Retinal Cell Biology in Health and Disease

Retinal Cell Biology in Health and Disease

Editor

Hossein Ameri

Basel • Beijing • Wuhan • Barcelona • Belgrade • Novi Sad • Cluj • Manchester

Editor
Hossein Ameri
University of Southern California
Los Angeles
USA

Editorial Office
MDPI
St. Alban-Anlage 66
4052 Basel, Switzerland

This is a reprint of articles from the Special Issue published online in the open access journal *Cells* (ISSN 2073-4409) (available at: https://i.mdpi.cn/team/training/manual/chapter/1501).

For citation purposes, cite each article independently as indicated on the article page online and as indicated below:

Lastname, A.A.; Lastname, B.B. Article Title. *Journal Name* **Year**, *Volume Number*, Page Range.

ISBN 978-3-7258-0759-8 (Hbk)
ISBN 978-3-7258-0760-4 (PDF)
doi.org/10.3390/books978-3-7258-0760-4

Cover image courtesy of Hossein Ameri

Contents

About the Editor

Hossein Ameri

Hossein Ameri is a clinician-scientist at the USC Roski Eye Institute of the University of Southern California and director of the USC Retinal Degeneration Center. Dr. Ameri's clinical practice comprises vitreoretinal surgery and medical retina. He has years of experience treating conditions such as diabetic retinopathy, age-related macular degeneration, and inherited retinal diseases. His research experience includes retinal prosthesis, retinal angiogenesis, retinal imaging, ocular gene therapy, and ocular drug delivery. He is an editorial board member of several journals, has patented many inventions, and has published numerous articles in peer-reviewed scientific journals.

cells MDPI

Editorial

Retinal Cell Biology in Health and Disease

Hossein Ameri

Department of Ophthalmology, USC Roski Eye Institute, Keck School of Medicine, University of Southern California, Los Angeles, CA 90033, USA; ameri@med.usc.edu

The intricate network of cells and processes that govern retinal health has long been a subject of fascination and intensive study within the scientific community. In recent breakthroughs, a series of papers have shed light on various aspects of ocular physiology, presenting findings that deepen our understanding of eye health and open new avenues for therapeutic interventions. From mitophagy in astrocytes to membrane attack complex-mediated cell death, these studies explore diverse facets of retinal cell biology, contributing to the broader landscape of vision research.

One of these compelling revelations comes from a study on mitophagy in astrocytes by Yazdankhah et al., which emphasizes the crucial role of mitophagy in preserving mitochondrial function and maintaining the health of the optic nerve [1]. The optic nerve, responsible for transmitting visual information from the retina to the brain, relies on the intricate interplay between cells, and the role of mitophagy in astrocytes highlights the importance of cellular maintenance mechanisms in preserving optic nerve function. This research not only enhances our comprehension of optic nerve health but also suggests potential therapeutic strategies for conditions affecting mitochondrial function such as Leber's hereditary optic neuropathy.

Another noteworthy discovery pertains to the interaction between Disheveled-1 and Claudin-5 in the context of Norrin-induced endothelial barrier restoration, which was studied by Diaz-Coranguez et al. [2]. Understanding the molecular mechanisms underlying endothelial barrier function is pivotal, especially in the context of vascular endothelial growth factor overexpression in conditions such as diabetic retinopathy and retinal vein occlusion. This research not only identifies a key player in the restoration process but also unveils potential targets for therapeutic intervention in conditions involving endothelial barrier dysfunction.

The synaptic ribbon of rod photoreceptor synapses takes center stage in the study of Rabconnectin-3α/DMXL2 enrichment by Dittrich et al. [3]. The local concentration of this protein at the synaptic ribbon underscores its significance in the intricate machinery governing neurotransmission in the retina. Insights into the molecular intricacies of photoreceptor synapses provide a foundation for understanding visual signal processing and may pave the way for novel therapeutic approaches in vision-related disorders.

Shifting the focus to cellular signaling pathways, the research on Peptide Lv by Pham et al. stands out due to its exploration of the trafficking and membrane insertion of $K_{Ca}3.1$ [4]. Unraveling the complex interplay between Protein Lv and MEK1–ERK and PI3K–Akt signaling pathways in this context not only deepens our understanding of cellular regulation but also presents potential targets for modulating ion channel activity in retinal vascular endothelial cells, opening new avenues for therapeutic intervention in conditions involving pathogenic angiogenesis such as wet age-related macular degeneration.

The anatomical success of reattaching a detached retina does not guarantee a full restoration of vision to its pre-detachment level. The visual recovery is inversely related to the duration of the detachment. The study conducted by Townes-Anderson et al. sheds light on the differential impact of retinal detachment on cone and rod synapses [5]. Furthermore, the protective role of Rho kinase inhibition highlights a potential avenue for the preservation of cone synapses and, consequently, visual function in chronic retinal detachment.

Citation: Ameri, H. Retinal Cell Biology in Health and Disease. *Cells* **2024**, *13*, 297. https://doi.org/10.3390/cells13040297

Received: 30 January 2024
Accepted: 5 February 2024
Published: 6 February 2024

Mitochondrial dysfunction takes center stage in Huang et al.'s study, who examined impaired antioxidant responses in retinal pigment epithelial cells associated with RCBTB1-related retinopathy [6]. This research not only provides insights into the pathophysiology of a specific retinal disorder but also underscores the broader significance of mitochondrial health in maintaining retinal function.

Lastly, the involvement of the membrane attack complex in retinal pigment epithelium cell death in Stargardt Macular Degeneration, studied by Ng et. al., unveils a critical aspect of the immune response in ocular degenerative diseases [7]. Understanding the mechanisms underlying cell death in these conditions is crucial for developing targeted therapies to halt or slow down disease progression.

In conclusion, the papers discussed here collectively contribute to a deeper understanding of ocular health, spanning various aspects of cellular and molecular biology. As we unravel the intricacies of mitophagy, synaptic ribbons, signaling pathways, and immune responses in the eye, we pave the way for innovative therapeutic strategies that may revolutionize the treatment of retinal disorders. These findings not only advance our scientific knowledge but also hold the promise of improving the lives of individuals affected by vision-related conditions. The journey towards comprehensive ocular health continues, fueled by the relentless pursuit of knowledge and the hope of transformative medical advancements.

Conflicts of Interest: The author declares no conflicts of interest.

References

1. Yazdankhah, M.; Ghosh, S.; Liu, H.; Hose, S.; Zigler, J.S.; Sinha, D. Mitophagy in Astrocytes Is Required for the Health of Optic Nerve. *Cells* **2023**, *12*, 2496. [CrossRef] [PubMed]
2. Díaz-Coránguez, M.; González-González, L.; Wang, A.; Liu, X.; Antonetti, D.A. Disheveled-1 Interacts with Claudin-5 and Contributes to Norrin-Induced Endothelial Barrier Restoration. *Cells* **2023**, *12*, 2402. [CrossRef] [PubMed]
3. Dittrich, A.; Ramesh, G.; Jung, M.; Schmitz, F. Rabconnectin-3α/DMXL2 Is Locally Enriched at the Synaptic Ribbon of Rod Photoreceptor Synapses. *Cells* **2023**, *12*, 1665. [CrossRef] [PubMed]
4. Pham, D.L.; Niemi, A.; Blank, R.; Lomenzo, G.; Tham, J.; Ko, M.L.; Ko, G.Y.-P. Peptide Lv Promotes Trafficking and Membrane Insertion of KCa3.1 through the MEK1-ERK and PI3K-Akt Signaling Pathways. *Cells* **2023**, *12*, 1651. [CrossRef] [PubMed]
5. Townes-Anderson, E.; Halász, É.; Sugino, I.; Davidow, A.L.; Frishman, L.J.; Fritzky, L.; Yousufzai, F.A.K.; Zarbin, M. Injury to Cone Synapses by Retinal Detachment: Differences from Rod Synapses and Protection by ROCK Inhibition. *Cells* **2023**, *12*, 1485. [CrossRef] [PubMed]
6. Huang, Z.; Zhang, D.; Chen, S.-C.; Huang, D.; Mackey, D.; Chen, F.K.; McLenachan, S. Mitochondrial Dysfunction and Impaired Antioxidant Responses in Retinal Pigment Epithelial Cells Derived from a Patient with RCBTB1-Associated Retinopathy. *Cells* **2023**, *12*, 1358. [CrossRef] [PubMed]
7. Ng, E.S.Y.; Kady, N.; Hu, J.; Dave, A.; Jiang, Z.; Pei, J.; Gorin, M.B.; Matynia, A.; Radu, R.A. Membrane Attack Complex Mediates Retinal Pigment Epithelium Cell Death in Stargardt Macular Degeneration. *Cells* **2022**, *11*, 3462. [CrossRef] [PubMed]

cells

MDPI

Article

Membrane Attack Complex Mediates Retinal Pigment Epithelium Cell Death in Stargardt Macular Degeneration

Eunice Sze Yin Ng [1,2,†], Nermin Kady [1,3,4,†], Jane Hu [1], Arpita Dave [1], Zhichun Jiang [1], Jacqueline Pei [1], Michael B. Gorin [1], Anna Matynia [1] and Roxana A. Radu [1,*]

1 UCLA Stein Eye Institute and Department of Ophthalmology, David Geffen School of Medicine at UCLA, University of California at Los Angeles, CA 90095, USA
2 Molecular Cellular and Integrative Physiology Interdepartmental Program, University of California, Los Angeles, CA 90095, USA
3 Department of Internal Medicine, Division of Hematology and Oncology, University of Michigan, Ann Arbor, MI 48109, USA
4 Clinical Pathology Department, Faculty of Medicine, Mansoura University, Mansoura 35516, Egypt
* Correspondence: radu@jsei.ucla.edu
† These authors contributed equally to this work.

Abstract: Recessive Stargardt disease (STGD1) is an inherited retinopathy caused by mutations in the *ABCA4* gene. The ABCA4 protein is a phospholipid-retinoid flippase in the outer segments of photoreceptors and the internal membranes of retinal pigment epithelial (RPE) cells. Here, we show that RPE cells derived via induced pluripotent stem-cell from a molecularly and clinically diagnosed STGD1 patient exhibited reduced ABCA4 protein and diminished activity compared to a normal subject. Consequently, STGD1 RPE cells accumulated intracellular autofluorescence-lipofuscin and displayed increased complement C3 activity. The level of C3 inversely correlated with the level of CD46, an early negative regulator of the complement cascade. Persistent complement dysregulation led to deposition of the membrane attack complex on the surface of RPE cells, decrease in transepithelial resistance, and subsequent cell death. These findings are strong evidence of complement-mediated RPE cell damage in STGD1, in the absence of photoreceptors, caused by reduced CD46 regulatory protein.

Keywords: recessive Stargardt disease (STGD1); retinal pigment epithelium (RPE); complement system; bisretinoid-lipofuscin; "disease-in-a-dish"; retinoids; macular degeneration; ABCA4

Citation: Ng, E.S.Y.; Kady, N.; Hu, J.; Dave, A.; Jiang, Z.; Pei, J.; Gorin, M.B.; Matynia, A.; Radu, R.A. Membrane Attack Complex Mediates Retinal Pigment Epithelium Cell Death in Stargardt Macular Degeneration. *Cells* 2022, *11*, 3462. https://doi.org/10.3390/cells11213462

Academic Editor: Hossein Ameri

Received: 8 October 2022
Accepted: 30 October 2022
Published: 2 November 2022

1. Introduction

Accumulation of toxic bisretinoid-lipofuscin material in the cells of the retinal pigment epithelium (RPE) is an age-dependent process, accelerated by loss or dysfunction of the ATP binding cassette subfamily A member 4 (ABCA4) protein. Mutations in the *ABCA4* gene are responsible for recessive Stargardt disease (STGD1), a juvenile maculopathy that shares many clinical and pathologic features with dry-form of age-related macular degeneration (AMD) [1,2]. Other mutations in *ABCA4* cause cone-rod dystrophy in approximately one-third of cases and serve as rare susceptibility loci for AMD [3,4]. None of these *ABCA4*-mediated blinding diseases are currently treatable. ABCA4 is a flippase for retinaldehyde conjugated to phosphatidylethanolamine (*N*-retinylidene-phosphatidylethanolamine, or *N*-ret-PE) to promote clearance of retinaldehyde and prevent the formation of retinaldehyde dimers (bisretinoids) [5,6]. Recently, the *ABCA4* was shown to be expressed in RPE cells, in addition to photoreceptors [7,8]. ABCA4 in RPE endo-lysosomal membranes is responsible for retinaldehyde recycling during proteolysis of visual pigments following the daily phagocytosis of photoreceptor outer segments (OS). The hypothesized function of ABCA4 in RPE cells was evaluated in vivo using a transgenic mouse line on the *Abca4*$^{-/-}$

background that expresses ABCA4 in RPE cells but not in photoreceptors [7]. These transgenic mice exhibited partial rescue of the photoreceptor degeneration and bisretinoid accumulation seen in non-transgenic *Abca4*$^{-/-}$ mice, suggesting that ABCA4 in RPE cells reduces the toxicity of retinaldehyde released during proteolysis of rhodopsin following phagocytosis [7]. More recently, we generated RPE cells from molecularly and clinically diagnosed STGD1 patients with altered levels of *ABCA4* transcript and protein as a new "disease-in-a-dish" biological model [9].

Bisretinoids and their oxidative products trigger strong complement reactivity in cultured RPE cells [10]. Further, accumulation of bisretinoids in the RPE of *Abca4*$^{-/-}$ mice was shown to activate complement, increase inflammatory markers, and cause oxidative stress [11]. Degeneration of photoreceptors in *Abca4*$^{-/-}$ mice and STGD1 patients is thought to result from chronic bisretinoid-mediated complement dysregulation and intracellular buildup of autofluorescent material in the RPE cells [11,12]. Notably, a much greater deposition of the membrane attack complex (MAC) on the basolateral membranes of RPE cells was also found in cadaveric donor eyes from STGD1 patients versus unaffected age-matched human controls [13,14]. In the current study, we used our newly developed "disease-in-a-dish" RPE model from a confirmed STGD1 patient to test the hypothesis that ABCA4 deficiency and complement dysregulation in RPE cells are coupled. We observed a strong correlation between time-dependent deposition of the MAC and RPE cell loss, which mirrored the RPE phenotype detected in *Abca4*$^{-/-}$ mice. These findings are the first indication of complement dysregulation and MAC-mediated damage in RPE cells of STGD1 in the absence of the photoreceptor cells.

2. Materials and Methods

2.1. Human-Derived RPE Cell Cultures

Intrinsically pluripotent stem cell lines (iPSC) from a clinically and molecularly characterized STGD1 patient (sample "H1" and "H3" in Matynia et al [9]) and an unaffected human donor (control and NDHF sample in Matynia et al. and Lowry et al. studies, respectively [9,15]) were differentiated into RPE cells following published methods [9,16–18]. Briefly, undifferentiated iPSCs colonies were passaged using ReLeSR (Stemcell Technologies, Kent, WA, USA) onto a matrigel (1:3 to 1:6, Corning Incorporated, Glenddale, AZ, USA) coated dish with a basal medium containing the DMEM/F12 with Glutamax, NEAA (Millipore Sigma, Burlington, MA, USA), B27, and N2 (Thermo Fisher Scientific, Waltham, MA, USA). The following growth factors were added to the basal medium for the first 48 h: Noggin (50 ng/mL, R & D Systems, Minneapolis, MN, USA), DKK1 (10 ng/mL, R & D Systems) and IGF1 (10 ng/mL, R & D Systems). For the next 48 h, the cells were grown in fresh medium with the above ingredients with the addition of bFGF (5 ng/mL, Thermo Fisher). Days five to six, the cells were grown in the basal medium with DKK1 and IGF1 (10 ng/mL, R & D Systems) and maintained in the basal medium with Activin (100 ng/mL, R & D Systems) and SU5402 (10 μm, Millipore Sigma) for days 7 to 14. At the end of day 14, the cells were switched to Miller medium [19] until the formation of pigmented RPE colonies. The pigmented RPE colonies were picked mechanically and dissociated with 0.25% Trypsin/EDTA (Thermo Fisher Scientific), and then placed onto laminin coated plates until they became confluent in the Miller medium [19]. At passage 2–5, about 2×10^5 total RPE cells (~600,000 cells/cm^2) were seeded on mouse laminin coated transwell inserts (Millicell-HA) with a pore size of 0.4 μm (Millipore Sigma), allowed to form a monolayer with developed junctional complexes which was assessed by measurement of the transepithelial resistance with an EVOM2 voltohmmeter (World Precision Instruments, Sarasota, FL, USA).

Experimental cultured human RPE cells originated from fibroblasts were collected from a clinically and molecularly characterized STGD1 patient (male) and an unaffected human donor (male). The iPSC lines were validated for viability and karyotype by the UCLA Broad Stem Cell Core using Cell Line Genetics as previously described in Lowry et al [15]. The RPE cells were authenticated using RNASeq to identify the loss of ABCA4 expression in patient cells and presence in Control cells, with key RPE genes comparable between the two cell lines [9]. Control and STGD1 RPE cells were genotyped for single nucleotide polymorphisms (SNPs) associated with AMD and neither of them contained the AMD risk haplotype (Table S1).

RPE cells were grown in culture using optimal media *without* retinoids for the initial characterization at two-, three-, and five-months. For longitudinal experiments, media was supplemented with bovine retinal extract, as a source of retinoid (~1.5 pmoles per feeding) for the RPE cells. Supplementation was initiated at two-months under our standard feeding protocol (twice a week) and RPE cells were maintained in culture up to 12-months. Bovine retinal extract was prepared following established protocols [20]. Briefly, retinas were isolated from fresh bovine eyes, sonicated in Ringer's buffer without calcium and magnesium, agitated overnight at 4 °C, and centrifuged for 20 min at 17,300× *g*. Supernatant was collected and stored at −80 °C. Retinoid content in retinal extract was quantified by normal phase high performance liquid chromatography (HPLC) as described previously [21] and detailed in methods Section 2.8 below.

2.2. Animals

Animals were housed under normal cyclic 12-h light/12-h dark conditions and fed *ad libitum*. Six-months old albino wild-type (BALB/c) and *Abca4*$^{-/-}$ on a BALB/c background mice (backcrossed at least eight times) were used for immunostaining and immunoblotting experiments. Equal numbers of male and female per genotype were used for all mouse studies.

2.3. Quantitative Real-Time PCR

Total RNA was extracted from RPE cell cultures using Qiagen RNeasy kit (Qiagen Inc., Hilden, Germany) according to the manufacturer's instructions. NanoDrop (Thermo Fisher) was used to determine total RNA concentration. cDNA was synthesized from one microgram of RNA using the superscript III first-strand synthesis system (Thermo Fisher Scientific). qRT-PCR was performed with SYBR Green (BIO-RAD, Hercules, CA, USA). Primer specificity for stem cell markers (OCT4, NANOG, PAX6, and OTX2), for melanogenesis markers (Tyrosinase, PMEL17 and pigment epithelium derived factor (PEDF)) were used along with housekeeping genes (glyceraldehyde phosphate dehydrogenase (GAPDH), peptidylprolyl isomerase A (PPIA), hydroxymethylbilane synthase (HMBS), and glucose phosphate isomerase 3′ (GPI)). Primer specificity was confirmed with melting temperature analysis and gel electrophoresis. Expression levels were normalized to the geometric mean of all four housekeeping genes: GAPDH, PPIA, HMBS, and GPI. Forward (F) and reverse (R) primer sequences are listed in Table S2.

2.4. In Situ Hybridization Assay

RPE cells were used for in situ hybridization assay with the RNAscope 2.5 HD Chromogenic Detection Kit (Advanced Cell Diagnostics, Hayward, CA, USA) according to the manufacturer's protocol. Cells were hybridized with human ABCA4 specific oligo probes (Advanced Cell Diagnostics), human RNA polymerase II subunit A probe (Polr2A) as a positive control probe, or bacterial dihydrodipicolinate reductase as a negative control probe, followed by amplification steps and chromogenic detection with Fast Red as previously described [7]. Briefly, Images were captured using a Zeiss Axiophot microscope equipped with a 40× oil-immersion objective lens and a CoolSNAP digital camera (Media Cybernetics, Silver Spring, MD, USA).

2.5. Western Electrophoresis System (WES)

For detection and quantification of ABCA4 protein in RPE cells, capillary electrophoresis was performed on the fully automated Western system (ProteinSimple, Minneapolis, MN, USA) following the manufacturer's recommendations. Briefly, 20 µg protein lysate were mixed with one µL of the 5× master mix containing sample buffer and DTT. The samples and protein ladder were heat denatured at 95 °C for five minutes. Three μL of the sample were added to the assay plate containing blocking buffer, primary and HRP-conjugated secondary anti-goat or anti-rabbit antibodies (Table S3), and wash buffer in independent wells, respectively for sequential processing. The plate was briefly centrifuged and loaded into the instrument for electrophoretic separation of proteins in capillary tubes containing a 66- to 440-kDa separation matrix. Protein levels were quantified using the automated Compass software according to the manufacturer's guidelines (ProteinSimple).

2.6. Western Blot (WB)

Mouse RPE along with Bruch's membrane and choroid were harvested from six-month-old albino wild-type and *Abca4*$^{-/-}$ mice, both on a BALB/c background. Human cultured RPE were grown on Millicell-HA for three months. Both mouse and cultured human RPE cells were homogenized in 1× PBS with Halt protease inhibitor mixture (Thermo Fisher Scientific). Protein samples were treated with benzonase nuclease (Millipore-Sigma) at room temperature (RT) for one hour and re-homogenized with 1% SDS. Cellular debris was removed by brief centrifugation, and protein concentration was determined using the Micro BCA Protein Assay Kit (Thermo Fisher Scientific). 25 μg of total protein of mouse homogenate or three μg of total protein of RPE homogenate or 40 μg cell concentrated media, respectively was then separated on 4–12% or 12% SDS/PAGE gels (Novex; Thermo Fisher Scientific). PVDF membranes were blocked with Odyssey Blocking Buffer (LI-COR Biosciences, Lincoln, NE, USA), followed by incubation at RT, probed with primary antibodies and cognate IR dye-labeled secondary antibodies (Table S3) [6,22–24]. Western blot analysis was done using the Odyssey CLx Infrared Imaging System (LI-COR). Colored images were converted to grayscale for viewing using Photoshop (Adobe, San Jose, CA, USA).

2.7. Exogenous Retinoid Uptake and Processing by RPE Cells

RPE cells with TER of >150 Ω/cm^2 grown in media *without* supplementation of bovine retinal extract for ~three-months were used in functional assays (visual cycle enzymatic steps and ABCA4-flippase activity). All experiments involving retinoids were conducted in a dark room under dim red light. To assay visual cycle activities, RPE cells were incubated overnight at 37 °C with 10 μM all-*trans*-retinol (MilliporeSigma) in DMEM (Thermo Fisher Scientific) supplemented with 1% BSA (bovine serum albumin, fatty acid–free) on the basal side. Apo inter-photoreceptor retinoid-binding protein (IRBP, 10 μM) was added to the apical medium to protect the released 11-*cis*-retinoids. IRBP was purified from freshly dissected bovine retinas and apo-IRBP was obtained after retinoids were degraded by exposure to UV-light [20]. To evaluate the ABCA4-flippase activity, we incubated overnight the RPE cells with bovine OS applied to the apical side (~75 × 10^6 OS per transwell; ~1.2 nmoles total retinaldehyde). Cultured cells on filters were washed with 1× PBS buffer and stored dry at −80 °C until further processing. Corresponding media samples were collected in tubes containing hydroxylamine (final concentration 200 mM) and processed immediately for retinoid extraction.

2.8. Retinoid Extraction and Normal-Phase Liquid Chromatography Analysis

RPE cells were homogenized in phosphate buffer containing 200 mM hydroxylamine. Cell homogenates and media samples containing hydroxylamine were mixed by vortexing and incubated at RT for ~20 min. Two mL of methanol was added, and retinoids were extracted twice with three mL of hexane followed by centrifugation at 3000 × g for five minutes. The organic phases were collected, dried under a stream of argon gas, and re-dissolved in 100 μL of hexane. Hexane solutions were analyzed by normal phase HPLC as previously described [21]. The identity of each retinoid-eluted peak was established by comparing the spectra and elution times with those of authentic retinoid standards. Retinoids standards were obtained from Sigma, and retinaldehyde oximes were rederived in house by hydroxylamine reduction and purified by HPLC [25]. Peak area for retinol and retinyl esters were acquired at 325 nm and retinaldehyde oximes at 350 nm, according to their maximum absorbance spectra. Sample peaks were quantitated by comparing peak areas to calibration curves established with retinoid standards using the published molar extinction coefficients [25,26].

2.9. Phagocytosis Assay

Pulse-chase phagocytosis assay was done with RPE cells grown as monolayers on culture inserts incubated with bovine photoreceptor OS (Invision BioResources, Seattle, WA, USA) (10 OS/cell; total ~2.5×10^6 OS per transwell) in the apical medium for two hours. After the initial OS challenge, cells were washed with Dulbecco's Phosphate Buffered Saline (DPBS), containing 0.9 mM Calcium Chloride and 0.49 mM Magnesium Chloride (DPBS-CM), and immediately processed for immunofluorescence (pulse), or incubated further for two hours and then processed for immunofluorescence (chase) as previously described [27]. A double immunofluorescence labeling strategy, using a primary antibody against rhodopsin (Rho), to distinguish between OS bound to the surface versus OS that has been internalized by the RPE cells. Briefly, cultures were fixed with 4% formaldehyde for ten minutes at RT and blocked with 1% BSA in DPBS-CM for 15 min. Surface-bound OS was labeled with the mouse anti-Rho 1D4 [27] conjugated with Alexa Fluor 488-nm goat anti-mouse secondary antibody (Table S3). After permeabilization with 50% ethanol in DPBS-CM for five minutes, OS was labeled with the same Rho 1D4 antibody conjugated with the Alexa Fluor 568-nm goat anti-mouse secondary antibody (Table S3). Finally, cells were washed with DPBS-CM and the membranes of the culture inserts were excised and mounted onto microscopy slides. Confocal z-stacks of randomly selected fields were captured using an Olympus FluoView FV1000 confocal microscope with a 60× NA1.4 oil objective. Surface-bound OS particles were labeled with both secondary antibodies (green and red), thereby appearing yellow. Internalized OS particles were labeled only with the Alexa Fluor 568-nm-conjugated secondary antibody (red only). For quantification, OS particle with a ≥0.5 μm diameter were counted from a total of five to ten fields of view using NIH Image J software (Rasband, W.S., ImageJ, U. S. National Institutes of Health, Bethesda, Maryland, USA; Available online https://imagej.nih.gov/ij/, accessed on 4 March 2019). Analysis of OS degradation (ingested) was estimated after subtracting the OS bound-only from the total number of OS particles after permeabilization of the cells.

2.10. Mouse Immunohistochemistry

Age-matched six-months albino wild-type and *Abca4*$^{-/-}$ were euthanized and the superior part of the eyeballs were marked with a cautery pen. The eyes were enucleated and fixed in 4% formaldehyde with 0.1 M sodium phosphate buffer ($NaPO_4$, pH 7.4) for 30 min at room temperature (RT) for OD eye and overnight for OS eye. After washing, the eyes were dissected in 0.1 M $NaPO_4$ and the anterior segment (cornea, lens, vitreous) and the neural retina were removed to obtain the RPE-choroid-scleral eyecups. The OD eyecup was flattened by cutting into eight leaflets to obtain an RPE-choroid-scleral flatmount. The OS eyecups were infiltrated with 10–30% sucrose for cryoprotection, embedded in Optimal Cutting Temperature embedding medium (OCT; Tissue-Tek, Torrance, CA, USA), and cut into 10-μm frozen sections. The RPE flatmounts and frozen sections were permeabilized with 1% Triton X-100 at RT for 1 hour and 10 min, respectively, then blocked with 1% BSA/5% goat serum/0.5% TritonX-100 at 4 °C overnight for flatmounts and one hour at RT for frozen sections. Primary antibodies were incubated at 4 °C for two days and overnight for RPE flatmounts and frozen sections, respectively. Following three washes with PBS/ 0.1% TritonX-100, secondary antibodies were applied at RT for two and one hour to flatmounts and frozen sections, respectively. The RPE cells were delineated by the phalloidin conjugated with Texas red, staining the actin filaments, with nuclei stained by DAPI (300 nM, Thermo Fisher Scientific), and then mounted with 5% n-propyl gallate in 100% glycerol. Images from superior temporal region, at similar distance from the optic nerve, of the RPE flatmounts and frozen sections were captured with an Olympus FluoView

FV1000 confocal microscope using a 60× oil-immersion objective. The z-orthogonal sections images were obtained using Imaris software (Bitplane Inc., Concord, MA, USA).

2.11. RPE Cell Immunocytochemistry (ICC)

Cultured RPE cells with their supporting filter were fixed in 4% formaldehyde and 0.1 M PBS for 30 min at RT. For confocal microscopy analysis of RPE monolayer cross-sections, the fixed cells along with their filter were embedded in agarose (Millipore-Sigma) and 50 μm sections were cut with a VT1000s vibratome (Leica Microsystem, Wetzlar, Germany) [28]. To visualize phosphatidylethanolamine in the cell membranes, RPE cells were incubated for two hours at RT with 1 μM Duramycin conjugated to Biotin (Molecular Target Technologies, West Chester, PA, USA) diluted from 200 μM stock solution in water containing 5% DMSO. For both wholemount and agarose sections of cultured RPE cells, primary antibodies were applied overnight at 4 °C followed by the secondary antibodies for 1 hour at RT using either goat anti-mouse IgG or goat anti-rabbit IgG, or streptavidin conjugated to Alexa Fluor dyes (Thermo Fisher Scientific). In addition, phalloidin and DAPI were used to delineate the cell boarder and nuclei, respectively. The cells were mounted with 5% n-propyl gallate in 100% glycerol with coverslips. All primary and secondary antibodies source and concentration were listed in Table S3. Images were captured with an Olympus FluoView FV1000 confocal microscope using a 60× oil-immersion objective. Images were visualized and analyzed by Imaris (Bitplane Inc.) or FluoView software (Olympus, Waltham, MA, USA).

2.12. Transmission Electron Microscopy

RPE cells grown on culture inserts (Millicell-HA) were fixed with 2% formaldehyde and 2.5% glutaraldehyde in 0.1 M PBS (pH 7.2), treated with 1% osmium tetroxide dissolved with 0.1 M PBS, then dehydrated in a graded series of alcohols and embedded in Araldite 502 (Electron Microscope Sciences, Hatfield, PA, USA). Ultrathin sections (at 70 nm in thickness) were cut on a Leica Ultracut microtome, collected on 200-mesh copper grids, and double-stained with uranyl acetate and lead citrate. The images (~12–18 per transwell) were collected in a JEM-1400 electron microscope (JEOL, Peabody, MA, USA) with ORIUS SC1000b Camera at 2000× and analyzed using Gatan Microscopy Suite Software (Gatan, Inc., Pleasanton, CA, USA).

2.13. Quantification and Statistical Analysis

Statistical analysis for each experiment is described in figure legends, where "n" represents biological replicates. For human studies, we used one line of Control and two lines of STGD1 RPE cells seeded and grown on filters in 24-transwell plates; one filter (with ~2×10^5 RPE cells per filter/transwell) represents an individual biological sample, unless otherwise specified. Every experiment with human RPE cells included a minimum of three independent biological samples and the experiments were repeated twice with triplicates (3 × single filter/transwell for a total of n = 6), or with duplicates (2 × single filter/transwell, for a total of n = 4) per group/genotype. The results were presented as means with standard deviation of a minimum of three to six animals or human biological samples per group/genotype. Data analyses were performed using GraphPad Prism 9.0 package (GraphPad by Dotmatics, San Diego, CA, USA). Comparisons between groups were evaluated using two-tailed Student's t-test with a statistical significance reported at * $p < 0.05$; ** $p < 0.005$; *** $p < 0.0005$; **** $p < 0.0001$.

3. Results

3.1. Control and STGD1 RPE Cultured Cells Have Similar Morphological Features

In vivo complement activation observed in *Abca4*$^{-/-}$ mice and excessive MAC deposition evidenced in STGD1 postmortem donor eyes compelled us to investigate the role of complement in STGD1 pathogenesis using patient-derived RPE cells [11–13]. For this study, we selected a STGD1 patient with two ABCA4 mutations on different alleles: (*i*) c.3386G > T; p.Arg1129Leu and (*ii*) c.[5461-10T > C;5603A > T]; p.[Thr1821Aspfs*6,Thr1821Valfs*13; (Asn1868Ile)] [29–31]. The R1129L substitution on the first allele is located within the nucleotide-binding domain 1 (NBD1) of ABCA4, previously shown to cause reduction in protein abundance and ATP-binding capacity when expressed in the 293T cells [32]. The second allele mutation was newly identified in this STGD1 patient and predicts a null protein [29]. The iPSC lines of control (normal) and clinically diagnosed STGD1 patient were derived into RPE cells according to standardized protocols [9,16,27] with their molecular characterization being recently published in Matynia et al [9]. In the current study, the RPE cells were seeded on a transwell filter and maintained in culture using a conditioned medium *without* retinoids for an initial morphological evaluation. By qRT-PCR analysis, mRNA for differentiation and pigmentation marker proteins were at similar levels in control and STGD1 RPE cells at three- and five-months in culture, while the mRNAs for pluripotency markers were insignificantly expressed (Figure S1A,B). By a chromogenic in situ hybridization assay, we observed a comparable distribution of ABCA4 mRNA expression in the STGD1 and control RPE cells (Figure S1C). Similar levels of ABCA4 mRNA in STGD1 and control RPE cells were found by qRT-PCR after two- and five-months in culture (Figure 1A), consistent with near-normal ABCA4 transcript levels found by RNASeq analysis [9]. At the protein level however, STGD1 RPE cells contained only ~15% of the ABCA4 found in the control RPE cells (Figures 1B and S1D). By immunohistochemistry, control RPE cells displayed a homogeneous membrane distribution of ABCA4 throughout, unlike the punctate profile observed in the STGD1 RPE cells (Figure 1C). Intracellular localization of normal ABCA4 protein in RPE cells was assessed with an antibody against the endosome marker EEA1 (Figure S1E), as previously shown in mouse RPE cells and fetal human RPE cultured cells [7]. In STGD1 RPE cells, the mutated ABCA4 protein seemed to colocalize with EEA1 (Figure S1E). Together, these findings suggest degradation and mislocalization of the mutated ABCA4 protein. At two-months in culture, control and STGD1 RPE cells had both grown to confluent monolayers with homogenous pigmentation and a typical RPE cobblestone appearance (Figure 1D). Transmission electron microscopy revealed polarized RPE cells with apical microvilli, basal infoldings, and intracellular pigment granules for both STGD1 and control RPE cells (Figures 1E and S1F). En-face confocal images of phalloidin staining for apically localized actin filaments showed well-defined hexagonal cell borders with similar cell counts for control and STGD1 RPE cells (Figure 1F,G). Since the formation of functional cellular junctions is crucial for both maintenance of epithelial integrity and the barrier role of RPE, we measured the transepithelial resistances (TER) in three-months cultures. STGD1 and control RPE cells exhibited similar TER of ~200 Ω/cm^2 (Figure 1H), comparable to the TER of human fetal RPE cells in culture [27,33,34].

3.2. OS Binding, Internalization, and Rhodopsin Degradation Are Normal in STGD1 RPE Cells

Phagocytosis of distal photoreceptor OS is an important function of RPE cells, which serves to remove toxic products and maintain photoreceptor health [35,36]. To determine if RPE cells are capable of phagocytosis, two-months cultured STGD1 and control RPE cells were incubated with purified bovine OS using an optimized pulse-chase assay [27], in which the amount of surface-bound Rho-containing OS (green channel) was determined before permeabilization of the cells, and the amount of ingested Rho-containing OS was determined after permeabilization of the cells as the difference between total Rho-containing OS (permeabilized cells, red channel) and bound Rho-containing OS (green channel). In the pulse phase (2 h, Figure S2A), staining with an antibody against rhodopsin (Rho) before (Figure S2A, Rho-bound) and after permeabilizing the cells (Figure S2A, Rho-total), we observed similar amounts of both bound (green arrows in all images) and ingested Rho-containing OS (white arrows in the red channel and merge images) in the STGD1 and control RPE cells (Figure S2A,C). In the chase phase (2 h after removal of bovine OS from the media), we observed degradation of internalized Rho and an overall decrease in Rho staining for both control and STGD1 RPE cells (Figure S2B). Quantification of total Rho showed no significant difference between control and STGD1 RPE cells, suggesting similar rates of OS uptake and Rho proteolysis (Figure S2D).

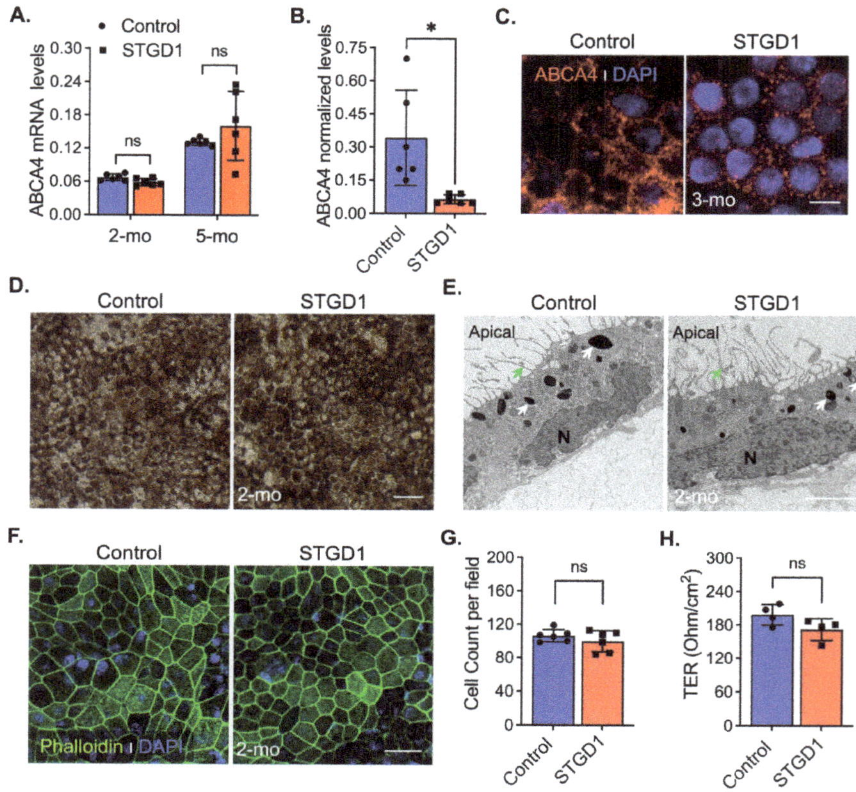

Figure 1. Molecular and morphological features of Control and STGD1 RPE cells. (**A**) Relative levels of ABCA4 mRNA by qRT-PCR from RPE cells for two (2)- and five (5)-months (mo) in culture. (**B**) Normalized ABCA4 protein levels of three-months control and STGD1 RPE cells homogenates (20 µg) by automated WES system (p = 0.025). STGD1 RPE cells have 85% reduction in ABCA4 levels compared to control. (**C**) Representative confocal microscopy images of three-months RPE cells fixed and stained for ABCA4 (red, custom-made antibody, generous gift from Dr. Hui Sun). DAPI (blue) stains the nuclei. Note a punctate distribution of ABCA4 throughout RPE cells of STGD1 compared to the control cells. Scale bar = 20 µm. (**D–F**) Representative images of two-months of control and STGD1 RPE cells by light (**D**), electron (**E**), and confocal (**F**) microscopy analysis. (**D**) RPE cells of both genotypes display similar cobblestone appearance and homogeneous pigmentation; Scale bar = 100 µm. (**E**) Control and STGD1 RPE cells developed as a monolayer and were polarized presenting apical microvilli (green arrow), melanosomes (white arrow), and nucleus (N). Scale bar = 2 µm. (**F**) Phalloidin staining F-actin filaments (green) evidences a well-defined hexagonal cell shape for both STGD1 vs control RPE cells. DAPI (blue) stains the nuclei. Scale bar = 20 µm. (**G**) Bar graph shows similar number of cells per view field (140 µm^2) between the control and STGD1

RPE cells (two-months in culture). (**H**) Histogram shows the transepithelial resistance (TER) measured across monolayers of control and STGD1 RPE cells (three-months in culture). Note: All RPE cells were grown in culture with conditioned medium *without* retinoids. In (**A**,**B**,**G**,**H**) data presented as mean ± SD; Statistics analysis using Student's *t*-test: ns = not significant; * $p < 0.05$; n = 4–6 filter/transwell per genotype.

3.3. ABCA4-Flippase Activity Is Reduced in STGD1 RPE Cells

To test for ABCA4-flippase activity, RPE cells grown in culture for three-months *without* exposure to retinoids were fed with a single dose of bovine OS, estimated to contain ~1200 pmoles 11-*cis*-retinaldehyde (11c-RAL) in the form of rhodopsin. Media and cells were harvested the following day (Figure 2A). Upon proteolytic digestion of OS proteins in RPE phagolysosomes, free 11c-RAL is released and thermally isomerized to all-*trans*-retinaldehyde (at-RAL) within acidic endolysosomes. ABCA4 was shown to 'flip' 11c- and at-RAL coupled to phosphatidylethanolamine (as N-ret-PE) across membranes with similar efficiencies [5]. The transfer of N-ret-PE from the lumenal to cytoplasmic face of the membranes, in photoreceptor OS and RPE endolysosomes, prevents the secondary condensation of N-ret-PE with another at-RAL to form a toxic bisretinoid. In RPE cells, translocation of N-ret-PE facilitates the recycling of at-RAL to 11c-RAL visual chromophore via the RPE visual cycle [5,7] (Figure 2B). After overnight incubation, we extracted retinoids from both the media and RPE cell-homogenates and analyzed them by HPLC. Untreated cells and their corresponding media showed *no* detectable retinoids (Figures 2C and S3A). The retinoid content of OS-supplemented media was similar for control- and STGD1-derived RPE cell cultures, mainly comprising 11c-RAL (Figure S3A,B). In contrast, HPLC analysis of the RPE cell homogenates showed approximately two-fold lower levels of all-*trans*-retinyl-palmitate (at-RP), an insoluble storage form of vitamin A, in STGD1 vs control RPE cells. This result is consistent with the reduced N-Ret-PE flippase activity of the mutated ABCA4 protein and impaired recycling of OS visual pigments (Figure 2C,D). PE phospholipid alone is also a substrate of ABCA4 that was shown to be actively flipped across the photoreceptor disc membranes in the same direction as N-Ret-PE [5]. Media of two-month-old cultured RPE cells was supplemented with bovine retina extract and the PE distribution was visualized by confocal microscopy at six-months after a short incubation with duramycin, a cyclic peptide that binds PE with high affinity and specificity [37]. We found that duramycin-bound PE was dispersed along both plasma and internal membranes of the control RPE cells (Figure 2E). In contrast, STGD1 RPE cells displayed intracellular PE-aggregates without clear PE-duramycin-association with the plasma membrane (Figures 2E and S3C). Quantification of pixel intensity from confocal images stained with PE-duramycin indicated ~50% increased levels in the RPE cells of STGD1 versus control (Figure 2F), which is further evidence of dysfunctional ABCA4 protein.

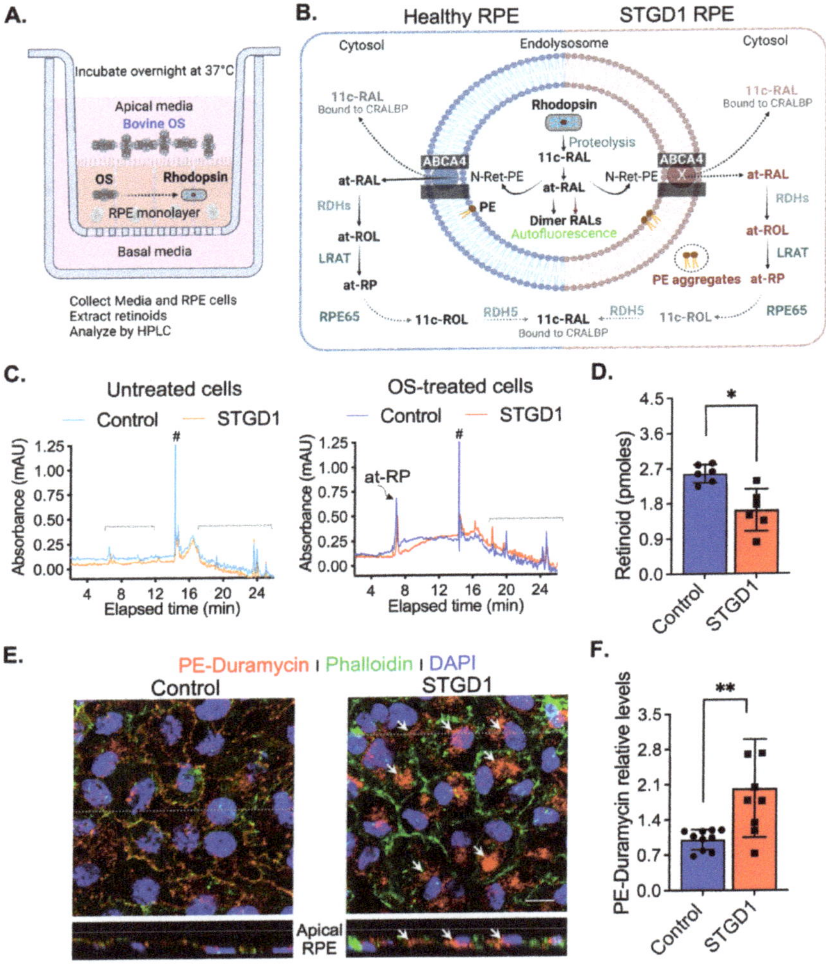

Figure 2. ABCA4-flippase activity is reduced in STGD1 RPE cells. (**A**) Diagram of a transwell with RPE cells grown on a filter to summarize our experimental setting to evaluate recycling of retinaldehyde following overnight incubation with bovine OS containing rhodopsin (retinaldehyde bound with opsin); retinoids were extracted in hexane from media and cell homogenate and analyzed by high-performance liquid chromatography (HPLC). (**B**) Diagram of proposed ABCA4-flippase activity in the endolysosomes of the RPE cell. Following rhodopsin proteolysis, 11-*cis*-retinaldehyde (11c-RAL) and its thermoisomerized form all-trans-RAL (at-RAL) condense with phosphatidylethanolamine (PE) in the endolysosomal membrane to form N-retinyledene-PE (N-Ret-PE). ABCA4 transporter translocates both N-Ret-PE and PE to the cytoplasmic side for recycling. Upon hydrolysis of N-Ret-PE, the 11c-RAL (bound to CRALBP) is readily available to reconstitute the visual pigment, while at-RAL must be reduced by RPE retinal dehydrogenases (RDHs) to all-*trans*-retinol (at-ROL) which is the substrate for lecithin-retinol-acyltransferase (LRAT), the major retinyl ester synthase in RPE cells. LRAT esterifies the at-ROL to all-*trans*-retinyl-palmitate (at-RP), an insoluble storage form of vitamin A, that can be stored or further used as the substrate of RPE65-isomerase to form 11-cis-retinol (11c-ROL) followed by RDH5-oxidation to 11c-RAL. (**C**) Representative HPLC chromatograms at 325 nm of hexane extracts of untreated (left) and OS-treated (right) of three-months control (cyan/blue traces) and STGD1 (orange/red traces) RPE cells. Note that untreated cells (left) have *no* detectable

retinoids and all-*trans*-retinyl palmitate (at-RP) is the only detected retinoid in the OS-treated cells (right). Brackets indicate non-retinoid peaks based on spectral analysis and (#) indicates change in solvent gradient. (**D**) Levels of at-RP in the STGD1 RPE cell homogenate were significantly lower vs control. Average data is presented as mean ± SD; Statistics analysis using Student's *t*-test; * $p < 0.05$ ($p = 0.0106$); n = 6 filter/transwell per genotype. (**E**) Representative confocal merge images of ~six-months control (left) and STGD1 (right) RPE cells that were treated with 1 µM Duramycin (red), specifically labeling the head-group of phosphatidylethanolamines (PE). Phalloidin (green) staining F-actin filaments and nuclei stained by DAPI (blue). Corresponding z-orthogonal representative confocal images are shown below the en-face images. Intracellular PE-aggregates (white arrows) are noticeable in both en-face and z-orthogonal images of STGD1 RPE cells versus control which display significant PE-Duramycin/Phalloidin co-localization (orange-like color area delineating cell boundaries). (**F**) Bar graph shows the relative PE-duramycin levels determined by measuring pixel intensity from acquired confocal images (n = 10 for Control and n = 9 for STGD1). Average data presented as mean ± SD; ** $p < 0.005$ ($p = 0.0048$); n = 3 filter/transwell per genotype; Scale bar = 10 µm. Note: RPE cells were grown in culture for three-months (**C**,**D**) *without* any retinoid supplementation and for six-months (**E**,**F**) with bovine retinal extract supplementation (~1.5 pmoles per feeding, twice per week). Drawings in (**A**) and (**B**) were created with BioRender.com.

3.4. LRAT and RPE65 Activities Are Similar in Control and STGD1 RPE Cells

Reduced levels of at-RP in homogenates of STGD1 versus control RPE cells (Figure 2D) could be caused by reduced LRAT protein levels instead of reduced at-ROL substrate for at-RP synthesis. To test this possibility, we performed quantitative immunoblotting and found similar LRAT levels in the control and STGD1 cell homogenates (Figure S4A,B). Immunoreactivity for LRAT antibody in agarose sections showed comparable staining in both STGD1 and control RPE cells (Figure S4C). Still another possibility is that RPE65 level and activity may be increased in STGD1 versus control cells. Since RPE65 is a retinoid isomerohydrolase that converts at-RP to 11c-ROL plus free palmitic acid [38], increased RPE65 could also cause reduced levels of at-RP in STGD1 RPE cells. We observed similar levels of the RPE65 protein in control and STGD1 RPE cells as determined by quantitative immunoblotting and confocal microscopy, respectively (Figure S4D,F). To evaluate the specific activities of LRAT and RPE65 in vivo, we incubated the RPE cells with 10 µM at-ROL in the basal media supplemented with 1% BSA (Figure S4G). Additionally, the apical medium was supplemented with 10 µM of apo-interphotoreceptor retinol-binding protein (apo-IRBP) to protect the released 11c-RAL generated after at-ROL was processed through the RPE visual cycle steps. IRBP has high affinity for 11c-RAL facilitating its transfer from the RPE to photoreceptors to regenerate the visual pigment by binding to the cone- and rod-opsins [39,40]. Extracts of the RPE cell homogenate and corresponding media were analyzed by HPLC to quantify the retinoid content. We found that levels of at-ROL and at-RP, the resultant product of LRAT activity, were similar in the homogenates of control and STGD1 RPE cells (Figure S4H,J). Also, no difference in the levels of at-ROL and 11c-RAL released in the media was observed in the control vs STGD1 RPE cells (Figure S4I,J). Taken together, these data suggest that the levels and activities of the key visual-cycle proteins, LRAT and RPE65, were similar in control versus STGD1 RPE cells supplied with at-ROL, and therefore not responsible for the reduced at-RP found in the STGD1 cells fed with OS containing retinaldehydes (Figure 2D).

3.5. Autofluorescent Material Accumulates in STGD1 RPE Cells

Impaired recycling of retinaldehyde, due to loss of the ABCA4 transporter function, leads to secondary condensation of *N*-ret-PE with at-RAL to form an autofluorescent bisretinoid such as A2E in the RPE cells of STGD1 donors and *Abca4*$^{-/-}$ mice [13,41]. Here, we acquired bisretinoid-autofluorescence images at 488nm of control and STGD1 RPE cells maintained in culture for three- or 12-months in the presence of bovine retinal extract (~1.5 pmoles RALs per feeding twice per week). Compared to RPE cells from unaffected human controls, STGD1 RPE cells showed slightly increased autofluorescence at three-months, and two-fold increased autofluorescence after 12-months in culture (Figure 3A,B). In *Abca4*$^{-/-}$ and STGD1 donor eyes, the presence of bisretinoids in the RPE caused cellular stress, evidenced by elevated lipid peroxidation products such as malondialdehyde (MDA) and 4-hydroxynonenal (4-HNE) [11,13]. Similarly, analysis of STGD1 RPE cells grown in culture for 12-months showed stronger immunoreactivity for the 4-HNE antibody compared to the control RPE cells (Figure 3C).

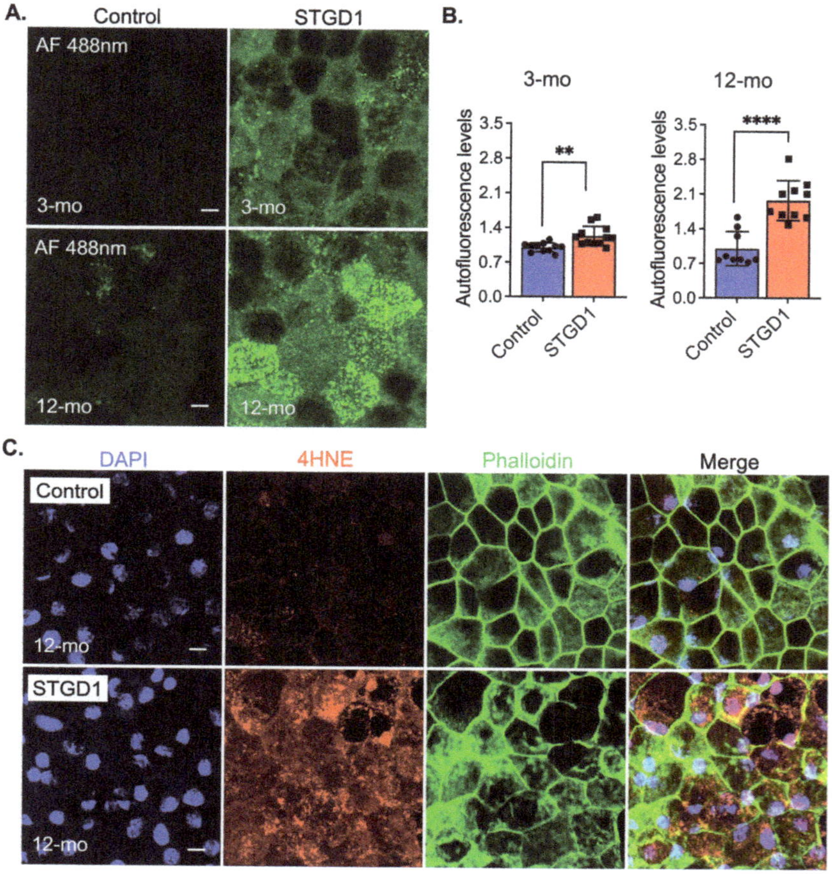

Figure 3. Age-dependent autofluorescent buildup and lipid peroxidation are amplified in STGD1 RPE cells. (**A**) Representative confocal images of autofluorescence (green) acquired with an excitation of 488nm (emission of 515nm) of control (left) and STGD1 (right) RPE cells in culture for three-months (3-mo, top row) and 12-months (12-mo, bottom row) in the presence of bovine retinal extract supplemented at two-months (~1.5 pmoles per feeding). Scale bar = 10 µm. (**B**) Bar graphs show the relative autofluorescence levels determined by measuring pixel intensity from images acquired at 3-mo (left) and 12-mo (right). Statistics analysis using Student's *t*-test; ** $p < 0.005$ ($p = 0.0012$) and **** $p < 0.0001$; n = 3 filter/transwell per genotype. (**C**) Representative confocal images of 12-months RPE cells fixed and stained with an antibody against 4-hydroxynonenal (4-HNE, red), a natural lipid peroxisomal marker. Phalloidin (green) stains the apical F-actin filaments and DAPI (blue) marks the cell nuclei. Note the intense immunoreactivity for the 4-HNE antibody on the STGD1 RPE cells. Scale bar = 10 µm; n = 3 filter/transwell per genotype.

3.6. Complement Dysregulation Is Evidenced in STGD1 RPE Cells

Stimulation of C3 activity by bisretinoids has been shown in vivo and in vitro [10,11,42]. Here, we assessed C3 components and the complement negative regulatory proteins, CD46,

CD59, and CFH by immunohistochemistry and immunoblotting at three-months in human-derived RPE cultured cells. On sections, C3 immunolabeling appeared throughout the RPE cells and was approximately two-fold higher in STGD1 versus control RPE cells (Figure 4A,B). Elevated C3b/iC3b breakdown fragments were further confirmed by quantitative immunoblot analysis (Figure 4C,D). Using fixed sections and RPE cell homogenates, C3a, a short-lived soluble fragment released after C3 fragmentation, was investigated with an antibody against its more stable form, C3a-desArg along with its C3a cognate membrane-bound receptor C3aR. While the C3a fragment was not detected in the RPE cells (Figure 4C), immunoblotting and immunostaining for its cognate receptor, C3aR, showed similar distribution and levels in control and STGD1 RPE cells (Figure S5A–C). C5aR levels were also similar in control and STGD1 RPE cells (Figure S5B,D), suggesting that the C5a/C5aR signaling pathway downstream of C3a/C3aR is also not activated. Next, we determined the levels of CD46, a membrane-bound, early-stage inhibitor of C3 convertase. CD46 appeared to localize predominantly on the basolateral side of the RPE cells. By confocal microscopy, CD46 immunostaining was notably diminished and pixel intensity quantification of CD46 showed approximately two-fold reduction in STGD1 versus human control RPE cells (Figure 4E,F). Further, CD46 levels in STGD1 RPE cells were only ~65% of the levels in control RPE cells by quantitative immunoblotting (Figure 4G,H). Unlike CD46, the cellular distribution and levels of CD59, another membrane-bound complement regulatory protein that inhibits the final step of the C3 cascade, were similar in the STGD1 and control RPE cells (Figure S5E,F). By quantitative immunoblotting, we measured levels of complement factor H (CFH), a major soluble complement regulatory protein, in the media of cultured RPE cells. Levels of CFH, along with unrelated secreted protein PEDF, were also similar between STGD1 and control RPE cells (Figure S5G,H). Taken together, these data suggest a primary deficiency of C3 convertase inhibition during the early complement amplification phase on the surface of STGD1 RPE cells.

3.7. Markedly Elevated MAC Deposition Is Present on the STGD1 RPE Cells

Excessive C3 hydrolysis and increased C3b opsonizing affinity on the plasma membrane due to C3 convertase activity are responsible for the assembly of terminal complement complex C5-C9 [43]. We assessed formation of this terminal complex on the surface of STGD1 and control RPE cells at different times in culture by confocal microscopy with an antibody against MAC (C5b-9 complex). We observed 1.2- and two-fold increased MAC levels on the STGD1 versus control RPE cells at three and 12 months in culture, respectively (Figures 5A,B and S6A). Orthogonal images obtained from z-stacked optical sections, from the apical side, showed that MAC deposition was predominantly on the basolateral poles of these cells and evidenced its internalization more abundantly in the STGD1 RPE cells (Figure 5A). Additionally, MAC localization was confirmed in the STGD1 and control RPE cells using an antibody against peropsin, an integral membrane protein expressed *only* in the apical microvilli of RPE cells (Figure S6B) [22]. Furthermore, the en-face view of phalloidin-staining apically distributed actin filaments showed an advanced dysmorphism (discontinued lines and wider cell size, indicated by stars, respectively) in the aged STGD1 versus control RPE cells (Figures 5A and S6A). These changes after 12-months in culture were mirrored by a decline in junctional complexes between cells, reflected by a ~35% loss of cell number and ~80% reduction of TER in the STGD1 versus control RPE cultures (Figure 5C,D).

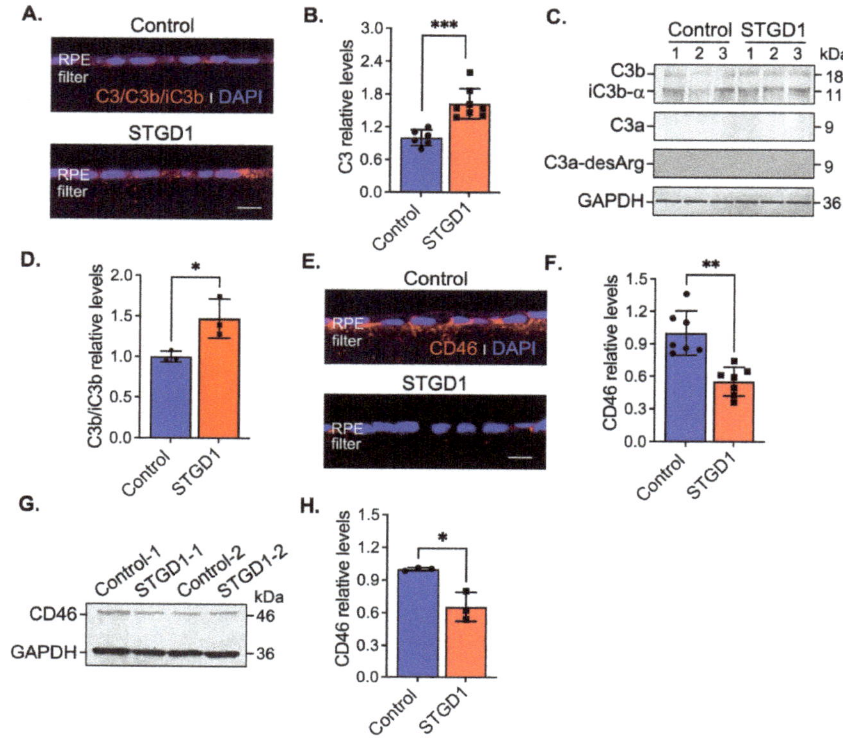

Figure 4. C3 complement dysregulation is observed in STGD1 RPE cells. (**A**) Representative confocal images of C3/C3b/iC3b (red) staining on sections of fixed and agarose embedded RPE cells of control (top) and STGD1 (bottom). (**B**) Histogram shows the relative levels of C3 based on pixel intensity acquired by confocal microscopy. Average data presented as mean ± SD; *** $p < 0.0005$ ($p = 0.0003$); n = 3 filter/transwell per genotype. (**C**) Representative immunoblots of equal amount of RPE cell homogenates of control and STGD1 using C3, C3a-desArg, and GAPDH antibodies. Under reducing conditions, C3 antibody recognizes C3b, iC3b, and C3a fragments. C3a fragment was not detected with either of the C3 or C3a-desArg antibodies. Numbers one to three on each blot represent independent biological samples for each genotype. (**D**) Histogram presents the relative levels of cumulative C3b and iC3b major fragments after normalizing to GAPDH loading control. Average data is presented as mean ± SD; * $p < 0.05$ ($p = 0.041$); n = 3 filter/transwell per genotype. (**E**) Representative confocal images of CD46 (red) staining on sections of RPE cells of control (top) and STGD1 (bottom). Noticeable diminished CD46 staining was observed in STGD1 RPE cells. (**F**) Histogram shows the relative levels of CD46 protein based on pixel intensity acquired by confocal microscopy. Average data is presented as mean ± SD; ** $p < 0.005$ ($p = 0.0013$); n = 3 filter/transwell per genotype. (**G**) Representative immunoblot of 10 µg total protein of RPE homogenates from two biological samples (1- and 2-) consisting of pooled two transwells of RPE cells of control and STGD1, respectively. (**H**) Histogram presents the relative levels of CD46 after normalizing to GAPDH loading control. Average data is presented as mean ± SD; * $p < 0.05$ ($p = 0.037$); n = 3 replicates per genotype. Note: All RPE cells were grown in culture for three-months in the presence of bovine retinal extract supplemented at two-months (~1.5 pmoles per feeding) at regular feeding regimen (twice a week). Immunohistochemistry staining and quantification of C3/C3b/iC3b (**A**,**B**) and CD46 (**E**,**F**) were done using sections from the same transwell (n = 3 filter/transwell per genotype). Experiment was repeated twice. In (**A**,**E**) nuclei were stained with DAPI (blue) and scale bar = 20 µm.

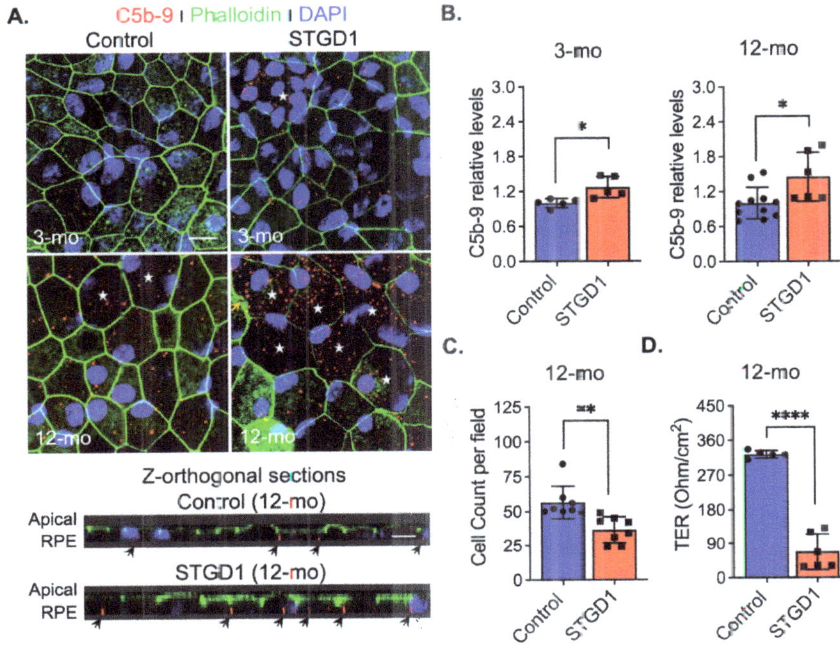

Figure 5. MAC (C5b-9) deposition induces cell damage in STGD1 RPE cells. (**A**) Representative confocal merge images of C5b-9 (red) and Phalloidin (green) acquired of control (left) and STGD1 (right) RPE cells in culture for three-months (3-mo, top row) and 12-months (12-mo, bottom row) in the presence of bovine retinal extract supplemented at two-months (twice per week). Corresponding z-orthogonal sections for 12-mo group are shown below the en-face confocal images. En-face images of 12-mo STGD1 RPE cells displayed numerous larger cells (white stars) delineated by the Phalloidin staining and co-localization of C5b-9 with Phalloidin is indicated by yellow arrows. Abundant C5b-9 immunoreactivity, with predominant basolateral distribution (indicated by the black arrows), is also evidenced in the z-orthogonal sections of 12-mo STGD1 RPE cells. Nuclei were stained with DAPI (blue). Scale bar = 10 µm. (**B**) Bar graphs show the relative C5b-9 levels determined by measuring pixel intensity from images acquired at 3-mo (left) and 12-mo (right). Average data presented as mean ± SD; * $p < 0.05$ ($p = 0.037$ for 3-mo and $p = 0.015$ for 12-mo); n = 3 filter/transwell per genotype. (**C**) Histogram shows average data presented as mean ± SD for total cell count of 12-mo control and STGD1 RPE cells; ** $p < 0.005$ ($p = 0.001$); n = 3 filter/transwell per genotype. (**D**) Transepithelial resistance measurements of 12-mo control and STGD1 RPE cells. Average data presented as mean ± SD; **** $p < 0.0001$; n = 5 filter/transwell for control and n = 6 filter/transwell for STGD1. Statistics analysis using Student's *t*-test.

3.8. MAC-Mediated RPE Cell Damage Is Evident in the Eyes of Abca4$^{-/-}$ Mice

Accumulation of toxic bisretinoids and elevated complement C3 activity were previously shown in the RPE of *Abca4*$^{-/-}$ mice at ~10-weeks, prior to the onset of photoreceptor degeneration [11,41]. Presumably, oxidized bisretinoids and C3b opsonization on the plasma membrane, along with the buildup of internalized C3b/iC3b fragments, led to a loss of RPE cells followed by degeneration of photoreceptors beginning at six-months in *Abca4*$^{-/-}$ mice [41,44,45]. Here, we evaluated the C3b/C3a breakdown fragments and their corresponding inactive forms (iC3b/C3a-desArg), by immunoblotting of RPE homogenates from six-month-old wild-type versus *Abca4*$^{-/-}$ mice (Figure S7A,B). We performed this analysis using a C3 antibody against the full-length protein and a C3a-desArg antibody specific for the smaller fragment C3a. Irrespective of the antibody and using reducing conditions to render all C3 fragments in the RPE cell homogenates, we could not detect the

C3a ~9 kDa band in the mouse RPE homogenates from either wild-type or *Abca4*$^{-/-}$ mice (Figure S7A,B). Moreover, immunostaining of mouse retina with C3a-desArg antibody displayed comparable C3a intensity for both wild-type and *Abca4*$^{-/-}$ mice (Figure S7D). At the same time, levels of the C3a receptor protein (C3aR) were similar in wild-type and *Abca4*$^{-/-}$ RPE homogenates by immunoblotting (Figure S7A,C), and in mouse retina sections by confocal immunofluorescence analysis (Figure S7D). To investigate the extent of the complement activation, we prepared RPE flatmounts from wild-type and *Abca4*$^{-/-}$ mouse RPE and performed immunohistochemistry with an antibody against C5b-9 terminal complement complex. We observed more intense immunolabeling of C5b-9 complex in RPE cells from *Abca4*$^{-/-}$ mice versus wild-type mice (Figure S7E,F). Further, we detected larger cells containing three or more nuclei, likely due to membrane fusion between cells, suggesting the formation of RPE syncytia in *Abca4*$^{-/-}$ compared to wild-type mouse flatmounts (Figure S7E). Orthogonal z-stacked optical sections, obtained from the apical side of the RPE flatmounts, revealed that MAC deposits were internalized by the RPE and were more pronounced on the basolateral plasma membrane of the *Abca4*$^{-/-}$ mice (Figure S7F). Cell count showed a two-fold increase in the multinucleated cells in the *Abca4*$^{-/-}$ vs wild-type RPE flatmount (Figure S7G). These data suggest that in the *Abca4*$^{-/-}$ mice, the C3 amplification loop is accompanied by an ongoing fully activated complement cascade and possible MAC-mediated lysis of RPE cellular plasma membrane.

4. Discussion

The RPE is an integral component of the blood-retina barrier and contributes to daily maintenance of the adjacent cellular layers [46]. The initiator of photoreceptor degeneration in STGD1, and in some cases of AMD that are associated with specific ABCA4 variants, is thought to be due to RPE dysfunction [3]. Rod and cone photoreceptors daily shed their distal OS, which are phagocytosed by the apical RPE as a mechanism to recycle OS components and to detoxify lipids that have undergone oxidation and other light-dependent modifications [47,48]. Following phagocytosis, rhodopsin and cone-opsin visual pigments undergo proteolysis, liberating their chromophore as free retinaldehyde, which is cytotoxic and highly reactive [49]. Until recently, little was known about the detoxification and recycling of retinaldehyde released into RPE phagolysosomes. When ABCA4 is missing, in *Abca4*$^{-/-}$ mice or STGD1 patients, fluorescent retinaldehyde adducts (bisretinoids including A2E) accumulate within RPE cells [41,50]. Recently, ABCA4 was shown to be present in RPE internal membranes, in addition to photoreceptor OS [7,8]. Together these findings suggest that ABCA4 in RPE serves to recycle retinaldehyde released during degradation of visual pigments by transferring *N*-ret-PE from the luminal to cytoplasmic face of endolysosomal membranes where the retinaldehyde can be reduced to retinol and re-enter the RPE visual cycle.

In the current study, we analyzed the role of ABCA4 in RPE cells from a clinically and genetically diagnosed STGD1 patient and an unaffected human control. Characterization of RPE cells from these individuals showed that the ABCA4 products of the mutant loci had no effect on the growth of fibroblasts, dedifferentiation of fibroblasts into iPSCs, or reprogramming of iPSCs into RPE cells (Figures 1 and S1, and [9]). Importantly, both control and STGD1 RPE cells matured into confluent monolayers at the same rate with similar trans-epithelium resistances and displayed normal ultrastructural features. Further, neither the control nor STGD1 RPE cells contained the AMD-associated SNP haplotypes (Table S1). Lastly, both the OS phagocytotic capacities and visual-cycle enzyme activities were similar in control and STGD1 RPE cells, allowing us to interrogate the coupling of ABCA4 deficiency and complement dysregulation as the driver of RPE cell death without the input from photoreceptor cells.

Although ABCA4 mRNA levels were comparable for STGD1 and control cells, levels of the ABCA4 protein were five-fold lower in the STGD1 cells signifying functional consequences [9,29,32]. We tested the ABCA4 N-Ret-PE flippase activity in the control and STGD1 RPE cells that were *never* previously exposed to visual retinoids, by feeding them with a single 'meal' of bovine OS. As predicted by the patient's ABCA4 variants, recycling of retinaldehyde was significantly reduced in the STGD1 RPE cells. ABCA4 also functions as a flippase for PE unconjugated to retinaldehyde [5], signaling a potential role in the cellular distribution of PE. To test this possibility, we investigated PE localization in RPE cells following treatment with duramycin, which binds the headgroup of PE [37]. Intracellular PE-aggregates and reduced PE in the plasma membrane in STGD1 RPE cells suggest that altered PE metabolism may play a role in STGD1 pathogenesis. This novel finding demands further investigation using additional RPE cells obtained from STGD1 patients with different ABCA4 mutations.

Patients with *ABCA4*-mediated retinopathies exhibit a strong correlation between visual loss and increased short wavelength (SW) autofluorescence in the RPE due to the presence of bisretinoids [2]. Bisretinoid levels are also several-fold higher in the RPE of *Abca4*$^{-/-}$ versus wild-type mice [7]. Although ABCA4 is far more abundant in photoreceptors versus RPE cells, bisretinoid levels are several-fold higher in the RPE versus neural retina [7]. Further, the major bisretinoid species, A2E, is undetectable in *Abca4*$^{-/-}$ neural retinas and only found in RPE cells. Here, RPE cells supplemented with bovine retinal extract showed much more 488-nm autofluorescent material in STGD1 versus control cells. *In vivo*, the bisretinoid chemical structure renders them susceptible to further modification and fragmentation, thereby generating reactive species responsible for increased oxidative stress and destabilization of cellular physiological activities [51–57]. Similar to STGD1 donor eyes [13], 12-month-old STGD1 RPE cells showed markedly increased levels of the lipid peroxidation marker, 4-HNE. Thus, RPE cells from the STGD1 patient recapitulates multiple features of the phenotypes seen in STGD1-human and *Abca4*$^{-/-}$ mouse eyes.

The complement system is a proteolytic cascade, activated through several initiating pathways, representing the first line of defense against an invading pathogen [43]. To prevent inappropriate inactivation and an attack on self, C3 activation, C3 hydrolysis (tick-over), and C3 degradation are all kept in check by multiple complement negative-regulatory proteins including CD46, CD59, and CFH. Both activating and regulatory proteins of the complement cascade were shown to be expressed in RPE cells, suggesting that the RPE protects the eye from blood-borne pathogens without dependence on systemic factors [58–60]. Disruption of RPE homeostasis can lead to C3-mediated attack on these cells Augmented autofluorescence build-up observed in STGD1 RPE cells triggered C3 activity in a similar fashion as was shown in the RPE cells of *Abca4*$^{-/-}$ mice and cultured human fetal RPE cells challenged with bisretinoids [11,61,62]. In a previous study, we showed that over-expressing the complement receptor 1-like protein y (CRRY), a major murine complement regulator and a functional homolog of human CD46, reduced complement attack on the RPE and rescued both bisretinoid accumulation and photoreceptor degeneration in *Abca4*$^{-/-}$ mice [12]. Here, increased C3 convertase activity and reduced levels of the CD46 negative-regulatory protein set up the RPE for attack by the complement system in STGD1 cells while CFH levels remain unchanged. These findings suggest that photoreceptor degeneration in *Abca4*$^{-/-}$ mice and STGD1 patients results from loss of RPE support due to chronic complement dysregulation in the early stage of the cascade.

The RPE's ability to internalize C3 and C5b-9 complex is another built-in defensive mechanism to prevent further C3b opsonizing fragments and formation of a cytolytic MAC on the plasma membrane [11,12,63]. Despite similar cell densities and morphologies, we observed substantial MAC deposition on the surface of STGD1 versus control RPE cells at three-months. MAC lytic effect is potentiated by the insertion of additional C9 molecules into the C5b-9 terminal complex blocked by CD59, another membrane-bound negative regulator [64]. Here, we found that CD59 expression was not altered in STGD1 RPE cells signifying further that the early inhibitory step of the complement cascade by CD46 is essential. When cells were maintained in culture for ~one year, we observed extensive MAC-mediated damage to membranes with ~35% loss of cells in the STGD1 RPE cultures. Finally, the barrier integrity of RPE cells derived from the STGD1 patient was severely compromised, indicated by a several-fold drop in TER. Comparable MAC-mediated pathological features were recently described in three donor eyes from STGD1 patients [13,14]. Notably, an inverse correlation between the basal laminar autofluorescent deposits in dysmorphic RPE cells and CD46 immunoreactivity was also evidenced in the AMD donor eyes [65]. STGD1 and AMD, despite superficially unrelated causes (loss of the ABCA4 transporter versus multigenic risk alleles including mutations in the genes for complement regulatory proteins), may actually both be caused by chronic inflammation of the RPE due to dysregulation of the complement system.

Taken together, these findings stress the importance of the CD46 regulator to limit the downstream C3 convertase activity that can lead to the harmful levels of complement terminal complex on the RPE cells. In STGD1, the ongoing intracellular PE-buildup and autofluorescent bisretinoid material, due to dysfunctional ABCA4 flippase, promotes the complement activity and drives the formation of the terminal complement complex C5b-9 (Figure 6). It appears that chronic deposition of the C5b-9 terminal complex, initially at a sub-lytic level, eventually destabilizes the plasma membrane by forming a lytic pore with loss of the RPE cells. Inefficient PE translocation by mutated ABCA4 may also compromise RPE cellular integrity due to an abnormal phospholipid composition of the plasma membrane. In STGD1 and sub-group of AMD patients, loss of RPE cells precedes the degeneration of photoreceptors. This may reflect an opportunity for therapeutic intervention window directed towards the RPE cells. Our current study supports the potential of the use of RPE cells as "disease-in-a-dish" biological system to test emerging therapies for *ABCA4*-mediated retinopathies.

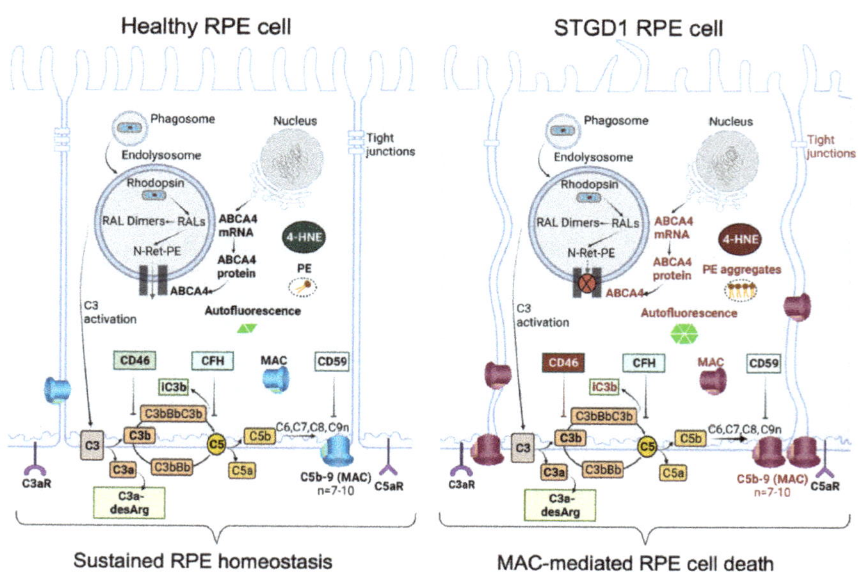

Figure 6. Schematic diagram summarizing our data of healthy RPE (**left**) and STGD1 RPE (**right**). ABCA4-deficiency in STGD1 RPE cells results in the ongoing intracellular accumulation of autofluorescent-lipofuscin material, including phosphatidylethanolamine (PE) aggregates, and augmented C3 complement cascade on the basolateral side of plasma membrane of RPE cells. An early key pathological feature of STGD1 RPE cells is the improper inhibition of C3 convertase by the CD46 complement negative regulator protein. Consequently, CD46 insufficiency in STGD1 RPE cells causes abnormal formation of complement terminal complex (C5b-9) on RPE plasma membrane. With more C9 addition to the C5b-9, the terminal complex destabilizes the plasma membrane by forming a lytic pore which causes loss of the RPE cells in the STGD1 patients. Diagram created with BioRender.com.

5. Conclusions

The RPE is thought to protect the retina from inappropriate attacks of the innate immune reactivity by expressing complement negative regulatory proteins. While genetic association studies have established the role of complement system in the etiology of maculopathies, such role for the RPE has never been shown in STGD1. Here, we recapitulated key pathological features of STGD1 in cultured RPE cells derived from a STGD1 patient STGD1 RPE cells showed lipofuscin-mediated complement activation and improper complement regulation. Decreased inhibition of the complement cascade by CD46 promoted the formation of the membrane attack complex on STGD1 cells causing loss of RPE homeostasis and later cell death. Therefore, mitigation of RPE local complement dynamics may slow disease progression in ABCA4-mediated retinopathies.

Supplementary Materials: The following supporting information can be downloaded at https //www.mdpi.com/article/10.3390/cells11213462/s1, Figure S1: Pluripotency, developmental, pigmentation markers, and ABCA4 expression and localization in human RPE cells cultured on 24-transwell inserts with growth medium without retinoids, Figure S2: Pulse-chase phagocytosis assay for human RPE cells, Figure S3: ABCA4-flippase activity in the RPE cells, Figure S4: Visual cycle proteins and activities in the RPE cells at three-months in culture, Figure S5: Complement C3a/C5a, CD59, and secreted proteins profile in human RPE cells, Figure S6: MAC (C5b-9) deposition accumulates in STGD1 RPE cells, Figure S7: Complement terminal complex destabilizes the RPE plasma membrane of the *Abca4*$^{-/-}$ mice, Table S1: Control and STGD1 patient variant information for additional SNPs correlated with AMD, Table S2: qRT-PCR primers information and Table S3: Antibodies information.

Author Contributions: Conceptualization, N.K., E.S.Y.N. and R.A.R.; methodology and investigation N.K., J.H., Z.J., J.P., A.M., E.S.Y.N. and R.A.R.; validation, N.K., E.S.Y.N. and R.A.R.; data curation N.K., E.S.Y.N. and R.A.R.; writing—original draft preparation N.K., E.S.Y.N. and R.A.R.; writing—review and editing, N.K., E.S.Y.N., A.M. and R.A.R.; visualization, N.K., J.H., Z.J., A.M., E.S.Y.N. A.D., M.B.G. and R.A.R.; supervision, R.A.R.; project administration, R.A.R.; funding acquisition R.A.R. All authors have read and agreed to the published version of the manuscript.

Funding: This research was funded by the National Eye Institute grants: R01 EY025002 (RAR), NRSA Vision Research Training Grant T32-EY007026 (ESYN), and P30-EY000331 Stein Eye Institute Core Grant for Vision Research. Additional support: Unrestricted Grant from Research to Prevent Blindness, Inc. (RPB, New York); Macula Vision Research Foundation Award #20142217 (RAR); Gerald Oppenheimer Family Foundation Center for Prevention of Eye Disease (RAR); the UCLA Eli and Edythe Broad Center of Regenerative Medicine and Stem Cell Research Rose Hills Foundation Innovator Grant Award #20172305 (RAR), the Daljit S. and Elaine Sarkaria Charitable Foundation (RAR, MBG); and Patricia and Joseph Yzurdiaga Vision Research Fund (RAR); RAR holds the Vernon O. Underwood Family Endowed Chair in Ophthalmology, UCLA David Geffen School of Medicine. The APC was funded by the UCLA Stein Eye Institute and Department of Ophthalmology unrestricted fund.

Institutional Review Board Statement: This study was conducted in accordance with the Declaration of Helsinki, approved by the Embryonic Stem Cell research Oversight and Institutional Review Board of UCLA (protocol#11-000020; approval: 08/25/21) and in accordance with regulations of the Health Insurance Portability and Accountability Act of 1996, with signed informed consent and specifying the use of the information for publications and research presentations. The animal study protocol was approved by the Animal Research Committee of UCLA (protocol#2012-057; approval: 5/17/2022) and performed in accordance with the Association for Research in Vision and Ophthalmology (ARVO) statement for the use of animal in ophthalmic and visual research.

Informed Consent Statement: Informed consent was obtained from all subjects involved in the study. All samples were de-identified before being used for our studies.

Data Availability Statement: The data presented in this study are available on request from R.A.R. (radu@jsei.ucla.edu).

Acknowledgments: We thank our human tissue donor participants, without whom this work would not be possible. We acknowledge Hui Sun, Gabe Travis (both at University of California, Los Angeles) and Joyce Tombran-Tink (Pennsylvania State University) for providing antibodies against human ABCA4, Peropsin, RPE65, and PDEF, respectively; Ashley Macabasco for support with tissue culture; Steven McMullen, Shannan Eddington, Marcia Lloyd, and Katherine Frei for their technical assistance with the microscopy studies. We thank Dean Bok, Anupama Dimashkie, and Gabe Travis for valuable comments on the manuscript.

Conflicts of Interest: The authors declare no conflict of interests. The funders had no role in the design of the study; in the collection, analyses, or interpretation of data; in the writing of the manuscript; or in the decision to publish the results.

References

1. Allikmets, R.; Shroyer, N.F.; Singh, N.; Seddon, J.M.; Lewis, R.A.; Bernstein, P.S.; Peiffer, A.; Zabriskie, N.A.; Li, Y.; Hutchinson, A.; et al. Mutation of the Stargardt disease gene (ABCR) in age-related macular degeneration. *Science* **1997**, *277*, 1805–1807. [CrossRef] [PubMed]
2. Burke, T.R.; Duncker, T.; Woods, R.L.; Greenberg, J.P.; Zernant, J.; Tsang, S.H.; Smith, R.T.; Allikmets, R.; Sparrow, J.R.; Delori, F.C. Quantitative fundus autofluorescence in recessive Stargardt disease. *Investig. Ophthalmol. Vis. Sci.* **2014**, *55*, 2841–2852. [CrossRef] [PubMed]
3. Fritsche, L.G.; Fleckenstein, M.; Fiebig, B.S.; Schmitz-Valckenberg, S.; Bindewald-Wittich, A.; Keilhauer, C.N.; Renner, A.B.; Mackensen, F.; Mossner, A.; Pauleikhoff, D.; et al. A subgroup of age-related macular degeneration is associated with mono-allelic sequence variants in the ABCA4 gene. *Investig. Ophthalmol. Vis. Sci.* **2012**, *53*, 2112–2118. [CrossRef] [PubMed]
4. Maugeri, A.; Klevering, B.J.; Rohrschneider, K.; Blankenagel, A.; Brunner, H.G.; Deutman, A.F.; Hoyng, C.B.; Cremers, F.P. Mutations in the ABCA4 (ABCR) gene are the major cause of autosomal recessive cone-rod dystrophy. *Am. J. Hum. Genet.* **2000**, *67*, 960–966. [CrossRef] [PubMed]
5. Quazi, F.; Lenevich, S.; Molday, R.S. ABCA4 is an N-retinylidene-phosphatidylethanolamine and phosphatidylethanolamine importer. *Nat. Commun.* **2012**, *3*, 925. [CrossRef]
6. Sun, H.; Nathans, J. Stargardt's ABCR is localized to the disc membrane of retinal rod outer segments. *Nat. Genet.* **1997**, *17*, 15–16. [CrossRef]
7. Lenis, T.L.; Hu, J.; Ng, S.Y.; Jiang, Z.; Sarfare, S.; Lloyd, M.B.; Esposito, N.J.; Samuel, W.; Jaworski, C.; Bok, D.; et al. Expression of ABCA4 in the retinal pigment epithelium and its implications for Stargardt macular degeneration. *Proc. Natl. Acad. Sci. USA* **2018**, *115*, E11120–E11127. [CrossRef]
8. Voigt, A.P.; Whitmore, S.S.; Lessing, N.D.; DeLuca, A.P.; Tucker, B.A.; Stone, E.M.; Mullins, R.F.; Scheetz, T.E. Spectacle: An interactive resource for ocular single-cell RNA sequencing data analysis. *Exp. Eye Res.* **2020**, *200*, 108204. [CrossRef]
9. Matynia, A.; Wang, J.; Kim, S.; Li, Y.; Dimashkie, A.; Jiang, Z.; Hu, J.; Strom, S.P.; Radu, R.A.; Chen, R.; et al. Assessing Variant Causality and Severity Using Retinal Pigment Epithelial Cells Derived from Stargardt Disease Patients. *Transl. Vis. Sci. Technol.* **2022**, *11*, 33. [CrossRef]
10. Zhou, J.; Kim, S.R.; Westlund, B.S.; Sparrow, J.R. Complement activation by bisretinoid constituents of RPE lipofuscin. *Investig. Ophthalmol. Vis. Sci.* **2009**, *50*, 1392–1399. [CrossRef]
11. Radu, R.A.; Hu, J.; Yuan, Q.; Welch, D.L.; Makshanoff, J.; Lloyd, M.; McMullen, S.; Travis, G.H.; Bok, D. Complement system dysregulation and inflammation in the retinal pigment epithelium of a mouse model for Stargardt macular degeneration. *J. Biol. Chem.* **2011**, *286*, 18593–18601. [CrossRef] [PubMed]
12. Lenis, T.L.; Sarfare, S.; Jiang, Z.; Lloyd, M.B.; Bok, D.; Radu, R.A. Complement modulation in the retinal pigment epithelium rescues photoreceptor degeneration in a mouse model of Stargardt disease. *Proc. Natl. Acad. Sci. USA* **2017**, *114*, 3987–3992. [CrossRef] [PubMed]
13. Hu, J.; Pauer, G.J.; Hagstrom, S.A.; Bok, D.; DeBenedictis, M.J.; Bonilha, V.L.; Hollyfield, J.G.; Radu, R.A. Evidence of complement dysregulation in outer retina of Stargardt disease donor eyes. *Redox Biol.* **2020**, *37*, 101787. [CrossRef] [PubMed]
14. Edwards, M.M.; Bonilha, V.L.; Bhutto, I.A.; Bell, B.A.; McLeod, D.S.; Hollyfield, J.G.; Lutty, G.A. Retinal Glial and Choroidal Vascular Pathology in Donors Clinically Diagnosed With Stargardt Disease. *Investig. Ophthalmol. Vis. Sci.* **2020**, *61*, 27. [CrossRef] [PubMed]
15. Lowry, W.E.; Richter, L.; Yachechko, R.; Pyle, A.D.; Tchieu, J.; Sridharan, R.; Clark, A.T.; Plath, K. Generation of human induced pluripotent stem cells from dermal fibroblasts. *Proc. Natl. Acad. Sci. USA* **2008**, *105*, 2883–2888. [CrossRef]
16. Buchholz, D.E.; Pennington, B.O.; Croze, R.H.; Hinman, C.R.; Coffey, P.J.; Clegg, D.O. Rapid and efficient directed differentiation of human pluripotent stem cells into retinal pigmented epithelium. *Stem Cells Transl. Med.* **2013**, *2*, 384–393. [CrossRef] [PubMed]
17. Karumbayaram, S.; Lee, P.; Azghadi, S.F.; Cooper, A.R.; Patterson, M.; Kohn, D.B.; Pyle, A.; Clark, A.; Byrne, J.; Zack, J.A.; et al. From skin biopsy to neurons through a pluripotent intermediate under Good Manufacturing Practice protocols. *Stem Cells Transl. Med.* **2012**, *1*, 36–43. [CrossRef]
18. Llorente, I.L.; Hatanaka, E.A.; Meadow, M.E.; Xie, Y.; Lowry, W.E.; Carmichael, S.T. Reliable generation of glial enriched progenitors from human fibroblast-derived iPSCs. *Stem Cell Res.* **2021**, *55*, 102458. [CrossRef]

19. Maminishkis, A.; Chen, S.; Jalickee, S.; Banzon, T.; Shi, G.; Wang, F.E.; Ehalt, T.; Hammer, J.A.; Miller, S.S. Confluent monolayers of cultured human fetal retinal pigment epithelium exhibit morphology and physiology of native tissue. *Investig. Ophthalmol. Vis. Sci.* **2006**, *47*, 3612–3624. [CrossRef]
20. Hu, J.; Bok, D. Culture of highly differentiated human retinal pigment epithelium for analysis of the polarized uptake, processing, and secretion of retinoids. In *Retinoids*; Springer: Berlin/Heidelberg, Germany, 2010; pp. 55–73.
21. Radu, R.A.; Hu, J.; Peng, J.; Bok, D.; Mata, N.L.; Travis, G.H. Retinal pigment epithelium-retinal G protein receptor-opsin mediates light-dependent translocation of all-trans-retinyl esters for synthesis of visual chromophore in retinal pigment epithelial cells. *J. Biol. Chem.* **2008**, *283*, 19730–19738. [CrossRef]
22. Sun, H.; Gilbert, D.J.; Copeland, N.G.; Jenkins, N.A.; Nathans, J. Peropsin, a novel visual pigment-like protein located in the apical microvilli of the retinal pigment epithelium. *Proc. Natl. Acad. Sci. USA* **1997**, *94*, 9893–9898. [CrossRef] [PubMed]
23. Mata, N.L.; Moghrabi, W.N.; Lee, J.S.; Bui, T.V.; Radu, R.A.; Horwitz, J.; Travis, G.H. Rpe65 is a retinyl ester binding protein that presents insoluble substrate to the isomerase in retinal pigment epithelial cells. *J. Biol. Chem.* **2004**, *279*, 635–643. [CrossRef] [PubMed]
24. Li, H.; Tran, V.V.; Hu, Y.; Mark Saltzman, W.; Barnstable, C.J.; Tombran-Tink, J. A PEDF N-terminal peptide protects the retina from ischemic injury when delivered in PLGA nanospheres. *Exp. Eye Res.* **2006**, *83*, 824–833. [CrossRef]
25. Garwin, G.G.; Saari, J.C. [19] High-performance liquid chromatography analysis of visual cycle retinoids. *Methods Enzymol.* **2000**, *316*, 313–324. [PubMed]
26. Yang, M.; Fong, H.K. Synthesis of the all-trans-retinal chromophore of retinal G protein-coupled receptor opsin in cultured pigment epithelial cells. *J. Biol. Chem.* **2002**, *277*, 3318–3324. [CrossRef] [PubMed]
27. Hazim, R.A.; Karumbayaram, S.; Jiang, M.; Dimashkie, A.; Lopes, V.S.; Li, D.; Burgess, B.L.; Vijayaraj, P.; Alva-Ornelas, J.A.; Zack, J.A.; et al. Differentiation of RPE cells from integration-free iPS cells and their cell biological characterization. *Stem Cell Res. Ther.* **2017**, *8*, 217. [CrossRef]
28. Hu, J.; Bok, D. The use of cultured human fetal retinal pigment epithelium in studies of the classical retinoid visual cycle and retinoid-based disease processes. *Exp. Eye Res.* **2014**, *126*, 46–50. [CrossRef]
29. Khan, M.; Cornelis, S.S.; Pozo-Valero, M.D.; Whelan, L.; Runhart, E.H.; Mishra, K.; Bults, F.; AlSwaiti, Y.; AlTalbishi, A.; De Baere, E.; et al. Resolving the dark matter of ABCA4 for 1054 Stargardt disease probands through integrated genomics and transcriptomics. *Genet. Med.* **2020**, *22*, 1235–1246. [CrossRef]
30. Zernant, J.; Lee, W.; Collison, F.T.; Fishman, G.A.; Sergeev, Y.V.; Schuerch, K.; Sparrow, J.R.; Tsang, S.H.; Allikmets, R. Frequent hypomorphic alleles account for a significant fraction of ABCA4 disease and distinguish it from age-related macular degeneration. *J. Med. Genet.* **2017**, *54*, 404–412. [CrossRef]
31. Runhart, E.H.; Sangermano, R.; Cornelis, S.S.; Verheij, J.; Plomp, A.S.; Boon, C.J.F.; Lugtenberg, D.; Roosing, S.; Bax, N.M.; Blokland, E.A.W.; et al. The Common ABCA4 Variant p.Asn1868Ile Shows Nonpenetrance and Variable Expression of Stargardt Disease When Present in trans With Severe Variants. *Investig. Ophthalmol. Vis. Sci.* **2018**, *59*, 3220–3231. [CrossRef]
32. Sun, H.; Smallwood, P.M.; Nathans, J. Biochemical defects in ABCR protein variants associated with human retinopathies. *Nat. Genet.* **2000**, *26*, 242–246. [CrossRef] [PubMed]
33. Sonoda, S.; Spee, C.; Barron, E.; Ryan, S.J.; Kannan, R.; Hinton, D.R. A protocol for the culture and differentiation of highly polarized human retinal pigment epithelial cells. *Nat. Protoc.* **2009**, *4*, 662–673. [CrossRef] [PubMed]
34. Ferrer, M.; Corneo, B.; Davis, J.; Wan, Q.; Miyagishima, K.J.; King, R.; Maminishkis, A.; Marugan, J.; Sharma, R.; Shure, M.; et al. A multiplex high-throughput gene expression assay to simultaneously detect disease and functional markers in induced pluripotent stem cell-derived retinal pigment epithelium. *Stem Cells Transl. Med.* **2014**, *3*, 911–922. [CrossRef] [PubMed]
35. Kamao, H.; Mandai, M.; Okamoto, S.; Sakai, N.; Suga, A.; Sugita, S.; Kiryu, J.; Takahashi, M. Characterization of human induced pluripotent stem cell-derived retinal pigment epithelium cell sheets aiming for clinical application. *Stem Cell Rep.* **2014**, *2*, 205–218. [CrossRef] [PubMed]
36. Young, R.W.; Bok, D. Participation of the retinal pigment epithelium in the rod outer segment renewal process. *J. Cell Biol.* **1969**, *42*, 392–403. [CrossRef] [PubMed]
37. Hou, S.; Johnson, S.E.; Zhao, M. A One-Step Staining Probe for Phosphatidylethanolamine. *Chembiochem* **2015**, *16*, 1955–1960. [CrossRef]
38. Jin, M.; Li, S.; Moghrabi, W.N.; Sun, H.; Travis, G.H. Rpe65 is the retinoid isomerase in bovine retinal pigment epithelium. *Cell* **2005**, *122*, 449–459. [CrossRef]
39. Nowak-Sliwinska, P.; van den Bergh, H.; Sickenberg, M.; Koh, A.H. Photodynamic therapy for polypoidal choroidal vasculopathy. *Prog. Retin. Eye Res.* **2013**, *37*, 182–199. [CrossRef]
40. Carlson, A.; Bok, D. Promotion of the release of 11-cis-retinal from cultured retinal pigment epithelium by interphotoreceptor retinoid-binding protein. *Biochemistry* **1992**, *31*, 9056–9062. [CrossRef]
41. Radu, R.A.; Yuan, Q.; Hu, J.; Peng, J.H.; Lloyd, M.; Nusinowitz, S.; Bok, D.; Travis, G.H. Accelerated accumulation of lipofuscin pigments in the RPE of a mouse model for ABCA4-mediated retinal dystrophies following Vitamin A supplementation. *Investig. Ophthalmol. Vis. Sci.* **2008**, *49*, 3821–3829. [CrossRef]
42. Kaur, G.; Tan, L.X.; Rathnasamy, G.; La Cunza, N.; Germer, C.J.; Toops, K.A.; Fernandes, M.; Blenkinsop, T.A.; Lakkaraju, A. Aberrant early endosome biogenesis mediates complement activation in the retinal pigment epithelium in models of macular degeneration. *Proc. Natl. Acad. Sci. USA* **2018**, *115*, 9014–9019. [CrossRef] [PubMed]

43. Anderson, D.H.; Radeke, M.J.; Gallo, N.B.; Chapin, E.A.; Johnson, P.T.; Curletti, C.R.; Hancox, L.S.; Hu, J.; Ebright, J.N.; Malek, G. et al. The pivotal role of the complement system in aging and age-related macular degeneration: Hypothesis re-visited. *Prog. Retin. Eye Res.* **2010**, *29*, 95–112. [CrossRef] [PubMed]
44. Radu, R.A.; Mata, N.L.; Bagla, A.; Travis, G.H. Light exposure stimulates formation of A2E oxiranes in a mouse model of Stargardt's macular degeneration. *Proc. Natl. Acad. Sci. USA* **2004**, *101*, 5928–5933. [CrossRef] [PubMed]
45. Wu, L.; Nagasaki, T.; Sparrow, J.R. Photoreceptor cell degeneration in Abcr−/−mice. In *Retinal Degenerative Diseases*; Springer: Berlin/Heidelberg, Germany, 2010; pp. 533–539.
46. Strauss, O. The retinal pigment epithelium in visual function. *Physiol. Rev.* **2005**, *85*, 845–881. [CrossRef] [PubMed]
47. LaVail, M.M. Rod outer segment disk shedding in rat retina: Relationship to cyclic lighting. *Science* **1976**, *194*, 1071–1074. [CrossRef]
48. Williams, D.; Anderson, D. *Rod and Cone Photoreceptor Cells: Outer-Segment Membrane Renewal*; Springer: Berlin/Heidelberg, Germany, 2010.
49. De, S.; Sakmar, T.P. Interaction of A2E with model membranes. Implications to the pathogenesis of age-related macular degeneration. *J. Gen. Physiol.* **2002**, *120*, 147–157. [CrossRef]
50. Sparrow, J.R.; Duncker, T. Fundus Autofluorescence and RPE Lipofuscin in Age-Related Macular Degeneration. *J. Clin. Med.* **2014**, 3, 1302–1321. [CrossRef]
51. Arunkumar, R.; Gorusupudi, A.; Li, B.; Blount, J.D.; Nwagbo, U.; Kim, H.J.; Sparrow, J.R.; Bernstein, P.S. Lutein and zeaxanthin reduce A2E and iso-A2E levels and improve visual performance in Abca4(-/-)/Bco2(-/-) double knockout mice. *Exp. Eye Res.* **2021**, *209*, 108680. [CrossRef]
52. Zhao, J.; Kim, H.J.; Ueda, K.; Zhang, K.; Montenegro, D.; Dunaief, J.L.; Sparrow, J.R. A vicious cycle of bisretinoid formation and oxidation relevant to recessive Stargardt disease. *J. Biol. Chem.* **2021**, *296*, 100259. [CrossRef]
53. Wu, Y.; Yanase, E.; Feng, X.; Siegel, M.M.; Sparrow, J.R. Structural characterization of bisretinoid A2E photocleavage products and implications for age-related macular degeneration. *Proc. Natl. Acad. Sci. USA* **2010**, *107*, 7275–7280. [CrossRef]
54. Finnemann, S.C.; Leung, L.W.; Rodriguez-Boulan, E. The lipofuscin component A2E selectively inhibits phagolysosomal degradation of photoreceptor phospholipid by the retinal pigment epithelium. *Proc. Natl. Acad. Sci. USA* **2002**, *99*, 3842–3847. [CrossRef] [PubMed]
55. Suter, M.; Remé, C.; Grimm, C.; Wenzel, A.; Jaäättela, M.; Esser, P.; Kociok, N.; Leist, M.; Richter, C. Age-related Macular Degeneration: The lipofuscin componentn-retinyl-n-retinylidene ethanolamine detaches proapoptotic proteins from mitochondria and induces apoptosis in mammalian retinal pigment epithelial cells. *J. Biol. Chem.* **2000**, *275*, 39625–39630. [CrossRef] [PubMed]
56. Schutt, F.; Davies, S.; Kopitz, J.; Holz, F.G.; Boulton, M.E. Photodamage to human RPE cells by A2-E, a retinoid component of lipofuscin. *Investig. Ophthalmol. Vis. Sci.* **2000**, *41*, 2303–2308.
57. Anderson, D.M.G.; Ablonczy, Z.; Koutalos, Y.; Hanneken, A.M.; Spraggins, J.M.; Calcutt, M.W.; Crouch, R.K.; Caprioli, R.M.; Schey, K.L. Bis(monoacylglycero)phosphate lipids in the retinal pigment epithelium implicate lysosomal/endosomal dysfunction in a model of Stargardt disease and human retinas. *Sci. Rep.* **2017**, *7*, 17352. [CrossRef] [PubMed]
58. Yang, P.; Tyrrell, J.; Han, I.; Jaffe, G.J. Expression and modulation of RPE cell membrane complement regulatory proteins. *Investig. Ophthalmol. Vis. Sci.* **2009**, *50*, 3473–3481. [CrossRef] [PubMed]
59. Pauly, D.; Agarwal, D.; Dana, N.; Schafer, N.; Biber, J.; Wunderlich, K.A.; Jabri, Y.; Straub, T.; Zhang, N.R.; Gautam, A.K.; et al. Cell-Type-Specific Complement Expression in the Healthy and Diseased Retina. *Cell Rep.* **2019**, *29*, 2835–2848. [CrossRef]
60. Hageman, G.S.; Anderson, D.H.; Johnson, L.V.; Hancox, L.S.; Taiber, A.J.; Hardisty, L.I.; Hageman, J.L.; Stockman, H.A.; Borchardt, J.D.; Gehrs, K.M.; et al. A common haplotype in the complement regulatory gene factor H (HF1/CFH) predisposes individuals to age-related macular degeneration. *Proc. Natl. Acad. Sci. USA* **2005**, *102*, 7227–7232. [CrossRef]
61. Jabri, Y.; Biber, J.; Diaz-Lezama, N.; Grosche, A.; Pauly, D. Cell-Type-Specific Complement Profiling in the ABCA4(-/-) Mouse Model of Stargardt Disease. *Int. J. Mol. Sci.* **2020**, *21*, 8468. [CrossRef]
62. Zhou, J.; Jang, Y.P.; Kim, S.R.; Sparrow, J.R. Complement activation by photooxidation products of A2E, a lipofuscin constituent of the retinal pigment epithelium. *Proc. Natl. Acad. Sci. USA* **2006**, *103*, 16182–16187. [CrossRef]
63. Georgiannakis, A.; Burgoyne, T.; Lueck, K.; Futter, C.; Greenwood, J.; Moss, S.E. Retinal Pigment Epithelial Cells Mitigate the Effects of Complement Attack by Endocytosis of C5b-9. *J. Immunol.* **2015**, *195*, 3382–3389. [CrossRef]
64. Morgan, B.P.; Boyd, C.; Bubeck, D. Molecular cell biology of complement membrane attack. *Semin. Cell Dev. Biol.* **2017**, *72*, 124–132. [CrossRef] [PubMed]
65. Vogt, S.D.; Curcio, C.A.; Wang, L.; Li, C.M.; McGwin, G., Jr.; Medeiros, N.E.; Philp, N.J.; Kimble, J.A.; Read, R.W. Retinal pigment epithelial expression of complement regulator CD46 is altered early in the course of geographic atrophy. *Exp. Eye Res.* **2011**, *93*, 413–423. [CrossRef] [PubMed]

cells

MDPI

Article

Mitophagy in Astrocytes Is Required for the Health of Optic Nerve

Meysam Yazdankhah [1,*], Sayan Ghosh [1], Haitao Liu [1], Stacey Hose [1], J. Samuel Zigler, Jr. [2] and Debasish Sinha [1,2]

1 Department of Ophthalmology, University of Pittsburgh School of Medicine, Pittsburgh, PA 15213, USA; sayang@pitt.edu (S.G.); hal140@pitt.edu (H.L.); stacey.hose@pitt.edu (S.H.); debasish@jhmi.edu (D.S.)

2 Department of Ophthalmology, The Wilmer Eye Institute, The Johns Hopkins School of Medicine, Baltimore, MD 21205, USA; szigler45@gmail.com

* Correspondence: meysam.yazdankhah@gmail.com; Tel.: +1-443-875-1871

Abstract: Mitochondrial dysfunction in astrocytes has been implicated in the development of various neurological disorders. Mitophagy, mitochondrial autophagy, is required for proper mitochondrial function by preventing the accumulation of damaged mitochondria. The importance of mitophagy specifically in the astrocytes of the optic nerve (ON), has been little studied. We introduce an animal model in which two separate mutations act synergistically to produce severe ON degeneration. The first mutation is in *Cryba1*, which encodes βA3/A1-crystallin, a lens protein also expressed in astrocytes, where it regulates lysosomal pH. The second mutation is in *Bckdk*, which encodes branched-chain ketoacid dehydrogenase kinase, which is ubiquitously expressed in the mitochondrial matrix and involved in the catabolism of the branched-chain amino acids. BCKDK is essential for mitochondrial function and the amelioration of oxidative stress. Neither of the mutations in isolation has a significant effect on the ON, but animals homozygous for both mutations (DM) exhibit very serious ON degeneration. ON astrocytes from these double-mutant (DM) animals have lysosomal defects, including impaired mitophagy, and dysfunctional mitochondria. Urolithin A can rescue the mitophagy impairment in DM astrocytes and reduce ON degeneration. These data demonstrate that efficient mitophagy in astrocytes is required for ON health and functional integrity.

Keywords: mitophagy; mitochondria; lysosome; BCKDK; βA3/A1-crystallin; astrocytes; autophagy; optic nerve

Citation: Yazdankhah, M.; Ghosh, S.; Liu, H.; Hose, S.; Zigler, J.S., Jr.; Sinha, D. Mitophagy in Astrocytes Is Required for the Health of Optic Nerve. *Cells* **2023**, *12*, 2496. https://doi.org/10.3390/cells12202496

Academic Editor: Hossein Ameri

Received: 31 August 2023
Revised: 17 October 2023
Accepted: 18 October 2023
Published: 20 October 2023

1. Introduction

Astrocytes are the most abundant glial cells in the central nervous system and play a crucial role in maintaining optic nerve health and function [1]. The retina and the optic nerve track are among the most metabolically active tissues, requiring significant energy resources [2,3]. Proper mitochondrial function in various ocular cells is essential to ensure an adequate energy supply for the transmission of light signals [4,5]. Irregularities in mitochondrial function are commonly observed in various ocular diseases, including glaucoma [6], age-related macular degeneration (AMD) [7,8], and diabetic retinopathy (DR) [9]. Mitochondrial dysfunction in astrocytes has also been implicated in the development of neurological disorders, including Alzheimer's disease (AD) [10], Parkinson's disease [11], and inflammatory diseases [12–14]. However, the impact of mitochondrial dysfunction in astrocytes on the pathogenesis of ocular diseases largely remains unknown. To investigate this, we developed a unique animal model with two spontaneous mutations.

We previously reported two spontaneous mutations that arose in the Sprague Dawley rat [15–17]. The first of these, called Nuc1, has an obvious eye-specific phenotype in which heterozygotes have dense nuclear cataracts, and homozygotes have a much more severe phenotype with microphthalmia and lens rupture prior to birth [16,18]. Gene linkage studies identified *Cryba1* as the affected gene and subsequent analysis demonstrated an insertion mutation in Exon 6 that replaced an absolutely conserved glycine with 10 novel amino acids [16,17]. *Cryba1* encodes βA3/A1-crystallin, two isoforms that are members of the β/γ-crystallin family of proteins, major structural proteins of the ocular lens essential to the establishment of the organ's critical attributes—refractivity and transparency [16]. Crystallins are known to have been recruited to the lens from proteins existing before the lens evolved with functions unrelated to their role in the lens. This gave rise to the concept of 'gene sharing' [19] or subsequently, 'moonlighting' proteins [20], where a single protein may have different functions when expressed in different cell types or under different conditions [20]. Our data suggest that, in Nuc1, the semi-dominant phenotype results from such a situation. In addition to being expressed at high abundance in the lens, *Cryba1* is also present at much lower levels in certain other ocular cell types, in particular astrocytes and retinal pigmented epithelium (RPE) cells [16]. While these cell types appear normal in Nuc1 heterozygotes, cells from homozygotes exhibit definitive defects in cell signaling networks particularly involving lysosomal function, autophagy, and the mTORC1 pathway [21]. Thus, Nuc1 is a recessive phenotype with respect to these cellular functions, but in the lens, the mutant protein is abundant and fails to fold properly, leading to aggregation and nuclear cataract formation even in heterozygotes. We have performed extensive investigations into the effects of the Nuc1 mutation in both astrocytes and RPE cells and the ultimate impact on the eye in vivo [15,22].

While maintaining a colony of Nuc1 rats, a second distinct phenotype appeared. In a litter of thirteen pups from a mating of two Nuc1 homozygotes, five pups displayed severe hindlimb dysfunction. The affected pups had a frog-like gait with hindlimb hypertonicity, hyperextension, and dorsal rotation. Their eyes did not appear different from their non-'frogleg' littermates. To determine whether we were dealing with a second de novo mutation or a new manifestation of Nuc1, we undertook a selective breeding approach to see if the two phenotypes could be separated. While this was complicated by the fact that we were unable to successfully breed 'frogleg' homozygotes with each other or with wild-type rats, we were eventually successful in separating the two phenotypes and establishing that 'frogleg' was the result of an independent mutation transmitted in an autosomal dominant fashion [15]. Using microsatellite markers spanning the rat genome, the affected gene was linked to rat chromosome 1q32-1q37; whole-genome sequencing within this linkage interval identified a missense mutation in the gene *Bckdk* (branched chain keto-acid dehydrogenase kinase) as the most likely cause of 'frogleg' [15]. BCKDK is ubiquitously expressed and is a critical component of the mitochondrial matrix, regulating levels of the branched-chain amino acids (BCAAs), leucine, isoleucine, and valine [23,24]. Specifically, BCKDK regulates the catabolism of the BCAAs by phosphorylating the rate-limiting enzyme in the pathway, BCKDH (branched-chain keto-acid dehydrogenase) [25]. When BCKDH is not phosphorylated at Ser293, its activity is unchecked, and levels of the BCAAs fall dramatically with significant pathological consequences [26]. We were able to demonstrate that the substitution of glutamine for a highly conserved glycine at position 369 of BCKDK was the responsible mutation by showing that, in 'frogleg' homozygotes, there was no detectable phosphorylation of Ser293 in BDKDH and that levels of the three BCAAs were markedly reduced, with no change in other amino acids [15]. A chance histological observation on the eyes of a rat homozygous for both mutations that had not been identified as such led us to the discovery that animals homozygous for both mutations (double mutant (DM)) exhibited severe degenerative effects on the inner retina and optic nerve. The optic nerve is composed of the axons of the retinal ganglion cells and is responsible for transmitting visual information from the retina to the brain. In the optic nerve, astrocytes express both *Cryba1* and *Bckdk*. Autophagy is a cellular process through which damaged cellular components, such as proteins and organelles, are delivered to lysosomes for degradation [27,28]. Selective mitochondrial autophagy, mitophagy, is a crucial quality control mechanism in the maintenance of proper mitochondrial function by preventing the accumulation of damaged mitochondria [29,30]. The importance of mitophagy, specifically in the astrocytes of the optic nerve, has been little studied. To address this question, here, we introduce a unique animal model in which two separate mutations (in *Cryba1* and *Bckdk*) act synergistically to produce severe optic nerve degeneration. While neither of the mutations in isolation has a significant effect on optic nerve structure or myelination, animals homozygous for both mutations exhibit very serious optic nerve degeneration and demyelination. We demonstrate that a simultaneous lack of functional βA3/A1-crystallin and BCKDK leads to lysosomal abnormality and dysfunctional mitochondria in astrocytes, which in turn leads to optic nerve degeneration. This finding highlights the importance of proper mitochondrial function and mitophagy in maintaining the health and function of astrocytes in the optic nerve.

2. Materials and Methods

2.1. Animals

All animal-related investigations adhered to the principles outlined in the Guide for the Care and Use of Animals (The National Academies Press, Washington, DC, USA) and received approval from the University of Pittsburgh Animal Care and Use Committee (Protocol Number: 20108281). Nuc1 was a spontaneous mutation in Sprague Dawley rats, and then the second mutation also arose spontaneously in our Nuc1 colony [15].

2.2. Astrocyte Cultures

The astrocytes from the optic nerve of 2-day-old rats were cultured following a previously established methodology [31].

2.3. Measurement of Autophagy Flux

To measure autophagy flux, cultured astrocytes from different genotypes were treated with bafilomycin A1 (BafA1) (Sigma-Aldrich, Burlington, MA, USA, B1793), an inhibitor of lysosomal acidification. The ratio of LC3-II/MAP1LC3B levels in the presence and absence of BafA1 was determined through immunoblotting, following a previously described method [31–33].

2.4. Measurement of Mitochondrial Membrane Potential and Superoxide Generation

To assess the mitochondrial membrane potential (MMP), we utilized tetramethylrhodamine methyl ester (TMRM; Thermo Fisher Scientific, Waltham, MA, USA, T668) as described before [33,34]. TMRM is a cationic dye that accumulates in the mitochondria in a membrane potential-dependent manner, making it useful for assessing mitochondrial health and function [35]. TMRM carries a positive charge, allowing it to accumulate within active mitochondria, which possess a negative charge. In contrast, depolarized or inactive mitochondria exhibit reduced membrane potential and do not retain TMRM as efficiently [36,37]. Additionally, mitochondrial superoxide levels were measured using the MitoSOX Red Mitochondrial Superoxide Indicator (Thermo Fisher Scientific, Waltham, MA, USA, M36008) according to the manufacturer's instructions. Cells from all experimental conditions were washed with PBS and incubated with either 5 μM MitoSOX or 400 nM TMRM for 15 min at 37 °C, protected from light. The cells were washed with PBS thrice, scraped from the plate, and analyzed using a flow cytometer, BD FACS Aria III (BD Bioscience, Franklin Lakes, NJ, USA). The data were analyzed using FlowJo software (v10.6.178). Alternatively, mitochondrial superoxide generation was determined using lucigenin (bis-N-methyl acridinium nitrate; Sigma-Aldrich, Burlington, MA, USA, M8010) as previously described [38]. Briefly, a solution of 5 mM lucigenin was prepared in Hanks' balanced salt solution (HBSS, Thermo Fisher Scientific, Waltham, MA, USA, 88284). Subsequently, 100 μL of pre-warmed lucigenin buffer was added to the culture plates and incubated at 37 °C and 5% CO_2 for 5 min. Luminescence readings were then obtained using a Glomax bioluminescence apparatus (Promega, Madison, WI, USA, E5311), with 5 consecutive readings per sample taken for 1 s each. The luminescence signal was adjusted based on the total protein content and expressed as arbitrary units (AUs) per milligram of protein.

2.5. Viral Delivery of COX8–EGFP–mCherry

For the generation of COX8–EGFP–mCherry, the pCLBW COX8 EGFP mCherry plas mid was procured from Addgene (Watertown, MA, USA, 78520), which was initiall deposited by Dr. David Chan's laboratory. Subsequently, the sequence encoding COX8 EGFP–mCherry was subcloned into an adenovirus vector by employing the services o Vector Biolabs, Malvern, PA, USA. To ensure the expression of the construct, cells wer infected with the adenovirus at a multiplicity of infection (MOI) of 50 and allowed t incubate for 24 h before commencing the experiments [33,39].

2.6. Western Blotting and Quantification

Western blotting and quantification were conducted following the previously estab lished protocols [33]. The signal intensity of the detected protein was normalized to tha of the housekeeping protein and analyzed using ImageJ analysis software (latest versio 1.53t). All targeted proteins and internal loading controls were detected and confirmed t be within the same linear range. The primary antibodies used for western blotting in thi study were as follows: MBP from Biolegend (San Diego, CA, USA (808401)); LC3, phosph P70S6K, P70S6K, phospho-S6, S6, and cleaved caspase 1 from Cell Signaling Technolog (Danvers, MA, USA (2775, 9204, 2708, 2211, 2217, and 89332)); SOD2, CNPase, and vinculi from Abcam (Waltham, MA, USA (ab13533, ab6319, and ab129002)); SQSTM1/p62 an NLRP3 from Novus Biologicals (Englewood, CO, USA (NBP1-42821 and NBP2-12446)); an β-Actin from Sigma-Aldrich (Burlington, MA, USA (A2066)). The secondary antibodie used were peroxidase-labeled goat anti-rabbit and anti-mouse antibodies from KPL (Sera Care, Milford, MA, USA, (074–1506 and 074–1806)).

2.7. Enzyme-Linked Immunosorbent Assay (ELISA)

ELISA was conducted utilizing lysates from 4-month-old optic nerves, following previously established procedures [40]. In brief, the optic nerves were homogenized i 300 μL of complete extraction buffer (Abcam, Waltham, MA, USA, ab193970). Subsequentl the ELISA was performed in 96-well microtiter plates (Sigma-Aldrich, Burlington, MA USA, M9410-1CS) coated with tissue lysates and then left to incubate overnight at 4 °C The plates were subsequently blocked with 5% BSA for 2 h. After PBS washing, 50 μL o IL1β antibody (Abcam, Waltham, MA, USA, ab9722) at a 1:1000 dilution was added to each well and allowed to incubate for 2 h at room temperature. The cytokines bound to the plate were detected using secondary IgG-HRP (KPL, Sera Care, Milford, MA, USA, 074–1506) TMB substrate solution (Thermo Fisher Scientific, Waltham, MA, USA, 34021) was used to develop color, and the reaction was stopped with 2 N H_2SO_4 solution (Thermo Fishe Scientific, Waltham, MA, USA, MK-H381-1). The absorbance was measured at 450 nm using a microplate reader [40].

2.8. Hematoxylin–Eosin Staining

The eyes of rats from different genotypes underwent the following procedure fo hematoxylin–eosin staining as described before [18]: The initial fixation of the eyes was accomplished using a solution containing 2.5% glutaraldehyde (Sigma-Aldrich, Burlington MA, USA, G5882) followed by formalin treatment. The fixed eyes were subsequently subjected to a series of graded ethanol solutions, progressively increasing in concentration This process aimed to facilitate dehydration. The eyes were then embedded in methyl methacrylate (Sigma-Aldrich, Burlington, MA, USA, M55909), and 1 μm sections were cut These sections were then subjected to hematoxylin and eosin staining (TissuePro Technology Gainesville, FL, USA, EY07-500R) and were examined utilizing a light microscope, enabling the detailed observation of the cellular structures and characteristics.

2.9. Electron Microscopy

The rat optic nerves from different genotypes underwent the following preparation steps for electron microscopy as explained previously [41]: The optic nerves were ini tially fixed in a solution containing 1.5% glutaraldehyde in 0.08 mol/L phosphate buffer (Sorenson's buffer, Thermo Fisher Scientific, Waltham, MA, USA, 50365831) at pH 7.2. This fixation process was conducted for a duration of 2 h at room temperature. After fixation, the nerves underwent two 10 min washes in the buffer. A post-fixation procedure was performed using 1% osmium tetroxide for 1 h. A series of alcohol washes were carried out to dehydrate the nerves. Subsequently, the nerves were transitioned to propylene oxide. The nerves were infiltrated with a mixture of 50% propylene oxide (Sigma-Aldrich, Burlington, MA, USA, 75569) and Epon resin (GracoRoberts, Arlington, TX, USA, V00298) overnight Ultrathin sections were prepared utilizing a Leica Ultramicrotome UCT (Leica Microsystems, Wetzlar, Germany). These sections were subsequently stained with uranyl acetate and lead citrate before being visualized using an H7600 transmission electron microscope (Hitachi, Tokyo, Japan).

2.10. Drug Delivery to Animals

UA (Sigma-Aldrich, Burlington, MA, USA, SML1791) was initially prepared at a concentration of 25 mg/mL in DMSO, dimethyl sulfoxide (Thermo Fisher Scientific, Waltham, MA, USA, J66650.AK). This concentrated solution was then appropriately diluted using PBS (pH = 7.4) to attain a final concentration of 0.25 mg/mL. The final DMSO concentration that was used as vehicle control in any treatment was 1%. DM rats were intraperitoneally (i.p.) treated with the prepared UA solution at a dose of 5 mg/kg body weight. In addition, 21-day-old WT and Nuc1 rats received intraperitoneal injections of UA or 1% DMSO (as the vehicle) three times a week over a span of three months. Following this treatment regimen, the optic nerves were fixed for immunohistochemical and immunoblotting analyses.

2.11. Tissue Preparation and Immunohistochemistry

Standard techniques were used for tissue preparation and subsequent immunohistochemistry analysis, following the established protocol as described [42,43]. Images from sections immunostained for Iba1 (Abcam, Waltham, MA USA, ab178846), RPBMS (Abcam, Waltham, MA USA, ab152101), and CD4 (Abcam, Waltham, MA USA, ab133616) were captured using a Zeiss confocal microscope. The obtained images were then subjected to analysis using the Zen Light Edition 2009 software (ZEN 2009-Zeiss Enhanced Navigation), which is a free digital image processing software offered by Zeiss for this purpose.

2.12. Statistical Analysis

GraphPad 6.0 software was utilized for conducting statistical analysis. The significance levels (p-values) were determined through 2-tailed Student's t-test and repeated-measure ANOVA tests, followed by Tukey's post hoc test. These analyses were based on data collected from a minimum of three distinct experiments. Statistical significance between the groups was established as follows: * for $p < 0.05$, ** for $p < 0.01$, and *** for $p < 0.001$. Data are expressed as mean ± SEM, and each biological replicate consists of a minimum of three technical replicates.

3. Results

3.1. Optic Nerve Degeneration in the DM Rats

Eyes and optic nerves were prepared for histological analysis from groups of wild-types (WT), Nuc1 homozygous (Nuc1), frogleg homozygous (FL), and double-mutant (DM) homozygous rats at 4 months of age. Optic nerve cross-sections for DM show massive degeneration on TEM images, while the Nuc1 and FL samples have an optic nerve structure similar to WT (Figure 1a). Likewise, cross-sections of the retinas clearly show the absence of the inner nuclear layer (ganglion cells) in the DM sample, while both single mutants appear essentially normal (Figure 1b). To corroborate these results and determine if the myelination defects seen in the optic nerve are selective for that tissue, extracts were prepared from optic nerves and the brainstem and cortex (other heavily myelinated tissues) in groups of WT and DM rats and analyzed via western blotting for markers of myelination (MBP-myelin basic protein and CNPase-2′,3′-Cyclic-nucleotide 3′-phosphodiesterase) [44,45]. Figure 1c shows that while both markers are highly significantly reduced in the DM optic nerve relative to WT, no such difference is seen in the brain stem or cortex (Figure 1d,e) at 4 months of age. Thus, the myelination defect seems to be specific to the optic nerve.

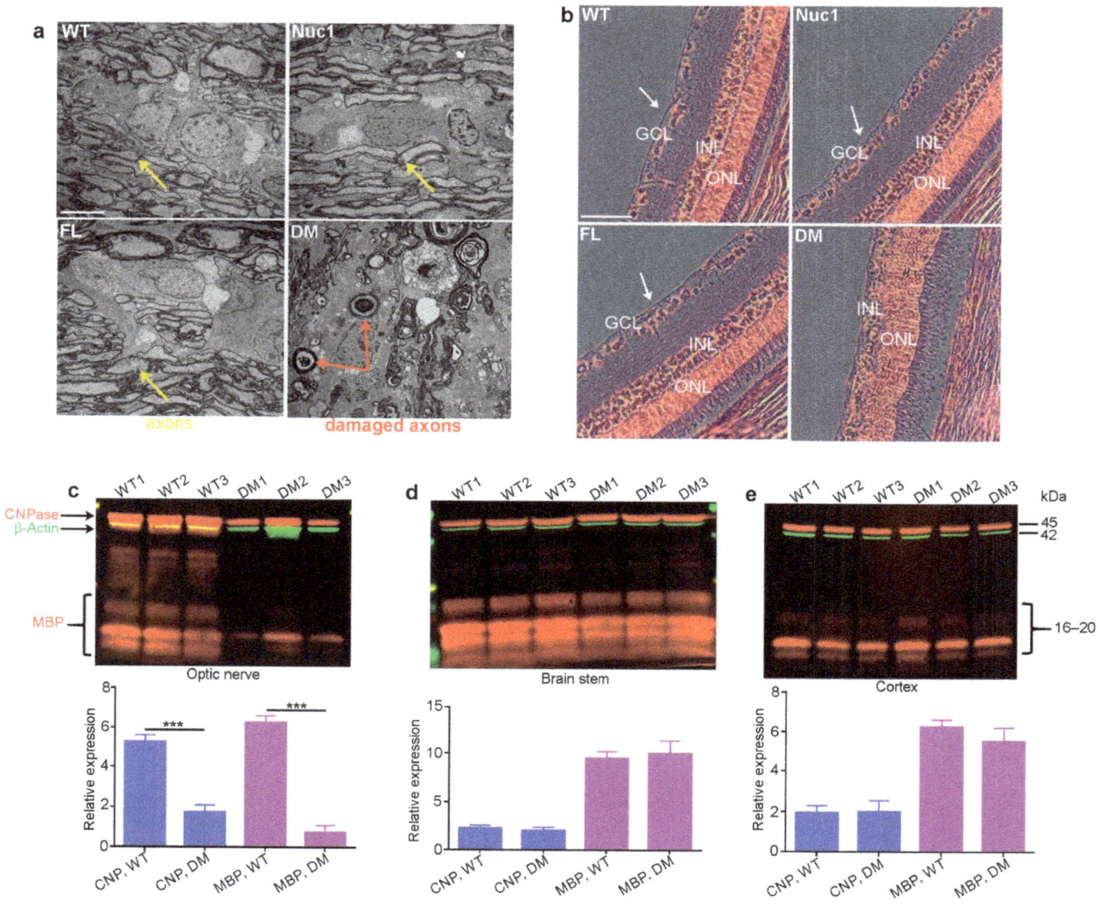

Figure 1. Degeneration of optic nerve in DM rats: (**a**) Transmission electron microscopy (TEM) revealed severe optic nerve degeneration in the DM rats compared to other genotypes at 4 months of age. Orange and red arrows show the axons and damaged axons, respectively. Bar: 5 μm. (**b**) H&E staining showing loss of RGCs (white arrows) in 4-month-old DM rats relative to other genotypes. GCL: ganglion cell layer; INL: inner nuclear layer; ONL: outer nuclear layer, Bar: 100 μm. (**c**) The levels of both CNPase and MBP are significantly decreased specifically in the optic nerves of DM rats at 4 months of age compared to WT, but (**d**) the brainstem and (**e**) cortex are not affected. Graphs represent mean ± SEM from at least 3 independent experiments repeated in triplicate, *** $p < 0.001$.

3.2. Optic Nerve Forms Normally in the DM and then Degenerates

In the rat optic nerve, myelination is generally completed by 3–4 weeks after birth [46]. To determine whether the myelination deficits seen in the DM optic nerves at 4 months reflected a failure to produce myelin or a degeneration of myelin after it was produced, we repeated the above analyses on 21-day-old rats. As seen in Figure 2a, at 21 days of age, all four genotypes have similar levels of both myelin markers. As could be expected, staining for retinal ganglion cells on retinal flat mounts from rats of all four genotypes at 21 days and 4 months of age showed cell loss only in the 4-month DM, with no apparent change at 21 days relative to the WT and single-mutant samples (Figure 2b). This is consistent with the great reduction in MBP in the optic nerve of 4-month DM rats, which is not observed in 21-day-old rats. These data demonstrate that the loss of myelin in the DM optic nerve results from degeneration after myelination is complete.

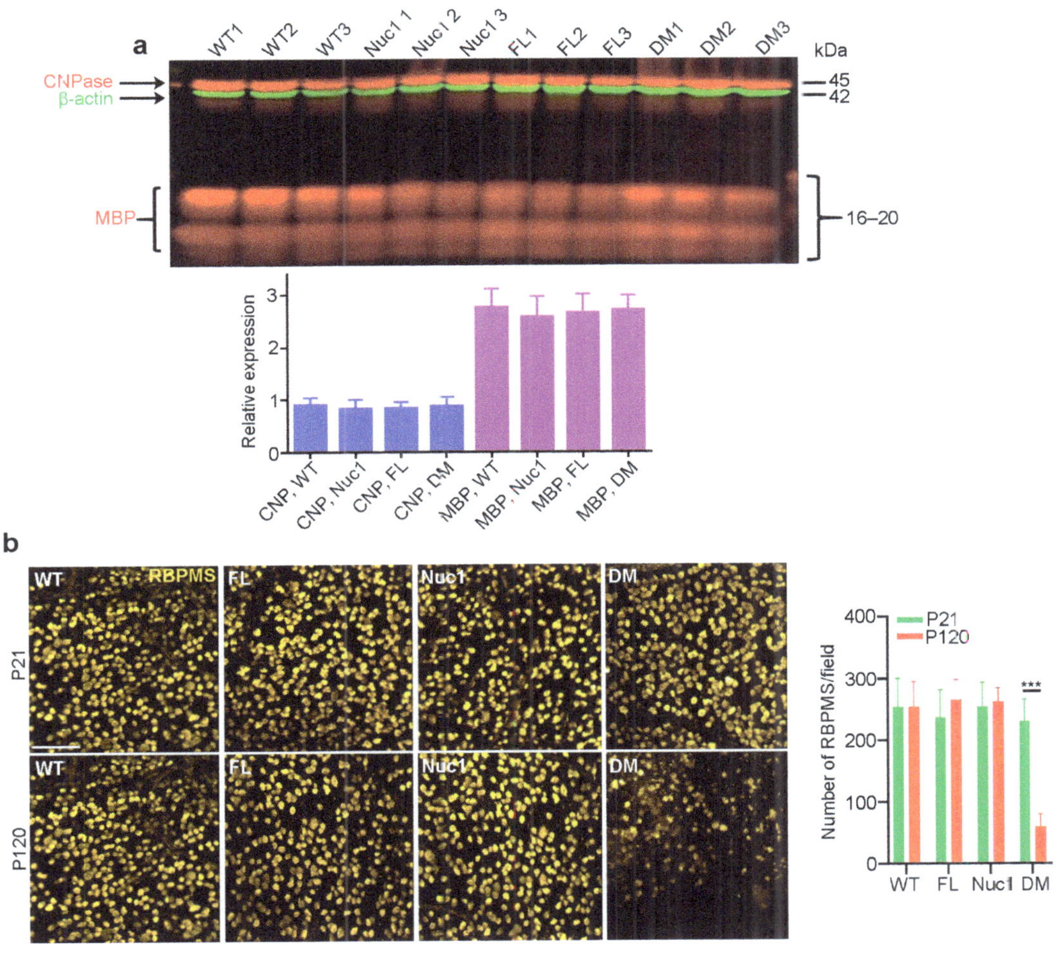

Figure 2. Optic nerves form normally in the DM and then degenerate: (**a**) In contrast to the changes shown in Figure 1c, at P21, immunoblotting showed that there were no differences in levels of CNPase or MBP in optic nerves of DM rats relative to the other genotypes. (**b**) Immunohistochemistry of retinal flat mounts against RBPMS (anti-RNA binding protein with multiple splicing) was used to detect RGCs. RBPMS staining was significantly reduced in DM retinas relative to other genotypes at 4 months of age. Bar: 50 μm. Graphs represent mean ± SEM from at least 3 independent experiments repeated in triplicate, *** $p < 0.001$.

3.3. *Optic Nerve Degeneration Is Associated with Inflammation*

The RNA-Seq analysis of optic nerves from 4-month-old rats revealed increases in the expression of several important inflammation-associated genes in DM compared to the other genotypes. However, the expression of myelin/oligodendrocytes-related genes (Mbp, Plp1, and Cspg4) was significantly reduced in DM (Figure 3a). To determine whether inflammation is involved in DM optic nerve degeneration, the microglia/macrophage pop ulations in the four different genotypes were characterized. To this end, optic nerves wer dissected from 4-month-old rats and analyzed via immunostaining and immunoblotting fo ionized calcium binding adaptor molecule 1 (Iba1, marker of microglia/macrophages) [47 DM rats demonstrated a drastic increase in the number of Iba1-positive cells, compared t the other genotypes (Figure 3b,c). Furthermore, the immunostaining of optic nerves fron different genotypes for CD4 led to the detection of CD4+ T lymphocytes [48] only in DN optic nerve (Figure S1). The activation of microglia/macrophages is also associated wit the activation of the inflammasome pathway, a critical component of the immune respons in the central nervous system. Its activation in both astrocytes and microglia can contribut to neuroinflammation and the pathogenesis of neurological disorders [49,50]. We foun that the expression of inflammasome markers was significantly increased only in the DN optic nerves (Figure 3d,e). The inflammasome pathway produces the bioactive form o IL-1β, a crucial cytokine for both immunoregulation and inflammation [51,52]. We foun that the expression of IL-1β is increased in DM optic nerves, indicating that degeneratio of the optic nerve in DM rats is associated with the activation of inflammation.

3.4. Excessive Mitochondrial Oxidative Stress in DM Astrocytes

Given that astrocytes are the only cells in the optic nerve that express both *Cryba1* an *Bckdk*, we investigated the synergistic effects of the two mutations on astrocyte functio Astrocytes were isolated from adult rats (all four different genotypes) and expande in culture. The MitoSox assay [33,53] demonstrated increased mitochondrial reactiv oxidative stress in FL and particularly DM astrocytes (Figure 4a). In addition, mitochondria superoxide anion levels were significantly increased in FL and DM astrocytes relative t WT and Nuc1 cells (Figure 4b). The expression of mitochondrial superoxide dismutas 2, (SOD2) [33,54], a mitochondrial antioxidant, was significantly increased in FL and DN astrocytes relative to other genotypes, consistent with increased oxidative stress in FL and DM cells (Figure 4c). In addition, the mitochondrial membrane potential (MMP) wa significantly reduced only in DM astrocytes relative to the other genotypes (Figure S2) These data collectively show that oxidative stress affects the function of mitochondria i DM astrocytes to a much greater extent than in astrocytes from single-mutant animals.

This information led us to speculate and envision the following mechanism for th generation of the astrocyte abnormality in the DM optic nerve (Figure 4d): In astrocyte derived from WT rats, we expect to see normal lysosomes and mitochondria, as well a basal levels of mitophagy and oxidative stress. In Nuc1 rats, the previously documente lysosomal abnormality is expected to result in impairment of autophagosome digestion leading to a reduction in autophagy flux, with the health of mitochondria being unaffecte In FL animals, oxidative stress is expected to result in mitochondrial damage [24] tha would be counteracted by increased mitophagy. However, in DM rats, the mitochondria damaged from the *Bckdk* mutation, are expected to accumulate due to the inability o abnormal lysosomes (caused by the *Cryba1* mutation) to complete the mitophagy proces The accumulation of damaged mitochondria would also be expected to significantly reduc MMP in DM astrocytes.

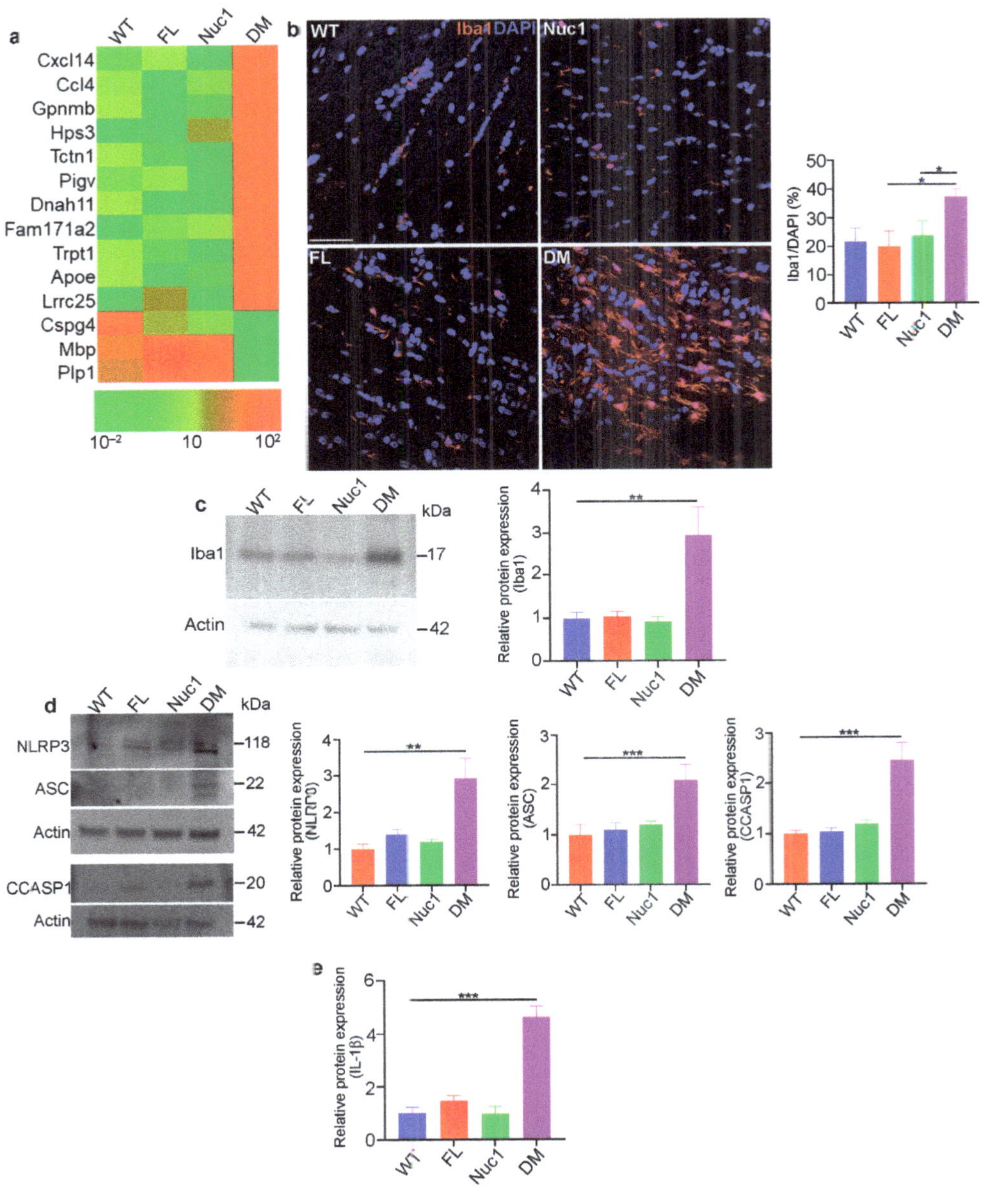

Figure 3. Inflammation in the optic nerve of DM rats: (**a**) RNA-Seq analysis of 4-month-old optic nerves revealed changes in the mRNA expression pattern of several inflammatory factors in DM relative to other genotypes. (**b**,**c**) Immunohistochemical staining and western blotting of optic nerve sections from 4-month-old rats against Iba1 (microglia/macrophages) showed increased staining and protein levels in DM rats relative to other genotypes. Bar: 100 µm. (**d**) Representative immunoblotting images demonstrate that expression of inflammasome markers including NLRP3, ASC, and cleaved caspase 1 were significantly increased in DM optic nerve relative to the other genotypes. (**e**) Measuring IL-1β using ELISA showed a significant increase in DM optic nerve relative to other genotypes. Values are plotted as mean ± SEM from 3 independent experiments repeated in triplicate, * $p < 0.05$; ** $p < 0.01$; *** $p < 0.001$.

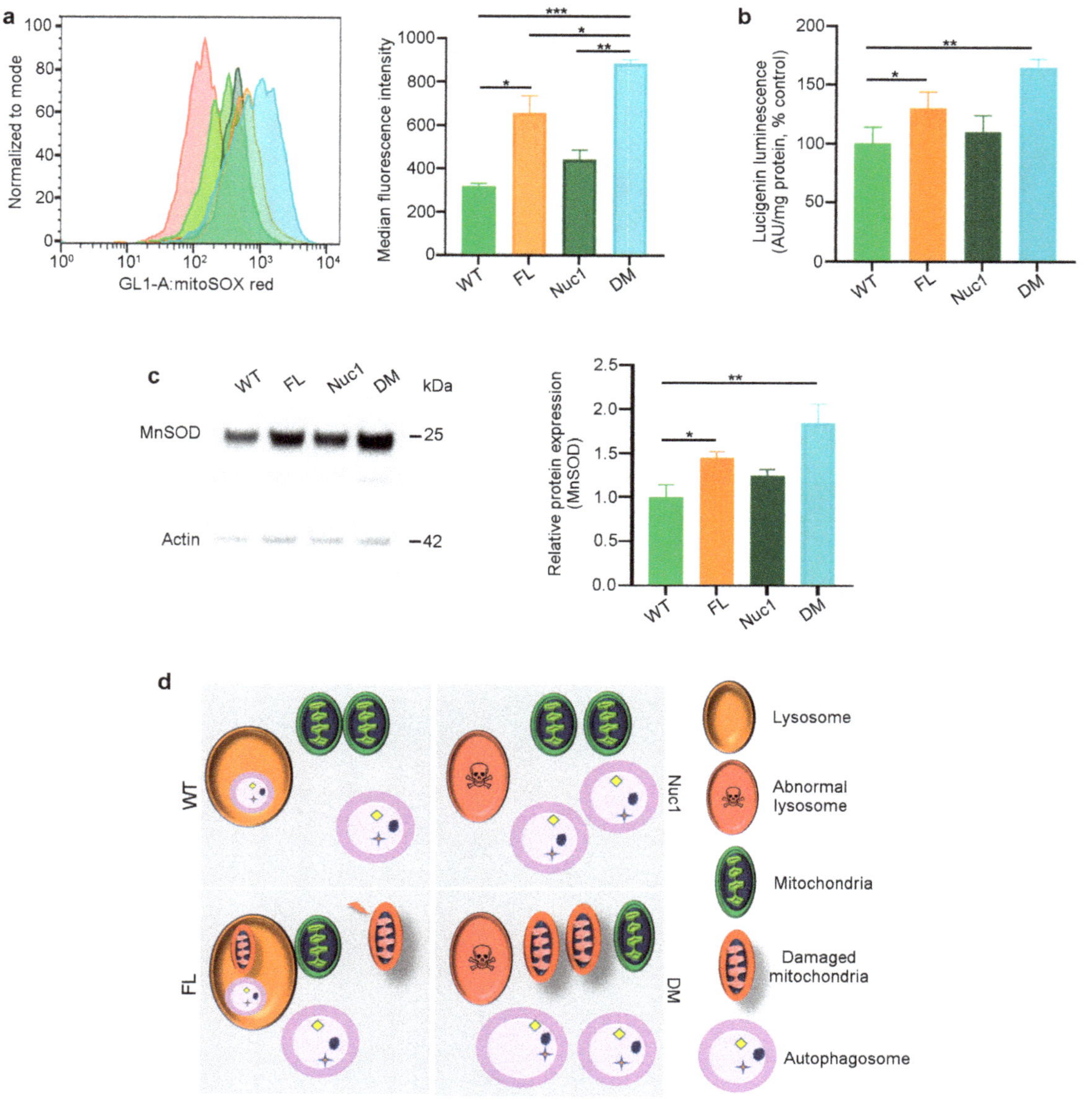

Figure 4. Mitochondrial superoxide production is increased in DM astrocytes: (**a**) Representative histograms of flow cytometric analysis of MitoSOX fluorescence in different experimental groups show that the expression of MitoSOX is increased in FL and DM astrocytes. (**b**) Quantitative analysis of superoxide release indicates that the level of superoxide is increased in FL and DM astrocytes relative to WT control cells. (**c**) Immunoblotting of proteins extracted from astrocytes showed elevated MnSOD expression in FL and DM genotypes relative to WT as control. (**d**) The proposed model of the astrocyte abnormality in DM rats. In WT astrocytes, mitochondria and lysosomes are healthy; thus, autophagosomes are digested in lysosomes. In Nuc1 astrocytes, lysosomes are abnormal, and autophagosome degradation is impaired. In FL astrocytes, oxidative stress leads to mitochondrial damage; however, lysosomes and the mitophagy process are normal. In DM astrocytes, lysosomes are abnormal, so mitophagy is not completed, leading to the accumulation of damaged mitochondria. Data are mean ± SEM, * $p < 0.05$; ** $p < 0.01$; *** $p < 0.001$.

3.5. Autophagy Flux Is Impaired in DM Astrocytes

Autophagy is a dynamic process, and its activity is described as 'autophagy flux' [55]. LC3, a key protein in the autophagy pathway, is proteolytically cleaved to form LC3I and then lipidated to generate LC3-II. LC3-II is a marker of autophagosomes and is degraded in lysosomes; therefore, LC3-II turnover is an indicator of autophagy flux [56,57]. A standard method of measuring autophagy flux is immunoblotting for LC3-II in the presence and absence of bafilomycin A1 (BafA1), a lysosomotropic agent that suppresses lysosomal acidification, thereby inhibiting lysosomal function. The LC3-II level in BafA1-treated cells is divided by the level of LC3-II in untreated cells [58]. Astrocytes from all four different genotypes were cultured in the presence and absence of 50 nM BafA1 for 3 h; then, the protein was extracted and analyzed for LC3-II via western blotting to determine autophagy flux. We found that the level of LC3-II accumulation in Nuc1 and DM astrocytes was significantly reduced relative to FL and WT cells, indicating an impairment in autophagy flux in these cells (Figure 5a). These data confirm our previous findings that demonstrated lysosomal abnormality and autophagy impairment in Nuc1 astrocytes [31,59]. In addition, to validate the autophagy impairment in DM astrocytes, these cells were transduced with a tandem repeat red fluorescent protein (RFP)–green fluorescent protein (GFP)–LC3B (microtubule-associated protein 1 light chain 3 beta) construct to show the levels of autolysosomes relative to autophagosomes. By fusing an acid-sensitive GFP with an acid-insensitive RFP, the transition from autophagosome (neutral pH) to autolysosome (acidic pH) was analyzed through the selective disappearance of GFP fluorescence [60]. The ratio of autolysosomes to autophagosomes, which is an indicator of autophagy flux [55], was significantly increased in FL astrocytes and reduced in both Nuc1 and DM astrocytes relative to WT (Figure 5b). This demonstrates that autophagy flux is significantly increased in FL astrocytes relative to astrocytes from the other genotypes. In contrast, the level of autophagy flux is greatly suppressed in DM.

Considering the fact that proper lysosomal function is essential for autophagy machinery and that the loss of *Cryba1* gene expression impairs lysosomal function in Nuc1 and DM astrocytes, we compared the function of lysosomes in the different genotypes. To measure lysosomal function, live cells were stained with LysoSensor as an indicator of lysosomal pH. The fluorescence of the LysoSensor dye increases in acidic environments [61]. We found that WT and FL astrocytes with normally acidic lysosomes displayed increased LysoSensor fluorescence. However, LysoSensor fluorescence intensity was significantly reduced in Nuc1 and DM astrocytes, indicating an abnormally elevated lysosomal pH in these cells (Figure 5c). In addition, the function of lysosomes was evaluated by measuring the activities of the lysosomal enzymes cathepsin D and L [62,63]. Cathepsins are a family of protease enzymes primarily found in lysosomes that are active in a low-pH environment [64]. We found that the activities of both cathepsin D and L were significantly reduced in DM and Nuc1 cells relative to WT and FL cells (Figure 5d,e).

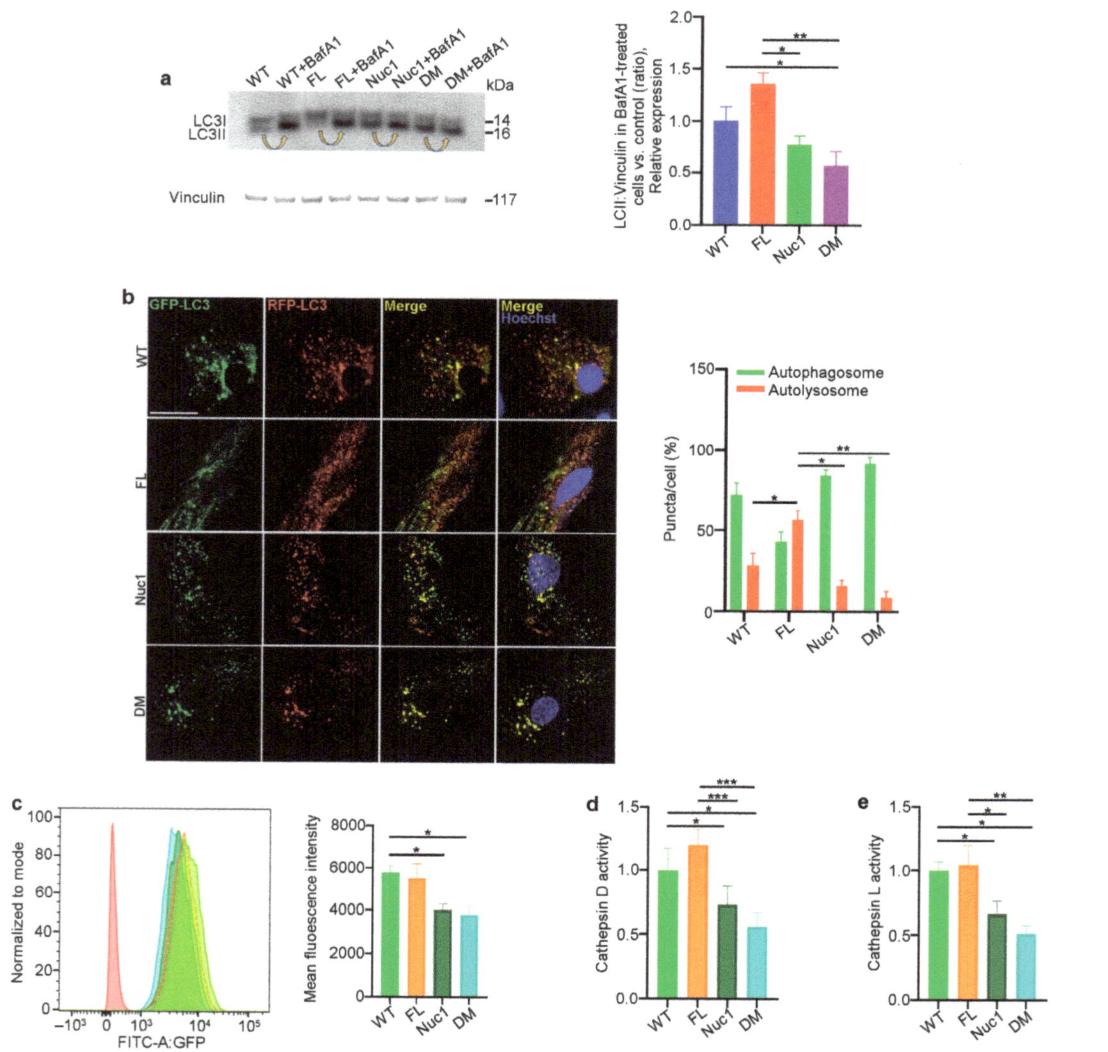

Figure 5. Autophagy is diminished in DM but not in FL astrocytes: (**a**) Astrocytes from the different genotypes were cultured in the presence or absence of the lysosomal inhibitor bafilomycin A1 (BafA1). After treatment, the cells were lysed, and protein lysates were analyzed for LC3 via immunoblotting. There was a significant decrease in the accumulation of the autophagosome-positive LC3-II isoform in DM cells compared to WT and FL cells after BafA1 treatment. Arrows show the accumulation of LC3-II after BafA1 treatment in each cell type. This suggests that the DM cells have a defect in autophagy, which may have caused the reduced accumulation of LC3-II. (**b**) Astrocytes from all genotypes were transduced using an RFP–GFP tandem fluorescent-tagged LC3 (RFP–GFP–LC3) construct. Yellow puncta indicate the presence of autophagosomal structures, as both GFP and RFP fluoresce at the cytoplasmic pH. In contrast, red puncta signify autolysosomes, as the acidity of lysosomes quenches the GFP fluorescence. Merged confocal images show that FL astrocytes have many red puncta, indicating the presence of autolysosomes, while Nuc1 and DM cells have mostly yellow puncta, indicating a blockage of autophagic flux. The numbers of autophagosomes and autolysosomes were quantified from 30 images per group for each experiment. (**c**) Representative

histograms of flow cytometric analysis of LysoSensor fluorescence in different experimental groups show that the fluorescent intensity of LysoSensor is reduced in DM astrocytes relative to WT cells. (**d**,**e**) The graphs show that the levels of active cathepsin D and L were significantly reduced in DM astrocytes. Data are mean ± SEM; n = 3, * $p < 0.05$. Scale bar: 20 µm. Data are mean ± SEM; n = 30 cells from three different experiments, * $p < 0.05$; ** $p < 0.01$; *** $p < 0.001$.

3.6. Mitophagy Is Impaired in DM Astrocytes

Since autophagy flux was dysregulated, and mitochondrial oxidative stress was increased in DM astrocytes relative to FL, we wondered if the clearance of abnormal mitochondria (mitophagy) is also impaired in the DM astrocytes. To measure the level of mitophagy, a fluorescent reporter COX8 (cytochrome c oxidase subunit 8)-enhanced GFP (EGFP)–mCherry fluorescence reporter that is targeted to the mitochondrial matrix was used [65]. Since EGFP is sensitive to pH, the red puncta correspond to mitochondria in lysosomes because EGFP is quenched by the acidity of lysosomes. The yellow puncta represent normal mitochondria because both EGFP and mCherry fluoresce at cytoplasmic pH. Therefore, the number of red puncta in each cell is a measure of mitophagy. We found that the number of red puncta was greatly reduced in DM astrocytes relative to FL. However, the number of red puncta was significantly increased in FL astrocytes relative to the other genotypes, suggesting that FL astrocytes contain many abnormal mitochondria that are being actively removed through mitophagy (Figure 6). We did not see a difference in mitophagy between WT and Nuc1 astrocytes, most likely because mitochondria are normal in these cells.

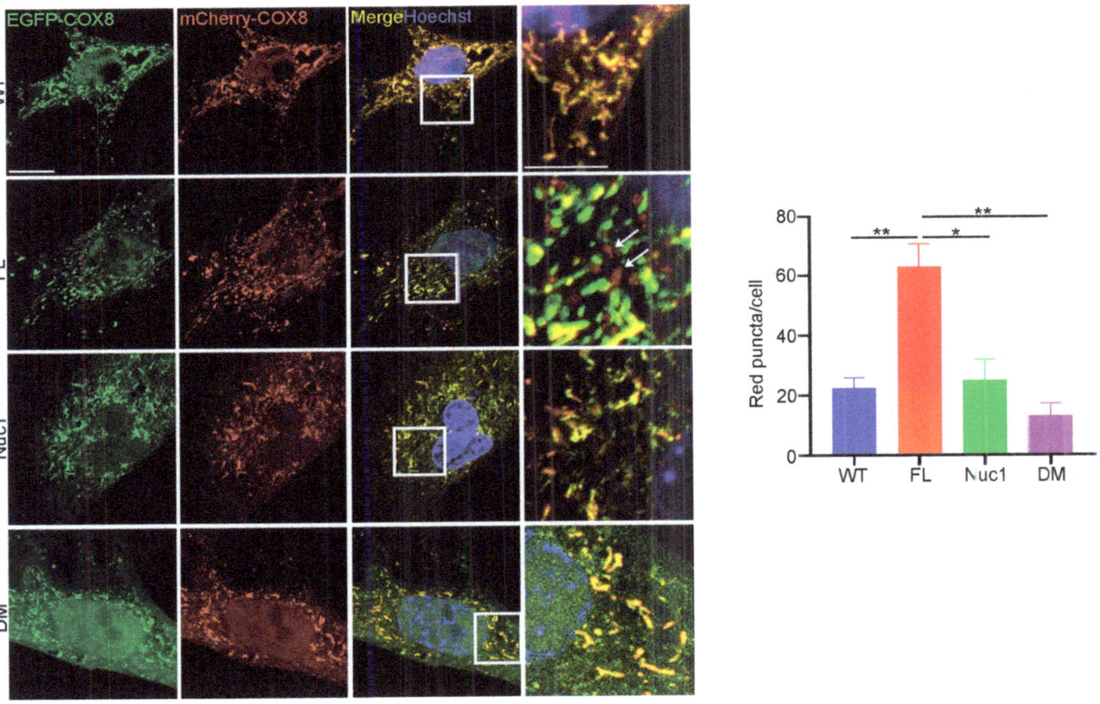

Figure 6. Mitophagy is impaired in DM astrocytes. Astrocytes from all genotypes were transduced by adenovirus–COX8–EGFP–mCherry to label mitochondria and observed under a confocal microscope. The number of acidic (red) mitochondria was significantly increased in FL astrocytes and greatly diminished in DM astrocytes relative to other genotypes. Scale bar: 20 µm. Data are mean ± SEM; n = 30 cells from three different experiments. * $p < 0.05$; ** $p < 0.01$

3.7. mTORC1 Is Overactivated in Cultured DM Astrocytes and Optic Nerve

We have previously shown that the loss of functional βA3/A1-crystallin in Nuc1 astrocytes leads to the upregulation of mTORC1, a negative regulator of autophagy [31]. In addition, BCKDK is also important in the regulation of cellular oxidative stress that in turn affects mTORC1 function. Therefore, we wanted to know if the lack of both functional *Cryba1* and *Bckdk* genes has a synergistic effect on the mTORC1 pathway in DM astrocytes. Consistent with our previous findings, we found that the level of mTORC1 signaling was increased in Nuc1 astrocytes. However, the level of mTORC1 activity was reduced in FL astrocytes, most likely because of dysregulation in the metabolism of BCAAs (Figure 7a,b). Interestingly, mTORC1 signaling in DM astrocytes was increased as in Nuc1 cells. Previous research has demonstrated that βA3/A1-crystallin controls the localization of mTORC1 to lysosomes by binding to V-ATPase, which is the crucial proton pump that acidifies lysosomes [16]. To confirm the in vitro observations, optic nerves from different genotypes were isolated and protein lysates were analyzed via immunoblotting. We found that the expression of SQSTM1 [66] was significantly increased in the Nuc1 and DM optic nerves, indicating the impairment of autophagy; however, it was reduced in FL optic nerves, indicating increased autophagic degradation. These data are consistent with in vitro data showing that autophagy flux is reduced in DM astrocytes. In addition, we also found that autophagy impairment is associated with mTORC1 overactivation in DM and Nuc1 optic nerves (Figure S3a,b).

3.8. Urolithin A (UA) Rescues Autophagy Impairment in DM Astrocytes

UA is produced by gut microflora from the transformation of ellagitannins that are found naturally in plants such as walnuts, pomegranates, raspberries, and strawberries [67,68]. UA treatment helps to maintain mitochondrial quality control by stimulating mitophagy and preventing the accumulation of damaged mitochondria [69,70]. It has been shown that UA induces autophagy by inhibiting the Akt/mTORC1 signaling pathway [71]. The treatment of DM astrocytes with UA leads to the inhibition of the mTORC1 signaling pathway and increased autophagy flux (Figure 7c and Figure S4). In addition, the mitophagy abnormality in DM astrocytes was rescued using UA treatment (Figure 7d).

3.9. UA Treatment Mitigates Optic Nerve Degeneration in DM Rats

UA can cross the blood–brain barrier [72] and has been shown to be a powerful neuroprotectant in vivo against various neurodegenerative diseases [73]. The intraperitoneal injection of UA guaranteed the presence of UA in circulation [74]. In this study, UA was administered within the time frame when optic nerve degeneration occurred in DM rats. UA (or vehicle) was injected intraperitoneally into DM rats three times per week starting at one month of age and continuing for three months. We found that UA treatment led to a reduction in mTORC1 pathway activity and decreased levels of SQSTM1, indicating increased autophagy in DM optic nerves (Figure 8a). In addition, the population of microglia/macrophages was significantly reduced by UA treatment, as was the expression of inflammasome pathway markers (Figure 8b–e). Finally, the level of MBP expression in the DM optic nerves was significantly increased through UA treatment (Figure 8f). Given the fact that MBP is significantly diminished as a result of optic nerve degeneration in DM rats, UA treatment led to a significant reduction in myelin loss and therefore the progression of optic nerve degeneration in DM rats.

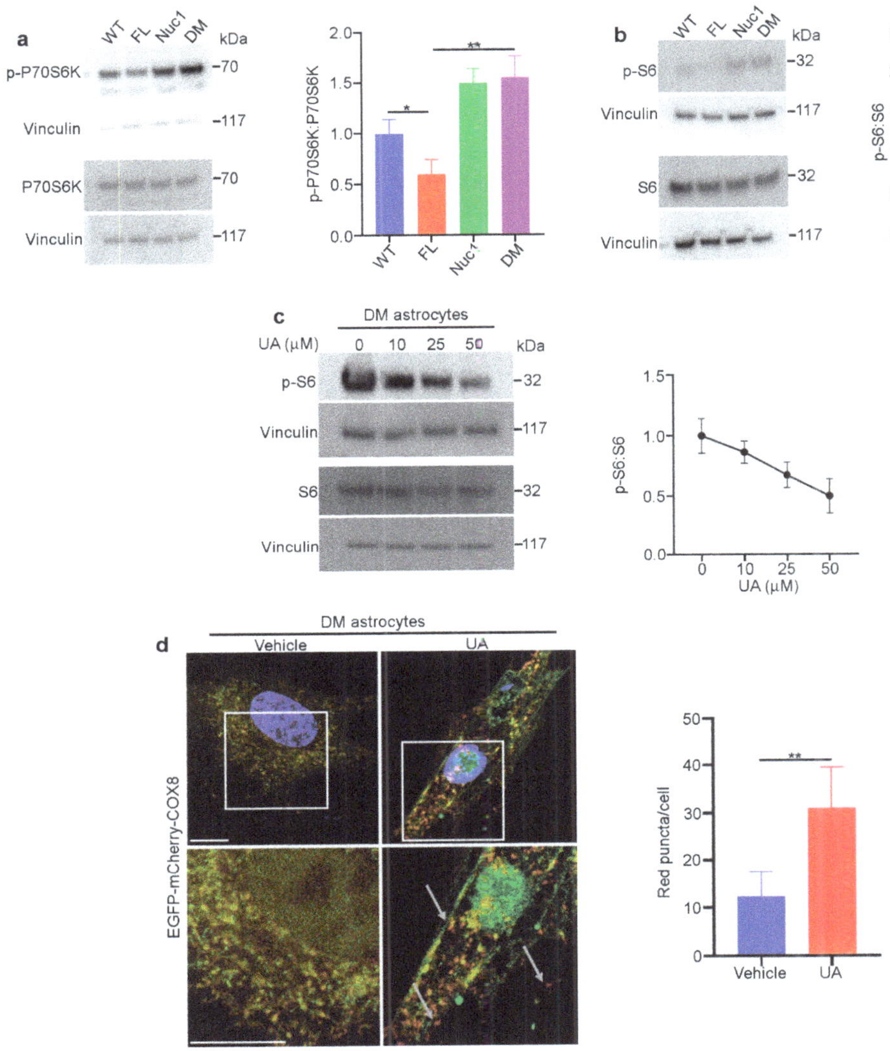

Figure 7. The mTORC1 pathway is overactivated in DM astrocytes: (**a**,**b**) Overactivation of the mTORC1 pathway in DM astrocytes. Protein lysates of cultured astrocytes were prepared and analyzed via immunoblotting for activation markers of the mTORC1 pathway. The DM and Nuc1 astrocytes showed a significant increase in two different phosphorylated (active) forms of RPS6KB1 and phosphorylated-RPS6 compared to the WT group. In contrast, the FL astrocytes showed a significant reduction in these activated molecules. The levels of total RPS6 were not significantly different among the different groups. The presented graphs show the ratios of phosphorylated proteins (active forms of RPS6KB1 and phosphorylated-RPS6) to total protein in the different genotypes. (**c**) Representative western blot images and densitometry showed that UA treatment reduced the ratio of PS6:S6 in DM astrocytes in a dose-dependent manner. Vinculin was used as a loading control. (**d**) DM astrocytes were transduced using a GFP–mCherry–COX8 construct and then treated with 100 μM UA for 24 h. The representative confocal images showed that the number of red puncta (mitochondria) increased with UA treatment (n = 30). White arrows show the red puncta. Values are plotted as mean ± SEM from 3 independent experiments repeated in triplicate. * $p < 0.05$; ** $p < 0.01$.

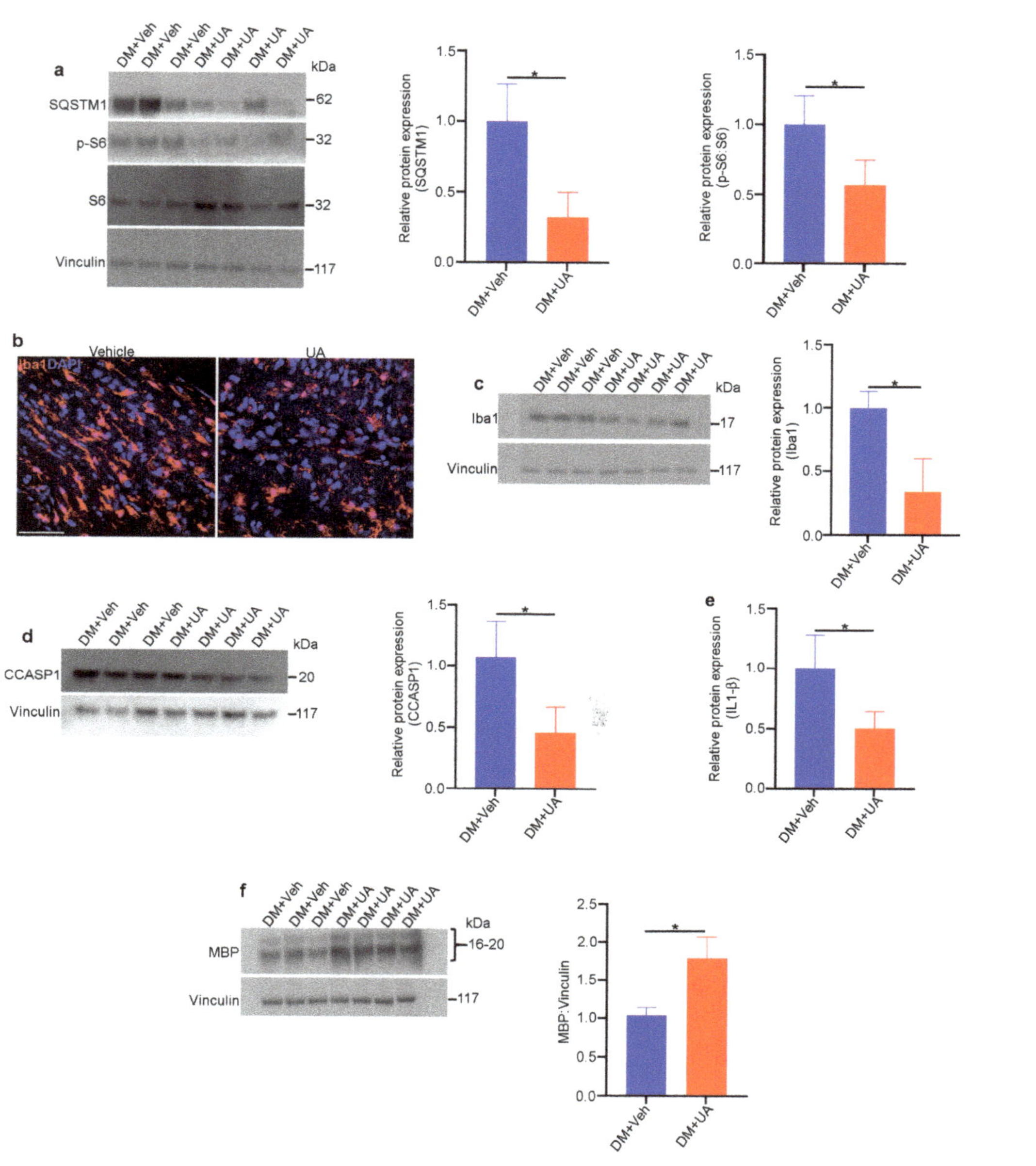

Figure 8. UA protects DM optic nerves against degeneration. Rats from different genotypes were treated with UA from P21 to P120. Then, optic nerve protein lysates were analyzed via immunoblotting: (**a**) Representative immunoblotting images showed that the expression of SQSTM1 and the ratio of PS6:S6 were reduced in the optic nerves of UA-treated DM rats. (**b**) Representative confocal microscopy images revealed that Iba1 staining was decreased via UA treatment. Scale bar: 100 µm (**c**) Immunoblotting of tissue lysates confirmed the reduction in Iba1. (**d**) Western blotting showed that the expression of cleaved caspase 1 (CCASP1) was reduced in UA-treated DM optic nerves (**e**) Analysis of IL-1β in optic nerve lysates indicated that IL-1β was reduced by UA treatment; n = 6 (**f**) Western blotting analysis showed that the expression of MBP was increased in DM optic nerves after UA treatment. Vinculin was used as a loading control. Values are given as mean ± SEM from 3 independent experiments, * $p < 0.05$.

4. Discussion

Our study shed light on the connection between mitophagy in astrocytes and optic nerve health. First, we introduced a rat animal model with mitochondrial abnormality in the optic nerve astrocytes. Second, we found that impairment in mitophagy and mitochondrial health in optic nerve astrocytes leads to optic nerve degeneration. Finally, UA rescues the autophagy/mitophagy impairment in optic nerve astrocytes and inhibits the degeneration of the optic nerve in DM rats. Therefore, our studies reveal that optic nerve health profoundly relies on mitochondrial health in astrocytes.

It has been shown that a lack of *Bckdk* causes mitochondrial abnormalities and the production of excessive oxidative stress [24]. Consistently, we also found that mitochondrial oxidative stress was increased in both FL and DM astrocytes. However, the mitochondrial oxidative stress was greater in DM astrocytes than in FL astrocytes. The oxidative stress was associated with a reduction in MMP in DM but not in FL astrocytes.

It has been previously shown that βA3/A1-crystallin regulates lysosomal function and modulates signaling molecules in the lysosomes of astrocytes [16]. We also found that a lack of functional *Cryba1* in Nuc1 astrocytes leads to lysosomal abnormality and autophagy impairment [31]. The elimination of damaged mitochondria through mitophagy is required for cellular homeostasis [75,76]. As our data demonstrate, in FL astrocytes, functional lysosomes and an efficient mitophagy process eliminate damaged mitochondria caused by the lack of *Bckdk*. However, in DM astrocytes, the process of mitophagy is not completed because of lysosomal abnormality, and it leads to the accumulation of damaged mitochondria and excessive mitochondrial oxidative stress (Figure 4d). Autophagy impairment does not lead to optic nerve degeneration in Nuc1 rats. However, the coincidence of mitochondrial abnormality with dysfunctional lysosomes in DM astrocytes leads to the accumulation of damaged mitochondria and excessive oxidative stress in astrocytes and consequently optic nerve degeneration. This evidence indicates that mitophagy in astrocytes is essential for the homeostasis of the optic nerve. Moreover, optic nerve degeneration in this animal model emphasizes that retinal ganglion cells are highly metabolic cells and very sensitive to inadequate astrocyte-derived nutrient supply [5].

Lysosomes were once thought to be simply a bag of waste disposal enzymes [77], but they are now recognized as a center for multiple signaling pathways, particularly the mTORC1 pathway, a master negative regulator of autophagy [78]. V-ATPase is a proton pump in lysosomes that is responsible for maintaining the acidic pH of the lysosomal lumen and plays an important role in the activation and deactivation of mTORC1 [16,79]. Furthermore, studies have shown that βA3/A1-crystallin interacts with V-ATPase and that the lack of *Cryba1* leads to lysosomal dysfunction and overactivation of the mTORC1 pathway [16]. In addition, it is known that mitochondrial oxidative stress inhibits mTORC1 activity [80,81], and this is likely the cause of the inhibited mTORC1 pathway and the increase in lysosomal function in FL astrocytes. In DM astrocytes, mTORC1 is overactivated, similar to Nuc1. This indicates that in DM astrocytes, lysosomal abnormality caused by the *Cryba1* deficiency prevents oxidative stress-mediated mTORC1 inhibition. Therefore, inhibiting the mTORC1 pathway in DM astrocytes can be a strategy for rescuing impaired autophagy and mitophagy.

It has been shown that UA induces autophagy/mitophagy by inhibiting the mTORC1 pathway in a variety of experimental settings [67,69,74]. Our data also demonstrated that the treatment of DM astrocytes with UA led to the inhibition of mTORC1 and the rejuvenation of impaired autophagy/mitophagy. Therefore, we postulate that mTORC1 inhibition by the mitophagy/autophagy inducer UA can be a strategy to inhibit optic nerve degeneration in DM rats. It has been shown that UA improves mitochondrial function, induces mitophagy, and reduces neurodegeneration in various animal models of AD [82,83]. Our data showed that UA inhibits the degeneration of DM optic nerves by inhibiting mTORC1 and inducing autophagy. In addition, it has been shown that UA reduces neuroinflammation and improves cognitive function in a mouse model of AD [84]. Our data showed that optic nerve degeneration in DM rats is associated with the activation of microglia and inflammation. We also found that UA targets inflammasome activation in microglia/macrophages and attenuates optic nerve degeneration in DM rats.

5. Conclusions

Mitochondrial dysfunction is linked to various ocular diseases, including autosomal dominant optic atrophy [85], Leber's hereditary optic neuropathy [86], glaucoma [87], age-related macular degeneration (AMD) [88], and diabetic retinopathy (DR) [89]. It is known that inflammation can be triggered by oxidative stress caused by mitochondrial damage [90]. Our data demonstrated the accumulation of a mitochondrial abnormality and in turn, excessive oxidative stress in DM astrocytes. However, further studies are required to understand the molecular mechanisms through which excessive oxidative stress in astrocytes promotes inflammation. Although UA restores the mitophagy abnormality in DM astrocytes in vitro, the therapeutic effects of UA in DM optic nerve degeneration might not be limited to this. UA might also be beneficial through its direct effects on immune cells. UA could potentially offer therapeutic benefits in the treatment of common ophthalmic disorders of aging associated with mitochondrial dysfunction, including diabetic retinopathy, age-related macular degeneration, and glaucoma. Taken together, our findings indicate that mitochondrial health in astrocytes is essential to the homeostasis of the optic nerve and that UA has potential as a therapeutic agent for optic nerve degenerative diseases.

Supplementary Materials: The following supporting information can be downloaded at: https://www.mdpi.com/article/10.3390/cells12202496/s1, Figure S1: Inflammation is activated in DM optic nerve. Immunohistochemical analysis of the DM optic nerve showed the presence of CD4-positive T lymphocytes (arrows). Scale bar: 50 μm.; Figure S2: The mitochondrial membrane potential is reduced in DM astrocytes. Representative TMRM fluorescence intensity histogram from a flow cytometric analysis shows significantly lower fluorescence intensity in DM astrocytes corresponding to lower membrane potential. Data are mean ± SEM. n = 3, ** $p < 0.01$; Figure S3: mTORC1 is overactivated in DM optic nerves. (a,b) Optic nerves from all 4 genotypes were dissected and protein lysates were analyzed by immunoblotting. The representative immunoblotting images showed that the level of SQSTM1/p62, and the ratio of PS6:S6 are increased in the optic nerves of DM rats. Data are mean ± SEM from 3 independent experiments repeated in triplicate. * $p < 0.05$, ** $p < 0.01$; Figure S4: UA treatment induces autophagy in DM astrocytes. DM astrocytes were treated with UA (25 μM) for 24 h. In the last 3 h of culture, 50 nM BafA1 was added and the levels of autophagosome-positive LC3II isoform were analyzed by immunoblotting. A significant increase in the accumulation of autophagosomes positive for the LC3II isoform was observed in UA-treated cells relative to those untreated after BafA1 treatment. * $p < 0.05$.

Author Contributions: D.S., J.S.Z.J. and M.Y. designed the study. S.G., H.L. and M.Y. conducted the experiments and analyzed the data. M.Y., S.H., J.S.Z.J. and D.S. wrote the paper. All authors have read and agreed to the published version of the manuscript.

Funding: This work was supported by a Knights Templar Eye Foundation Career-Starter Research Grant (MY), The University of Pittsburgh start-up funds (DS), the Jennifer Salvitti Davis, M.D. Chair Professorship in Ophthalmology (DS), P30 core award EY08098 from the National Eye Institute, NIH (to the University of Pittsburgh Department of Ophthalmology), and unrestricted funds from The Research to Prevent Blindness Inc., NY (to the University of Pittsburgh Department of Ophthalmology and the Wilmer Eye Institute).

Institutional Review Board Statement: The animal study protocol was approved by the Institutional Animal Care and Use Committee of the University of Pittsburgh for studies involving animals.

Informed Consent Statement: Not applicable.

Data Availability Statement: All data generated or analyzed during this study are included in this published article (and its Supplementary Materials files).

Acknowledgments: The authors would like to thank Peter Calabresi and members of his laboratory at the Johns Hopkins University School of Medicine for their help with some of the experiments reported in this study. The authors would also like to acknowledge the UPMC Flow Cytometry and Imaging Cores in the Department of Pediatrics, for their valuable contributions and support.

Conflicts of Interest: The authors declare no conflict of interest.

References

1. García-Bermúdez, M.Y.; Freude, K.K.; Mouhammad, Z.A.; van Wijngaarden, P.; Martin, K.K.; Kolko, M. Glial Cells in Glaucoma: Friends, Foes, and Potential Therapeutic Targets. *Front. Neurol.* **2021**, *12*, 624983. [CrossRef]
2. Wangsa-Wirawan, N.D.; Linsenmeier, R.A. Retinal Oxygen: Fundamental and Clinical Aspects. *Arch. Ophthalmol.* **2003**, *4*, 547–557. [CrossRef] [PubMed]
3. Perge, J.A.; Koch, K.; Miller, R.; Sterling, P.; Balasubramanian, V. How the Optic Nerve Allocates Space, Energy Capacity, and Information. *J. Neurosci.* **2009**, *29*, 7917–7928. [CrossRef]
4. Carelli, V.; Ross-Cisneros, F.N.; Sadun, A.A. Mitochondrial dysfunction as a cause of optic neuropathies. *Prog. Retin. Eye Res.* **2004**, *23*, 53–89. [CrossRef]
5. Ito, Y.A.; Di Polo, A. Mitochondrial Dynamics, Transport, and Quality Control: A Bot-tleneck for Retinal Ganglion Cell Viability in Optic Neuropathies. *Mitochondrion* **2017**, *36*, 186–192. [CrossRef] [PubMed]
6. Kong, G.Y.; Van Bergen, N.J.; Trounce, I.A.; Crowston, J.G. Mi-tochondrial Dysfunction and Glaucoma. *J. Glaucoma* **2009**, 18.
7. Kenney, M.C.; Riazi-Esfahani, M.; Kuppermann, B.D. The role of mitochondria in AMD: Current knowledge and future applications. *J. Ophthalmic Vis. Res.* **2017**, *12*, 424–428. [CrossRef]
8. Ferrington, D.; Fisher, C.R.; Kowluru, R.A. Mitochondrial Defects Drive Degenerative Retinal Diseases. *Trends Mol. Med.* **2020**, *26*, 105–118. [CrossRef] [PubMed]
9. Miller, D.J.; Cascio, M.A.; Rosca, M.G. Diabetic Retinopathy: The Role of Mitochondria in the Neural Retina and Microvascular Disease. *Antioxidants* **2020**, *9*, 905. [CrossRef]
10. Lampinen, R.; Belaya, I.; Boccuni, I.; Kanninen, T.M. Mitochondrial Function in Alzheimer's Disease: Focus on Astrocytes. In *Astrocyte*; Teresa, G.M., Colucci, D.L., Eds.; IntechOpen: Rijeka, Croatia, 2017.
11. Booth, H.D.; Hirst, W.D.; Wade-Martins, R. The Role of Astrocyte Dysfunction in Parkinson's Disease Pathogenesis. *Trends Neurosci.* **2017**, *40*, 358–370. [CrossRef]
12. Bantle, C.M.; Hirst, W.D.; Weihofen, A.; Shlevkov, E. Mitochondrial Dysfunction in Astrocytes: A Role in Parkinson's Disease? *Front. Cell Dev. Biol* **2020**, *8*, 608026. [CrossRef] [PubMed]
13. Gollihue, J.; Norris, C. Astrocyte mitochondria: Central players and potential therapeutic targets for neurodegenerative diseases and injury. *Ageing Res. Rev.* **2020**, *59*, 101039. [CrossRef]
14. Borsche, M.; König, I.R.; Delcambre, S.; Petrucci, S.; Balck, A.; Brüggemann, N.; Zimprich, A.; Wasner, K.; Pereira, S.L.; Avenali, M.; et al. Mitochondrial damage-associated inflammation highlights biomarkers in PRKN/PINK1 parkinsonism. *Brain* **2020**, *143*, 3041–3051. [CrossRef] [PubMed]
15. Zigler, J.S.; Hodgkinson, C.A.; Wright, M.; Klise, A.; Sundin, O.; Broman, K.W.; Hejtmancik, F.; Huang, H.; Patek, B.; Sergeev, Y.; et al. A Spontaneous Missense Mutation in Branched Chain Keto Acid Dehydrogenase Kinase in the Rat Affects Both the Central and Peripheral Nervous Systems. *PLoS ONE* **2016**, *11*, e0160447. [CrossRef] [PubMed]
16. Zigler, J.S., Jr.; Sinha, D. Ba3/A1-Crystallin: More Than a Lens Protein. *Prog. Retin. Eye Res.* **2015**, *44*, 62–85. [CrossRef]
17. Sinha, D.; Klise, A.; Sergeev, Y.; Hose, S.; Bhutto, I.A.; Hackler, L., Jr.; Malpic-Llanos, T.; Samtani, S.; Grebe, R.; Goldberg, M.F.; et al. betaA3/A1-crystallin in astroglial cells regulates retinal vascular remodeling during development. *Mol. Cell. Neurosci.* **2008**, *37*, 85–95. [CrossRef]
18. Sinha, D.; Hose, S.; Zhang, C.; Neal, R.; Ghosh, M.; O'Brien, T.P.; Sundin, O.; Goldberg, M.F.; Robison, W.G.; Russell, P.; et al. A spontaneous mutation affects programmed cell death during development of the rat eye. *Exp. Eye Res.* **2005**, *80*, 323–335. [CrossRef]
19. Cvekl, A.; Zheng, D. Gene sharing and evolution. *Hum. Genom.* **2009**, *4*, 66–67. [CrossRef]
20. Jeffery, C.J. Protein moonlighting: What is it, and why is it important? *Philos. Trans. R. Soc. Lond. Ser. B Biol. Sci.* **2018**, *373*, 20160523. [CrossRef]
21. Valapala, M.; Wilson, C.; Hose, S.; Bhutto, I.A.; Grebe, R.; Dong, A.; Greenbaum, S.; Gu, L.; Sengupta, S.; Cano, M.; et al. Lysosomal-Mediated Waste Clearance in Retinal Pigment Epithelial Cells Is Regulated by Cryba1/Ba3/A1-Crystallin Via V-Atpase-Mtorc1 Signaling. *Autophagy* **2014**, *10*, 480–496. [CrossRef]
22. Zigler, J.S.; Zhang, C.; Grebe, R.; Sehrawat, G.; Hackler, L.; Adhya, S.; Hose, S.; McLeod, D.S.; Bhutto, I.; Barbour, W.; et al. Mutation in the Ba3/A1-Crystallin Gene Impairs Phagosome Degradation in the Retinal Pigmented Epithelium of the Rat. *J. Cell Sci.* **2011**, 523–531. [CrossRef] [PubMed]
23. Cook, K.G.; Bradford, A.P.; Yeaman, S.J.; Aitken, A.; Fearnley, I.M.; Walker, J.E. Regulation of Bovine Kidney Branched-Chain 2-Oxoacid Dehydrogenase Complex by Reversible Phosphorylation. *Eur. J. Biochem.* **1984**, *145*, 587–591. [CrossRef] [PubMed]

24. Oyarzabal, A.; Bravo-Alonso, I.; Sánchez-Aragó, M.; Rejas, M.T.; Merinero, B.; García-Cazorla, A.; Artuch, R.; Ugarte, M Rodríguez-Pombo, P. Mitochondrial Response to the Bckdk-Deficiency: Some Clues to Understand the Positive Dietary Respons in This Form of Autism. *Biochim. Biophys. Acta* **2016**, *1862*, 592–600. [CrossRef]
25. Joshi, M.A.; Jeoung, N.H.; Obayashi, M.; Hattab, E.M.; Brocken, E.G.; Liechty, E.A.; Kubek, M.J.; Vattem, K.M.; Wek, R.C.; Harris R.A. Impaired Growth and Neurological Abnormalities in Branched-Chain Alpha-Keto Acid Dehydrogenase Kinase-Deficien Mice. *Biochem. J.* **2006**, *400*, 153–162. [CrossRef] [PubMed]
26. Popov, K.; Zhao, Y.; Shimomura, Y.; Kuntz, M.; Harris, R. Branched-chain alpha-ketoacid dehydrogenase kinase. Molecula cloning, expression, and sequence similarity with histidine protein kinases. *J. Biol. Chem.* **1992**, *267*, 13127–13130. [CrossRef]
27. Sung, K.; Jimenez-Sanchez, M. Autophagy in Astrocytes and Its Implications in Neurodegeneration. *J. Mol. Biol.* **2020**, *43* 2605–2621. [CrossRef]
28. Parzych, K.R.; Klionsky, D.J. An Overview of Autophagy: Morphology, Mechanism, Regulation. *Antioxid. Redox Signal.* **2014**, *2* 460–473. [CrossRef]
29. Ding, W.X.; Yin, X.M. Mitophagy: Mechanisms, Pathophysiological Roles, and Anal-ysis. *Biol. Chem.* **2012**, *393*, 547–56 [CrossRef]
30. Chen, G.; Kroemer, G.; Kepp, O. Mitophagy: An Emerging Role in Aging and Age-Associated Diseases. *Front. Cell Dev. Biol.* **202** *8*, 200. [CrossRef]
31. Yazdankhah, M.; Shang, P.; Ghosh, S.; Bhutto, I.A.; Stepicheva, N.; Grebe, R.; Sinha, D. Modulating Egfr-Mtorc1-Autophagy as Potential Therapy for Persistent Fetal Vasculature (Pfv) Disease. *Autophagy* **2020**, *16*, 1130–1142. [CrossRef]
32. Loos, B.; Du Toit, A.; Hofmeyr, J.-H.S. Defining and measuring autophagosome flux—Concept and reality. *Autophagy* **2014**, *1* 2087–2096. [CrossRef] [PubMed]
33. Yazdankhah, M.; Ghosh, S.; Shang, P.; Stepicheva, N.; Hose, S.; Liu, H.; Chamling, X.; Tian, S.; Sullivan, M.L.G.; Calderon M.J.; et al. Bnip3l-Mediated Mitophagy Is Required for Mitochondrial Remodeling During the Differentiation of Optic Nerv Oligodendrocytes. *Autophagy* **2021**, *17*, 3140–3159. [CrossRef]
34. Joshi, D.C.; Bakowska, J.C. Determination of Mitochondrial Membrane Potential and Reactive Oxygen Species in Live Rat Cortica Neurons. *J. Vis. Exp.* **2011**, *51*, e2704.
35. Chazotte, B. Labeling Mitochondria with Tmrm or Tmre. *Cold Spring Harb. Protoc.* **2011**, *2011*, 895–897. [PubMed]
36. Ehrenberg, B.; Montana, V.; Wei, M.-D.; Wuskell, J.; Loew, L. Membrane potential can be determined in individual cells from th nernstian distribution of cationic dyes. *Biophys. J.* **1988**, *53*, 785–794. [CrossRef]
37. Perry, S.W.; Norman, J.P.; Barbieri, J.; Brown, E.B.; Gelbard, H.A. Mitochondrial membrane potential probes and the proton gradient: A practical usage guide. *BioTechniques* **2011**, *50*, 98–115. [CrossRef]
38. Kuribayashi, F.; Tsuruta, S.; Yamazaki, T.; Nunoi, H.; Imajoh-Ohmi, S.; Kanegasaki, S.; Nakamura, M. Cell adhesion markedly increases lucigenin-enhanced chemiluminescence of the phagocyte NADPH oxidase. *Genes Cells* **2008**, *13*, 1249–1256. [CrossRef
39. Ma, X.; Ding, W.-X. A fluorescence imaging based-assay to monitor mitophagy in cultured hepatocytes and mouse liver. *Liver Res* **2021**, *5*, 16–20. [CrossRef]
40. Ghosh, S.; Padmanabhan, A.; Vaidya, T.; Watson, A.M.; Bhutto, I.A.; Hose, S.; Shang, P.; Stepicheva, N.; Yazdankhah, M.; Weiss, J. et al. Neutrophils homing into the retina trigger pathology in early age-related macular degeneration. *Commun. Biol.* **2019**, *2*, 34 [CrossRef]
41. Prayer, D.; Barkovich, A.J.; Kirschner, D.A.; Prayer, L.M.; Roberts, T.P.; Kucharczyk, J.; Moseley, M.E. Visualization of Nonstruc tural Changes in Early White Matter Development on Diffusion-Weighted MR Images: Evidence Supporting Premyelination Anisotropy. *Am. J. Neuroradiol.* **2001**, *22*, 1572–1576.
42. Battistini, J.I.; Mastrorilli, V.; Nicolis di Robilant, V.; Saraulli, D.; Marinelli, S.; Farioli Vecchioli, S. Role of Running-Activated Neural Stem Cells in the Anatomical and Functional Re-covery after Traumatic Brain Injury in P21 Knock-out Mice. *Int. J. Mo Sci.* **2023**, 24.
43. Petrella, C.; Strimpakos, G.; Torcinaro, A.; Middei, S.; Ricci, V.; Gargari, G.; Mora, D.; De Santa, F.; Farioli-Vecchioli, S. Proneuro genic and neuroprotective effect of a multi strain probiotic mixture in a mouse model of acute inflammation: Involvement of th gut-brain axis. *Pharmacol. Res.* **2021**, *172*, 105795. [CrossRef] [PubMed]
44. Kuhn, S.; Gritti, L.; Crooks, D.; Dombrowski, Y. Oligodendrocytes in Development, Myelin Generation and Beyond. *Cells* **2019**, *8* 1424. [CrossRef] [PubMed]
45. Gargareta, V.I.; Reuschenbach, J.; Siems, S.B.; Sun, T.; Piepkorn, L.; Mangana, C.; Werner, H.B. Conservation and Divergence o Myelin Proteome and Oligodendrocyte Transcriptome Profiles between Humans and Mice. *Elife* **2022**, *11*, e77019. [CrossRef [PubMed]
46. Han, W.; Pan, Y.; Han, Z.; Cheng, L.; Jiang, L. Advanced Maternal Age Impairs Myelination in Offspring Rats. *Front. Pediatr.* **2022** *10*, 850213. [CrossRef]
47. Lier, J.; Streit, W.J.; Bechmann, I. Beyond Activation: Characterizing Microglial Functional Phenotypes. *Cells* **2021**, *10*, 2236 [CrossRef]
48. Luckheeram, R.; Zhou, R.; Verma, A.D.; Xia, B. $Cd4^+$ T Cells: Differentiation and Functions. *Clin. Dev. Immunol.* **2012**, *2012* 925135. [CrossRef]
49. Subhramanyam, C.S.; Wang, C.; Hu, Q.; Dheen, S.T. Microglia-mediated neuroinflammation in neurodegenerative diseases *Semin. Cell Dev. Biol.* **2019**, *94*, 112–120. [CrossRef]

50. Shao, F.; Wang, X.; Wu, H.; Wu, Q.; Zhang, J. Microglia and Neuroinflammation: Crucial Pathological Mechanisms in Traumatic Brain Injury-Induced Neurodegeneration. *Front. Aging Neurosci.* **2022**, *14*, 825086. [CrossRef]
51. She, N.; Shi, Y.; Feng, Y.; Ma, L.; Yuan, Y.; Zhang, Y.; Cao, Z.; Chen, X.; Zhao, B.; Liu, H.; et al. Nlrp3 Inflammasome Regulates Astrocyte Transformation in Brain Injury Induced by Chronic Intermittent Hypoxia. *BMC Neurosci.* **2022**, *23*, 70. [CrossRef]
52. Voet, S.; Srinivasan, S.; Lamkanfi, M.; van Loo, G. Inflammasomes in neuroinflammatory and neurodegenerative diseases. *EMBO Mol. Med.* **2019**, *11*, e10248. [CrossRef]
53. Kauffman, M.E.; Kauffman, M.K.; Traore, K.; Zhu, H.; Trush, M.A.; Jia, Z.; Li, Y.R. Mi-tosox-Based Flow Cytometry for Detecting Mitochondrial Ros. *React. Oxyg. Species Apex* **2016**, *2*, 361–370.
54. Flynn, J.M.; Melov, S. SOD2 in mitochondrial dysfunction and neurodegeneration. *Free Radic. Biol. Med.* **2013**, *62*, 4–12. [CrossRef] [PubMed]
55. Mizushima, N.; Murphy, L.O. Autophagy Assays for Biological Discovery and Therapeutic Development. *Trends Biochem. Sci.* **2020**, *45*, 1080–1093. [CrossRef] [PubMed]
56. Tanida, I.; Ueno, T.; Kominami, E. LC3 conjugation system in mammalian autophagy. *Int. J. Biochem. Cell Biol.* **2004**, *36*, 2503–2518. [CrossRef]
57. Zhang, X.J.; Chen, S.; Huang, K.X.; Le, W.D. Why Should Autophagic Flux Be Assessed? *Acta Pharmacol. Sin.* **2013**, *10*, 595–599. [CrossRef]
58. Yoshii, S.R.; Mizushima, N. Monitoring and Measuring Autophagy. *Int. J. Mol. Sci.* **2017**, *18*, 1865. [CrossRef] [PubMed]
59. Valapala, M.; Hose, S.; Gongora, C.; Dong, L.; Wawrousek, E.F.; Zigler, J.S., Jr.; Sinha, D. Impaired Endolysosomal Function Disrupts Notch Signalling in Optic Nerve Astrocytes. *Nat. Commun.* **2013**, *4*, 1629. [CrossRef]
60. Maulucci, G.; Chiarpotto, M.; Papi, M.; Samengo, D.; Pani, G.; De Spirito, M. Quantitative Analysis of Autophagic Flux by Confocal Ph-Imaging of Autophagic Intermediates. *Autophagy* **2015**, *10*, 1905–1916. [CrossRef]
61. Albrecht, L.; Tejeda-Muñoz, N.; De Robertis, E. Protocol for Probing Regulated Lysosomal Activity and Function in Living Cells. *STAR Protoc.* **2020**, *1*, 100132. [CrossRef]
62. Lu, S.; Sung, T.; Lin, N.; Abraham, R.T.; Jessen, B.A. Lysosomal adaptation: How cells respond to lysosomotropic compounds. *PLoS ONE* **2017**, *12*, e0173771. [CrossRef] [PubMed]
63. Baranski, T.J.; Faust, P.L.; Kornfeld, S. Generation of a lysosomal enzyme targeting signal in the secretory protein pepsinogen. *Cell* **1990**, *63*, 281–291. [CrossRef] [PubMed]
64. Čaval, T.; Hecht, E.S.; Tang, W.; Uy-Gomez, M.; Nichols, A.; Kil, Y.J.; Sandoval, W.; Bern, M.; Heck, A.J.R. The lysosomal endopeptidases Cathepsin D and L are selective and effective proteases for the middle-down characterization of antibodies. *FEBS J.* **2021**, *288*, 5389–5405. [CrossRef] [PubMed]
65. Rojansky, R.; Cha, M.Y.; Chan, D.C. Elimination of Paternal Mitochondria in Mouse Embryos Occurs through Autophagic Degradation Dependent on Parkin and Mul1. *Elife* **2016**, *5*, e17896. [CrossRef]
66. Kumar, A.V.; Mills, J.; Lapierre, L.R. Selective Autophagy Receptor P62/Sqstm1, a Pivotal Player in Stress and Aging. *Front. Cell Dev. Biol.* **2022**, *10*, 793328. [CrossRef]
67. D'Amico, D.P.; Andreux, A.; Valdés, P.; Singh, A.; Rinsch, C.; Auwerx, J. Impact of the Natural Compound Urolithin a on Health, Disease. Aging. *Trends Mol. Med.* **2021**, *27*, 687–699 [CrossRef]
68. Nishimoto, Y.; Fujisawa, K.; Ukawa, Y.; Kudoh, M.; Funahashi, K.; Kishimoto, Y.; Fukuda, S. Effect of urolithin A on the improvement of vascular endothelial function depends on the gut microbiota. *Front. Nutr.* **2023**, *9*, 1077534. [CrossRef]
69. Dongryeol, R.; Mouchiroud, L.; Andreux, P.A.; Katsyuba, E.; Moullan, N.; Nicolet-dit-Félix, A.A.; Williams, E.G.; Jha, P.; Sasso, G.L.; Huzard, D.; et al. Urolithin a Induces Mitophagy and Prolongs Lifespan in C. Elegans and Increases Muscle Function in Rodents. *Nat. Med.* **2016**, *22*, 879–888.
70. Denk, D.; Petrocelli, V.; Conche, C.; Drachsler, M.; Ziegler, P.K.; Braun, A.; Kress, A.; Nicolas, A.M.; Mohs, K.; Becker, C.; et al. Expansion of T Memory Stem Cells with Superior Anti-Tumor Immunity by Urolithin a-Induced Mitophagy. *Immunity* **2022**, *55*, 2059–2073.e8. [CrossRef]
71. Zhang, Y.; Jiang, L.; Su, P.; Yu, T.; Ma, Z.; Liu, Y.; Yu, J. Urolithin A suppresses tumor progression and induces autophagy in gastric cancer via the PI3K/Akt/mTOR pathway. *Drug Dev. Res.* **2022**, *84*, 172–184. [CrossRef]
72. Tan, S.; Tong, W.H.; Vyas, A. Urolithin-A attenuates neurotoxoplasmosis and alters innate response towards predator odor. *Brain Behav. Immun. Health* **2020**, *8*, 100128. [CrossRef]
73. Andreux, P.; Blanco-Bose, W.; Ryu, D.; Burdet, F.; Ibberson, M.; Aebischer, P.; Auwerx, J.; Singh, A.; Rinsch, C. The Mitophagy Activator Urolithin a Is Safe and Induces a Molecular Signature of Improved Mitochondrial and Cellular Health in Humans. *Nat. Metab.* **2019**, *1*, 595–603. [CrossRef] [PubMed]
74. Savi, M.; Bocchi, L.; Mena, P.; Dall'asta, M.; Crozier, A.; Brighenti, F.; Stilli, D.; Del Rio, D. In vivo administration of urolithin A and B prevents the occurrence of cardiac dysfunction in streptozotocin-induced diabetic rats. *Cardiovasc. Diabetol.* **2017**, *16*, 1–13. [CrossRef] [PubMed]
75. Jetto, C.T.; Nambiar, A.; Manjithaya, R. Mitophagy and Neurodegeneration: Between the Knowns and the Unknowns. *Front. Cell Dev. Biol.* **2022**, *10*, 837337. [CrossRef] [PubMed]
76. Onishi, M.; Yamano, K.; Sato, M.; Matsuda, N.; Okamoto, K. Molecular mechanisms and physiological functions of mitophagy. *EMBO J.* **2021**, *40*, e104705. [CrossRef] [PubMed]

77. Tancini, B.; Buratta, S.; Delo, F.; Sagini, K.; Chiaradia, E.; Pellegrino, R.M.; Emiliani, C.; Urbanelli, L. Lysosomal Exocytosis: The Extracellular Role of an Intracellular Organelle. *Membranes* **2020**, *10*, 406. [CrossRef]
78. Settembre, C.; Fraldi, A.; Medina, D.L.; Ballabio, A. Signals from the lysosome: A control centre for cellular clearance and energy metabolism. *Nat. Rev. Mol. Cell Biol.* **2013**, *14*, 283–296. [CrossRef]
79. Pamarthy, S.; Kulshrestha, A.; Katara, G.K.; Beaman, K.D. The curious case of vacuolar ATPase: Regulation of signaling pathways. *Mol. Cancer* **2018**, *17*, 1–9. [CrossRef]
80. Wang, J.; Yang, X.; Zhang, J. Bridges between Mitochondrial Oxidative Stress, Er Stress and Mtor Signaling in Pancreatic B Cells. *Cell Signal.* **2016**, *28*, 1099–1104. [CrossRef]
81. Dermit, M.; Casado, P.; Rajeeve, V.; Wilkes, E.H.; Foxler, D.E.; Campbell, H.; Critchlow, S.; Sharp, T.V.; Gribben, J.G.; Unwin, R.; et al. Oxidative stress downstream of mTORC1 but not AKT causes a proliferative defect in cancer cells resistant to PI3K inhibition. *Oncogene* **2016**, *36*, 2762–2774. [CrossRef]
82. Ballesteros-Álvarez, J.; Nguyen, W.; Sivapatham, R.; Rane, A.; Andersen, J.K. Urolithin A reduces amyloid-beta load and improves cognitive deficits uncorrelated with plaque burden in a mouse model of Alzheimer's disease. *GeroScience* **2022**, *45*, 1095–1113. [CrossRef]
83. Fang, E.F.; Hou, Y.; Palikaras, K.; Adriaanse, B.A.; Kerr, J.S.; Yang, B.; Lautrup, S.; Hasan-Olive, M.M.; Caponio, D.; Dan, X.; et al. Mitophagy Inhibits Amyloid-B and Tau Pathology and Reverses Cognitive Deficits in Models of Alzheimer's Disease. *Nat. Neurosci.* **2019**, *22*, 401–412. [CrossRef]
84. Gong, Z.; Huang, J.; Xu, B.; Ou, Z.; Zhang, L.; Lin, X.; Ye, X.; Kong, X.; Long, D.; Sun, X.; et al. Urolithin A attenuates memory impairment and neuroinflammation in APP/PS1 mice. *J. Neuroinflamm.* **2019**, *16*, 1–13. [CrossRef]
85. Strachan, E.L.; Mac White-Begg, D.; Crean, J.; Reynolds, A.L.; Kennedy, B.N.; O'sullivan, N.C. The Role of Mitochondria in Optic Atrophy With Autosomal Inheritance. *Front. Neurosci.* **2021**, *15*, 784987. [CrossRef] [PubMed]
86. Peverelli, L.; Catania, A.; Marchet, S.; Ciasca, P.; Cammarata, G.; Melzi, L.; Bellino, A.; Fancellu, R.; Lamantea, E.; Capristo, M.; et al. Leber's Hereditary Optic Neuropathy: A Report on Novel Mtdna Pathogenic Variants. *Front. Neurol.* **2021**, *12*, 657317. [CrossRef]
87. Kuang, G.; Halimitabrizi, M.; Edziah, A.-A.; Salowe, R.; O'brien, J.M. The potential for mitochondrial therapeutics in the treatment of primary open-angle glaucoma: A review. *Front. Physiol.* **2023**, *14*, 1184060. [CrossRef] [PubMed]
88. Cano, M.; Wang, L.; Wan, J.; Barnett, B.P.; Ebrahimi, K.; Qian, J.; Handa, J.T. Oxidative stress induces mitochondrial dysfunction and a protective unfolded protein response in RPE cells. *Free Radic. Biol. Med.* **2014**, *69*, 1–14. [CrossRef]
89. Alka, K.; Kumar, J.; Kowluru, R.A. Impaired mitochondrial dynamics and removal of the damaged mitochondria in diabetic retinopathy. *Front. Endocrinol.* **2023**, *14*, 1160155. [CrossRef] [PubMed]
90. Marchi, S.; Guilbaud, E.; Tait, S.W.G.; Yamazaki, T.; Galluzzi, L. Mitochondrial control of inflammation. *Nat. Rev. Immunol.* **2022**, *23*, 159–173. [CrossRef]

cells

Article

Rabconnectin-3α/DMXL2 Is Locally Enriched at the Synaptic Ribbon of Rod Photoreceptor Synapses

Alina Dittrich [1], Girish Ramesh [1,2], Martin Jung [3] and Frank Schmitz [1,*]

1 Institute of Anatomy and Cell Biology, Saarland University, 66421 Homburg, Germany; alina-dittrich@gmx.de (A.D.); girishr.19@gmail.com (G.R.)
2 Institute of Biophysics, Saarland University, 66421 Homburg, Germany
3 Institute of Medical Biochemistry and Molecular Biology, Saarland University, 66421 Homburg, Germany; martin.jung@uks.eu
* Correspondence: frank.schmitz@uks.eu

Abstract: Ribbon synapses reliably transmit synaptic signals over a broad signalling range. Rod photoreceptor ribbon synapses are capable of transmitting signals generated by the absorption of single photons. The high precision of ribbon synapses emphasizes the need for particularly efficient signalling mechanisms. Synaptic ribbons are presynaptic specializations of ribbon synapses and are anchored to the active zone. Synaptic ribbons bind many synaptic vesicles that are delivered to the active zone for continuous and faithful signalling. In the present study we demonstrate with independent antibodies at the light- and electron microscopic level that rabconnectin-3α (RC3α)—alternative name Dmx-like 2 (DMXL2)—is localized to the synaptic ribbons of rod photoreceptor synapses in the mouse retina. In the brain, RC3α-containing complexes are known to interact with important components of synaptic vesicles, including Rab3-activating/inactivating enzymes, priming proteins and the vesicular H^+-ATPase that acidifies the synaptic vesicle lumen to promote full neurotransmitter loading. The association of RC3α/DMXL2 with rod synaptic ribbons of the mouse retina could enable these structures to deliver only fully signalling-competent synaptic vesicles to the active zone thus contributing to reliable synaptic communication.

Keywords: retina; ribbon synapse; rabconnectin-3α; DMXL2; vesicular H^+-ATPase; Rab3

Citation: Dittrich, A.; Ramesh, G.; Jung, M.; Schmitz, F. Rabconnectin-3α/DMXL2 Is Locally Enriched at the Synaptic Ribbon of Rod Photoreceptor Synapses. *Cells* 2023, *12*, 1665. https://doi.org/10.3390/cells12121665

Academic Editor: Hossein Ameri

Received: 28 April 2023
Revised: 8 June 2023
Accepted: 17 June 2023
Published: 19 June 2023

1. Introduction

Ribbon synapses are specialized, continuously active synapses built in the retina, pineal gland and inner ear [1–7]. In the retina, rod and cone photoreceptors and bipolar cells form ribbon synapses. Ribbon synapses faithfully transmit a broad range of stimulus intensities by computing graded changes of membrane potential. Rod ribbon synapses can even reliably transmit signals generated by the detection of single photons [8–19]. The highly reliable synaptic transmission at this type of synapse requires structural and functional specializations. The synaptic ribbon is the characteristic presynaptic structural specialization of ribbon synapses. Synaptic ribbons bind many synaptic vesicles and deliver them to the active zone to promote continuous synaptic transmission in a precise and largely indefatigable manner at high temporal resolution. RIBEYE is the main building block of synaptic ribbons and is essential to make the synaptic ribbon [20–26]. RIBEYE consists of a unique amino-terminal A-domain. The carboxyterminal B-domain of RIBEYE is identical to CtBP2 except for the first 20 amino-terminal amino acids [20]. CtBP2 functions as an NAD(H)-binding nuclear co-repressor and is highly homologous to CtBP1. The latter proteins evolved from D-isomer-specific 2-hydroxyacid dehydrogenases [27–30].

Intense synaptic vesicle trafficking events occur at the synaptic ribbon [7,31,32]. Synaptic vesicles bind to the synaptic ribbon and translocate along the ribbon to the active zone at which synaptic vesicle fusion occurs [33–37]. At many synapses, the synaptic vesicle-associated small GTP-binding protein Rab3A [38,39] is important for the recruitment of synaptic vesicles to the active zones, thereby also mediating aspects o synaptic plasticity [40–45]. Rab3A interconverts between a GTP-bound active state and GDP-bound inactive state [40,46]. At ribbon synapses, Rab3A most likely plays a particu larly prominent role [34,47,48]. Rab3A mediates vesicle delivery to the synaptic ribbon [34 and antibodies against Rab3 immunolabel synaptic ribbons in hair cells [48]. Bindin and release of synaptic vesicles depends upon a GTP/GDP cycle [34] emphasizing th importance of proteins that regulate the nature of the guanine nucleotide bound to Rab3A Rabconnectin-3α (RC3α)—alternative name Dmx-like 2 (DMXL2) [49]—serves, togethe with rabconnectin-3β, as a scaffold for the GDP/GTP exchange factor (GEF) protein and th GTPase activating protein (GAP) of Rab3A [50,51]. Therefore, we aimed to analyse the distr bution of rabconnectin-3α/DMXL2 (RC3α/DMXL2), in retinal ribbon synapses and focuse on photoreceptor ribbon synapses that are characterized by particularly large synaptic rib bons and intense synaptic vesicle trafficking. In the present study, we found RC3α/DMXL located to the synaptic ribbon in rod photoreceptor synapses of the mouse retina (a ro photoreceptor-dominated retina) using different antibodies against RC3α/DMXL2 an light and electron microscopic immunolabelling techniques suggesting the importance o proteins that regulate the nucleotide binding state of Rab3A for ribbon-associated vesicl trafficking and related events at the ribbon synapse.

2. Materials and Methods

2.1. Materials

2.1.1. Mice

All mouse care/organ dissection procedures were approved by the responsible loca authorities (Landesamt für Verbraucherschutz; Geschäftsbereich 4; 66115 Saarbrücken Germany; GB 3-2.4.2.2-25-2020). Prior to organ collection, C57BL/6J mice were deeply anaesthetized with isoflurane. Mice were killed by cervical dislocation. Organ isolation was performed within 5 min *post-mortem*.

2.1.2. Primary Antibodies

Antibodies against Rabconnectin3α/DMXL2

Two mouse monoclonal antibodies against RC3α/DMXL2 (clones 2G2 and 12D8 were generated and used in the present study for the determination of the localization of RC3α/DMXL2 in the retina. Mouse monoclonal antibodies were raised against re combinant, bacterially expressed and purified GST fusion protein in which the following 110 amino acid long peptide stretch from mouse RC3α/DMXL2 (from N- to C-terminal KKDQLDSVSGRMENGPSESKPVSRSDGGSGADWSAVTSSQFDWSQPMVTVDEEPLRL DWGDDHDGALEEDDGGGLVMKTTDAKKAGQEQSASDPRALLTPQDEECADGDTE was fused to the carboxy-terminus of GST using standard DNA cloning techniques. Fu sion protein expression and purification, immunization, hybridoma screening, hybridoma sub-cloning, counter-screening against GST and antibody isotyping was performed by Absea (Beijing, China). Two antibody hybridoma clones against RC3α/DMXL2 were used in the present study (2G2 and 12D8; both IgG2a immunoglobulins). For peptide array experiments the antibodies were used in a 1:20,000 dilution (~10 ng/mL final immunoglob- ulin concentration) and for immunofluorescence (IF) and post-embedding immunogold microscopy in a 1:10 dilution (~20 μg/mL final immunoglobulin concentration).

Furthermore, a commercially available rabbit polyclonal antibody against RC3α/DMXL2 (HPA039375; Sigma, Taufkirchen, Germany) was also used in this study. The affinity-purified antibody has been raised against a recombinant protein fragment of 92 amino acids length of human RC3α/DMXL2: (from N- to C-terminal: TKTSALSAKKDQPDFISHRMDDVP-SHSKALSDGNGSSGIEWSNVTSSQYDWSQPIVKVDEEPLNLDWGEDHDSA LDEEED-DAVGLVMKSTDA). This peptide sequence from human DMXL2 largely, though not com- pletely, corresponds to the peptide region of mouse RC3α/DMXL2 that was used for the generation of the monoclonal antibodies 2G2/12D8 (see first paragraph in Section 2.1.2). The affinity-purified polyclonal antibody against RC3α/DMXL2 was used for WB in a

1:1000 dilution; for peptide array experiments in a 1:20,000; for IF in a 1:300 dilution. Further primary antibodies used in the present study have been summarized in Table 1.

Table 1. Further primary antibodies.

Antibody	Source	Reference	Dilution
RIBEYE(B) U2656, rabbit polyclonal	Lab-made	[20]	1:10,000 (IF)
RIBEYE(B) (2D9), mouse monoclonal	Lab-made	[52]	1:200 (IF) 1:400 (EM)
PSD-95 (postsynaptic density protein-95), rabbit polyclonal	Gift Dr. T.C. Südhof	[53]	1:1000 (IF)
CSP (cysteine-string protein), rabbit polyclonal	Lab-made	Raised against recombinant full-length mouse CSP	1:500 (IF)
Cav1.4 Nterm, rabbit polyclonal	Lab-made	[54]	1:500 (IF)
GST, mouse monoclonal	Sigma-Aldrich, G1160	[55]	1:10,000 (WB)

2.1.3. Secondary Antibodies

All secondary antibodies used in the present study have been summarized in Table 2.

Table 2. Secondary Antibodies.

Antibody	Source	Dilution
Donkey anti-mouse Alexa 488	Invitrogen; Karlsruhe, Germany; A-21202	1:1000 (IF)
Chicken anti-mouse DyLight 488	Jackson ImmunoResearch; 715485150	1:1000 (IF)
Chicken anti-rabbit Alexa 488	Invitrogen; Karlsruhe, Germany; A-21441	1:1000 (IF)
Chicken anti-rabbit Alexa 568	Invitrogen; Karlsruhe, Germany; A-10042	1:1000 (IF)
Chicken anti-mouse Alexa 488	ThermoFisher; Karlsruhe, Germany; 10114192	1:1000 (IF)
Goat anti-mouse peroxidase-conjugate (POX)	Sigma; Taufkirchen, Germany; A3673	1:5000 (WB)
Goat anti-mouse conjugated to 1.4 nm Nanogold	Nanoprobes/Biotrend, Cologne, Germany, #N-2001	1:100 (EM)

Abbreviations: Immunofluorescence (IF), Western blot (WB) with enhanced chemiluminescence (ECL) detection, Electron microscopy (EM).

2.1.4. Additional Materials

Silver Enhancement Kit (Nanoprobes 2012 45ML, Nanoprobes, Inc. 95 Horseblock Road, Unit 1 Yaphank, NY, USA).

HiMark^{TM} Pre-stained protein standard (Invitrogen LC 5699).

Roti-Mark Standard for SDS-PAGE.

2.1.5. Plasmids

pGEX-mouse RC3α/DMXL2 (Absea), was cloned in pGEX-KG via BamHI/XhoI using synthetic DNA encoding aa 1919–aa 2029 of mouse RC3α/DMXL2 (NP_766359) using standard methods.

2.2. Methods

2.2.1. Embedding of Mouse Retinas and Immunohistochemistry on 0.5 µm Thin Resin Sections

Mouse retinas were processed for immunofluorescence (IF) microscopy on 0.5 µm thin resin sections as described [25,52,54–59]. Semithin sections provide a better resolution than cryostat sections for immunolabelling [60] and are capable of resolving single rod terminals and single rod active zones [52,54,56–58]. Eyes were isolated within 5 min *post-mortem* and the dissected posterior eyecups were flash-frozen in liquid nitrogen-cooled isopentane. Lyophilization of the tissue was performed at ~10^{-7} mbar for ~48 h. During the lyophilization, the tissue was continuously cooled by liquid nitrogen. Freeze-dried samples were equilibrated to room temperature, infiltrated with Epon resin at 28 °C on a rotor (at 2 rpm, for ~24 h), degassed in a vacuum chamber and subsequently polymerized at 60 °C for ~24 h, as described [25,52,54–59].

Semi-thin sections (0.5 µm thin) were cut from the polymerized tissue blocks with a Reichert ultramicrotome, as described [56,57]. Sections were collected on glass coverslips. Epon resin was removed from the sections by incubating sections with sodium methylate (30% *w*/*v* in methanol (Sigma-Aldrich #8.18194), 10 min); xylene/methanol (1:1 *v*/*v*, 10 min); acetone (2 × 10 min), H_2O (10 min) and PBS (10 min), as previously described [25,52,54–57]. All resin removal steps were performed at room temperature (RT).

Next, sections were incubated simultaneously with the indicated primary antibodies overnight at 4 °C, as also previously described [25,52,55–57]. Following the incubation in the primary antibody dilutions, sections were washed several times with PBS and incubated with the corresponding fluorophore-conjugated secondary antibodies (1 h at RT). After several washes with PBS, immunolabelled sections were mounted with an N-propyl gallate-containing anti-fading solution, as previously described [25,52,55–57]. Immunolabelling experiments were performed with three different sets of embedded mouse retinas.

In control experiments, sections were incubated also without primary antibody; all other steps of the immunolabelling protocol remained the same. Additional controls were performed for double immunolabelling experiments by setting individual laser power lines to zero. The detection settings remained unchanged. These controls were performed to make sure that the immunosignals in the respective detection channel do not result from signals of the "neighbouring" detection channel ("bleed-through controls"). Immunolabelled retina sections were analysed by confocal microscopy, as described in the next paragraph.

2.2.2. Confocal Microscopy of Immunolabelled Sections

We used an A1R laser scanning microscope (Nikon, Düsseldorf, Germany) for confocal microscopy, as previously described [25,52,55–57,61]. Images were acquired with 60×/1.40 N.A. oil objective and the 488 nm and 568 nm laser excitation lines. Image acquisition was performed with the NIS Elements software (NIS Elements AR 3.2, 64 bit; Nikon, Düsseldorf, Germany).

2.2.3. Preparation and Immunolabelling of Retinal Cryostat Sections

The posterior eyecups were dissected within 5 min of *post-mortem* and flash-frozen in liquid nitrogen-cooled isopentane, as previously described [20,62]. Cryostat sections of 8 µm were cut from these samples with a Leica cryostat CM950. Cryosections were heat-fixed by putting them on a heating pad (30 min at 60 °C). Incubation of the heat-fixed cryosections with primary and secondary antibodies, negative and positive controls as well as bleed-through controls was performed as described above for semi-thin sections. Immunolabelled sections were analysed by confocal microscopy, as described above.

2.2.4. Embedding of Retinas in LR Gold for Post-Embedding Immunogold Electron Microscopy

Mouse retinas were processed for post-embedding immunogold labelling as previously described [20,25,56]. Retinas, dissected as described above, were fixed overnight in 2% freshly depolymerized paraformaldehyde in PBS (pH 7.4) at 4 °C. Afterward, samples were dehydrated with ethanol (30% ethanol (4 °C, 10 min); next with ethanol concentrations of 50%, 70%, 80% to 99% ethanol (20 min each step, at −20 °C with mild agitation using an overhead rotator). Samples were infiltrated with increasing concentrations of LR Gold (ethanol/LR-Gold: 2/1, 1/1, 1/2 (*v*/*v*); 1 h each, at −20 °C) as described [20,25,56]. Samples were transferred to pure LR Gold resin (overnight at −20 °C) and finally infiltrated with LR Gold containing 0.1% benzil (*w*/*v*). Polymerization was performed for ≈48 h at −20 °C with UV light. Ultrathin sections (≈70 nm thin) were cut with a Reichert–Jung ultramicrotome and collected on 100 mesh gold grids. Please note that no OsO_4 can be used for lipid-contrasting in post-embedding immunogold electron microscopy. Therefore, membranes, e.g., synaptic vesicle membranes, are only weakly visible.

2.2.5. Post-Embedding Immunogold Labelling with Ultrasmall Immunogold Particles and Subsequent Silver Intensification

Post-embedding immunogold labelling was performed largely as previously described with some modifications [20,25,56]. Ultrathin sections were obtained from LR Gold-embedded tissue and treated with blocking buffer, containing 1% bovine serum albumin (BSA) in PBS, pH 7.4 (1 h, at RT) to saturate unspecific protein binding sites [20,25,56]. Then, sections were incubated with RC3α/DMXL2 2G2 mouse monoclonal antibody diluted 1:10 in blocking buffer (overnight, at 4 °C). Following several washes with blocking buffer, ultrathin sections were incubated with goat anti-mouse secondary antibody (1:100 dilution in blocking buffer, 90 min, at RT). The secondary antibody was conjugated to ultrasmall gold particles (~1.4 nm in diameter). After incubation with the secondary antibody, sections were washed several times with PBS and treated with 2.5% glutaraldehyde in PBS (15 min, at RT). Next sections were washed with H_2O and the immunolabeled sections were silver-intensified according to the manufacturer's instructions in the dark (4 min, at RT). This enhancement procedure was done to improve sensitivity. Following silver enhancement, sections were washed three times with H_2O and contrasted with 2% uranyl acetate (in H_2O, 10 min, at RT). As negative controls in these immunolabelling experiments, incubations in which no primary antibody was applied were used. All other steps of the immunogold labelling procedure remained identical in these negative control experiments. Immunolabelling experiments were performed with three sets of LR Gold-embedded mouse retinas. Immunolabelled ultrathin sections were analysed with a Tecnai Biotwin 12 transmission electron microscope (FEI, Eindhoven, The Netherlands) operated at 100 kV [25,56]. Images were acquired with a Megaview III digital camera (Gatan, Unterschleissheim, Germany) under the control of the iTEM acquisition software (Olympus; Hamburg, Germany).

2.2.6. Peptide Arrays for Antibody Epitope Mapping

We performed epitope mapping of all three RC3α/DMXL2 antibodies applied in the present study using peptides immobilized on cellulose membranes. For antibody epitope mapping of RC3α/DMXL2 antibodies, peptides of mouse RC3α/DMXL2 covering aa 1919–aa 2029 of mouse DMXL2 (NP_766359) were analysed. Peptides with a length of 20 amino acids each (overlap of 10 amino acids) were synthesized on the membrane. Peptide synthesis was accomplished on hardened cellulose membranes using a ResPepSL-Synthesizer (Intavis Bioanalytical Instruments; Cologne, Germany) [63–65]. Peptide arrays were processed for epitope mapping as previously described [25,66]. The membrane was activated with methanol (1 min, at RT). Next, the membrane with the peptide arrays was briefly washed with H_2O and incubated for 2 h with binding buffer (50 mM Tris-HCl, pH 7.5, 150 mM NaCl, 0.1% Triton X-100) with mild shaking at RT. Unspecific protein binding sites were saturated by incubating the membrane in blocking buffer (1 µM BSA in binding buffer; 1 h, at RT). Following incubation in blocking buffer, membranes were incubated with the primary antibodies that are indicated in the respective experiments (2G2 and 12D8 RC3α/DMXL2 mouse monoclonal; RC3α/DMXL2 rabbit polyclonal (Sigma-Aldrich; Taufkirchen, Germany); all in a 1:20,000 dilution in blocking buffer, overnight at 4 °C). Next, the membrane was washed 3 × 10 min with blocking buffer and incubated with goat anti-mouse antibody conjugated to horseradish peroxidase (HRP) (1:10,000 in blocking buffer) for 1 h at RT on a shaker. Antibody binding was visualized by enhanced chemiluminescence (ECL) with a ChemiDoc™ XRS Gel Doc system (Bio-Rad, Feldkirchen, Germany). After ECL detection, the locations of all peptide spots were visualized by UV illumination, as described [25,66].

2.3. Miscellaneous Methods

2.3.1. SDS-PAGE and Western Blotting

Retinas were isolated within 5 min *post-mortem* and dissolved in 200 µL hot Laemmli buffer [25]. The samples were solubilized by homogenization by up/down pipetting in a 100 µL tip and heated at 96 °C for 10 min [25]. The protein concentration of these retina samples was determined as described [67]. Retinal lysates (~50 µg total protein per lane) were separated by 5% acrylamide SDS-PAGE. Proteins were electro-transferred to nitrocellulose membrane (Protran 0.45 µm) (at 40 V, 10 h, at 4 °C). Unspecific protein binding sites were saturated by incubation in 5% skimmed milk powder in PBS (45 min RT) followed by overnight incubation in primary antibody (4 °C). Binding of the primary antibody was detected by the respective peroxidase-conjugated secondary antibodies and analysed by enhanced chemiluminescence (ECL) using a ChemiDoc™ XRS GelDoc system (Bio-Rad, Feldkirchen, Germany).

2.3.2. Expression and Purification of GST-Tagged Fusion Proteins

BL21(DE3) bacteria were transformed with the respective pGEX plasmids by electroporation and plated on Ampicillin plates. Induction, expression with IPTG and purification of recombinant GST-tagged fusion protein was performed with standard methods as previously described [20,55,68,69].

3. Results

We first applied a commercially available affinity-purified polyclonal rabbit antibody against DMXL2 to determine the localization of RC3α/DMXL2 in the retina. The polyclonal antibody detected a single high molecular weight band in WB analyses at the characteristic running position of ~340 kDa in mouse retinal lysates (Figure 1A). These Western blotting data clearly demonstrated that RC3α/DMXL2 is expressed in the retina. With peptide arrays that covered the entire region used for immunization we determined the precise epitopes of RC3α/DMXL2 which the polyclonal antibody reacts with (Figure 1B). The polyclonal antibody reacts with aa1949 to aa1988 of mouse RC3α/DMXL2 (NP_766359.2). This peptide region corresponds to aa1950 to aa1989 of human RC3α/DMXL2 (AAL93215) and is highly conserved between mouse and human DMXL2 (Figure 1C).

Next, we used the polyclonal RC3α/DMXL2 antibody for immunolabelling of 0.5 µm thin sections obtained from mouse retina (Figure 2). In cross-sections of the retina, we observed strong RC3α/DMXL2 immunosignals in the outer plexiform layer (OPL) in which the photoreceptor ribbon synapses are located (Figure 2A). We found the inner plexiform layer only weakly, if at all, immunolabeled (Figure 2A). Therefore, we focused on the localization of RC3α/DMXL2 in photoreceptor synapses of the OPL. Photoreceptor synapses in the mouse retina are predominantly rod photoreceptors synapses [7,70]. Rod photoreceptor synapses typically possess a single, large active zone with a single and large horseshoe-shaped synaptic ribbon [7,70]. The entire presynaptic terminal is filled with many highly motile synaptic vesicles that can bind to the synaptic ribbon [71].

Double-immunolabelling with antibodies against RIBEYE confirmed the synaptic localization of DMXL2 in the OPL (Figure 2B,C). Double immunolabelling with antibodies against RIBEYE showed partial co-localization of RC3α/DMXL2 with RIBEYE and suggested that a significant portion of RC3α/DMXL2 could be localized to the synaptic ribbon (Figure 2(B1–B3,C1–C3)). "Bleed-through" controls demonstrated that the RC3α/DMXL2 immunosignals at the synaptic ribbon are not influenced by RIBEYE immunosignals from the neighbouring detection channel but completely persist if the excitation for the RIBEYE channel is completely switched off (Figure 3). These data show that the RC3α/DMXL2 immunosignals do not result from a "bleed-through" from the RIBEYE immunosignals.

Unfortunately, the rabbit RC3α/DMXL2 antibody was not suitable for electron microscopic analyses and the ultrastructural distribution of RC3α/DMXL2 in rod photoreceptor synapses could thus not be resolved with this antibody.

In order to also resolve the ultrastructural distribution of RC3α/DMXL2 in rod photoreceptor synapses, we generated novel monoclonal antibodies against RC3α/DMXL2. Two monoclonal RC3α/DMXL2 antibodies, 2G2 and 12D8, were raised against a GST fusion protein containing a 110 amino acid long peptide stretch carboxyterminal of the central Rav1P_C domains of RC3α/DMXL2 (Figure 4A,B). Both monoclonal RC3α/DMXL2 antibodies strongly reacted with the RC3α/DMXL2-GST fusion protein in Western blot (WB) analyses (Figure 4(C1) lane 3; Figure 4(D1), lane 5), but not with GST alone (Figure 4(C1) lane 4; Figure 4(D1), lane 6). The same WB membranes that were first incubated with RC3α/DMXL2 antibodies were then re-probed with GST antibodies to analyse equal loading of the respective fusion proteins (Figure 4(C2,D2)). With peptide arrays we determined the precise binding epitopes of the RC3α/DMXL2 antibodies 2G2 and 12D8. The 2G2 antibody strongly reacted with the RC3α/DMXL2 peptide KKDQLDSVSGRMENGPSESK (Figure 4(E2),G) whereas 12D8 reacted with the RC3α/DMXL2 peptide ADWSAVTSSQFDWSQPMVTV (Figure 4(F2),G). UV illumination was used to determine the localization of the peptide spots (Figure 4(E1,F1)).

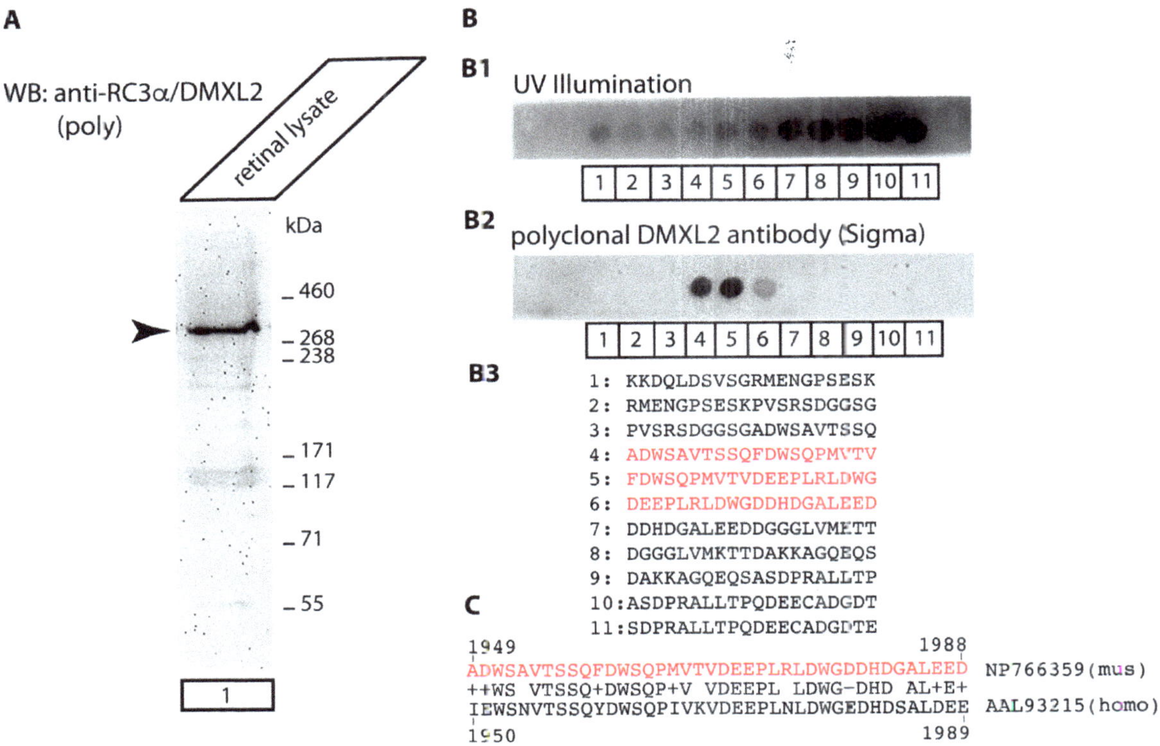

Figure 1. RC3α/DMXL2 expression in the mouse retina. (**A**) Retinal lysate from wild-type mice was probed by Western blot (WB) with affinity-purified rabbit polyclonal anti-RC3α/DMXL2. A high molecular weight band at ≈340 kDa was detected by the antibody in retinal lysates. (**B**) Peptide arrays that correspond to the RC3α/DMXL2 region against which this polyclonal antibody was generated, were incubated with affinity-purified rabbit polyclonal anti-RC3α/DMXL2 antibody. (**B1**) The location of all peptide spots was visualized by UV illumination. (**B2**) shows the result of the immunolabelling of the peptide array with affinity-purified rabbit polyclonal anti-RC3α/DMXL2 (ECL detection). Peptide spots #4, #5 and #6 strongly reacted with the anti-RC3α/DMXL2 polyclonal antibody. (**B3**) The amino acid sequence of spots #4, #5 and #6, that strongly reacted with the polyclonal RC3α/DMXL2 antibody, are highlighted in red. (**C**) The peptide sequence of mouse RC3α/DMXL2 (NP766359) that reacted with the affinity-purified rabbit polyclonal anti-RC3α/DMXL2 (highlighted in red) was aligned with the corresponding sequence from human RC3α/DMXL2 (AAL93215). The corresponding sequences are highly conserved between mouse and human RC3α/DMXL2 (73% amino acid identities).

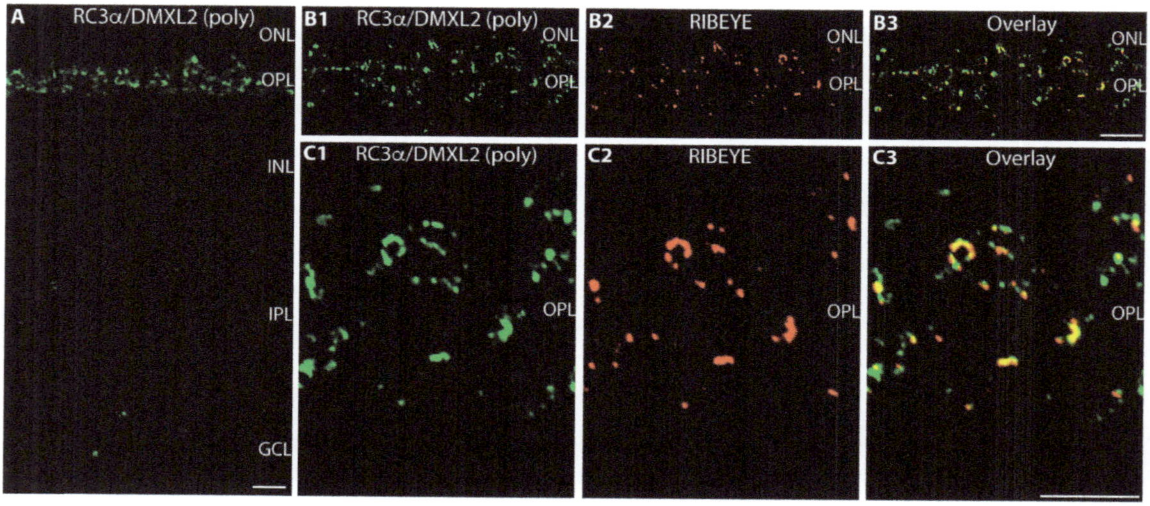

Figure 2. RC3α/DMXL2 is highly expressed in photoreceptor synapses in the OPL in close vicinity to synaptic ribbons. (**A**) 0.5 µm-thin retina sections incubated with affinity-purified rabbit polyclonal antibody against RC3α/DMXL2. RC3α/DMXL2 immunosignals are strongly enriched in the OPL in which photoreceptor ribbon synapses are found. (**B1–B3**,**C1–C3**) 0.5 µm-thin sections of the retina double-immunolabelled with rabbit anti-RC3α/DMXL2 (green channel) and with mouse anti-RIBEYE (2D9) (red channel). Signals from green and red channels (**B1**,**C1**/**B2**,**C2**) were overlaid in (**B3**/**C3**). Abbreviations: ONL, outer nuclear layer; OPL, outer plexiform layer; INL, inner nuclear layer; IPL, inner plexiform layer; GCL, ganglion cell layer. Scale bars: 5 µm.

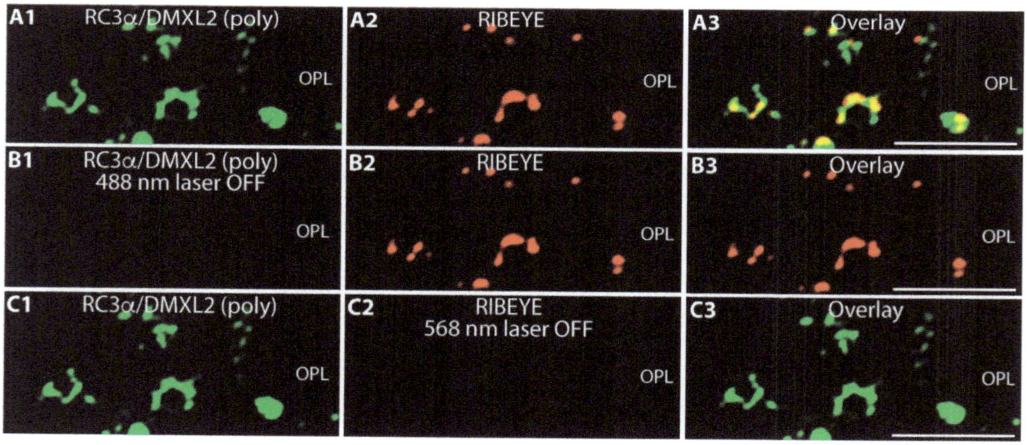

Figure 3. Control exposures. (**A1–A3**) 0.5 µm-thin retina sections were immunolabelled with rabbit anti-RC3α/DMXL2 (green channel) and with mouse anti-RIBEYE (2D9) (red channel), as in Figure 2. (**B1–B3**) The same immunolabelled retina section as shown in (**A1–A3**), but with the 488 nm laser turned off (with all PMT detection settings remaining unchanged). (**C1–C3**) Same double-immunolabelled retina section as shown in (**A1–A3**), but with the 568 nm laser turned off (with all PMT detection settings remaining unchanged). Signals from green and red channels (**A1**,**B1**,**C1**/**A2**,**B2**,**C2**) were overlaid in (**A3**,**B3**,**C3**). Abbreviations: OPL, outer plexiform layer. Scale bars: 5 µm.

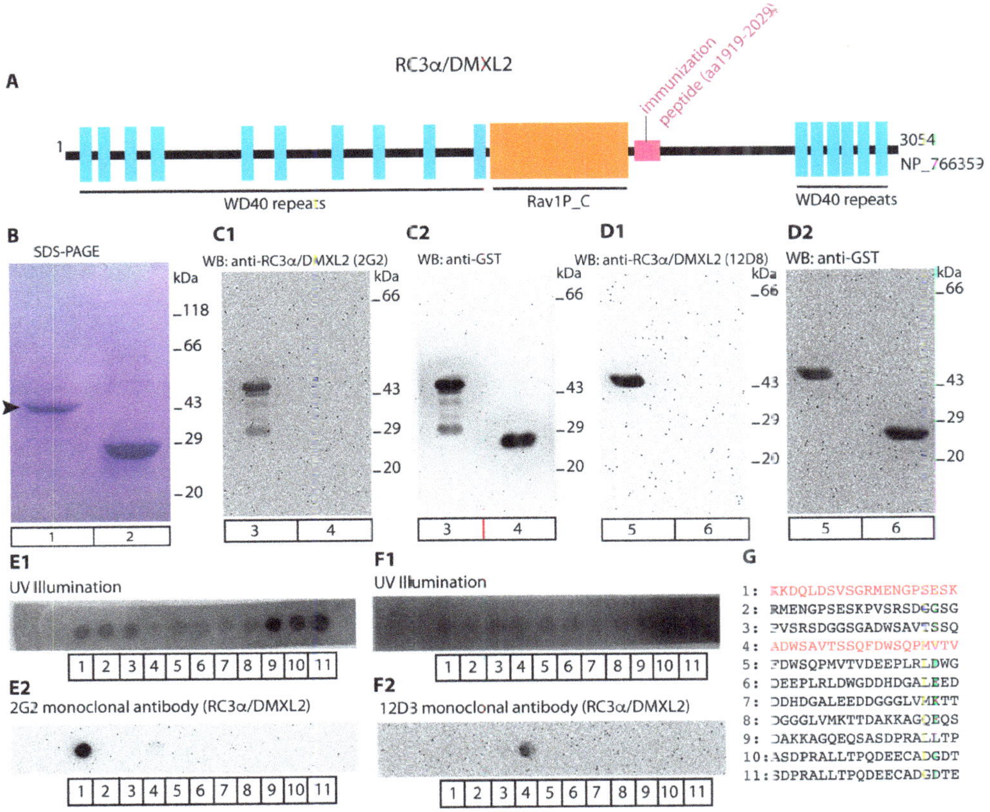

Figure 4. (**A**) Schematic domain structure of RC3α/DMXL2. Amino- and carboxyterminal WD40 repeats (blue boxes) and the central Rav1p_C domain (orange box) are schematically depicted. Monoclonal antibodies (2G2 and 12D8) were generated against a peptide stretch downstream of the Rav1p_C-domain (aa1919–aa2029, highlighted in pink). (**B**) SDS-PAGE of the purified GST-tagged fusion proteins (Coomassie Blue stained gel). Lane 1: RC3α/DMXL2 (aa1919–aa2029)-GST, lane 2: GST alone. (**C1**,**D1**) WB analyses of purified GST fusion proteins probed with RC3α/DMXL2 monoclonal antibodies. RC3α/DMXL2 (aa1919–aa2029)-GST was applied in lanes 3 and 5; GST in lanes 4 and 6 of (**C1**,**C2**,**D1**,**D2**). In (**C1**), RC3α/DMXL2-GST and GST were probed with anti-RC3α/DMXL2 (2G2). In (**C2**) the same blot was re-probed with anti-GST to verify equal loading. In (**D1**), RC3α/DMXL2 (aa1919–2029)-GST and GST were probed with anti-RC3α/DMXL2 (12D8). In (**D2**) the same blot was re-probed with anti-GST to verify equal loading. (**E1**,**E2**,**F1**,**F2**) show peptide arrays that cover the protein region of RC3α/DMXL2 against which the antibodies were generated (aa1919–aa2029). These peptide arrays were probed with the indicated monoclonal anti-RC3α/DMXL2 antibodies to determine the precise binding epitope of the antibodies. (**E1**,**F1**) UV light was used to visualize the location of all peptide spots. (**E2**) shows the peptide array that was incubated with monoclonal anti-RC3α/DMXL2 2G2; (**F2**) shows the peptide array incubated with monoclonal anti-RC3α/DMXL2 12D8. Peptide spot #1 (KKDQLDSVSGRMENGPSESK) strongly reacted with the 2G2 monoclonal antibody and peptide spot #4 (ADWSAVTSSQFDWSQPMVTV) with the 12D8 monoclonal antibody. (**G**) Amino acid sequences of all peptide spots. The peptide sequences of peptide spots #1 and #4, that strongly reacted with anti-RC3α/DMXL2 monoclonal antibodies 2G2 and 12D8, are highlighted in red.

Both novel monoclonal RC3α/DMXL2 antibodies 2G2 and 12D8 confirmed the previously observed synaptic localization of RC3α/DMXL2 (Figure 5) that has been obtained with the affinity-purified, rabbit polyclonal RC3α/DMXL2 antibody (Figures 2 and 3). Both antibodies (2G2 and 12D8) generated strong RC3α/DMXL2 immunosignals in the OPL (Figure 5A,C,E,F). The RC3α/DMXL2 antibodies 2G2 and 12D8 worked both on cryostat sections (Figure 5A,C) and on semi-thin sections (Figure 5E,F). The resolution of semi-thin sections was better than the resolution of cryostat sections (Figure 5A,C vs. Figure 5E,F).

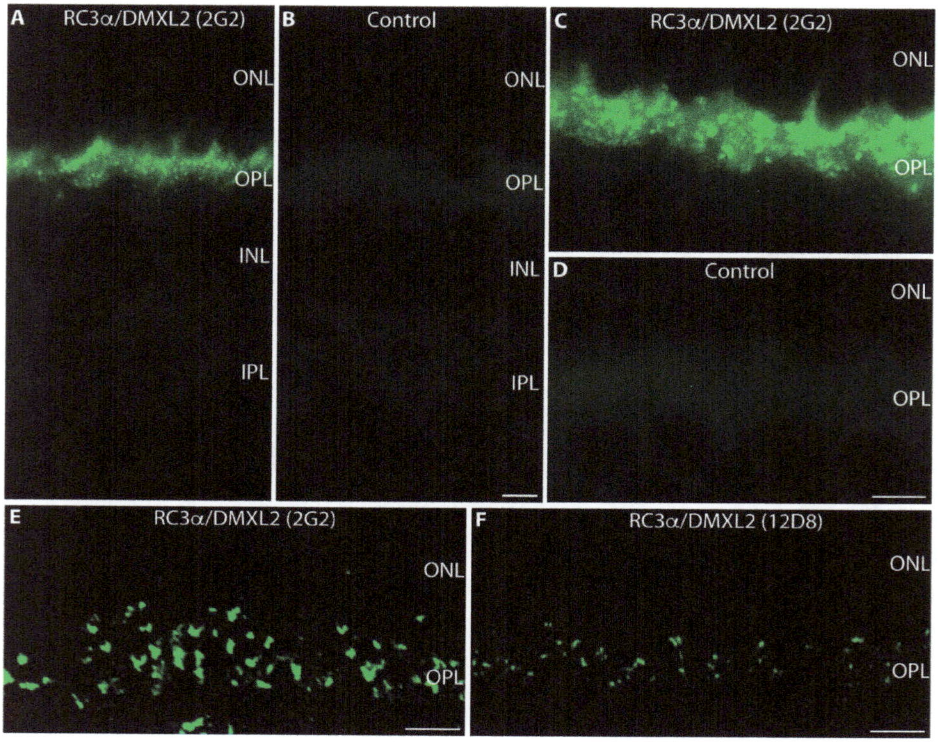

Figure 5. Single immunolabelling of mouse retina sections with the monoclonal RC3α/DMXL2 antibodies 2G2 and 12D8. In (**A**,**C**), 10 μm—thick cryostat sections of the mouse retina were immunolabelled with the indicated antibodies; in (**E**,**F**) 0.5 μm thin resin sections of the mouse retina. (**B**,**D**) represent control incubations in which the primary antibody was omitted. All other steps of the immunolabelling procedure were identical. Abbreviations: ONL, outer nuclear layer; OPL, outer plexiform layer; INL, inner nuclear layer; IPL, inner plexiform layer. Scale bars: 5 μm.

The 2G2 and 12D8 RC3α/DMXL2 mouse monoclonal antibodies produced discrete punctate immunosignals in the OPL, that partly appeared horseshoe-shaped (Figure 5A,C,E,F). The 2G2 antibody generated stronger immunosignals in the OPL and thus appeared more suitable for further immunocytochemical analyses.

The RC3α/DMXL2 (2G2) immunosignals were located within the photoreceptor presynaptic terminals as judged by double-immunolabelling with antibodies against cysteine-string protein (CSP), a synaptic vesicle protein of the presynaptic photoreceptor terminal (Figure 6(A1–A3)). As mentioned above, the entire large presynaptic photoreceptor terminal is occupied by many synaptic vesicles that contain CSP [72]. Double-immunolabelling with anti-PSD-95 revealed that RC3α/DMXL2 is located within the presynaptic terminal close to the presynaptic plasma membrane (Figure 6(B1–B3)). Please note that PSD-95 (postsynaptic density protein-95) is a presynaptic protein in photoreceptor synapses [73]. In photoreceptor synapses, PSD-95 is located beneath the presynaptic plasma membrane of the presynaptic terminal [73]. Immunosignals for RC3α/DMXL2 and voltage-gated

Cav1.4 calcium channels, which are located at the presynaptic active zone close to the synaptic ribbon [58,74], showed a similar localization/distribution at the light microscopic level (Figure 6(C1–C3)). Double-immunolabelling with antibodies against RIBEYE further indicated an enrichment of RC3α/DMXL2 at the synaptic ribbons (Figure 6(D1–D3,E1–E3)).

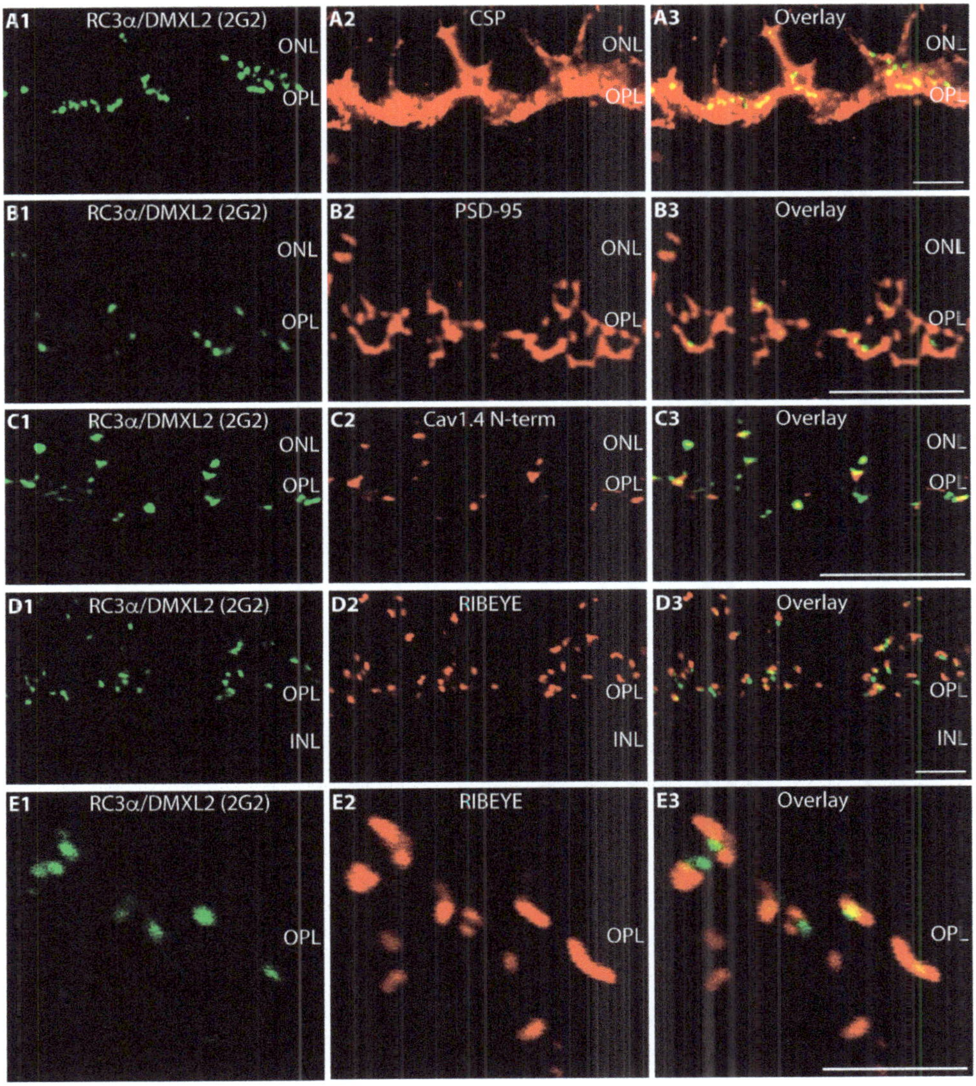

Figure 6. RC3α/DMXL2 is strongly expressed in photoreceptor synapses of the OPL in close vicinity to the synaptic ribbons. (**A1–A3**,**B1–B3**,**C1–C3**,**D1–D3**,**E1–E3**) Retina sections double-immunolabelled with mouse anti-RC3α/DMXL2 2G2 (**A1**,**B1**,**C1**,**D1**,**E1**) and with rabbit antibodies against CSP (**A2**), PSD-95 (**B2**), Cav1.4 (**C2**) and RIBEYE (**D2**,**E2**). Signals from green channels (**A1**,**B1**,**C1**,**D1**,**E1**) and red channels (**A2**,**B2**,**C2**,**D2**,**E2**) were overlaid in (**A3**,**B3**,**C3**,**D3**,**E3**). In (**A1–A3**,**D1–D3**,**E1–E3**), 10 µm-thick cryostat sections were used for immunolabelling; in (**B1–B3**,**C1–C3**) 0.5 µm—thin resin sections. Abbreviations: ONL, outer nuclear layer; OPL, outer plexiform layer. Scale bars: 5 µm.

The RC3α/DMXL2 2G2 mouse monoclonal antibody against RC3α/DMXL2 was suitable for post-embedding electron microscopic immunolabelling analyses (Figure 7). Post-embedding immunogold electron microscopy demonstrated that the synaptic ribbon is strongly decorated by the RC3α/DMXL2 (2G2) antibody. These data demonstrate also at the ultrastructural level that the synaptic ribbon is associated with RC3α/DMXL2, confirming the light microscopic immunolabelling data obtained with the polyclonal and monoclonal RC3α/DMXL2 antibodies. Other components of the presynaptic terminal were not strongly immunolabelled in our post-embedding immunogold labelling approach.

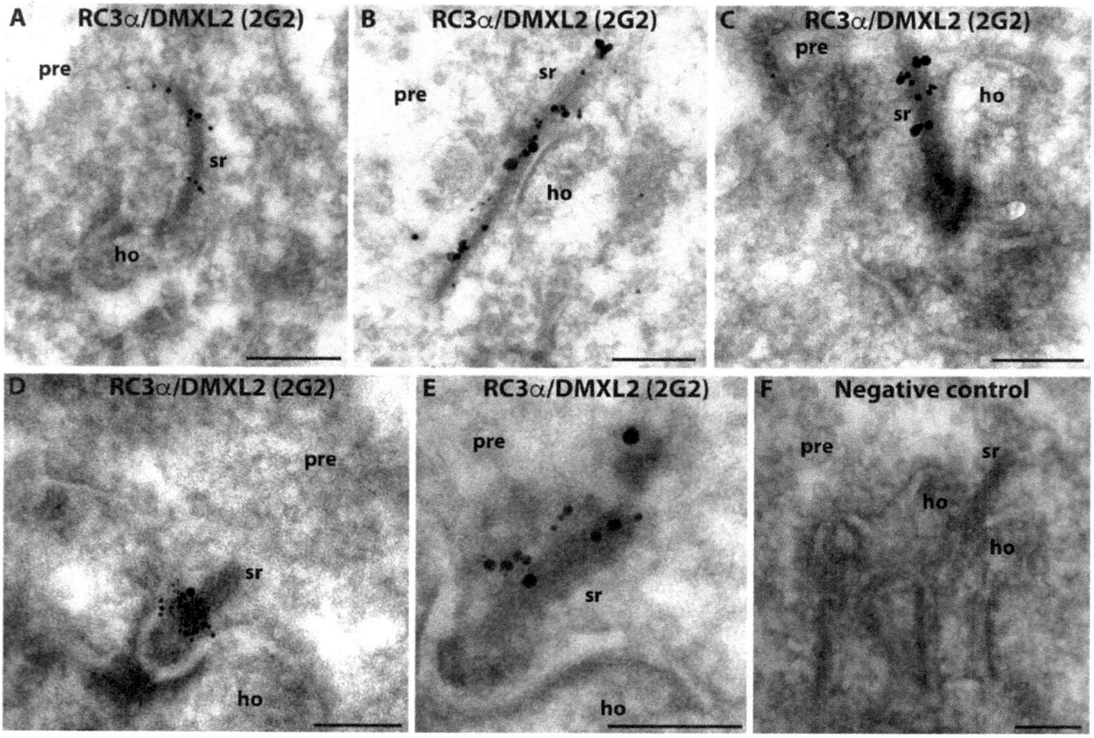

Figure 7. Post-embedding immunogold labelling of ultrathin LR Gold sections from the mouse retina (**A–F**). Rod photoreceptor synapses were immunolabelled with monoclonal anti-RC3α/DMXL2 2G2 antibody (**A–E**). (**F**) shows a representative negative control incubation in which the primary antibody was omitted. All other steps of the immunolabelling procedure remained the same. Please note that membrane contrast of membranes is limited because a post-embedding approach (without usage of OsO_4) was applied. Secondary goat anti-mouse antibodies were conjugated to ultrasmall (1.4 nm diameter) gold particles that were subsequently silver-intensified. Abbreviations: sr, synaptic ribbon; pre, presynaptic; ho, postsynaptic dendrites of horizontal cells. Scale bars: 300 nm (**A–F**).

4. Discussion

In the present study we have shown that rabconnectin3α (RC3α)/DMXL2 is localized to the synaptic ribbon in rod photoreceptor synapses of the mouse retina. The presence of RC3α/DMXL2 at the synaptic ribbon was consistently shown with three independent RC3α/DMXL2 antibodies at the light microscopic level using high resolution confocal microscopy. The localization at the synaptic ribbon was also confirmed at the ultrastructural level. Post-embedding immunogold electron microscopy demonstrated the presence of RC3α/DMXL2 at the synaptic ribbon in rod photoreceptor synapses. The retina of mice is a rod-dominated retina; more than 95% of photoreceptor synapses are made by rod photoreceptors [7]. Rod synapses have a very characteristic morphology at the light- and electron microscopic level [7]. Whether cone synaptic ribbons are also associated with RC3a/DMXL2 remains to be shown by future investigations. In the inner plexiform layer, we did not observe an obvious RC3α/DMXL2 immunosignal which might be based on the much smaller size of synaptic ribbons in the IPL in comparison to the OPL [7]. Thus, RC3α/DMXL2 immunosignals in the IPL might be under the detection limit for immunofluorescence microscopy on semi-thin sections.

Based on its subcellular localization at rod photoreceptor synaptic ribbons identified in the present study, RC3α/DMXL2 might serve as an acceptor complex for synaptic vesicles at the synaptic ribbon. As mentioned, RC3α/DMXL2 serves as a scaffold that binds GAP and GEF proteins that interact with Rab3A and determine the nature of its bound nucleotide (GDP vs. GTP) and the activity status. Rab3A is a component of synaptic vesicles, also at retinal ribbon synapses [75]. Thus, the previously observed binding of Rab3-containing synaptic vesicles to the synaptic ribbon [34,48] could be mediated by Rab3A interacting proteins, such as Rab3GEF/GAP, that are recruited via RC3α/DMXL2 to the synaptic ribbon. Electron-dense connections ("tethers") between synaptic vesicles and synaptic ribbons have been identified by electron microscopy [76]. Rab3A and Rab3A effectors have been previously proposed to be components of these tethers [34,48]. The known association of RC3α/DMXL2 with both GAP and GEF proteins of Rab3A [50,51] suggests that GTP/GDP exchange processes of Rab3A occur at the synaptic ribbon. Rab3A interacts with important Rab3A effectors in a GTP/GDP-dependent manner [40,46,47]. Interestingly, GTP/GDP-dependent interactors of Rab3A are enriched at the synaptic ribbon complex [20,47,74,75,77,78]. These include the RIM family of active zone proteins, which are important effectors of depolarization-evoked synaptic vesicle fusion at photoreceptor ribbon synapses [47,77]. Therefore, these GTP/GDP exchange processes will likely be functionally relevant for the intense vesicle trafficking events associated with the synaptic ribbon.

In the brain, RC3α/DMXL2 was purified from a crude synaptic vesicle fraction [50,51,79] and RC3α/DMXL2 has been localized to synaptic vesicles in the brain by a pre-embedding immunolabelling approach [50]. In the mouse retina, we did not find a RC3α/DMXL2 signal on most of the synaptic vesicles present in rod photoreceptor presynaptic terminals. We do not want to exclude that RC3α/DMXL2 is also present on non-ribbon-associated synaptic vesicles in ribbon synapses. The epitopes of RC3α/DMXL2 might be blocked/inaccessible on synaptic vesicles in post-embedding immunogold labelling procedures or the amount of RC3α/DMXL2 on vesicles in ribbon synapses could be less compared to the amount of RC3α/DMXL2 at the synaptic ribbon and too low to be detected by our antibodies. Further investigations are needed to analyse these possibilities.

RC3α/DMXL2 is a large protein with multiple amino- and carboxyterminal WD40 repeats and a central Rav1P_C domain. Interestingly, RC3α/DMXL2 not only interacts with GAP/GEF proteins of Rab3 but also with vesicular protein H^+-ATPases [80–83]. RC3α/DMXL2 is homologous to the yeast Rav1 protein, a central component of the yeast RAVE (Regulator of H^+-ATPase of vacuolar and endosomal membranes) complex [83–85]. The rabconnectin3 complex (RC3α/DMXL2 and RC3β) in higher eucaryotes/RAVE complex in yeast was shown to interact with components of vesicular H^+-ATPases [82–84,86,87]. In hair cells of zebrafish, RC3α/DMXL2 promotes the assembly of the vesicular H^+-ATPase and its functional activity that results in the acidification of synaptic vesicles [88]. RC3α/DMXL2 has been localized by immunofluorescence microscopy to the basal portion of hair cells [88]. At this location, ribbon synapses are found in hair cells [88].

Acidification of synaptic vesicles and a proton electrochemical gradient generated by the activity of the vesicular H^+-ATPase drive the loading of neurotransmitter into the synaptic vesicles (for review, [89,90]). Full acidification allows complete neurotransmitter loading of the synaptic vesicles, which was shown to be relevant for synaptic signalling [89–91]. Furthermore, full acidification of the synaptic vesicle lumen leads to the dissociation of V_o/V_1 complexes of the vesicular H^+-ATPase and to a change (increase) in fusion competence of the respective vesicles [91]. Of note, the vesicular H^+-ATPase has been found in protei complexes that were immunopurified with RIBEYE antibodies [48]. Thus, the association o RC3α/DMXL2 with the synaptic ribbon might ensure that only fully signalling-competen synaptic vesicles, i.e., synaptic vesicles that are completely filled with neurotransmitter, wi be made available to the active zone. The association of RC3α/DMXL2 with the synapti ribbon could help to prevent synaptic transmission failures that might result from th fusion of synaptic vesicles that are not or incompletely filled with neurotransmitter. T prevent failures in synaptic transmission seems particularly important for rod synapse that can faithfully transmit even very weak signals, e.g., tiny membrane potential change caused by the absorption of a single photon. A proton electrochemical gradient-dependen loading with glutamate has been identified as an important determinant of synaptic vesicl quantal size (amount of neurotransmitter in a synaptic vesicle) [92,93].

Interestingly, the rabconnectin3 complex, consisting of RC3α/DMXL2 and RC3β, ha also been reported to interact with CAPS1 (via rabconnectin-3β; [94]). CAPS1 has bee characterized in the brain as a synaptic protein with a dual role in vesicle priming an neurotransmitter filling. The localization and function of CAPS1 in the retina remains t be elucidated. Furthermore, RC3α/DMXL2 was found to interact with voltage-gated Cav channels in conventional synapses [95]. These data propose a central role of RC3α/DMXL for synaptic signalling.

The association of RC3α/DMXL2 with human diseases, e.g., Ohtahara syndrome (syndromic deafness-associated disease with mutations in the RC3α/DMXL2 gene) an some non-syndromic hearing losses with sensorineural impairment [96–98], emphasize the importance to further explore its function at ribbon synapses, including ribbon synapse of the retina.

Author Contributions: A.D., G.R. and F.S. performed experiments. M.J. provided important experi mental tools. A.D. and F.S. wrote the manuscript. F.S. designed the study. All authors have read an agreed to the published version of the manuscript.

Funding: Work of the authors was supported by a research grant from the Rolf M. Schwiet foundation (grant 2021-022), SFB894 (TP A7 and TP P3, Deutsche Forschungsgemeinschaft) an DFG Schm797/8-1.

Institutional Review Board Statement: The study was conducted according to the guideline of the Declaration of Helsinki and approved by the Institutional Review Board (Landesamt fü Verbraucherschutz, Geschäftsbereich 3; 66115 Saarbrücken, Germany; GB 3-2.4.1.1-K110/180-07).

Informed Consent Statement: Not applicable.

Data Availability Statement: All data are presented in the main manuscript and the manuscript figure

Acknowledgments: Work of the authors was supported by a research grant from the Rolf M. Schwiet foundation (grant 2021-022) and SFB894 (TP A7 and TP P3, Deutsche Forschungsgemeinschaft). W thank Gabi Kiefer for expert technical assistance; Shweta Suiwal and Soni Shankhwar for help wit immunoEM; Karin Schwarz for help with the formatting of the manuscript references.

Conflicts of Interest: The authors declare no conflict of interest.

References

1. Sjöstrand, F.S. Ultrastructure of retinal rod synapses of the guinea pig eye as revealed by three-dimensional reconstructions from serial sections. *J. Ultrastruct. Res.* **1958**, *2*, 122–170. [CrossRef]
2. Vollrath, L.; Huss, H. The synaptic ribbons of the guinea-pig pineal gland under normal and experimental conditions. *Z. Zellforsch Mikrosk. Anat.* **1973**, *139*, 417–429. [CrossRef] [PubMed]
3. Nakajima, Y.; Wang, D.W. Morphology of afferent and efferent synapses in hearing organ of goldfish. *J. Comp. Neurol.* **1974** *156*, 403–416. [CrossRef]
4. Krstic, R. Ultracytochemistry of the synaptic ribbons in the rat pineal organ. *Cell Tissue Res.* **1976**, *166*, 135–143. [CrossRef]
5. McNulty, J.A. Ultrastructural observations on synaptic ribbons in the pineal organ of the goldfish. *Cell Tissue Res.* **1980**, *210*, 249–25 [CrossRef] [PubMed]

6. Matthews, G.; Fuchs, P. The diverse roles of ribbon synapses in sensory neurotransmission. *Nat. Rev. Neurosci.* **2010**, *11*, 812–822. [CrossRef] [PubMed]
7. Moser, T.; Grabner, C.P.; Schmitz, F. Sensory processing at ribbon synapses in the retina and the cochlea. *Physiol. Rev.* **2020**, *100*, 103–144. [CrossRef]
8. Hecht, S.; Shlaer, S.; Pirenne, M.H. Energy, quanta, and vision. *J. Gen. Physiol.* **1942**, *25*, 819–840. [CrossRef]
9. Bulmer, M.G.; Howarth, C.I. Noise and the visual threshold. *Nature* **1957**, *180*, 1403–1404. [CrossRef]
10. Sakitt, B. Counting every quantum. *J. Physiol.* **1972**, *223*, 131–150. [CrossRef]
11. Baylor, D.A.; Nunn, B.J.; Schanpf, J.L. The photocurrent, noise and spectral sensitivity of rods of the monkey macaca fascicularis. *J. Physiol.* **1984**, *357*, 575–607. [CrossRef]
12. Schneeweis, D.M.; Schnapf, J.L. Photovoltage of rods and cones in the macaque retina. *Science* **1995**, *268*, 1053–1056. [CrossRef]
13. Rieke, F.; Baylor, D.A. Origin of reproducibility in the responses of retinal rods to single photons. *Biophys. J.* **1998**, *75*, 1836–1857. [CrossRef]
14. Field, G.D.; Rieke, F. Mechanisms regulating variability of the single photon responses of mammalian rod photoreceptors. *Neuron* **2002**, *35*, 733–747. [CrossRef]
15. Schein, S.; Ahmad, K.M. A clockwork hypothesis: Synaptic release by rod photoreceptors must be regular. *Biophys. J.* **2005**, *89*, 3931–3949. [CrossRef]
16. Gross, O.P.; Pugh, E.N.; Burns, M.E. cGMP in mouse rods: The spatiotemporal dynamics underlying single photon responses. *Front. Mol. Neurosci.* **2015**, *8*, 6. [CrossRef] [PubMed]
17. Reingruber, J.; Holcman, D.; Fain, G.L. How rods respond to single photons: Key adaptations of a G-protein cascade that enable vision at the physical limit of perception. *BioEssays* **2015**, *37*, 1243–1252. [CrossRef]
18. Hays, C.L.; Sladek, A.L.; Field, G.D.; Thoreson, W.B. Properties of multivesicular release from mouse rod photoreceptors support transmission of single-photon responses. *eLife* **2021**, *10*, e67446. [CrossRef] [PubMed]
19. Field, G.D.; Sampath, A.P.; Rieke, F. Retinal processing near absolute threshold: From behavior to mechanism. *Annu. Rev. Physiol.* **2005**, *67*, 491–514. [CrossRef]
20. Schmitz, F.; Königstorfer, A.; Südhof, T.C. RIBEYE, a component of synaptic ribbons. A protein's journey through evolution provides insight into synaptic ribbon function. *Neuron* **2000**, *28*, 852–872.
21. Maxeiner, S.; Luo, F.; Tan, A.; Schmitz, F.; Südhof, T.C. How to make a synaptic ribbon: RIBEYE deletion abolishes ribbons in retinal synapses and disrupts neurotransmitter release. *EMBO J.* **2016**, *35*, 1098–1114. [CrossRef]
22. Lv, C.; Stewart, W.J.; Akanyeti, O.; Frederick, C.; Zhu, J.; Santos-Sacchi, J.; Sheets, L.; Liao, J.C.; Zenisek, D. Synaptic ribbons require Ribeye for electron density, proper synaptic localization, and recruitment of calcium channels. *Cell Rep.* **2016**, *15*, 2784–2795. [CrossRef] [PubMed]
23. Jean, P.; de la Marina, D.L.; Michanski, S.; Tobón, L.M.J.; Chakrabarti, R.; Picher, M.M.; Neef, J.; Jung, S.; Gültas, M.; Maxeiner, S.; et al. The synaptic ribbon is critical for sound encoding at high rates with temporal precision. *eLife* **2018**, *7*, e29275. [CrossRef]
24. Becker, L.; Schnee, M.E.; Niwa, M.; Sun, W.; Maxeiner, S.; Talaei, S.; Kachar, B.; Rutherford, M.A.; Ricci, A.J. The presynaptic ribbon maintains vesicle populations at the hair cell afferent fiber synapse. *eLife* **2018**, *7*, e30241. [CrossRef] [PubMed]
25. Shankhwar, S.; Schwarz, K.; Katiyar, R.; Jung, M.; Maxeiner, S.; Südhof, T.C.; Schmitz, F. RIBEYE B-domain is essential for RIBEYE A-domain stability and assembly of synaptic ribbons. *Front. Mol. Neurosci.* **2022**, *15*, 838311. [CrossRef]
26. Mesnard, C.S.; Barta, C.L.; Sladek, A.L.; Zenisek, D.; Thoreson, W.B. Eliminating synaptic ribbons from rods and cones halves the releasable vesicle pools and slows down vesicle replenishment. *Int. J. Mol. Sci.* **2022**, *23*, 6429. [CrossRef]
27. Goldberg, J.D.; Yoshida, T.; Brick, P. Crystal structure of a NAD-dependent D-glycerate dehydrogenase at 2.4A resolution. *J. Mol. Biol.* **1994**, *236*, 1123–1140. [CrossRef]
28. Piatigorski, J. Dual use of the transcriptional repressor (CtBP2)/ribbon synapse (RIBEYE) gene: How prevalent are multifunctional genes? *Trends Neurosci.* **2001**, *24*, 555–557. [CrossRef]
29. Chinnadurai, G. CtBP, an unconventional transcriptional corepressor in development and oncogenesis. *Mol. Cell* **2002**, *9*, 213–224. [CrossRef]
30. Chinnadurai, G. CtBP family proteins: More than transcriptional repressors. *Bioessays* **2003**, *25*, 9–12. [CrossRef] [PubMed]
31. Heidelberger, R.; Thoreson, W.B.; Witkovsky, P. Synaptic transmission at retinal ribbon synapses. *Prog. Ret. Eye Res.* **2005**, *24*, 682–720. [CrossRef]
32. Thoreson, W.B. Transmission at rod and cone ribbon synapses in the retina. *Europ. J. Physiol.* **2021**, *473*, 1469–1491. [CrossRef] [PubMed]
33. Zenisek, D.; Steyer, J.A.; Almers, W. Transport, capture and exocytosis of single synaptic vesicles at active zones. *Nature* **2000**, *406*, 849–854. [CrossRef] [PubMed]
34. Tian, M.; Xu, S.; Montpetit, R.; Kramer, R.H. Rab3A mediates vesicle delivery at photoreceptor ribbon synapses. *J. Neurosci.* **2012**, *32*, 6931–6936. [CrossRef]
35. Van Hook, M.J.; Parmelee, C.M.; Chen, M.; Cork, K.M.; Curto, C.; Thoreson, W.B. Calmodulin enhances ribbon replenishment and shapes filtering of synaptic transmission by cone photoreceptors. *J. Gen. Physiol.* **2014**, *144*, 357–378. [CrossRef]
36. Vaithianathan, T.; Wollmuth, L.P.; Henry, D.; Zenisek, D.; Matthews, G. Tracking newly released synaptic vesicle proteins at ribbon active zones. *iScience* **2019**, *17*, 10–23. [CrossRef]

37. Joselevitch, C.; Zenisek, D. Direct observation of vesicle transport on the synaptic ribbon provides evidence that vesicles are mobilized and prepared rapidly for release. *J. Neurosci.* **2020**, *40*, 7390–7404. [CrossRef]
38. Matsui, Y.; Kikuchi, A.; Kondo, J.; Hishida, T.; Teranishi, Y.; Takai, Y. Nucleotide and deduced amino acid sequences of a GTP-binding protein family with molecular weights of 25,000 from bovine brain. *J. Biol. Chem.* **1988**, *263*, 11071–11074. [CrossRef]
39. Fischer von Mollard, G.; Mignery, G.A.; Baumert, M.; Perin, M.S.; Hanson, T.J.; Burger, P.M.; Jahn, R.; Südhof, T.C. Rab3 is a small GTP-binding protein exclusively localized to synaptic vesicles. *Proc. Natl. Acad. Sci. USA* **1990**, *87*, 1988–1992. [CrossRef] [PubMed]
40. Takai, Y.; Sasaki, T.; Shirataki, H.; Nakanishi, H. Rab3 small GTP binding protein in Ca^{2+}-dependent exocytosis. *Genes Cells* **1996**, *1*, 615–632. [CrossRef]
41. Geppert, M.; Bolshakov, V.Y.; Siegelbaum, S.A.; Takei, K.; De Camilli, P.; Hammer, R.E.; Südhof, T.C. The role of Rab3A in neurotransmitter release. *Nature* **1994**, *369*, 493–497. [CrossRef] [PubMed]
42. Geppert, M.; Goda, Y.; Stevens, C.F.; Südhof, T.C. The small GTP-binding protein Rab3A regulates a late step in synaptic vesicle fusion. *Nature* **1997**, *387*, 810–814. [CrossRef]
43. Leenders, A.G.M.; Lopes da Silva, F.H.; Ghijsen, W.E.J.M.; Verhage, M. Rab3A is involved in transport of synaptic vesicles to the active zone in mouse brain nerve terminals. *Mol. Biol. Cell* **2001**, *12*, 3095–3102. [CrossRef]
44. Schlüter, O.M.; Schmitz, F.; Jahn, R.; Südhof, T.C. A complete genetic analysis of neuronal Rab3 function. *J. Neurosci.* **2004**, *24*, 6629–6637. [CrossRef]
45. Schlüter, O.M.; Baus, J.; Südhof, T.C.; Rosenmund, C. Rab3 superprimes synaptic vesicles for release: Implications for short-term synaptic plasticity. *J. Neurosci.* **2006**, *26*, 1239–1246. [CrossRef] [PubMed]
46. Südhof, T.C. Function of Rab3 GDP-GTP exchange. *Neuron* **1997**, *18*, 519–522. [CrossRef]
47. Wang, Y.; Okamoto, M.; Schmitz, F.; Hofmann, K.; Südhof, T.C. RIM is a putative Rab3 effector in regulating synaptic-vesicle fusion. *Nature* **1997**, *388*, 593–598. [CrossRef] [PubMed]
48. Uthaiah, R.C.; Hudspeth, A.J. Molecular anatomy of the hair cell's ribbon synapse. *J. Neurosci.* **2010**, *30*, 12387–12399. [CrossRef]
49. Kraemer, C.; Weil, B.; Christmann, M.; Schmidt, E.R. The new gene DmX from Drosophila melanogaster encodes a novel WD-repeat protein. *Gene* **1998**, *216*, 267–276. [CrossRef]
50. Nagano, F.; Kawabe, H.; Nakanishi, H.; Shinohara, M.; Deguchi-Tawarada, M.; Takeuchi, M.; Sasaki, T.; Takai, Y. Rabconnectin-3, a novel protein that binds both GD/GTP exchange protein and GTPase-activating protein for Rab3 small G protein family. *J. Biol. Chem.* **2002**, *277*, 9629–9632. [CrossRef]
51. Kawabe, H.; Sakisaka, T.; Yasumi, M.; Shingai, T.; Izumi, G.; Nagano, F.; Deguchi-Tawarada, M.; Takeuchi, M.; Nakanishi, H.; Takai, Y. A novel Rabconnectin-3-binding protein that directly binds a GDP/GTP exchange protein for Rab3A small G protein implicated in Ca^{2+}-dependent exocytosis of neurotransmitter. *Genes Cells* **2003**, *8*, 537–546. [CrossRef]
52. Dembla, M.; Kesharwani, A.; Natarajan, S.; Fecher-Trost, C.; Fairless, R.; Williams, S.K.; Flockerzi, V.; Diem, R.; Schwarz, K.; Schmitz, F. Early auto-immune targeting of photoreceptor ribbon synapses in mouse models of multiple sclerosis. *EMBO Mol. Med.* **2018**, *10*, e8926. [CrossRef]
53. Irie, M.; Hata, Y.; Takeuchi, M.; Ichtchenko, K.; Toyoda, A.; Hirao, K.; Takai, Y.; Rosahl, T.W.; Südhof, T.C. Binding of neuroligins to PSD-95. *Science* **1997**, *277*, 1511–1515. [CrossRef] [PubMed]
54. Mukherjee, A.; Katiyar, R.; Dembla, E.; Dembla, E.; Kumar, P.; Belkacemi, A.; Jung, M.; Beck, A.; Flockerzi, V.; Schwarz, K.; et al. Disturbed presynaptic Ca^{2+} signaling in photoreceptors in the EAE mouse model of multiple sclerosis. *iScience* **2020**, *23*, 101830. [CrossRef]
55. Dembla, M.; Wahl, S.; Katiyar, R.; Schmitz, F. ArfGAP3 is a component of the photoreceptor synaptic ribbon complex and forms an NAD(H)-regulated, redox-sensitive complex with RIBEYE that is important for endocytosis. *J. Neurosci.* **2014**, *34*, 5245–5260. [CrossRef]
56. Wahl, S.; Katiyar, R.; Schmitz, F. A local, periactive zone endocytic machinery at photoreceptor synapses in close vicinity to synaptic ribbons. *J. Neurosci.* **2013**, *33*, 10278–10300. [CrossRef] [PubMed]
57. Wahl, S.; Magupalli, V.G.; Dembla, M.; Katiyar, R.; Schwarz, K.; Köblitz, L.; Alpadi, K.; Krause, E.; Rettig, J.; Sung, C.H.; et al. The Disease Protein Tulp1 Is Essential for Periactive Zone Endocytosis in Photoreceptor Ribbon Synapses. *J. Neurosci.* **2016**, *36*, 2473–2493. [CrossRef] [PubMed]
58. Dembla, E.; Dembla, M.; Maxeiner, S.; Schmitz, F. Synaptic ribbons foster active zone stability and illumination-dependent active zone enrichment of RIM2 and Cav1.4. *Sci. Rep.* **2020**, *10*, 5957. [CrossRef]
59. Kesharwani, A.; Schwarz, K.; Dembla, E.; Dembla, M.; Schmitz, F. Early changes in exo- and endocytosis in the EAE mouse model of multiple sclerosis correlate with decreased synaptic ribbon size and reduced ribbon-associated vesicle pools in rod photoreceptor synapses. *Int. J. Mol. Sci.* **2021**, *221*, 789. [CrossRef]
60. Punge, A.; Rizzoli, S.O.; Jahn, R.; Wildanger, J.D.; Meyer, L.; Schönle, A.; Kastrup, L.; Hell, S.W. 3D reconstruction of high-resolution STED microscope images. *Microsc. Res. Tech.* **2008**, *71*, 644–650. [CrossRef]
61. Eich, M.L.; Dembla, E.; Wahl, S.; Schwarz, K.; Schmitz, F. The calcineurin binding, activity-dependent splice variant dynamin1xb is highly enriched in synapses in various regions of the central nervous system. *Front. Mol. Neurosci.* **2017**, *10*, 230. [CrossRef]
62. Schmitz, F.; Bechmann, M.; Drenckhahn, D. Purification of synaptic ribbons, structural components of the active zone complex of photoreceptor synapses. *J. Neurosci.* **1996**, *16*, 7109–7116. [CrossRef]

3. Ronald, F. The SPOT-synthesis technique. Synthetic peptide arrays on membrane supports-principles and applications. *J. Immunol. Meth.* **2002**, *267*, 13–26.
4. Hilpert, K.; Winkler, D.F.H.; Hancock, R.E.W. Peptide arrays on cellulose support: SPOT synthesis, a time and cost-efficient method for synthesis of large numbers of peptides in a parallel and addressable fashion. *Nat. Protocols* **2007**, *2*, 1333–1349. [CrossRef] [PubMed]
5. Harsman, A.; Kopp, A.; Wagner, R.; Zimmermann, R.; Jung, M. Calmodulin regulation of the calcium-leak channel Sec61 is unique to vertebrates. *Channels* **2011**, *5*, 293–298. [CrossRef]
6. Suiwal, S.; Dembla, M.; Schwarz, K.; Katiyar, R.; Jung, M.; Carius, Y.; Maxeiner, S.; Lauterbach, M.A.; Lancaster, C.R.D.; Schmitz, F. Ciliary proteins repurposed by the synaptic ribbon: Trafficking myristoylated proteins at rod photoreceptor synapses. *Int. J. Mol. Sci.* **2022**, *23*, 7135. [CrossRef] [PubMed]
7. Dieckmann-Schuppert, A.; Schnittler, H.J. A simple assay for quantification of protein in tissue sections, cell cultures and cell homogenates and of protein immobilized on solid surfaces. *Cell Tissue Res.* **1997**, *288*, 119–126. [CrossRef]
8. Magupalli, V.G.; Schwarz, K.; Alpadi, K.; Natarajan, S.; Seigel, G.M.; Schmitz, F. Multiple RIBEYE-RIBEYE interactions create a dynamic scaffold for the formation of the synaptic ribbon. *J. Neurosci.* **2008**, *28*, 7954–7967. [CrossRef]
9. Alpadi, K.; Magupalli, V.G.; Käppel, S.; Köblitz, L.; Schwarz, K.; Seigel, G.M.; Sung, C.H.; Schmitz, F. RIBEYE recruits Munc118, the mammalian ortholog of the Caenorhabditis elegans protein unc119 to synaptic ribbons of photoreceptor synapses. *J. Biol. Chem.* **2008**, *283*, 26461–26467. [CrossRef] [PubMed]
0. Fredericks, C.E.; Zenisek, D. Ribbon synapses and retinal disease: Review. *Int. J. Mol. Sci.* **2023**, *24*, 5090. [CrossRef]
1. Holt, M.; Cooke, A.; Neef, A.; Lagnado, L. High mobility of vesicles supports continuous exocytosis at a ribbon synapse. *Curr. Biol.* **2004**, *14*, 173–178. [CrossRef] [PubMed]
2. Schmitz, F.; Tabares, L.; Khimich, D.; Strenzke, N.; de la Villa-Polo, P.; Castellano-Munoz, M.; Bulankina, A.; Moser, T.; Fernandez-Chacon, R.; Südhof, T.C. CSPalpha-deficiency causes massive and rapid photoreceptor degeneration. *Proc. Natl. Acad. Sci. USA* **2006**, *103*, 2926–2931. [CrossRef]
3. Koulen, P.; Fletcher, E.L.; Craven, S.E.; Wässle, H. Immunocytochemical localization of the postsynaptic density protein PSD-95 in the mammalian retina. *J. Neurosci.* **1998**, *18*, 10136–101349. [CrossRef]
4. tom Dieck, S.; Altrock, W.D.; Kessels, M.M.; Qualmann, B.; Regus, H.; Brauner, D.; Fejtova, A.; Bracko, O.; Gundelfinger, E.D.; Brandstätter, J.H. Molecular dissection of the photoreceptor ribbon synapse: Physical interaction of Bassoon and RIBEYE is essential for the assembly of the synaptic ribbon complex. *J. Cell Biol.* **2005**, *168*, 825–836. [CrossRef] [PubMed]
5. von Kriegstein, K.; Schmitz, F.; Link, E.; Südhof, T.C. Distribution of synaptic vesicle proteins in the mammalian retina identifies obligatory and facultative components of ribbon synapses. *Eur. J. Neurosci.* **1999**, *11*, 1335–1348. [CrossRef] [PubMed]
6. Usukura, J.; Yamada, E. Ultrastructure of the synaptic ribbons in photoreceptor cells of Rana catesbaiana revealed by freeze-etching and freeze-substitution. *Cell Tissue Res.* **1987**, *247*, 483–488. [CrossRef]
7. Grabner, C.P.; Gandini, M.A.; Rehak, R.; Le, Y.; Zamponi, G.W.; Schmitz, F. RIM1/2-mediated facilitation of Cav1.4 channels opening is required for Ca^{2+}-stimulated release in mouse rod photoreceptors. *J. Neurosci.* **2015**, *35*, 13133–13147. [CrossRef]
8. Löhner, M.; Babai, N.; Müller, T.; Gierke, K.; Atorf, J.; Joachimsthaler, A.; Peukert, A.; Martens, H.; Feigenspan, A.; Kremers, J.; et al. Analysis of RIM expression and function at mouse photoreceptor ribbon synapses. *J. Neurosci.* **2015**, *37*, 7848–7363. [CrossRef]
9. Sakisata, T.; Takai, Y. Purification and properties of Rabconnectin-3. *Methods Enzymol.* **2005**, *403*, 401–407.
0. Yan, Y.; Denef, N.; Schupbach, T. The vacuolar proton pump, V-ATPase, is required for notch signaling and endosomal trafficking in Drosophila. *Dev. Cell* **2009**, *17*, 387–402. [CrossRef]
1. Li, K.W.; Chen, N.; Klemmer, P.; Koopmans, F.; Karupothula, R.; Smit, A.B. Identifying true protein complex constituents in interaction proteomics: The example of the DMXL2 protein complex. *Proteomics* **2012**, *12*, 2428–2432.
2. Merkulova, M.; Paunescu, T.G.; Azroyan, A.; Marshansky, V.; Breton, S.; Brown, D. Mapping the H^+ (V)-ATPase interactome: Identification of proteins involved in the trafficking, folding, assembly and phosphorylation. *Sci. Rep.* **2015**, *5*, 14827. [CrossRef]
3. Jaskolka, M.C.; Kane, P.C. Interaction between the yeast RAVE complex and Vph1-containing Vo sectors is a central glucose-sensitive interaction required for v-ATPase reassembly. *J. Biol. Chem.* **2020**, *295*, 2259–2269. [CrossRef]
4. Smardon, A.M.; Tarsio, M.; Kane, P.M. The RAVE complex is essential for stable assembly of the yeast V-ATPase. *J. Biol. Chem.* **2002**, *277*, 13831–13839. [CrossRef] [PubMed]
5. Sipos, G.; Brickner, J.H.; Brace, E.J.; Chen, L.; Rambourg, A.; Kepes, F.; Fuller, R.S. Soi3p/Rav1p functions at the early endosome to regulate endocytic trafficking to the vacuole and localization of trans-Golgi network transmembrane proteins. *Mol. Biol. Cell* **2004**, *15*, 3196–3209. [CrossRef]
6. Seol, J.H.; Shevchenko, A.; Deshaies, R.J. Skp1 forms multiple protein complexes, including RAVE, a regulator of V-ATPase assembly. *Nat. Cell Biol.* **2001**, *3*, 384–391. [CrossRef]
7. Gowrisankaran, S.; Milosevic, I. Regulation of synaptic vesicle acidification at the neuronal synapse. *IUBMB Life* **2020**, *7*, 568–576. [CrossRef]
8. Einhorn, Z.; Trapani, J.G.; Liu, Q.; Nicolson, T. Rabconnectin3a promotes stable activity of the H^+-pump on synaptic vesicles in hair cells. *J. Neurosci.* **2012**, *32*, 11144–11156. [CrossRef]
9. Eriksen, J.E.; Li, F.; Edwards, R.H. The mechanism and regulation of vesicular glutamate transport: Coordination with the synaptic vesicle cycle. *Biochem. Biophys. Acta Biomembr.* **2020**, *1862*, 183259. [CrossRef]

90. Pietrancosta, N.; Djibo, M.; Daumas, S.; Mestikawy, S.E.; Erickson, J.D. Molecular, structural, functional, and pharmacologica sites for vesicular glutamate transporter regulation. *Mol. Neurobiol.* **2020**, *57*, 3118–3142. [CrossRef]
91. Bodzeta, A.; Kahms, M.; Klingauf, J. The presynaptic v-ATPase reversibly disassembles and thereby modulates exocytosis but i not part of the fusion machinery. *Cell Rep.* **2017**, *20*, 1348–1359. [CrossRef] [PubMed]
92. Edwards, R.H. The neurotransmitter cycle and quantal size. *Neuron* **2007**, *55*, 835–857. [CrossRef]
93. Takamori, S. Presynaptic molecular determinants of quantal size. *Front. Syn. Neurosci.* **2016**, *8*, 2. [CrossRef]
94. Crummy, E.; Mani, M.; Thellman, J.C.; Martin, T.F. The priming factor CAPS1 regulates dense core vesicle acidification b interacting with rabconnectin3b/WDR7 in neuroendocrine cells. *J. Biol. Chem.* **2019**, *294*, 9402–9415. [CrossRef] [PubMed]
95. Gandini, M.A.; Souza, I.A.; Fan, J.; Li, K.; Wang, D.; Zamponi, G.W. Interactions of Rabconnectin-3 with Cav2 calcium-channel *Mol. Brain* **2019**, *12*, 62. [CrossRef]
96. Chen, D.-Y.; Liu, X.-F.; Lin, X.-J.; Zhang, D.; Chai, Y.-C.; Yu, D.-H.; Sun, C.-L.; Wang, X.-L.; Zhu, W.-D.; Chen, Y.; et al. A dominan variant in DMXL2 is linked to nonsyndromic hearing loss. *Genet. Med.* **2017**, *19*, 553–558. [CrossRef] [PubMed]
97. Esposito, A.; Falace, A.; Wagner, M.; Gal, M.; Mei, D.; Conti, V.; Pisano, T.; Aprile, D.; Cerullo, M.S.; De Fusco, A.; et al. Bialleli DMXL2 mutations impair autophagy and cause Ohtahara syndrome with progressive course. *Brain* **2019**, *142*, 3876–389 [CrossRef]
98. Wonkam-Tingang, E.; Schrauwen, I.; Esoh, K.K.; Bharadwaj, T.; Nouel-Saied, A.; Acharya, A.; Nasir, A.; Leal, S.M.; Wonkam A. A novel variant in DMXL2 gene is associated with autosomal dominant non-syndromic hearing impairment (DFNA71) in cameroonian family. *Exptl. Biol. Med.* **2021**, *246*, 1524–1532. [CrossRef]

cells

MDPI

Article

Injury to Cone Synapses by Retinal Detachment: Differences from Rod Synapses and Protection by ROCK Inhibition

Ellen Townes-Anderson [1,*,†], Éva Halász [1,†], Ilene Sugino [2], Amy L. Davidow [3], Laura J. Frishman [4], Luke Fritzky [5], Fawad A. K. Yousufzai [5] and Marco Zarbin [2]

1 Department of Pharmacology, Physiology and Neuroscience, Rutgers New Jersey Medical School, 185 South Orange Avenue, Newark, NJ 07103, USA; eh394@njms.rutgers.edu
2 Institute of Ophthalmology and Visual Science, Rutgers New Jersey Medical School, 90 Bergen Street, Newark, NJ 07103, USA; suginoik@kajohnson.com (I.S.); zarbin@earthlink.net (M.Z.)
3 Department of Biostatistics, New York University School of Global Public Health, 708 Broadway, New York, NY 10003, USA; davida02@nyu.edu
4 Department of Vision Sciences, College of Optometry, University of Houston, Martin Luther King Blvd, Houston, TX 77204, USA; lfrishma@central.uh.edu
5 Cellular Imaging and Histology Core, Rutgers New Jersey Medical School, 205 South Orange Avenue, Newark, NJ 07103, USA; fritzklf@njms.rutgers.edu (L.F.); fawad.y@rutgers.edu (F.A.K.Y.)
* Correspondence: andersel@njms.rutgers.edu
† These authors contributed equally to this work.

Abstract: Attachment of a detached retina does not always restore vision to pre-injury levels, even if the attachment is anatomically successful. The problem is due in part to long-term damage to photoreceptor synapses. Previously, we reported on damage to rod synapses and synaptic protection using a Rho kinase (ROCK) inhibitor (AR13503) after retinal detachment (RD). This report documents the effects of detachment, reattachment, and protection by ROCK inhibition on cone synapses. Conventional confocal and stimulated emission depletion (STED) microscopy were used for morphological assessment and electroretinograms for functional analysis of an adult pig model of RD. RDs were examined 2 and 4 h after injury or two days later when spontaneous reattachment had occurred. Cone pedicles respond differently than rod spherules. They lose their synaptic ribbons, reduce invaginations, and change their shape. ROCK inhibition protects against these structural abnormalities whether the inhibitor is applied immediately or 2 h after the RD. Functional restoration of the photopic b-wave, indicating cone-bipolar neurotransmission, is also improved with ROCK inhibition. Successful protection of both rod and cone synapses with AR13503 suggests this drug will (1) be a useful adjunct to subretinal administration of gene or stem cell therapies and (2) improve recovery of the injured retina when treatment is delayed.

Keywords: cone cell; photoreceptor; synapse; retinal detachment; Rho kinase; synaptic ribbons; pedicle; spherule; STED microscopy; pig

Citation: Townes-Anderson, E.; Halász, É.; Sugino, I.; Davidow, A.L.; Frishman, L.J.; Fritzky, L.; Yousufzai, F.A.K.; Zarbin, M. Injury to Cone Synapses by Retinal Detachment: Differences from Rod Synapses and Protection by ROCK Inhibition. *Cells* 2023, *12*, 1485. https://doi.org/10.3390/cells12111485

Academic Editor: Hossein Ameri

Received: 15 April 2023
Revised: 22 May 2023
Accepted: 24 May 2023
Published: 27 May 2023

1. Introduction

Rod photoreceptors evolved from cone photoreceptors [1,2] to give us the capability to see over a wide range of luminosity with sensitivity to low light provided by rod cells and constant activity at ambient light levels provided by cone cells. The evolutionary link between rod and cone cells indicates numerous similarities. There are, however, also many differences in anatomy and physiology and, at the molecular level, in mechanisms such as light transduction and synaptic transmission. It is no surprise, therefore, that the responses of cone and rod cells to mechanical injury and retinal detachment display many differences. Fisher and colleagues [3–6] were the first to describe some of these differences concerning the synaptopathy of retinal injury. Retinal detachment results in the retraction of rod spherules from their bipolar and horizontal cell partners that, in many cases, do not repair even after long-term reattachment. In contrast, cone synapses do not retract in response to detachment, although their pedicles change shape, invaginations flatten, and ribbons can disappear. Data about cone synapses, however, remain limited in part because the synaptic cone changes are arguably more subtle and because cone markers disappear with increased time after detachment [7].

Because of the importance of the cone pathway to visual acuity and color vision, we undertook to expand the knowledge of cone pedicle structure after injury. We used quantitative analysis of confocal images and the new microscopic technique, stimulated emission depletion (STED), applied to a porcine model of retinal detachment. Moreover, because we reported that rod synaptic change occurs within hours of the injury [8], we have examined early time points (2–4 h) after detachment, times for which there is presently no information, as well as times after spontaneous retinal reattachment.

We have previously reported that the rapid changes in rod spherules, i.e., retraction, can be significantly reduced by the use of inhibitors to the RhoA pathway [8–13]. RhoA is a key signaling molecule in pathways that control the cytoskeleton. Its activation is known to occur with mechanical stimulation of cell membranes [14] and to play a role in both the development of synapses as well as synaptic plasticity [15,16]. Moreover, RhoA is present in both rod and cone synaptic terminals [9,10]. Most recently, we reported that the Rho kinase (ROCK) inhibitor AR13503 reduces spherule retraction and improves scotopic full-field electroretinogram (ffERG) responses two days after iatrogenic detachment followed by spontaneous reattachment [17]. The improvement in scotopic responses was highly correlated with a reduction of synaptic retraction, indicating that cytoskeletal control after an injury can reduce cellular dysfunction. Thus, in addition to examining the synaptic structure, we have assessed the responses of cone cells in experiments that repeat the conditions previously used for rod cells by testing cone photopic ERG responses two days after detachment and spontaneous reattachment. We show that AR13503 can improve cone synaptic function even though the pattern of damage in cone pedicles is distinct from rod spherule damage.

In these two-day experiments, iatrogenic retinal detachment, followed by spontaneous reattachment, follows a scenario similar to a surgical subretinal application or insertion of a viral vector or stem cell transplant [18,19]. ROCK inhibition, therefore, may be relevant for iatrogenic detachments created for these purposes. Moreover, the experiments were done on an adult pig, a diurnal animal whose eyes are similar to human eyes in many respects: the retina is similar in size, vascularization, and cellular structure, having both rod and cone cells and an area of high cone density, and the area centralis, similar in function to the human macula [20–22]. Porcine and human ERGs are similar as well [23,24]. Using porcine retinal detachment as a model of retina injury, therefore, enhances the translational significance of the results. Additionally, to assess whether ROCK inhibition can be applied not only at the time of detachment but also sometime after injury, we have tested whether treatment with ROCK inhibition can be delayed.

Our results indicate that ROCK inhibition protects both rod and cone synapses from injury, and thus its application may improve recovery from retinal detachment and reduce the likelihood of unsatisfactory outcomes reported after surgical retinal reattachment procedures [25–32].

2. Materials and Methods

2.1. Animals

Three-month-old male or female Yorkshire pigs weighing 30 kg were obtained from Animal Biotech Industries (Danboro, PA, USA) and kept on a 12 h light/12 h dark cycle. Animals were housed in an Association for Accreditation and Assessment of Laboratory Animal Care (AAALAC)-accredited pathogen-free facility, one animal to a pen. They were subject to overnight fasting with access to water ad libitum before surgery. Experimental procedures and methods of euthanasia were approved by the New Jersey Medical School Institutional Animal Care and Use Committee (IACUC ID 202000121, Protocol

PROTO999900995, 30 November 2020) and adhered to the ARVO Statement for the Use of Animals in Ophthalmic and Vision Research. A total of 19 animals and 38 eyes were used. Further descriptions of the animals can be found in Table A1 (Appendix A).

2.2. Retinal Detachment and Experimental Design

For consistency, retinal detachments were always created between 9 a.m. and 12 p.m. in the morning. Animals were injected with atropine (0.02 mg/kg; VetUS, Henry Schein, Dublin, OH, USA) and sedated with an injection of ketamine (20 mg/kg; Mylan Institutional LLC, Galway, Ireland) or telazol (3–4 mg/kg; Zoetis, Parsippany-Troy Hills, NJ, USA) and xylazine (2.2 mg/kg; Lloyd Lab., Shenandoah, IA, USA), all administered intramuscularly. After 5–10 min, a peripheral venous catheter was inserted through the auricular vein, and the animal was intubated with an endotracheal tube. In some cases, to deepen anesthesia, propofol (3 mL; Zoetis, Parsippany-Troy Hills, NJ, USA) was administered. To maintain anesthesia, the animals were supplied with 0.5% to 3.0% isoflurane in oxygen using a ventilator. Lactated Ringer's solution was infused intravenously at a rate of 8 mL/kg/h. Vital signs (oxygen saturation, heart rate, and body temperature) were monitored and maintained within the normal range throughout the experiment.

For surgery, pupils were dilated with a topical application of 1% Tropicamide (Bausch & Lomb, Tampa, FL, USA) and 2.5% phenylephrine (Paragon Bioteck, Portland, OR, USA). A standard 3-port vitrectomy was performed using 20 g instrumentation. The posterior hyaloid was detached over the area centralis using active suction, and a core vitrectomy was completed. During and after vitrectomy, the vitreous cavity of the eye was perfused with a balanced salt solution (BSS; Alcon, Fort Worth, TX, USA) containing 2 µg/mL epinephrine (Henry Schein, Dublin, OH, USA). To mitigate variability in anesthesia levels or other possible biological variables, each animal served as its own control, with one eye receiving BSS and the fellow eye the ROCK inhibitor. A 33 g metal cannula was used to inject BSS subretinally to create a retinal detachment (~10–15 mm in diameter) in the inferior nasal quadrant below the area centralis (Figure 1). For subretinal drug administration, the Rho kinase inhibitor AR13503 (Aerie Pharmaceuticals, Durham, NC, USA) was dissolved in the BSS to make the detachment. Immediately after the procedure, the sclerotomies were closed with a 7-0 vicryl suture. For intravitreal administration of a drug, 150 µL of 10 or 15 µm AR13503 dissolved in BSS was injected with a 30-gauge needle into the vitreous cavity (entering ~3 mm posterior to the limbus) with the fellow eye receiving an equal volume injection of BSS alone. The volume of the vitreous cavity was estimated to be 3 mL, and the final intravitreal concentrations were estimated to be 0.5 and 0.75 µm, respectively.

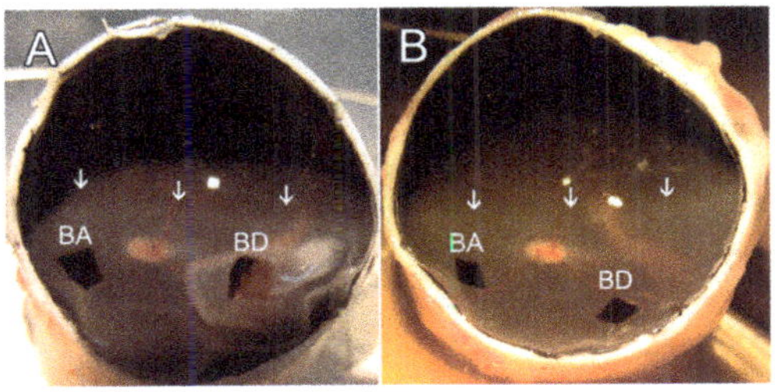

Figure 1. Pig eyes after enucleation and fixation. (**A**) A right eye 2 h after the formation of a detachment. The retinal detachment is present in the nasal inferior retina, below the area centralis (arrows). The boxes indicate the location of samples removed for sectioning. BD, sample of detached retina; BA, sample of attached retina. (**B**) A right eye two days after formation of a detachment. The detachment has spontaneously reattached. BD, sample of formerly detached retina; BA, sample of attached retina. Arrows indicate the area centralis, which extends from nasal to temporal retina.

After RDs were created, the animals survived for an additional 2 to 4 h or for two days. For the 2–4 h procedures, animals were kept under anesthesia after detachments were made and then euthanized with 7 mL intravenous Euthasol (Vibrac AH, Fort Worth TX, USA) for enucleation. For the longer survivals, the conjunctiva was sutured (7-0 vicryl after the sclerotomies were closed, 1.6 mg (0.4 mL) Dexamethasone (Fresenius Kabi, Lake Zurich, IL, USA) and 0.1 g (0.5 mL) Cefazolin (WG Critical Care, LLC, Paramus, NJ, USA were injected subconjunctivally, and Tobradex ointment (Alcon, Fort Worth, TX, USA) was applied topically. Once the animals had recovered, they were maintained in their cage with constant monitoring, for an additional two days. Animals were administered pre and post-operative intramuscular injections of buprenorphine (0.01–0.05 mg/kg; Reckitt Benckiser HealthCare, Hull, England) and enrofloxacin (10 mg/kg; Bayer HealthCare Shawnee, KS, USA). At the two-day time point, the animals were again anesthetized, using the previous protocol, for ERG recording and structural analysis by fundus photography and optical coherence tomography (OCT), as described below, before being euthanized with 7 mL intravenous Euthasol for enucleation.

2.3. Full-Field Flash Electroretinogram (ffERG), Fundus Photography, Optical Coherence Tomography (OCT)

The procedures for recording ffERGs, fundus photography, and OCT were done under general anesthesia, as described above. For all three procedures, pupils were dilated and accommodation was relaxed with topical application of 1% Tropicamide and 2.5% phenylephrine hydrochloride drops. Adjustable lid specula were used to keep the eyelids separated. ERGs were recorded in animals that had two-day survivals both before retinal surgery and two days after surgery. Fundus photography and OCT were performed in the animals two days after retinal surgery to confirm the status of the retina.

During electroretinography, flashes were produced, and responses were recorded using a UTAS ERG system with a BigShot Ganzfeld stimulator (LKC Technologies, Inc. Gaithersburg, MD, USA). The pig's head was placed inside the ganzfeld bowl, and bilateral ERGs were recorded simultaneously using ERG-Jet electrodes (Fabrinal SA, La Chaux-de Fonds, Switzerland) placed on the cornea. The cornea was kept moist with a hypromellose ophthalmic demulcent solution of 2.5% (Akorn, Inc., Lake Forest, IL, USA). The reference electrode was placed at the midline of the forehead, about the same distance from both eyes. The ground electrode was placed on the right shoulder of the animal. The stimulus protocol was based on the International Society for Clinical Electrophysiology of Vision (ISCEV) standard for clinical ffERG [33]. Briefly, to isolate the cone a- and b-waves, the animal was light-adapted for 10 min to a 20 cd/m^2 background. The ffERG was recorded to strobe flash intensities of 3 cd.s/m^2 (0 db) with an interstimulus interval (ISI) of 0.5 s (15 samples). A notch filter (60 Hz) and 85 Hz low pass filter were applied during data analysis using Matlab® (The Mathworks, Natick, MA, USA) to eliminate noise. The a wave amplitudes were measured from the beginning of the a-wave to the a-wave trough and a-wave implicit times were measured from stimulus onset to a-wave peak. The b-wave amplitudes were measured from the a-wave trough to the b-wave peak, and b-wave implicit times were measured from stimulus onset to the b-wave peak. Data points were automatically identified, and values were calculated by custom-made script in Matlab® (The Mathworks). Individual responses were analyzed, and aberrant waveforms were rejected before averaging.

2.4. Sample Preparation and Immunohistochemistry

After enucleation, a 5 mm slit was made at the ora serrata, and the eyes were immersed in 4% paraformaldehyde (EMS, Hatfield, PA, USA) for 15 min. The eyes were then opened the anterior segment and any remaining vitreous humor were removed, and eyecups were fixed overnight at 4 °C. Samples were collected from areas of the retina that had been detached and from areas of the retina that had not been detached as illustrated in Figure 1. Retinas were immersed in 30% sucrose overnight at 4 °C. On the following day specimens were embedded in OCT compound (Sakura Finetek, Torrance, CA, USA) at room temperature for 2 h, then frozen and cut into 15-µm-thick sections using a cryostat, as described previously [8].

Procedures for immunolabeling have been previously described [8]. Briefly, sections were washed two times with 0.3% Triton X-100 in PBS (PBS-T) and blocked with 10% blocking buffer (10% normal donkey serum, 33% PBS-T, and 57% PBS) for 1 h at room temperature. For visualization of cone synapses with confocal microscopy, sections were incubated in antibodies for CtBP2 (1:100, Fisher Scientific, Pittsburgh, PA, USA) and PSD-95 (1:100, Cell Signaling Technologies, Danvers, MA, USA) and peanut agglutinin (PNA) conjugated to FITC (1:100, Vector Labs, Burlingame, CA, USA) either overnight at room temperature or for 36 h at 4 °C. The following day the sections were washed three times with 0.3% Triton-100 in PBS and incubated with secondary antibodies conjugated to Alexa Fluor 546 or 647 (1:100, Life Technologies, Norwalk, CT, USA) and again with PNA-FITC for 90 min at room temperature or overnight at 4 °C. For STED microscopy, the primary antibody for CtBP2 was the same, but the secondary antibody was conjugated to STAR 635P (1:100, Abberior GmbH, Gottingen, Germany), whereas the PNA label used was conjugated to Alexa Fluor 594 (1:50, Life Technologies). In addition, incubation with the primary antibody and PNA was for 36 h, and with the secondary antibody and PNA, 24 h, at 4 °C. The resulting stronger labeling was useful for the increased intensity of the lasers needed to visualize the specimen with STED. For visualization of rod synapses with confocal microscopy, sections were incubated in an antibody to SV2 (1:100 dilution, Developmental Studies Hybridoma Bank, Iowa City, IA, USA) overnight at 4 °C. The next day, the sections were washed three times with 0.3% Triton-100 in PBS and incubated with a secondary antibody conjugated to Alexa Fluor 488 (1:100, Life Technologies) for 90 min at room temperature, followed by nuclear staining with 1 µg/mL propidium iodide (PI, 1:100, Sigma-Aldrich, St. Louis, MO, USA) or TO-PRO3 (1:500; Life Technologies) for 5 min at room temperature. After two washes with 0.3% Triton-100 in PBS, sections were covered with Fluoromount-G medium (Southern Biotech, Birmingham, AL, USA) or Diamond Antifade 5 (Fisher Scientific, Boston, MA, USA) and preserved under coverslips sealed with nail polish. For all immunohistochemistry, sections from retinal areas to be compared were placed on a single slide so that they were labeled together under the same conditions; control sections were also processed simultaneously with experimental sections but without primary antibodies.

2.5. Quantification of Cone Pedicles

All data were collected by a person masked to the sample identifications. Sections were examined using confocal microscopy (Nikon A1R+ HD Confocal Microscope, Melville, NY, USA) with a 60× oil immersion objective N.A. 1.4. Laser power and gain were set to obtain unsaturated images. In sections, locations for imaging were selected at random, and stacks of approximately 40 images of 0.20 µm optical thickness to allow for adequate overlap were created. Laser power, gain, and scanning rate were unchanged during the collection of images from a single experiment. Images were denoised before analysis. Enhancements in brightness and contrast were performed (Photoshop 7.0 software; Adobe, CA, USA) only for presentation purposes.

Two samples (BSS-treated attached retina, BA; BSS-treated detached retina, BD; AR13503-treated attached retina, AA; AR13503 Detached retina, AD) from each eye, four samples per animal, were obtained (Figure 1); two to three stacks of images were taken of each retinal sample. Cone pedicles were determined by the PNA label, which, in the outer plexiform layer (OPL), only labels the base of cone synapses. Stacks were scanned to follow intact ribbons, i.e., where both the beginning and the end of the ribbon were apparent in the stack, and ribbon length was measured. About 12–50 ribbons in each sample were measured. The number of ribbons varied because, in some conditions, ribbons were lost. The measurements are reported in micrometers.

2.6. Quantification of Rod Synapses

As mentioned above, data were collected by a person masked to the sample identifications. Sections were examined using confocal microscopy (model LSM510; Carl Zeiss Microscopy, Jena, Germany or Nikon A1R+ HD Confocal Microscope) by scanning 1 μm optically thick sections with a 63× N.A. 1.4 or 60× oil immersion objective N.A. 1.4, respectively. Laser power and gain were set to obtain unsaturated images. Laser power, gain, and scanning rate were unchanged throughout a single experiment. Enhancements in brightness and contrast were performed (Photoshop 7.0 software) only for presentation purposes.

As above, two samples (BA, BD or AA, AD) from each eye, four samples per animal, were obtained (Figure 1); 30–45 images were taken of each retina sample, and data were collected from 2 to 4 sections per sample, examining at least three different areas of each section. SV2 immunolabeling in the ONL was analyzed as described [8]. Briefly, a binary mask was created for each image, the ONL was outlined, and the labeled pixels in the ONL were counted using ImageJ software (version 1.45s; NIH). The measurements are reported as pixels per micrometer of ONL length.

2.7. Stimulated Emission Depletion (STED) Microscopy and 3D Imaging

Sections immunolabeled for STED microscopy were examined with a Stellaris 8 tau-STED Super Resolution Microscope (Leica Microsystems, Wetzlar, Germany) using a 93× glycerol-immersion objective lens N.A. 1.3, Leica's white laser for excitation, and a 755 nm laser for depletion. Two to three stacks of 0.14 μm optical sections were collected from each sample, with approximately 160 images per stack.

Stacks were viewed with Imaris software (v 10.0; Imaris for Core Facilities, Oxford Instruments, Tubney Woods, UK). CtBP2 immunolabel, for ribbons, and PNA label, indicative of the base of the cone pedicle, were viewed as 3D images by surface rendering of the fluorescent images. In some cases, all remaining fluorescent labels and CtBP2 immunolabels not associated with the PNA marker, diagnostic for rod ribbons, were removed to increase the clarity of the cone pedicles and associated ribbons for presentation.

Analysis of ribbons per pedicle was made from the rendered CtBP2 label; the surface area of the base of the pedicles was measured from en face views of the rendered PNA-labeled base of the pedicles using NIH ImageJ 1.5t (https://imagej.nih.gov/ij/index.html accessed on 3 January 2023). The perimeters of basal PNA labeling were outlined using the freehand selection tool, and area measurements were performed on ellipses fitted to PNA perimeters.

2.8. Statistical Analysis

For statistical analyses, the negative-binomial model with random effects, a generalized estimating equation (GEE) [34], or a linear model with random intercept was applied. Eyes within animals were randomized for BSS or AR13503 treatment. GEE and random intercept models were applied to estimate the parameters of a linear regression model while adjusting for the intrinsic correlation between repeated outcomes on the same subject, e.g., eyes within animals, multiple areas within eyes, and so on.

Statistical analysis was performed with SAS (version 9.4). The graphics were produced using GraphPad Prism 5.1 and Matlab® (The Mathworks, Natick, MA, USA). Box plots created with GraphPad show the median as a bar and the mean as a plus sign; the whiskers represent 1.5X the interquartile range (IQR) with dots as outliers. Figure 13 was done with Matlab. The whiskers represent the standard deviation, the bar is the mean, and the dots represent all data. We set alpha (type I error rate) at 0.05. Reported *p*-values were obtained via GEE analysis or the linear model unless otherwise noted.

3. Results

3.1. Normal Cone Synapses

Cone synapses were examined from normal, unoperated animals with confocal and STED microscopy to allow for comparison with cone pedicles after a detachment injury with or without treatment with the ROCK inhibitor AR13503. All images were from paraformaldehyde-fixed and frozen tissue. Sections for conventional confocal microscopy were triple-labeled for (see graphic abstract and Figure 2) (1) CtBP2, a homolog of the B domain of the synaptic ribbon protein RIBEYE, present in rod and cone terminals [35], and identical to a transcription repressor leading to the label of retinal nuclei in the sections as well [36]; (2) PSD-95, a scaffolding protein that coats the inner surface of both the rod and cone presynaptic terminals [37,38] excluding the base of the synapses; and (3) PNA, a lectin which lines the membranes only of cone cells extracellularly and thus is a cone-specific marker [39]. PNA is bright along the cone's inner and outer segments, lightly labels the cone cell body and axon, and then becomes bright again at the base of the cone pedicles; for STED microscopy, only CtBP2 and PNA were labeled.

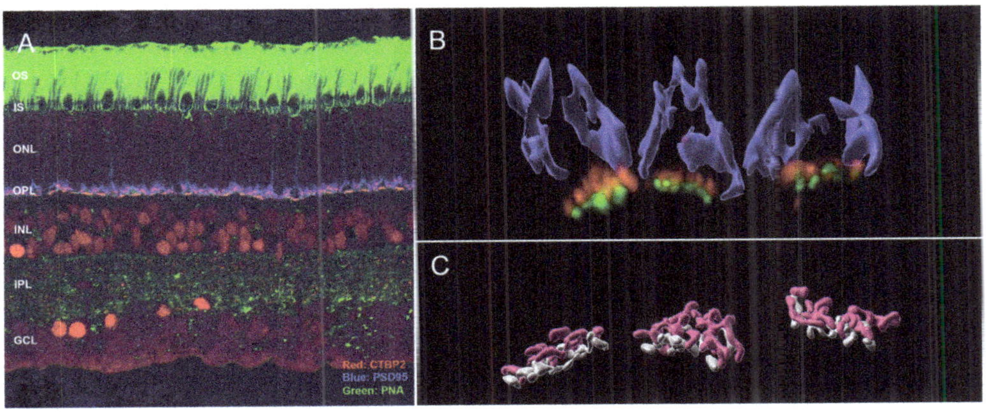

Figure 2. Normal porcine retina. (**A**) Low magnification of the retinal layers. CtPB2 (red) labels the ribbons in the outer plexiform layer (OPL) and is prominent in some nuclei. PSD-95 (blue) labels all photoreceptor synapses in the OPL. PNA (green) labels cone cells. It is present at the outer (OS) and inner (IS) segments, the cone axon, and the base of the pedicles. PNA is also present in the inner plexiform layer (IPL). ONL, outer nuclear layer; INL, inner nuclear layer; GCL, retinal ganglion cell layer. (**B**) High magnification of cone pedicles as seen from a stack of images taken with conventional confocal microscopy. PSD-95 has been surfaced rendered and labels the inside surface of the pedicles. PNA appears as discreet spots, and ribbons (CtBP2) as elongated spots. (**C**) Cone pedicles as seen from a stack of images taken with STED microscopy. The ribbon (pink) and PNA (gray) labels have been surface rendered. STED microscopy with surface rendering shows additional detail in the ribbon structure and PNA aggregates.

As in most mammalian retinas, the cone pedicles in the porcine retina form a narrow band in the outer plexiform layer (OPL) facing the bipolar dendritic processes (Figure 2). The rod spherules lie at multiple layers in the OPL distal to the cone synapses, and the rod ribbons, therefore, occupy a band about 5 μms wide. With conventional confocal microscopy, the PSD-95 label outlines the shape of the cone pedicle, which usually appears somewhat triangular (Figures 2 and 3). CtBP2 does not label the entire height of the plate-such as an electron-dense ribbon structure that is seen in electron micrographs (EM), as it is only labeling one component of RIBEYE, which itself makes up only about 67% of the mature ribbon [40]. However, the label extends the entire length of the ribbon. The cone terminal has multiple ribbons in contrast to rod spherules, which have a single ribbon. Cone ribbons often appeared as a row of elongated points, just above (distal) a row of PNA dots which are presumably labeling the base of the pedicle between synaptic invaginations (Figures 2 and 3). In conventional confocal microscopy, approximately 4–6 complete cone ribbons were seen in our image stacks of normal pedicles (Note: the number of complete ribbons visible in these stacks does not reflect the total number of ribbons present in a pedicle). Cone ribbons averaged 1.1 μms long in both the nasal inferior and temporal inferior quadrants that were analyzed (Figure 3). In some stacks, their horseshoe- or u shape is apparent. Consistent with other mammalian retinas, the cone ribbons are shorter than rod ribbons; rod ribbons averaged 1.8 μms in length in nasal and temporal inferior quadrants (Figure 3).

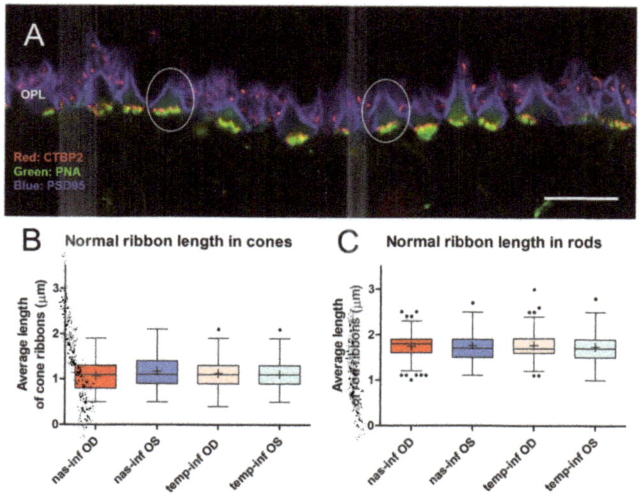

Figure 3. (**A**) Normal retina seen with conventional confocal microscopy. Cone pedicles (ovals) lie amongst rod spherules. Cone ribbons lie just above the PNA, whereas rod ribbons occupy a band about 5 μms wide and surround the pedicles. A stack of similar images was used to quantitate the length of photoreceptor ribbons. (**B**) The length of cone ribbons. The size of ribbons is relatively consistent across the inferior nasal and temporal retina. n = 2 animals. (**C**) The length of rod ribbons. Rod ribbons are longer than cone ribbons but also consistent in size across the inferior retina. n = 2 animals. OS, left eye OD, right eye. For box plots, whiskers are 1.5X IQR; dots are outliers. Bar = 10 μm.

With the 3D reconstruction of STED images and surface rendering of the fluorescent label, the PNA label for an individual pedicle appears as a loosely connected aggregation of islands of labels of various sizes (Figure 4). Cone ribbons are distributed uniformly above (distal to) the PNA label and generally are horseshoe- or u-shaped as in conventional confocal microscopy (Figure 4). The "u" shape is a result of the fact that the ribbons parallel the surface of the synaptic invagination where the processes of bipolar and horizontal cells lie. Additionally, in some cases, the ribbons may exhibit knobs or slight swellings at their ends (Figure 4). Finally, some ribbons have a y- or x-shape (Figure 4), as if there had been anastomoses of two or more ribbons. This finding has been previously illustrated in serial block face scanning EM in the mouse retina [41].

The aggregates of the PNA label are usually oval in shape and vary in size from 11.2 to 33.8 μm^2 with an average area of 24.1 μm^2 (n = 43 pedicles from two animals). We suggest that the "breaks" in the PNA label indicate the openings of synaptic invaginations. There is some difference of opinion about whether PNA is within the invagination or not. Unlike a previous report on the primate retina [42], we have never seen co-localization of PNA and Goalpha in the porcine retina (Townes-Anderson, personal observations). Instead, in STED images, we observe PNA on either end of horseshoe-shaped ribbons (Figure 4E), indicating its presence at the base where OFF bipolar processes contact the pedicle. Thus, our images support the report in mice by Sherry et al. [43].

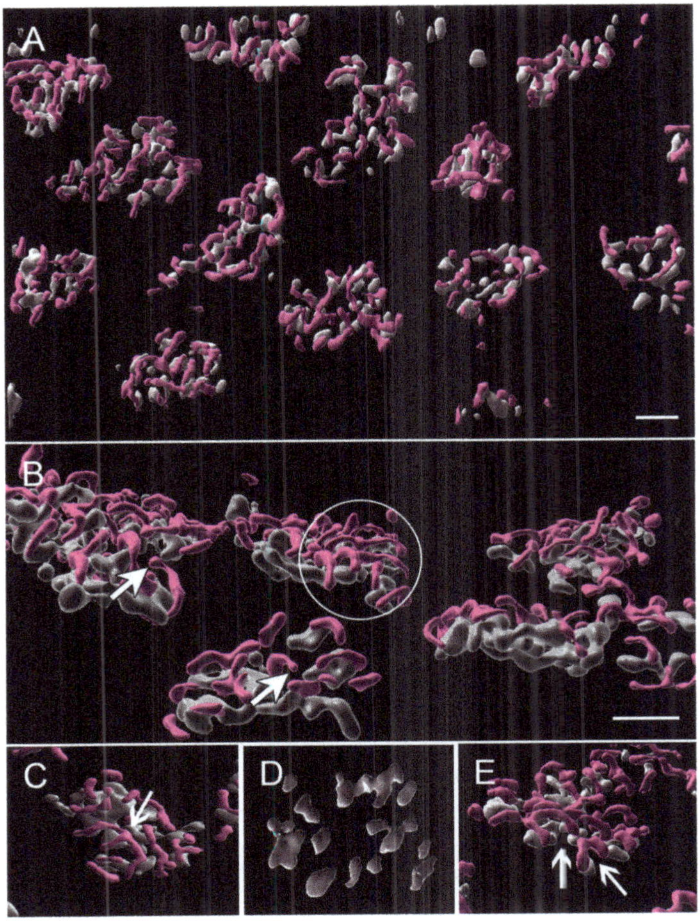

Figure 4. Normal retina from a stack of images taken with STED microscopy; fluorescent label was surface rendered. (**A**) The cone pedicles are regularly arrayed in this en face view. The PNA (gray) creates loose aggregates or islands of labels outlining the base of the pedicle. Ribbons (pink) appear as a loose network over the PNA label. (**B**) Most ribbons are single with a distinct horseshoe curvature (circle). Occasional swellings or knobs (arrows) appear along the ribbons. (**C**) Ribbons can also be in more complex branched configurations (arrow). (**D**) The aggregates of PNA are usually circular or oval. The spaces between the PNA are likely to be openings to synaptic invaginations. (**E**) Synaptic invaginations (arrows), defined by the curved ribbons, are bordered by a PNA label at the base of the pedicle. Bars = 2 μm.

A 3D reconstruction of STED images allows assessment of the total number of ribbons per pedicle. There were from 4 to 17 ribbons per pedicle, depending on the size of the PNA-labeled synaptic base, with an average of 12.3 ribbons per pedicle (n = 21 pedicles from two animals).

3.2. Intravitreal Administration of AR13503 Decreased the Loss of Cone Ribbons in 2-h Detachments

We have previously reported on the effects of detachment to the rod spherule. We discovered that rod spherules retract rapidly toward their cell bodies as soon as 2 h after detachment [8], so we initially examined cone synapses at a similar 2 h point after detachment. Additionally, as before, we tested the effects of ROCK inhibition on injury-induced synaptic changes. The ROCK inhibitor used in these experiments was AR13503, a ROCK inhibitor that we have shown to be more efficacious in reducing rod photoreceptor damage after detachment than other ROCK inhibitors [17]. AR13503 is the active metabolite of the FDA-approved netarsudil (AR13324) developed by Aerie Pharmaceuticals Inc. (Durham, NC, USA); the sustained release implant form is currently in clinical trials for diabetic retinopathy and age-related macular degeneration (phase 1, https://clinicaltrials.gov, (accessed on 1 May 2022)). AR13503 has a Ki of 0.2 nM for both ROCK 1 and 2 and Ki's of 1 nM and 27 nM for PKA and PKC, respectively [44]. AR13503 reduces the amount of rod spherule retraction induced by retinal detachment both by subretinal injection during the creation of a detachment or intravitreal injection following detachment creation [17]. Here, we examined sections from experiments in which the drug was applied intravitreally immediately following detachment. Some specimens had been previously examined for rod spherule retraction, and some came from newly performed experiments following the exact same protocols.

We created retinal detachments in both eyes of a single animal and injected one eye with the ROCK inhibitor diluted in BSS (to reach 0.75 μm AR13503 in the vitreous) and the fellow eye with BSS alone. (Note: although previous studies showed 0.5 μm AR13503 gave the best outcome for protection against rod retraction [17], we had had some success with 0.75 μm AR13503 as well and had material that could be used for investigation of cone synapses from these experiments.) Once created, detachments remained for 2 h before euthanasia and enucleation. The presence of retinal detachment was confirmed after fixation and bisection of the eyes. Specimens for microscopic examination came from the center of the detachment and attached regions of the retina approximately 1.5 cm further temporally. Both locations were inferior to the area centralis, the region of high cone cell density (Figure 1). The pig is a dichromat. The S and M cones are found throughout the retina, with S cones constituting about 13% of the cones. The nasal and temporal inferior porcine retina, from which we have taken our samples, have a similar distribution of cone cell types [45]. The distribution of S cones in the porcine retina has no consistent topographical pattern and can vary from the existence of small clusters of S cones to small areas lacking S cones [45]. Although S cones are reported to be more sensitive to retinal detachment in some species (human, primate, and rat [46–48]), we have not attempted to distinguish the two cone cell types, and thus our description and analyses related to all cones collectively.

In conventional confocal microscopy of triple-labeled sections, cone pedicles looked relatively normal (Figure 5). The cone ribbons formed a single layer facing the bipolar dendrites. However, some decrease in the number of ribbons was apparent in the detached retina from a cursory examination. Quantitating the length of cone ribbons demonstrated that the untreated, detached retina had significantly shorter ribbons than the detached retina that had been treated with AR13503 by about 40% (BD vs. AD, $p < 0.0001$; Figure 5). The number of ribbons observed in stacks of conventional confocal images was also reduced in the untreated detached retina compared to the treated, detached retina (e.g., 3.0 vs. 4.8 ribbons/pedicle in 1 animal). The ribbons in the attached retina from untreated and treated eyes were similarly examined since we had previously observed disruption of the rod synapse in the attached retina [8]. However, there were no significant differences in the length of cone ribbons in untreated and treated retina (BA vs. AA Figure 5) in areas temporal to the retinal detachment.

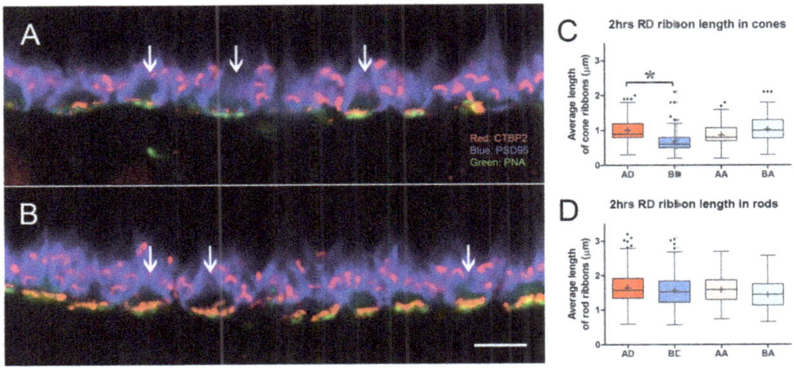

Figure 5. 2 h after detachment. (**A**) Detached retina, untreated. Although ribbons are present in most pedicles, the arrows indicate pedicles with reduced and shortened ribbons. Rod ribbons are present between the pedicles. (**B**) Detached retina treated with AR13503. All pedicles have ribbons (arrows). (**C**) Length of cone ribbons decreased significantly in the untreated retina (BD) compared to the treated retina (AD), * $p < 0.0001$. There was no significant difference in ribbon length in attached retinas whether treated or not (BA vs. AA). n = 3 animals. (**D**) Length of rod ribbons showed no differences in the treated vs. untreated eyes in detached and attached retinas. n = 3 animals. For box plots, whiskers are 1.5X IQR; dots are outliers. Bar = 5 μm.

With STED imaging and Imaris rendering, some pedicles of untreated retina appeared to be sparsely covered with ribbons. Cone ribbons also looked reduced in length with STED microscopy as quantitatively described above; the anastomoses of ribbons seen in the normal retina were less frequent, and the arched shape of the ribbons appeared more shallow (Figure 6). In a normal retina, the ribbon follows the curvature of the synaptic invagination; thus, the flattening of the ribbons strongly suggests there was some flattening of the invagination. Flattening of synaptic invaginations in cone pedicles has been previously described in detached retinas, but the data come only from detachments of three days duration or longer [4]. In contrast, in the treated, detached retina, most ribbons were normally arched, although some loss of the characteristic horseshoe shape in the treated retina could be found (Figure 6). Finally, the bases of the pedicles were significantly smaller in area, as determined by the PNA label, in the treated retina compared to the untreated retina (negative-binomial model, $p = 0.0004$, Figure 7) and compared to normal ($p = 0.003$). Immunohistochemical expression of PNA is relatively stable in detached rat retina compared to S cone opsin [46]; in the detached feline retina, the PNA label does not decrease for several days and remains at the synaptic terminal even after the labeling in the interphotoreceptor matrix has disappeared [7]. Changes in measurements based on the PNA label, therefore, are unlikely to be due to changes in PNA expression.

Thus, detachment results in the shortening, flattening, and disappearance of cone ribbons. ROCK inhibition applied at the time of detachment reduces these injury-induced changes. Unexpectedly, treated retinas had pedicles with a smaller presynaptic base. Both rounding and flattening of the shape of the feline pedicle have been described in longer-term detachments in vivo [4,49]. In vitro, we have seen rounding of porcine cone pedicles after 2 h of culture and spreading and flattening after 24 h [10]. It is possible that untreated pedicles were beginning to round up and lose their invaginations, whereas those treated with the ROCK inhibitor were not and were held in a more elongated configuration leading to a smaller base as well as better retention of the synaptic invaginations. Examination of the PSD-95 label confirmed that differences in pedicle shape existed between the untreated and treated retina (Figure 7).

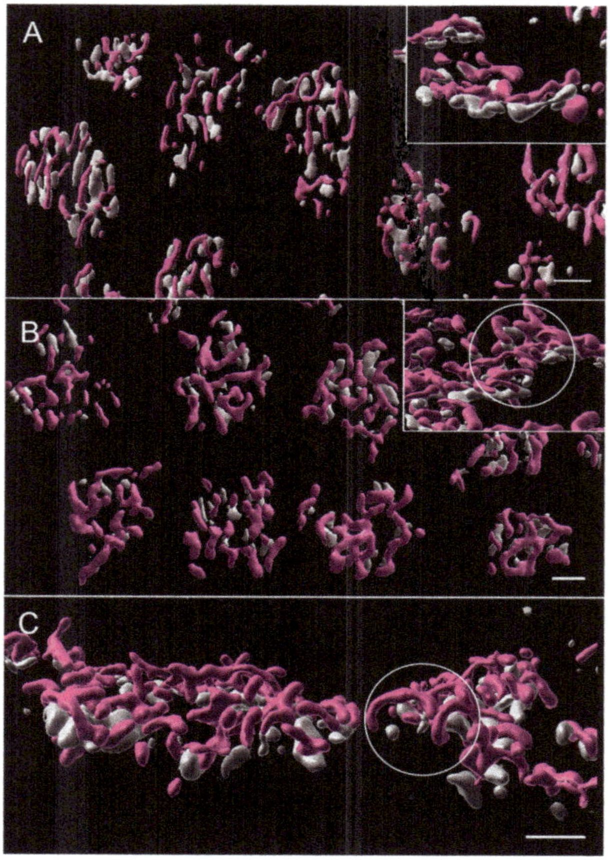

Figure 6. 2 h after detachment. (**A**) Detached retina, untreated. The array of pedicles, highlighted by PNA label (gray), is less organized than the normal retina. The network of ribbons (pink) also looks less dense, indicating a loss of ribbons. Branched ribbons were uncommon. In many cases, the ribbons have lost their horseshoe curvature (inset). (**B**) Detached retina, treated. The array of pedicles looks more normal, as does the network of ribbons overlying the PNA aggregates. In many cases, the ribbons have the characteristic U-shape curvature (inset, circle). (**C**) Attached retina, untreated. The ribbons looked densely distributed over the PNA label, and the horseshoe shape was consistently present (circle). STED microscopy with surface rendering. Bars = 2 μm.

For rod cells, although rod spherules retract into the outer nuclear layer after retinal detachment, there is no change in length from rod ribbon length in the normal retina (Figure 3). In addition, an analysis of the total CtBP2-labeled pixels in the rod zone of the OPL of untreated versus treated detached retina showed no difference (data not shown), suggesting there is no loss of ribbons. Thus, rod ribbons appear unaffected by detachment or ROCK inhibition.

Since some of the specimens had previously been examined for rod spherule retraction (2 h data in [17]), we can conclude that (1) rod spherule retraction and cone ribbon shortening occur together in the detached, untreated retina and (2) both occur quickly, within hours of detachment. However, in the attached retina temporal to the detachment, where rod spherule retraction is present, albeit to a lesser degree than in the detached retina [8], significant cone ribbon shortening or loss of ribbons was not seen. These structural changes in the cone synapses, therefore, did not spread to the attached retina.

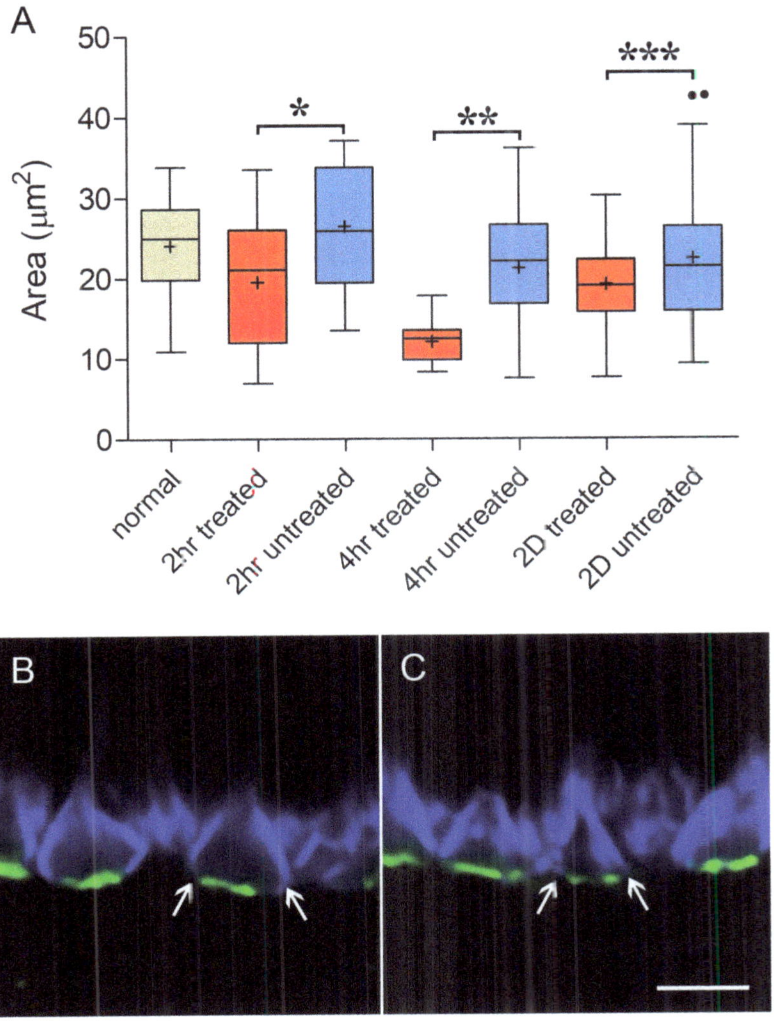

Figure 7. Size and shape of pedicles. (**A**) Area of pedicle base, measured from PNA label (see methods), at different times after detachment (2 and 4 h) and two days after detachment and reattachment (2D). Pedicle bases from 2 h, 4 h, and two-day treated retina are significantly smaller than those in untreated retina. For 2 h n = 41 pedicles from 1 animal * $p = 0.0004$; for 4 h n = 28 pedicles from 1 animal ** $p = 0.007$; for 2D n = 72 pedicles from 2 animals, *** $p = 0.049$. Pedicles from 4 h treated ($p < 0.001$) and untreated retina ($p = 0.042$) and 2 h ($p = 0.003$) and 2D treated retina ($p = 0.0028$) are smaller than normal; for normal n = 42 pedicles from 2 animals. For box plots, whiskers are 1.5X IQR; dots are outliers. (**B**) 2 h after detachment. Detached retina, untreated. Pedicles look slightly larger and rounder than in normal retinas. PSD-95 (blue); PNA (green). The PNA label indicates that the presynaptic terminals are from cone cells. (**C**) 2 h after detachment. Detached retina, treated. The cone pedicles look slightly smaller and triangular in shape, and their sides are straighter than the untreated retina. Arrows point to the edges of the PSD-95 label. Bar = 5 μm.

3.3. Delayed Intravitreal Administration of AR13503 Decreased Injury-Induced Changes in Both Cone and Rod Synapses

We have reported that the activation of RhoA increases by over 1.5-fold 2 h after retinal detachment in vivo [8]. For in vitro preparations of pig retina, activation remained high for at least 24 h [8]; in vivo, RhoA-GTP levels are higher 24 h after detachment compared to the control retina, but this increase did not reach statistical significance (n = 4 animals, *t*-test, $p = 0.07$, unpublished data). These data suggest that RhoA activation dissipates with time but may be present long after an injury has occurred.

In order to determine if ROCK inhibition could produce protection for photoreceptor synapses not just if initiated immediately during an injury but also if initiated a short time after injury, we applied the drug 2 h after detachment. Retinal detachments were made in both eyes of animals by subretinal injection of BSS. After 2 h, AR13503 was injected intravitreally to reach a concentration of 0.5 µm. Fellow eyes received an intravitreal BSS injection. The pigs survived an additional 2 h before euthanasia and enucleation. Thus, the total time from the original injury was 4 h.

Cone ribbons in conventional confocal microscopy were shorter in the untreated detached retina compared to the treated detached retina (BD vs. AD, $p < 0.05$, Figure 8). Although the difference in the data between treated and untreated detached retinas was significant, the variability between animals was striking. In one animal, the difference between untreated and treated retina was negligible, whereas, in another animal, there were numerous empty pedicles (i.e., no ribbons) in the confocal stacks of the detached untreated retina compared to the treated eye. Variability between animals is not unexpected, as individuals would be expected to show differences. However, with detachment, there is also variability within the detached retina. We have observed differences between sections analyzed at different points in a single retinal sample and have therefore used multiple animals and multiple images for statistical analyses. Our observations are supported by others who report a high level of variability in the number of TUNEL-positive cells across detached rat retina [46] as well as in outer segment length, outer segment/retinal pigmented epithelial interface, mislocalization of rod opsin, and rod neuritic sprouting into the inner retina after reattachment of feline retinas [4,50,51]. In some cases, these variables contribute to a morphological "patchwork" appearance of photoreceptor recovery [50]. In the untreated detached retina, cone ribbons were shorter than in the normal retina (Mann–Whitney $p = 0.009$). In contrast, there were no apparent differences in cone ribbon length or number between the untreated and treated attached retina, and the cone ribbon lengths appeared normal (BA vs. AA, Figure 8).

With STED microscopy, the visualization of the morphological differences in untreated and treated detached retinas in the 2 h delay experiments was enhanced: without ROCK inhibition, there appeared to be fewer, shorter ribbons, and many were flattened (Figure 9). With ROCK inhibition, there were more ribbons, which appeared longer and frequently had some curvature (Figure 9). The general flattening at the base of the pedicle, therefore, was pronounced with a longer (4 h) detachment, whereas the loss of synaptic invaginations was reduced or reversed with ROCK inhibition. Additionally, there was a "knobbiness" to the ribbons from the treated retina; in other words, skinny ribbon segments alternating with slight swellings or protrusions (Figure 9). Although these swellings were occasionally seen in other specimens, including in the normal retina, as discussed, the treated retina in these delayed-treatment experiments showed the knobs more frequently. There were also occasional aggregates of ribbon labels that were unassociated with PNA-labeled terminals (Figure 9). Finally, the PNA-labeled base of the pedicles was smaller for both untreated and treated detached retina than normal retina (negative-binomial model, $p < 0.05$, details in legend for Figure 7), whereas in the treated retina, as at 2 h, the PNA-labeled base was significantly smaller than in the untreated retina (negative-binomial model, $p = 0.007$, Figure 7).

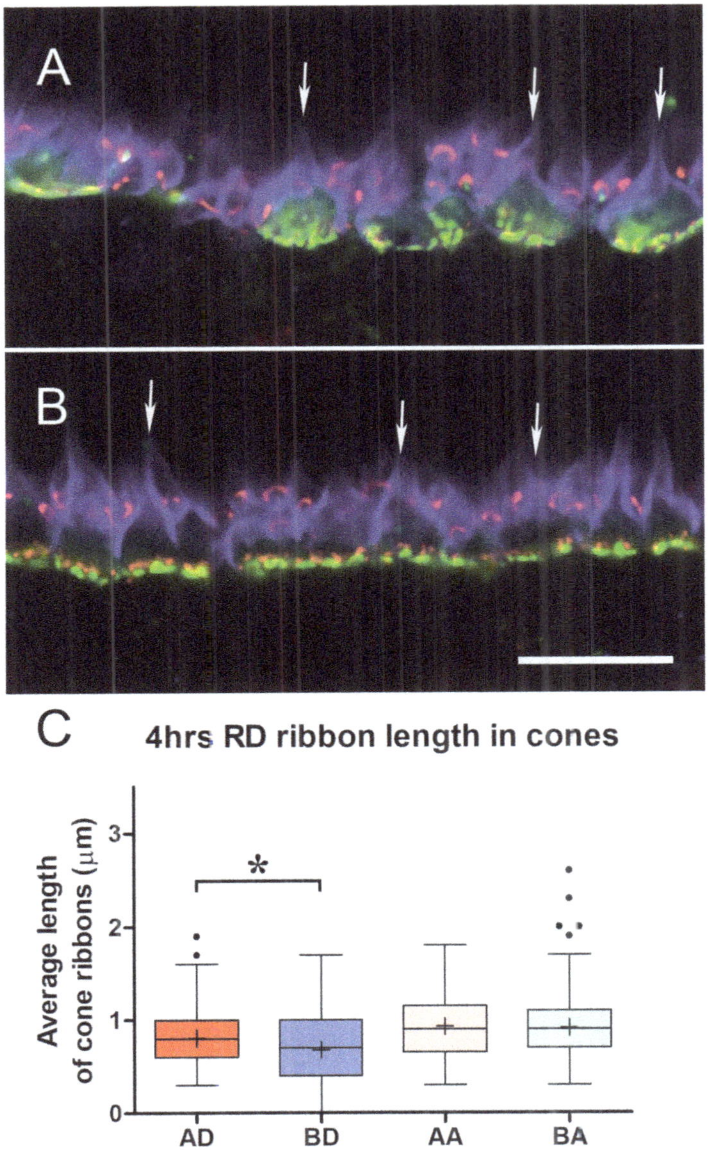

Figure 8. 4 h after detachment. (**A**) Detached retina, untreated. Pedicles are rounded in shape (arrows), and cone ribbons are short. Rod ribbons are present above (distal to) the base of the pedicles. (**B**) Detached retina treated with AR13503 after a two-hour delay. Pedicles have the more usual triangular shape (arrows), and ribbons appear longer. (**C**) Length of cone ribbons in the untreated detached retina (BD) is significantly smaller than those in the treated retina (AD), * $p < 0.05$. Length of ribbons in attached, untreated, and treated retinas (BA vs. AA) showed no significant differences. n = 4 animals. For box plots, whiskers are 1.5X IQR; dots are outliers. Bar = 5 μm.

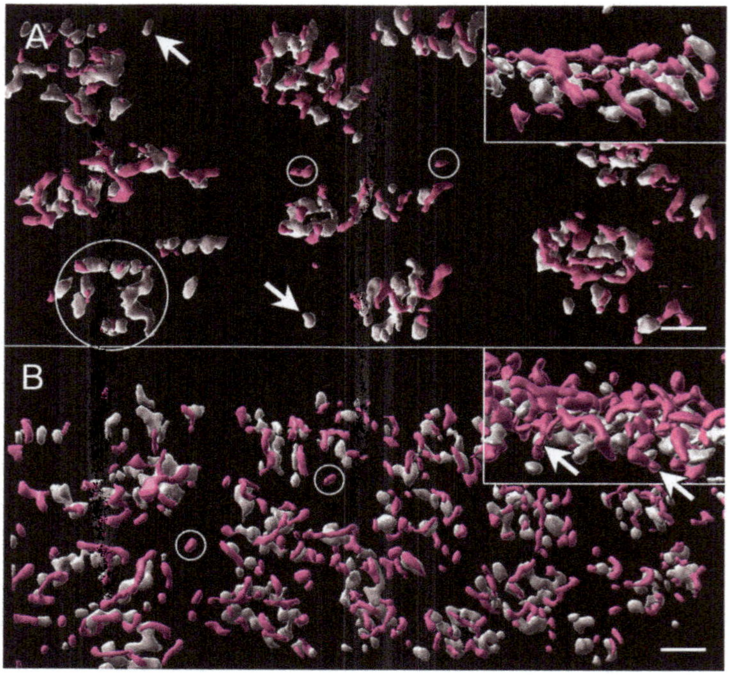

Figure 9. 4 h after detachment. (**A**) Detached retina, untreated. PNA aggregates (gray) have reduced numbers of ribbons (pink), which appear generally shorter than normal and, in some cases, thinner. Some pedicle bases are nearly devoid of ribbons (circle). Isolated PNA labels (arrows) and ribbons (small circles) are present. Many ribbons are flattened against the pedicle base (inset). (**B**) Detached retina, treated. Pedicles appear more tightly arrayed than in the untreated retina. There are more ribbons associated with the PNA label but also some isolated ribbon fragments (circles). Some ribbons are arched, and many have "knobs" along their lengths and at their ends (inset arrows). STED microscopy with surface rendering. Bars = 2 µm.

Because ROCK inhibition seemed to be useful as a delayed treatment to prevent loss of ribbons in cone synapses, we also examined rod spherule retraction in the specimens from (1) some of the same animals or (2) animals subject to the same experimental conditions. The retraction was most severe in the untreated, detached retina but was reduced by 37.6% with ROCK inhibition (BD vs. AD, p = 0.03, Figure 10). As with the cone data, there was considerable variability; however, all four animals showed less retraction with drug treatment than without. There was no difference in retraction in untreated versus treated attached retina (BA vs. AA, Figure 10).

In a previous study with the ROCK inhibitor fasudil [11], we had not seen any reduction of rod retraction after a 2 h delayed treatment. Thus, AR13503 was more effective than fasudil in reducing synaptic injury to rod cells. Moreover, ROCK inhibition with AR13503 was effective in reducing injury for both cone and rod cells.

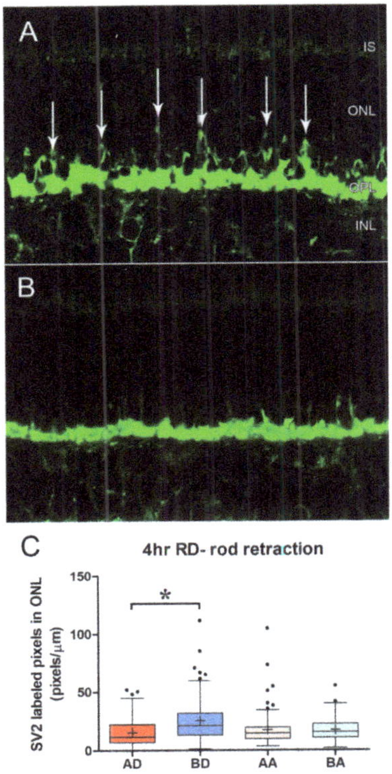

Figure 10. 4 h after detachment. (**A**) Detached retina, untreated. Synaptic vesicles are labeled for SV2 (green). There are many retracted rod spherules (arrows). (**B**) Detached retina treated with AR13503 after a two-hour delay. Very few retracted rod spherules are present. (**C**) There is significantly more retraction of rod spherules in the untreated detached retina (BD) than in the treated detached retina (AD) * $p = 0.03$. There is no difference in the amount of retraction in attached, untreated (BA) vs. treated (AA) retinas. n = 4 animals. For box plots, whiskers are 1.5X IQR; dots are outliers.

3.4. Structural and Functional Changes in Cone Synapses after Retinal Detachment and Spontaneous Retinal Reattachment

To assess the effects of reattachment on cone synaptic structure, we used a protocol that we had previously used for the examination of rod synapses [17]. Two days after detachment, most detachments have reattached spontaneously. Thus, we could examine cone synapses after reattachment as well as the effects of iatrogenic detachments followed by reattachment on cone function.

For these experiments, ffERG baseline recordings were made before surgery, and then retinal detachments were made in both eyes. In one eye, the subretinal injection included 0.5 µm AR13503; in the fellow eye, the detachment was made with BSS only. After two days, ERG recordings were made again, the eyes were examined with fundus photography and OCT to determine if the retina had reattached, and the animal was then euthanized and enucleated. Only animals with reattachments were used in our analysis (Figure 1).

With conventional confocal microscopy, no differences in cone ribbon length were apparent between untreated vs. treated retina in the detached or attached retina (Figure 11). Ribbon lengths were consistent with those of normal retinas. With STED microscopy and Imaris rendering, the cone pedicles in untreated and treated retina looked similar after reattachment: Most PNA-labeled presynaptic bases were covered with multiple ribbons (Figure 12). In the untreated and treated retina, there were 11.9 (n = 14 pedicles from 2 animals) and 10.3 (n = 14 pedicles from 2 animals) ribbons per pedicle, respectively. Thus, the abundance of ribbons was comparable to the normal retinas, although occasionally, the number of ribbons was reduced. The complexity of the ribbons in terms of the number of anastomoses between ribbons was also similar to that in the normal retina. Notably, however, the curvature of the ribbons in both treated and untreated retina was reduced (Figure 12). Many ribbons appeared as "shallow" arches, unlike the typical u-shaped morphology seen in the normal retina. Some pedicles contained knobby ribbons as well. An additional abnormality was the size of the PNA-labeled base of the pedicles. Consistent with the 2 h and 4 h experiments, the treated retina had a PNA-labeled area that was slightly but significantly smaller than the untreated retina (negative-binomial model, $p = 0.003$) and normal (negative-binomial model, $p = 0.049$, Figure 7). Thus, even though cone ribbon length and number returned to normal after reattachment, synaptic invaginations, as detected by ribbon arching, were not as deep as in normal retinas. Some variability in the distribution of ribbons and a reduced area at the base of the pedicle also suggested that recovery at this time point was not complete throughout the areas of reattachment.

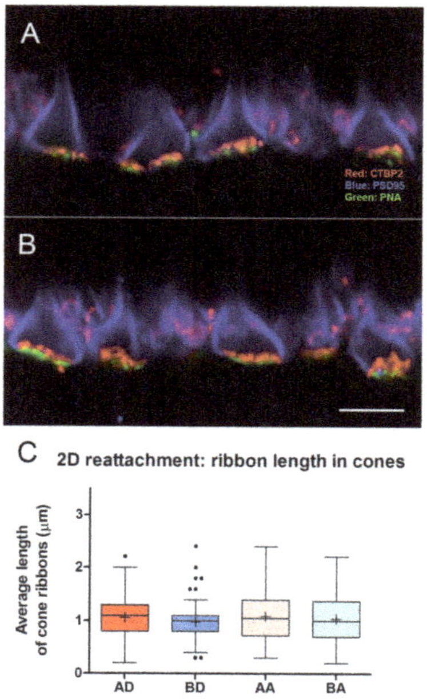

Figure 11. Two days after detachment and with reattachment. (**A**) Detached retina, untreated. (**B**) Detached retina treated with AR13503. Pedicles in both retinas show a normal distribution of ribbons. (**C**) Cone ribbons lengths showed no differences in all groups, AD, treated detached, AA, treated attached retina, BD, untreated detached, BA, untreated attached, n = 3 animals. For box plots, whiskers are 1.5X IQR; dots are outliers. Bar = 5 μm.

ERG recordings revealed that the function of the cone pedicles was not completely normal as well. Previously, to test for the synaptic function of rod spherules, we recorded and evaluated the scotopic b-wave [17]. We found that untreated retinas had lower amplitudes in their rod-specific, scotopic, b-wave. Here we examined the cone-specific, photopic b-wave using a similar analytical approach. To account for variability between animals, amplitudes were normalized as a percent of the baseline for each eye. We found that untreated retina had lower amplitudes than treated retina ($p = 0.0004$, Figure 13). The difference in the photopic b-wave was not due to transduction differences, as the a-waves were no different for untreated and treated retinas (Figure 13). The data indicate that treatment with AR13503 results in better synaptic transmission between cone photoreceptors and ON bipolar cells after reattachment.

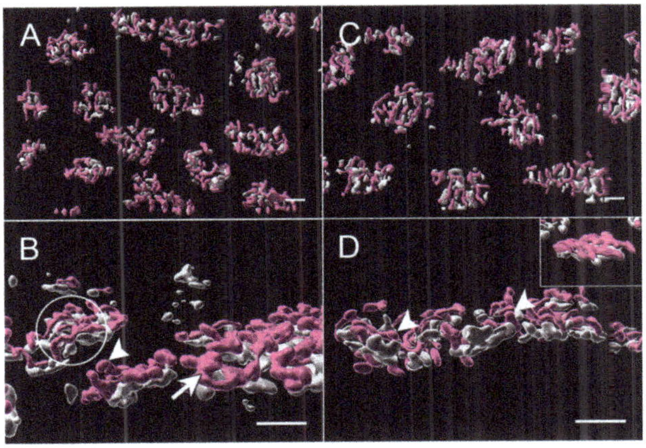

Figure 12. Two days after detachment and with reattachment. (**A**,**B**) Detached retina, untreated. The array of cone pedicles looks normal, and the density of the ribbon network also appears normal. However, there is considerable variability in the shapes of ribbons: ribbons with knobs (arrowhead) and ribbons with more complex shapes (arrow) are found on adjacent pedicles. Depths of ribbon arching are also variable, with some ribbons arched as in the normal retina and other ribbons shallower (circle). Some ribbons are completely flattened onto the pedicle base. (**C**,**D**) Detached retina treated with AR13503. The array of pedicles and density of the ribbons network is similar to the untreated retina. Many ribbons have knobs (arrowheads), similar to the untreated retinas; some ribbons appear flattened unto the pedicle base (inset). STED microscopy with surface rendering. Bars = 2 µm.

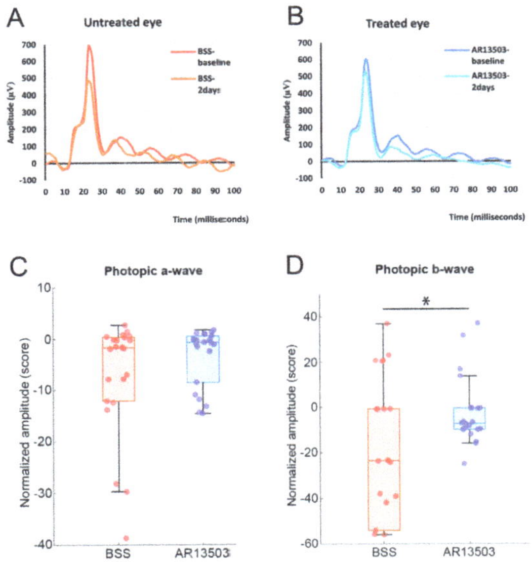

Figure 13. ERG recordings two days after detachment and with reattachment. (**A**,**B**) Representative waves of cone photopic b-wave for a BSS-untreated eye and an AR13503-treated eye. (**C**) Box plot showing normalized amplitude of photopic a-wave. There is no significant difference in a-wave amplitudes at two days from baseline recordings for both untreated and treated eyes. (**D**) Box plot showing normalized amplitude of photopic b-wave. BSS-untreated eyes have lower amplitudes at 2 days compared to baseline than AR13503 treated eyes, * $p = 0.0004$, n = 5 animals, an average of 5 recordings, repeated three times per animal. For box plots, whiskers are the standard deviation; dots represent all data points.

3.5. Summary

We have now examined cone and rod presynaptic terminals 2–4 h after detachment an two days after detachment, followed by spontaneous reattachment. Both rod [17] and con cells show a reduction in photoreceptor-bipolar transmission 2 days after a detachmen even though the retina has reattached. The similarity in functional injury is not paralle by similar morphological patterns of injury. Indeed, the cone and rod photoreceptor show distinct structural changes after detachment. We have summarized our findings i a table to facilitate a comparison of these morphological distinctions (Table 1). Wherea reduced rod spherule retraction correlated with improved scotopic b-wave amplitud no structural change in cone pedicles seemed correlated with the improved photopic b wave associated with ROCK inhibition. Nonetheless, AR13503 reduced the morphologica and functional changes caused by detachment in both cone and rod cells suggesting th potential usefulness of ROCK inhibition in the treatment of the retinal injury.

Table 1. Synaptic Structural Changes in Untreated Detached Retina.

	Cone Pedicles		
	2 h after Detachment	**4 h after Detachment**	**2 days after Detachment and with Reattachment**
Reduction in ribbon			
Length	xx	xx	o
Number	x	xx	o
Curvature	x	xx	x
Reduction in area of PNA-label	o	x	o
		Rod Spherules	
Retraction of terminal	xx *	xx	x *

x or xx—relative difference from normal; o—no different than normal; * Data from [17].

4. Discussion

Inhibition of activity in the RhoA pathway reduces synaptic damage to rod pho toreceptors after injury, both in vitro and in vivo [8–13,17]. This current work is our firs analysis of the effects of ROCK inhibition on cone synapses in vivo. We had, howeve previously examined isolated cone cells from the tiger salamander and found that blockin the RhoA pathway with the ROCK inhibitor Y27632 stimulated neuritic process growt and formation of presynaptic varicosities in culture [9]. Since detachment in vivo doe not produce neuritic sprouting in the time frame that we are examining, the in vitro wor was not predictive of what we observed in vivo. However, we did observe in vitro tha cone and rod salamander cells react differently to RhoA antagonists. Cone cells responde significantly faster and more robustly with the growth of processes and varicosities i the presence of Y27632, compared to rod cells [9]. One suggested conclusion was tha RhoA activity levels and regulation are different in the two photoreceptor cell types. Thi suggestion is consistent with our current results, as cone and rod cells show quite differen structural responses to detachment and, thus, presumably, to activation of RhoA signalin

For cone cells in this porcine model of retinal detachment, there was a significant decreas in ribbon length, changes in ribbon morphology, and loss of ribbons within hours afte detachment; changes in the length and number of ribbons were transient, being reversed wit reattachment. In contrast, the ribbons in rod spherules seemed fairly stable, even thoug spherules retracted from their postsynaptic partners. They were clearly unaltered 2 h afte detachment, the time point we examined quantitatively when cone pedicles have alread shown some ribbon loss. In detached feline retinas, however, rod ribbons do eventuall shorten after days of detachment in retracted and non-retracted spherules [4].

Under certain circumstances, both cone and rod ribbons are known to be plasti changing shape from elongated plates, as seen with EM, to plate-like structures wit swellings along the edges which serve to remove material from the ribbon to spherica electron-dense bodies that no longer appear as ribbons although they continue to be coated with attached synaptic vesicles. Spherical ribbons are usually detached from the active zone and found floating in the cytoplasm of the presynaptic terminal. These changes in size and shape have been reported to occur during the diurnal light phase, most notably in some strains of mice, with hibernation and a decrease in synaptic activity [52]. Moreover, the changes are reversible, with the greatest ribbon lengths achieved just after dark, along with increased visual sensitivity [53]. The knobbiness of porcine cone ribbons, as observed in 3D STED stacks of images, likely represents either (1) the process of reduction in ribbon size as spherical aggregates from elongated ribbons detach during periods of inactivity caused by detachment or (2) the reformation of ribbons after reattachment. In our specimens, the highest degree of knobbiness was seen in treated retinas, where the treatment had been delayed by 2 h. Additionally, small aggregates of ribbon labels were present unassociated with PNA-labeled synaptic bases. In the untreated fellow eyes, ribbon lengths were the shortest we observed. Treatment, therefore, may have halted the process of reduction mid-stream, leaving misshapen ribbons and free-floating ribbon spheres. Some knobs or protrusions are also present even after retinal reattachment.

Presynaptic calcium concentration appears to be a key determinant of ribbon size in photoreceptors. When photoreceptors are hyperpolarized in the light, calcium channels are closed, and the number of elongated plate-like ribbons is reduced [53–56]. Both EGTA and BAPTA, calcium chelators, likewise reduce the number of elongated ribbons and increase the number of spherical-shaped ribbon materials, whereas high levels of calcium reduce the number of spherical ribbons [56,57]. Interestingly, cone ribbons display higher sensitivity to calcium reduction than rod ribbons [57]. Most cone cell types have different L-type calcium channels at their terminals than rod cells [58], and cone terminals depend primarily on the Na^+/Ca^+ exchanger for maintaining calcium homeostasis, whereas rod synapses use primarily plasma membrane calcium ATPase (PMCA). These differences contribute to differing calcium dynamics in cone and rod synaptic terminals [59]. In the light, for instance, the estimated free [Ca^{2+}] in cone pedicles is 3.2-fold lower than in rod spherules [59]. The mechanisms to lower calcium levels in cone pedicles with light-induced hyperpolarization may contribute to the increased sensitivity of cone ribbons in detachment-induced hyperpolarization compared to rod ribbons. Furthermore, cone ribbons, unlike rod ribbons, do not depend on piccolino, one of the molecular constituents of ribbons along with ribeye/CtBP2, to maintain an elongate shape [60]. Intraterminal free [Ca^{2+}] may also contribute to the lack of ribbon changes in cone pedicles in the attached retina.

Rod spherules retract after detachment in both the detached retina and throughout the retina in the areas of the attached retina [17]. Spherule retraction throughout the retina is consistent with increased RhoA activity in both detached and attached retina [8]. In contrast, cone terminal morphology seems normal in the attached portions of the retina after detachment. Cone cells in the attached retina would not be hyperpolarized as they are in the detached retina. In the dark, cones cells are known to have rapid neurotransmission, faster than synaptic kinetics at rod synapses. Thoreson [58] suggested that the differences in synaptic kinetics are due to differences in intraterminal calcium handling. Cone cells have higher calcium levels than rod cells in the dark [61], with particularly high calcium hots spots at the cone ribbon [62]. Thus, in the attached retina, where photoreceptors are generally in a depolarized state and not in prolonged detachment-induced hyperpolarization, high calcium in cone pedicles would favor the presence of mature elongate ribbons.

Since the addition of a ROCK inhibitor reduced ribbon loss in cone pedicles after detachment, it is reasonable to suggest that there is a connection between RhoA activity and cone ribbon plasticity. Ribbons are not thought to have connections with the actin cytoskeleton; however, photoreceptor terminals contain myosin Va, an actin-based motor molecule linked to control of the size and shape of photoreceptor ribbons [63]. Cytoskeletal movement, in turn, is controlled by Rho signaling, but exactly how that results in changes in cone ribbons is currently unknown. Alternatively, the effect of ROCK on cone ribbons may come indirectly through interactions with voltage-gated calcium channels. ROCK interacts with smooth muscle L- and neuronal N-type calcium channels, causing reduced and increased calcium currents, respectively [64]. ROCK also interacts with T-type calcium channels. The kinase decreases currents in $Ca_v3.1$ and 3.3 and increases calcium currents in $Ca_v3.2$ channels [65]. Just recently, T-type calcium cells have been reported in mouse retinas at release sites exclusively in cone pedicles [66]. In the mouse, they are the $Ca_v3.2$ type of channel. ROCK would increase calcium current if the cone pedicle regulation by ROCK were the same as dorsal root ganglion cells [65]. However, because of the highly variable effects of ROCK on voltage-gated calcium channels and the high specificity of regulation mechanisms for T-type channels [67], it is difficult to predict how ROCK inhibition may affect calcium current amplitude in pedicles. The connection between RhoA activity and ribbon plasticity warrants future investigation.

The normal arching of cone ribbons was observed to flatten after detachment. This flattening indicates the flattening or disappearance of synaptic invaginations. After detachment, the loss of invaginations seemed to occur simultaneously with changes in pedicle shape. Fisher and colleagues [4,49] reported that feline cone pedicles could either round up or flatten days after detachment. In vitro, we observed porcine pedicles to round up 2 h after detachment and then appear flattened and spread out by 24 h in vitro [10]. With EM, after several days of detachment, invaginations have completely disappeared [4]. These shape changes would necessarily involve the cytoskeleton, and it is therefore not surprising that loss of synaptic invaginations, as seen by flattening of ribbons, and rounding up, as seen with PSD-95 label, are reduced with ROCK inhibition. Similar to ribbon knobbiness, the depth of synaptic invaginations did not return to normal for many active zones after reattachment at two days. The lack of discernable differences between untreated and treated retina in arching of ribbons after detachment and spontaneous reattachment suggest there may be residual RhoA activity days after injury and that pedicle structural changes are no longer inhibited by the drug AR13503. In traumatic brain injury, activated RhoA is detected for months after injury [68]. Loss of synaptic invaginations could change the effectiveness or properties of synaptic transmission at the photoreceptor synapse since glutamate diffusion and, therefore, glutamate levels, as well as the local ionic environment in the synaptic cleft, may be affected [69,70] and the relationships of dendrites, with their postsynaptic receptors, to sites of glutamate release may alter [71]. The differences between the functional restoration of untreated and treated reattached retina may lie in part with subtle differences in the geometry of the synaptic invaginations, differences that we are not able to distinguish with current morphological techniques.

In addition to the connection between the cytoskeletal control of presynaptic shape and synaptic invaginations and RhoA signaling, cytoskeleton and RhoA signaling also play a role in exo- and endocytosis at the synapse. ROCK signaling, for instance, reduced the size of the readily releasable pool of synaptic vesicles and increased the rate of synaptic vesicle endocytosis in central nervous system neurons [72,73]. Any imbalance in exo- and endocytosis adds to or removes plasmalemma at the active zone and can change synaptic efficacy. Reduction in endocytosis at photoreceptor synapses, for instance, slows transmission as calcium channels become more distant from vesicle release sites [70,74]. Cone synapses have a unique polyphosphoinositide phosphatase, Synaptojanin 1, which participates in clathrin-mediated endocytosis [75]. The presence of this unique protein indicates that the process of synaptic vesicle recycling differs in cone vs. rod cells. In addition, amphiphysin, also associated with clathrin-mediated endocytosis, is reduced in cones compared to rod terminals in mice [76]. It is possible that Rho-kinase-stimulated endocytosis may contribute to the rapid loss of synaptic invaginations in cones but not rod presynaptic terminals. Rod spherules do lose their invaginations in a feline model of retinal detachment but only slowly over a period of days [4]. Thus, Rho-kinase activity may alter cone presynaptic morphology and function after detachment directly by its effect on the cytoskeleton and indirectly through possible interaction with voltage-gated calcium channels and synaptic vesicle recycling. With reattachment, subtle differences in cone presynaptic structure and synaptic processes may still remain for the untreated and treated retina. Functional differences, however, could also be due to changes in molecular constituents postsynaptically.

Congenital stationary blindness (CSNB) has an ERG phenotype in which the a-wave is normal, and the b-wave is missing. Multiple genes are known to cause this retinal disease. Many mouse mutants share a no-b-wave ERG phenotype. They are known collectively as *nob* mice. The mutated genes are expressed in the ribbon synapse either presynaptically, affecting glutamate release, or postsynaptically affecting the depolarization of ON bipolar cells. Significantly, when the problem is postsynaptic, the morphology of the ribbon synapse can appear normal both with light and electron microscopy [77]. After detachment and reattachment, we found that untreated and treated retinas did not appear different morphologically. There was a reduction in ribbon arching and the presence of knobs on ribbons in both untreated and treated retinas; otherwise, the synapses appeared normal. The abnormalities in ribbon arching and knobbiness observed with STED microscopy would most likely not be seen in light or conventional confocal microscopy or even in electron microscopy, where 3D reconstruction is rarely done and shows only a small portion of the active zone. So with conventional techniques, the synapses might be described as normal. Thus, we suggest that the functional problem, in the form of a lower photopic b-wave amplitude in the untreated porcine retina with a normal a-wave, may be due to some postsynaptic component of the synapses. In CSNB, mutations have been found in the metabotropic glutamate receptor (mGLUR6), the signaling components Goalpha and Gβ5, and in nyctalopin, an extracellular protein located at the base of the synapse. In cone synapses, nyctalopin appears distal to the ribbons, near the PNA label, and is necessary for synaptic transmission due to its association with the transient receptor potential melastatin channel, TRPM1 [78,79]. If this suggestion is correct, reduced cone b-wave amplitude would be due, at least in part, to reduced mGLUR6 signaling. Interestingly, in a report on the cone-specific ELFN2 synaptic adhesion protein, a binding partner of mGLUR6, a strong correlation was observed between reduced b-wave amplitude and reduced mGluR6 synaptic content [80]. Clearly, this hypothesis remains to be proven, but it would be consistent with the morphological and functional description of *CSNB1* and *nob* mice. For cone cells, the loss of postsynaptic signaling, however, would not be due to genetic abnormalities but rather to the injury response and would likely occur in a limited area. In contrast, for rod cells, we have shown a high correlation between increased rod axon retraction and reduced scotopic b-wave amplitudes [17]. Thus, the causes of reduced photoreceptor-bipolar transmission after retinal detachment are disparate for the two photoreceptor cell types.

It is likely that every cell type in the eye contains RhoA and that ROCK inhibitors, when injected subretinally or intravitreally, could potentially influence activity in many cell types within the eye. For instance, the retinal vasculature responds to ROCK inhibition by relaxation [81]. Mechanosensitive TPRV4 channels on Müller cells cause increased RhoA activity resulting in increased GFAP expression and release of MCP-1, which, in turn, can cause photoreceptor apoptosis in retinal detachment [82]. ROCK inhibition can reduce glial reactivity [83]. Additionally, RPE structure and barrier function is restored in diabetic retinopathy with ROCK inhibition [81]. So AR13503 injections may have multiple effects intraocularly. Nonetheless, damage to the first synapse in the visual system may be the most negative feature of the retinal injury response to detachment and protection of this synapse, both its presynaptic and postsynaptic components, the most important therapeutic role for treatment with ROCK inhibitors. We have suggested, based on observation of rod synapses, that ROCK inhibitors could be included in iatrogenic procedures for gene and stem cell therapies where subretinal injections are used [17]. The additional information on cone synapses presented here reinforces this suggestion. In this report, we also demonstrate that damage to both cone and rod synapses is reduced even when the treatment is administered 2 h after retinal detachment. This response indicates that post-injury injections may also be useful for retinal trauma. Moreover, it may be beneficial to include a ROCK inhibitor during surgical reattachment as a mechanical perturbation, which can stimulate RhoA activity, is involved. Indeed, as we have previously mentioned (see Section 1), anatomically successful surgical reattachment in patients does not guarantee a return to normal vision. Functional and structural consequences of RhoA activity likely contribute to remaining visual losses. Our own and others' work has shown, in humans [84] and animal models [4,17,51], that abnormal neuronal networks remain after reattachment, whereas rod neuritic sprouting only appears with retinal reattachment [4]. Application of a ROCK inhibitor during or after retinal injury may improve patient outcomes by reducing synaptic abnormalities.

Author Contributions: Conceptualization, E.T.-A., É.H., I.S. and M.Z.; methodology, E.T.-A., É.H., I.S., M.Z., L.F., F.A.K.Y., A.L.D. and L.J.F.; validation, E.T.-A., É.H., I.S., M.Z., A.L.D. and L.J.F.; formal analysis, É.H., I.S. and A.L.D.; investigation, E.T.-A., É.H., I.S., M.Z., L.F. and F.A.K.Y.; resources, L.F. and F.A.K.Y.; data curation, É.H. and I.S.; writing—original draft preparation, E.T.-A.; writing—review and editing, E.T.-A., É.H., I.S., M.Z., L.F., F.A.K.Y., A.L.D. and L.J.F.; visualization, E.T.-A., É.H. and I.S.; supervision, E.T.-A.; project administration, E.T.-A. and É.H.; funding acquisition, E.T.-A., M.Z. and L.F. All authors have read and agreed to the published version of the manuscript.

Funding: This work was supported by the Department of Defense Research Award W81XWH1910819, Joseph J. and Marguerite DiSepio Retina Research Fund, New Jersey Lions Eye Research Foundation, Eng Family Foundation, and NIH High-End Instrumentation Grant S10 OD25182.

Institutional Review Board Statement: The animal study protocol was approved by the Institutional Animal Care and Use Committee of New Jersey Medical School (protocol 999900995, 30 November 2020).

Informed Consent Statement: Not applicable.

Data Availability Statement: Please contact the corresponding author for information regarding data.

Acknowledgments: We thank Qian Sun and the staff of the New Jersey Medical School Animal Facilities for technical support during the surgeries. We also are grateful to Sun for excellent immunocytochemistry. AR13503 was generously supplied by Aerie Pharmaceuticals Inc.

Conflicts of Interest: The authors declare no conflict of interest. The funders had no role in the design of the study; in the collection, analyses, or interpretation of data; in the writing of the manuscript; or in the decision to publish the results.

Appendix A

Table A1. All animals used for this study. Although only pigs with dark, pigmented irises were accepted in this study, to minimize the possibility of depigmentation in the posterior segment, some areas of depigmentation were present in the fundus, as noted. (RD: retinal detachment; os: oculus sinister; od: oculus dexter; ou: oculus uterque) * Rod data from these animals on spherule retraction were used for a previous study [17]. ** Rod data from these animals on spherule retraction and rod ERGs were used for a previous study [17].

Animal ID	Sex	Notes	Depigmentation
Control (terminal)			
DOD6	F	Data not used because enucleation was several hours after euthanasia	
DOD16	M	Cone and rod data	
DOD21	M	Cone and rod data	
Intravitreal 0.75 μm, 2 h RD (terminal)			
69	F	Cone and rod data *	depigmentation in od
70	F	Cone and rod data *	
DOD14	M	Cone and rod data	

Table A1. *Cont.*

Animal ID	Sex	Notes	Depigmentation
Intravitreal 0.5 μm, 4 h RD (terminal)			
DOD2	M	Cone and rod data	depigmentation in ou
DOD3	M	Rod data	
DOD4	M	Rod data	
DOD8	M	Cone data	
DOD12	M	Cone and rod data	depigmentation in os
DOD20	M	Cone data	
Subretinal 0.5 μm, 2 Days (survival)			
45B	F	Cone ERG data **	depigmentation in ou
48	F	Cone ERG data **	intraretinal hemorrhage in od(BSS)
49	F	Cone ERG data **	
50	F	Cone ERG data **	depigmentation in ou
74	F	Cone morphological and ERG data **	depigmentation in ou
DOD17	M	Cone morphological data	
DOD18	M	Cone morphological data	depigmentation in ou

References

1. Lamb, T.D.; Collin, S.P.; Pugh, E.N., Jr. Evolution of the vertebrate eye: Opsins, photoreceptors, retina and eye cup. *Nat. Rev. Neurosci.* **2007**, *8*, 960–976. [CrossRef]
2. Morshedian, A.; Fain, G.L. The evolution of rod photoreceptors. *Philos. Trans. R. Soc. Lond. B Biol. Sci.* **2017**, *372*. [CrossRef]
3. Erickson, P.A.; Fisher, S.K.; Anderson, D.H.; Stern, W.H.; Borgula, G.A. Retinal detachment in the cat: The outer nuclear and outer plexiform layers. *Investig. Ophthalmol. Vis. Sci.* **1983**, *24*, 927–942.
4. Fisher, S.K.; Lewis, G.P.; Linberg, K.A.; Verardo, M.R. Cellular remodeling in mammalian retina: Results from studies of experimental retinal detachment. *Prog. Retin. Eye Res.* **2005**, *24*, 395–431. [CrossRef]
5. Lewis, G.P.; Linberg, K.A.; Fisher, S.K. Neurite outgrowth from bipolar and horizontal cells after experimental retinal detachment. *Investig. Ophthalmol. Vis. Sci.* **1998**, *39*, 424–434.
6. Linberg, K.A.; Lewis, G.P.; Fisher, S.K. Retraction and remodeling of rod spherules are early events following experimental retinal detachment: An ultrastructural study using serial sections. *Mol. Vis.* **2009**, *15*, 10–25.
7. Rex, T.S.; Fariss, R.N.; Lewis, G.P.; Linberg, K.A.; Sokal, I.; Fisher, S.K. A survey of molecular expression by photoreceptors after experimental retinal detachment. *Investig. Ophthalmol. Vis. Sci.* **2002**, *43*, 1234–1247.
8. Wang, J.; Zarbin, M.; Sugino, I.; Whitehead, I.; Townes-Anderson, E. RhoA Signaling and Synaptic Damage Occur Within Hours in a Live Pig Model of CNS Injury, Retinal Detachment. *Investig. Ophthalmol. Vis. Sci.* **2016**, *57*, 3892–3906. [CrossRef]
9. Fontainhas, A.M.; Townes-Anderson, E. RhoA and its role in synaptic structural plasticity of isolated salamander photoreceptors. *Invesigt. Ophthalmol. Vis. Sci.* **2008**, *49*, 4177–4187. [CrossRef]
10. Fontainhas, A.M.; Townes-Anderson, E. RhoA inactivation prevents photoreceptor axon retraction in an in vitro model of acute retinal detachment. *Investig. Ophthalmol. Vis. Sci.* **2011**, *52*, 579–587. [CrossRef]
11. Townes-Anderson, E.; Wang, J.; Halasz, E.; Sugino, I.; Pitler, A.; Whitehead, I.; Zarbin, M. Fasudil, a Clinically Used ROCK Inhibitor, Stabilizes Rod Photoreceptor Synapses after Retinal Detachment. *Transl. Vis. Sci. Technol.* **2017**, *6*, 22. [CrossRef]
12. Wang, W.; Halasz, E.; Townes-Anderson, E. Actin Dynamics, Regulated by RhoA-LIMK-Cofilin Signaling, Mediates Rod Photoreceptor Axonal Retraction After Retinal Injury. *Investig. Ophthalmol. Vis. Sci.* **2019**, *60*, 2274–2285. [CrossRef]
13. Wang, W.; Townes-Anderson, E. LIM Kinase, a Newly Identified Regulator of Presynaptic Remodeling by Rod Photoreceptors After Injury. *Investig. Ophthalmol. Vis. Sci.* **2015**, *56*, 7847–7858. [CrossRef]
14. Lessey, E.C.; Guilluy, C.; Burridge, K. From mechanical force to RhoA activation. *Biochemistry* **2012**, *51*, 7420–7432. [CrossRef]
15. Martin-Camara, O.; Cores, A.; Lopez-Alvarado, P.; Menendez, J.C. Emerging targets in drug discovery against neurodegenerative diseases: Control of synapsis disfunction by the RhoA/ROCK pathway. *Eur. J. Med. Chem.* **2021**, *225*, 113742. [CrossRef]
16. Tolias, K.F.; Duman, J.G.; Um, K. Control of synapse development and plasticity by Rho GTPase regulatory proteins. *Prog. Neurobiol.* **2011**, *94*, 133–148. [CrossRef]
17. Halasz, E.; Zarbin, M.A.; Davidow, A.L.; Frishman, L.J.; Gombkoto, P.; Townes-Anderson, E. ROCK inhibition reduces morphological and functional damage to rod synapses after retinal injury. *Sci. Rep.* **2021**, *11*, 692. [CrossRef]

18. Maguire, A.M.; Simonelli, F.; Pierce, E.A.; Pugh, E.N., Jr.; Mingozzi, F.; Bennicelli, J.; Banfi, S.; Marshall, K.A.; Testa, F.; Surace, E.M.; et al. Safety and efficacy of gene transfer for Leber's congenital amaurosis. *N. Engl. J. Med.* **2008**, *358*, 2240–2248. [CrossRef]
19. Schwartz, S.D.; Hubschman, J.P.; Heilwell, G.; Franco-Cardenas, V.; Pan, C.K.; Ostrick, R.M.; Mickunas, E.; Gay, R.; Klimanskaya, I.; Lanza, R. Embryonic stem cell trials for macular degeneration: A preliminary report. *Lancet* **2012**, *379*, 713–720. [CrossRef]
20. Braekevelt, C.R. Fine structure of the retinal rods and cones in the domestic pig. *Graefes Arch. Clin. Exp. Ophthalmol.* **1983**, *220*, 273–278. [CrossRef]
21. Gerke, C.G., Jr.; Hao, Y.W.; Wong, F. Topography of rods and cones in the retina of domestic the pig. *Hong Kong Med. J.* **1995**, *1*, 302–308.
22. Simoens, P.; De Schaepdrijver, L.; Lauwers, H. Morphologic and clinical study of the retinal circulation in the miniature pig. A: Morphology of the retinal microvasculature. *Exp. Eye Res.* **1992**, *54*, 965–973. [CrossRef]
23. Fernandez de Castro, J.P.; Scott, P.A.; Fransen, J.W.; Demas, J.; DeMarco, P.J.; Kaplan, H.J.; McCall, M.A. Cone photoreceptors develop normally in the absence of functional rod photoreceptors in a transgenic swine model of retinitis pigmentosa. *Investig. Ophthalmol. Vis. Sci.* **2014**, *55*, 2460–2468. [CrossRef]
24. Petters, R.M.; Alexander, C.A.; Wells, K.D.; Collins, E.B.; Sommer, J.R.; Blanton, M.R.; Rojas, G.; Hao, Y.; Flowers, W.L.; Banin, E.; et al. Genetically engineered large animal model for studying cone photoreceptor survival and degeneration in retinitis pigmentosa. *Nat. Biotechnol.* **1997**, *15*, 965–970. [CrossRef]
25. Burton, T.C. Recovery of visual acuity after retinal detachment involving the macula. *Trans. Am. Ophthalmol. Soc.* **1982**, *80*, 475–497.
26. Campo, R.V.; Sipperley, J.O.; Sneed, S.R.; Park, D.W.; Dugel, P.U.; Jacobsen, J.; Flindall, R.J. Pars plana vitrectomy without scleral buckle for pseudophakic retinal detachments. *Ophthalmology* **1999**, *106*, 1811–1815; discussion 1816. [CrossRef]
27. Murtagh, P.J.; Stephenson, K.A.; Rhatigan, M.; McElnea, E.M.; Connell, P.P.; Keegan, D.J. Rhegmatogenous retinal detachments: Primary reattachment rates and visual outcomes over a 4-year period. *Ir. J. Med. Sci.* **2020**, *189*, 355–363. [CrossRef]
28. Ozgür, S.; Esgin, H. Macular function of successfully repaired macula-off retinal detachments. *Retina* **2007**, *27*, 358–364. [CrossRef]
29. Ross, W.H.; Kozy, D.W. Visual recovery in macula-off rhegmatogenous retinal detachments. *Ophthalmology* **1998**, *105*, 2149–2153. [CrossRef]
30. Salicone, A.; Smiddy, W.E.; Venkatraman, A.; Feuer, W. Visual recovery after scleral buckling procedure for retinal detachment. *Ophthalmology* **2006**, *113*, 1734–1742. [CrossRef]
31. Wykoff, C.C.; Smiddy, W.E.; Mathen, T.; Schwartz, S.G.; Flynn, H.W., Jr.; Shi, W. Fovea-sparing retinal detachments: Time to surgery and visual outcomes. *Am. J. Ophthalmol.* **2010**, *150*, 205–210.e202. [CrossRef]
32. Zabel, P.; Zabel, K.; Kazmierczak, K.; Stankiewicz, M.; Jaworski, D.; Suwala, K.; Buszko, K.; Stafiej, J.; Malukiewicz, G.; Kaluzny, J.J. Vascular density and macular sensitivity in eyes after scleral buckling surgery for macula-on rhegmatogenous retinal detachment. *PLoS ONE* **2023**, *18*, e0279683. [CrossRef]
33. Marmor, M.F.; Fulton, A.B.; Holder, G.E.; Miyake, Y.; Brigell, M.; Bach, M. ISCEV Standard for full-field clinical electroretinography (2008 update). *Doc. Ophthalmol.* **2009**, *118*, 69–77. [CrossRef]
34. Liang, K.-Y.; Zeger, S.L. Longitudinal data analysis using generalized linear models. *Biometrika* **1986**, *73*, 13–22. [CrossRef]
35. Schmitz, F.; Königstorfer, A.; Südhof, T.C. RIBEYE, a component of synaptic ribbons: A protein's journey through evolution provides insight into synaptic ribbon function. *Neuron* **2000**, *28*, 857–872. [CrossRef]
36. tom Dieck, S.; Altrock, W.D.; Kessels, M.M.; Qualmann, B.; Regus, H.; Brauner, D.; Fejtova, A.; Bracko, O.; Gundelfinger, E.D.; Brandstatter, J.H. Molecular dissection of the photoreceptor ribbon synapse: Physical interaction of Bassoon and RIBEYE is essential for the assembly of the ribbon complex. *J. Cell. Biol.* **2005**, *168*, 825–836. [CrossRef] [PubMed]
37. Blackmon, S.M.; Peng, Y.W.; Hao, Y.; Moon, S.J.; Oliveira, L.B.; Tatebayashi, M.; Petters, R.M.; Wong, F. Early loss of synaptic protein PSD-95 from rod terminals of rhodopsin P347L transgenic porcine retina. *Brain Res.* **2000**, *885*, 53–61. [CrossRef]
38. Koulen, P.; Fletcher, E.L.; Craven, S.E.; Bredt, D.S.; Wässle, H. Immunocytochemical localization of the postsynaptic density protein PSD-95 in the mammalian retina. *J. Neurosci.* **1998**, *18*, 10136–10149. [CrossRef]
39. Blanks, J.C.; Johnson, L.V. Specific binding of peanut lectin to a class of retinal photoreceptor cells. A species comparison. *Investig. Ophthalmol. Vis. Sci.* **1984**, *25*, 546–557.
40. Zenisek, D.; Horst, N.K.; Merrifield, C.; Sterling, P.; Matthews, G. Visualizing synaptic ribbons in the living cell. *J. Neurosci.* **2004**, *24*, 9752–9759. [CrossRef]
41. Kerov, V.; Laird, J.G.; Joiner, M.L.; Knecht, S.; Soh, D.; Hagen, J.; Gardner, S.H.; Gutierrez, W.; Yoshimatsu, T.; Bhattarai, S.; et al. $\alpha_2\delta$-4 Is Required for the Molecular and Structural Organization of Rod and Cone Photoreceptor Synapses. *J. Neurosci.* **2018**, *38*, 6145–6160. [CrossRef] [PubMed]
42. Haverkamp, S.; Grünert, U.; Wässle, H. The synaptic architecture of AMPA receptors at the cone pedicle of the primate retina. *J. Neurosci.* **2001**, *21*, 2488–2500. [CrossRef] [PubMed]
43. Sherry, D.M.; Mitchell, R.; Standifer, K.M.; du Plessis, B. Distribution of plasma membrane-associated syntaxins 1 through 4 indicates distinct trafficking functions in the synaptic layers of the mouse retina. *BMC Neurosci.* **2006**, *7*, 54. [CrossRef] [PubMed]
44. Lin, C.W.; Sherman, B.; Moore, L.A.; Laethem, C.L.; Lu, D.W.; Pattabiraman, P.P.; Rao, P.V.; deLong, M.A.; Kopczynski, C.C. Discovery and Preclinical Development of Netarsudil, a Novel Ocular Hypotensive Agent for the Treatment of Glaucoma. *J. Ocul. Pharmacol. Ther.* **2018**, *34*, 40–51. [CrossRef] [PubMed]

5. Hendrickson, A.; Hicks, D. Distribution and density of medium- and short-wavelength selective cones in the domestic pig retina. *Exp. Eye Res.* **2002**, *74*, 435–444. [CrossRef]
6. Chidlow, G.; Chan, W.O.; Wood, J.P.M.; Casson, R.J. Differential Effects of Experimental Retinal Detachment on S- and M/L-Cones in Rats. *Mol. Neurobiol.* **2022**, *59*, 117–136. [CrossRef]
7. Nork, T.M. Acquired color vision loss and a possible mechanism of ganglion cell death in glaucoma. *Trans. Am. Ophthalmol. Soc.* **2000**, *98*, 331–363. [PubMed]
8. Nork, T.M.; Millecchia, L.L.; Strickland, B.D.; Linberg, J.V.; Chao, G.M. Selective loss of blue cones and rods in human retinal detachment. *Arch. Ophthalmol.* **1995**, *113*, 1066–1073. [CrossRef]
9. Fisher, S.K.; Lewis, G.P. Muller cell and neuronal remodeling in retinal detachment and reattachment and their potential consequences for visual recovery: A review and reconsideration of recent data. *Vis. Res.* **2003**, *43*, 887–897. [CrossRef]
0. Fisher, S.K.; Lewis, G.P. Cellular effects of detachemnt and reattachment onthe neural retina and the retinal pigment epithelium. In *Retina*, 4th ed.; Ryan, S.J., Wilkinson, C.P., Eds.; Sirgical Retina; Mosby: Maryland Heights, MO, USA, 2006; Volume 3, pp. 1991–2012.
1. Lewis, G.P.; Charteris, D.G.; Sethi, C.S.; Leitner, W.P.; Linberg, K.A.; Fisher, S.K. The ability of rapid retinal reattachment to stop or reverse the cellular and molecular events initiated by detachment. *Investig. Ophthalmol. Vis. Sci.* **2002**, *43*, 2412–2420.
2. Voorn, R.A.; Vogl, C. Molecular Assembly and Structural Plasticity of Sensory Ribbon Synapses—A Presynaptic Perspective. *Int. J. Mol. Sci.* **2020**, *21*, 8758. [CrossRef] [PubMed]
3. Balkema, G.W.; Cusick, K.; Nguyen, T.H. Diurnal variation in synaptic ribbon length and visual threshold. *Vis. Neurosci.* **2001**, *18*, 789–797. [CrossRef]
4. Adly, M.A.; Spiwoks-Becker, I.; Vollrath, L. Ultrastructural changes of photoreceptor synaptic ribbons in relation to time of day and illumination. *Investig. Ophthalmol. Vis. Sci.* **1999**, *40*, 2165–2172.
5. Fuchs, M.; Sendelbeck, A.; Atorf, J.; Kremers, J.; Brandstatter, J.H. Strain differences in illumination-dependent structural changes at mouse photoreceptor ribbon synapses. *J. Comp. Neurol.* **2013**, *521*, 69–78. [CrossRef]
6. Spiwoks-Becker, I.; Glas, M.; Lasarzik, I.; Vollrath, L. Mouse photoreceptor synaptic ribbons lose and regain material in response to illumination changes. *Eur. J. Neurosci.* **2004**, *19*, 1559–1571. [CrossRef]
7. Regus-Leidig, H.; Specht, D.; Tom Dieck, S.; Brandstätter, J.H. Stability of active zone components at the photoreceptor ribbon complex. *Mol. Vis.* **2010**, *16*, 2690–2700. [PubMed]
8. Thoreson, W.B. Kinetics of synaptic transmission at ribbon synapses of rods and cones. *Mol. Neurobiol.* **2007**, *36*, 205–223. [CrossRef] [PubMed]
9. Johnson, J.E., Jr.; Perkins, G.A.; Giddabasappa, A.; Chaney, S.; Xiao, W.; White, A.D.; Brown, J.M.; Waggoner, J.; Ellisman, M.H.; Fox, D.A. Spatiotemporal regulation of ATP and Ca^{2+} dynamics in vertebrate rod and cone ribbon synapses. *Mol. Vis.* **2007**, *13*, 887–919. [PubMed]
0. Muller, T.M.; Gierke, K.; Joachimsthaler, A.; Sticht, H.; Izsvak, Z.; Hamra, F.K.; Fejtova, A.; Ackermann, F.; Garner, C.C.; Kremers, J.; et al. A Multiple Piccolino-RIBEYE Interaction Supports Plate-Shaped Synaptic Ribbons in Retinal Neurons. *J. Neurosci.* **2019**, *39*, 2606–2619. [CrossRef]
1. Sheng, Z.; Choi, S.Y.; Dharia, A.; Li, J.; Sterling, P.; Kramer, R.H. Synaptic Ca^{2+} in darkness is lower in rods than cones, causing slower tonic release of vesicles. *J. Neurosci.* **2007**, *27*, 5033–5042. [CrossRef]
2. Choi, S.Y.; Jackman, S.; Thoreson, W.B.; Kramer, R.H. Light regulation of Ca^{2+} in the cone photoreceptor synaptic terminal. *Vis. Neurosci.* **2008**, *25*, 693–700. [CrossRef] [PubMed]
3. Libby, R.T.; Lillo, C.; Kitamoto, J.; Williams, D.S.; Steel, K.P. Myosin Va is required for normal photoreceptor synaptic activity. *J. Cell. Sci.* **2004**, *117*, 4509–4515. [CrossRef] [PubMed]
4. Piccoli, G.; Rutishauser, U.; Bruses, J.L. N-cadherin juxtamembrane domain modulates voltage-gated Ca^{2+} current via RhoA GTPase and Rho-associated kinase. *J. Neurosci.* **2004**, *24*, 10918–10923. [CrossRef]
5. Iftinca, M.; Hamid, J.; Chen, L.; Varela, D.; Tadayonnejad, R.; Altier, C.; Turner, R.W.; Zamponi, G.W. Regulation of T-type calcium channels by Rho-associated kinase. *Nat. Neurosci.* **2007**, *10*, 854–860. [CrossRef] [PubMed]
6. Davison, A.; Lux, U.T.; Brandstätter, J.H.; Babai, N. T-Type Ca^{2+} Channels Boost Neurotransmission in Mammalian Cone Photoreceptors. *J. Neurosci.* **2022**, *42*, 6325–6343. [CrossRef] [PubMed]
7. Wolfe, J.T.; Wang, H.; Howard, J.; Garrison, J.C.; Barrett, P.Q. T-type calcium channel regulation by specific G-protein betagamma subunits. *Nature* **2003**, *424*, 209–213. [CrossRef]
8. Brabeck, C.; Beschorner, R.; Conrad, S.; Mittelbronn, M.; Bekure, K.; Meyermann, R.; Schluesener, H.J.; Schwab, J.M. Lesional expression of RhoA and RhoB following traumatic brain injury in humans. *J. Neurotrauma* **2004**, *21*, 697–706. [CrossRef]
9. Haverkamp, S.; Grünert, U.; Wässle, H. The cone pedicle, a complex synapse in the retina. *Neuron* **2000**, *27*, 85–95. [CrossRef]
0. Thoreson, W.B. Transmission at rod and cone ribbon synapses in the retina. *Pflug. Arch.* **2021**, *473*, 1469–1491. [CrossRef]
1. Sterling, P.; Matthews, G. Structure and function of ribbon synapses. *Trends Neurosci.* **2005**, *28*, 20–29. [CrossRef]
2. González-Forero, D.; Montero, F.; García-Morales, V.; Domínguez, G.; Gómez-Pérez, L.; García-Verdugo, J.M.; Moreno-López, B. Endogenous Rho-kinase signaling maintains synaptic strength by stabilizing the size of the readily releasable pool of synaptic vesicles. *J. Neurosci.* **2012**, *32*, 68–84. [CrossRef] [PubMed]
3. Taoufiq, Z.; Eguchi, K.; Takahashi, T. Rho-kinase accelerates synaptic vesicle endocytosis by linking cyclic GMP-dependent protein kinase activity to phosphatidylinositol-4,5-bisphosphate synthesis. *J. Neurosci.* **2013**, *33*, 12099–12104. [CrossRef] [PubMed]

74. Mercer, A.J.; Szalewski, R.J.; Jackman, S.L.; Van Hook, M.J.; Thoreson, W.B. Regulation of presynaptic strength by controlling Ca^{2} channel mobility: Effects of cholesterol depletion on release at the cone ribbon synapse. *J. Neurophysiol.* **2012**, *107*, 3468–347 [CrossRef] [PubMed]
75. Holzhausen, L.C.; Lewis, A.A.; Cheong, K.K.; Brockerhoff, S.E. Differential role for synaptojanin 1 in rod and cone photoreceptor *J. Comp. Neurol.* **2009**, *517*, 633–644. [CrossRef]
76. Sherry, D.M.; Heidelberger, R. Distribution of proteins associated with synaptic vesicle endocytosis in the mouse and goldfis retina. *J. Comp. Neurol.* **2005**, *484*, 440–457. [CrossRef]
77. McCall, M.A.; Gregg, R.G. Comparisons of structural and functional abnormalities in mouse b-wave mutants. *J. Physiol.* **2008**, *58* 4385–4392. [CrossRef]
78. Gregg, R.G.; Kamermans, M.; Klooster, J.; Lukasiewicz, P.D.; Peachey, N.S.; Vessey, K.A.; McCall, M.A. Nyctalopin expressio in retinal bipolar cells restores visual function in a mouse model of complete X-linked congenital stationary night blindnes *J. Neurophysiol.* **2007**, *98*, 3023–3033. [CrossRef]
79. Pearring, J.N.; Bojang, P., Jr.; Shen, Y.; Koike, C.; Furukawa, T.; Nawy, S.; Gregg, R.G. A role for nyctalopin, a small leucine-ric repeat protein, in localizing the TRP melastatin 1 channel to retinal depolarizing bipolar cell dendrites. *J. Neurosci.* **2011**, *3* 10060–10066. [CrossRef]
80. Cao, Y.; Wang, Y.; Dunn, H.A.; Orlandi, C.; Shultz, N.; Kamasawa, N.; Fitzpatrick, D.; Li, W.; Zeitz, C.; Hauswirth, W.; et a Interplay between cell-adhesion molecules governs synaptic wiring of cone photoreceptors. *Proc. Natl. Acad. Sci. USA* **2020**, *11* 23914–23924. [CrossRef]
81. Rothschild, P.R.; Salah, S.; Berdugo, M.; Gelize, E.; Delaunay, K.; Naud, M.C.; Klein, C.; Moulin, A.; Savoldelli, M.; Bergin, C.; et a ROCK-1 mediates diabetes-induced retinal pigment epithelial and endothelial cell blebbing: Contribution to diabetic retinopath *Sci. Rep.* **2017**, *7*, 8834. [CrossRef]
82. Guarino, B.D.; Paruchuri, S.; Thodeti, C.K. The role of TRPV4 channels in ocular function and pathologies. *Exp. Eye Res.* **2020**, *20* 108257. [CrossRef] [PubMed]
83. Tura, A.; Schuettauf, F.; Monnier, P.P.; Bartz-Schmidt, K.U.; Henke-Fahle, S. Efficacy of Rho-kinase inhibition in promoting ce survival and reducing reactive gliosis in the rodent retina. *Investig. Ophthalmol. Vis. Sci.* **2009**, *50*, 452–461. [CrossRef] [PubMed
84. Sethi, C.S.; Lewis, G.P.; Fisher, S.K.; Leitner, W.P.; Mann, D.L.; Luthert, P.J.; Charteris, D.G. Glial remodeling and neural plasticit in human retinal detachment with proliferative vitreoretinopathy. *Investig. Ophthalmol. Vis. Sci.* **2005**, *46*, 329–342. [CrossRef [PubMed]

cells

MDPI

Article

Peptide Lv Promotes Trafficking and Membrane Insertion of $K_{Ca}3.1$ through the MEK1–ERK and PI3K–Akt Signaling Pathways

Dylan L. Pham [1], Autumn Niemi [1], Ria Blank [1], Gabriella Lomenzo [1], Jenivi Tham [1], Michael L. Ko [1,2] and Gladys Y.-P. Ko [1,3,*]

1 Department of Veterinary Integrative Biosciences, School of Veterinary Medicine and Biomedical Sciences, Texas A&M University, College Station, TX 77843, USA; phamdarch@tamu.edu (D.L.P.) autumnniemi@tamu.edu (A.N.); kmblank246@tamu.edu (R.B.); g_lomenzo11@tamu.edu (G.L.); jenivit@tamu.edu (J.T.); michael.ko@blinn.edu (M.L.K.)
2 Department of Biology, Division of Natural and Physical Sciences, Blinn College, Bryan, TX 77802, USA
3 Texas A&M Institute for Neuroscience, Texas A&M University, College Station, TX 77843, USA
* Correspondence: gko@cvm.tamu.edu

Abstract: Peptide Lv is a small endogenous secretory peptide that is proangiogenic through hyperpolarizing vascular endothelial cells (ECs) by enhancing the current densities of $K_{Ca}3.1$ channels. However, it is unclear how peptide Lv enhances these currents. One way to enhance the current densities of ion channels is to promote its trafficking and insertion into the plasma membrane. We hypothesized that peptide Lv-elicited $K_{Ca}3.1$ augmentation occurs through activating the mitogen-activated protein kinase kinase 1 (MEK1)-extracellular signal-regulated kinase (ERK) and phosphoinositide 3-kinase (PI3K)–protein kinase B (Akt) signaling pathways, which are known to mediate ion channel trafficking and membrane insertion in neurons. To test this hypothesis, we employed patch-clamp electrophysiological recordings and cell-surface biotinylation assays on ECs treated with peptide Lv and pharmaceutical inhibitors of ERK and Akt. Blocking ERK or Akt activation diminished peptide Lv-elicited EC hyperpolarization and increase in $K_{Ca}3.1$ current densities. Blocking PI3K or Akt activation decreased the level of plasma membrane-bound, but not the total amount of $K_{Ca}3.1$ protein in ECs. Therefore, the peptide Lv-elicited EC hyperpolarization and $K_{Ca}3.1$ augmentation occurred in part through channel trafficking and insertion mediated by MEK1–ERK and PI3K–Akt activation. These results demonstrate the molecular mechanisms of how peptide Lv promotes EC-mediated angiogenesis.

Keywords: angiogenesis; potassium channel; endothelial cell; peptide Lv; protein trafficking; signaling pathway

Citation: Pham, D.L.; Niemi, A.; Blank, R.; Lomenzo, G.; Tham, J.; Ko, M.L.; Ko, G.Y.-P. Peptide Lv Promotes Trafficking and Membrane Insertion of $K_{Ca}3.1$ through the MEK1–ERK and PI3K–Akt Signaling Pathways. *Cells* 2023, *12*, 1651. https://doi.org/10.3390/cells12121651

Academic Editor: Hossein Ameri

Received: 23 May 2023
Revised: 9 June 2023
Accepted: 15 June 2023
Published: 17 June 2023

1. Introduction

Peptide Lv is a newly discovered endogenous secretory peptide with about 40–50 amino acids depending on species, is highly conserved across species [1], and is expressed in various organs and tissues including the retina and vascular endothelium [1,2]. Functionally, peptide Lv is able to augment ion channel currents in neurons, cardiomyocytes, and vascular endothelial cells (ECs) [1–4], as it was first discovered to augment L-type voltage-gated calcium channels in retinal photoreceptors [1,2], thus the name peptide "Lv". Peptide Lv is also proangiogenic through stimulating EC proliferation, migration, and sprouting, and it can elicit vasodilation [2,3]. Peptide Lv is upregulated in the retinas of patients with early proliferative diabetic retinopathy, diabetic animals, and mice with oxygen-induced retinopathy [3]. Thus, peptide Lv may contribute to pathological neovascularization.

The angiogenic activity of peptide Lv occurs in part through its ability to bind and activate vascular endothelial growth factor receptor 2 (VEGFR2) [2,3]. It also resembles VEGF and elicits vasodilation of coronary and retinal arterioles [3,5]. VEGF-elicited vasodilation is dependent on nitric oxide (NO), since N(G)-nitro-L-arginine methyl ester (L-NAME), a NO synthase inhibitor, completely blocks VEGF-elicited vasodilation [5]. However, peptide Lv-elicited vasodilation is only partially blocked by L-NAME [3], indicating that peptide Lv has a VEGF/VEGFR2-independent mechanism for vasodilation and possibly for angiogenesis. The hyperpolarization of ECs in blood vessels can lead to vasodilation and angiogenesis, which can be independent of NO [6–10]. Opening of the potassium channels mediates EC hyperpolarization, leading to smooth muscle cell hyperpolarization and relaxation, which dilates blood vessels [6–8,11,12]. Calcium-dependent potassium (K_{Ca}) channels, in particular the small-conductance K_{Ca} ($K_{Ca}2.3$) and intermediate-conductance K_{Ca} ($K_{Ca}3.1$) channels, are two of the major potassium channels involved in EC hyperpolarization [8,13,14].

We recently showed that peptide Lv hyperpolarizes cultured ECs by augmenting $K_{Ca}3.1$, and blocking $K_{Ca}3.1$ prevents peptide Lv-elicited proliferation of ECs [4]. This finding suggests that $K_{Ca}3.1$ may play a role in VEGF-independent angiogenesis, as EC proliferation is a major process in angiogenesis [15–17]. We found that peptide Lv increases the expression of $K_{Ca}3.1$ [4], but increased ion channel expression in itself does not necessarily increase the channel's current densities [18–20]. Ion channel proteins must be properly shuttled and inserted into the plasma membrane to be physiologically functional [18–20]. As peptide Lv augments the current density of $K_{Ca}3.1$ in ECs, it is possible that peptide Lv promotes the trafficking and membrane insertion of $K_{Ca}3.1$. In neurons, exogenous trophic factors increase K_{Ca} currents by increasing protein expression and promoting trafficking and membrane insertion of the channel complex [18,20]. The mitogen-activated protein kinase kinase (MEK1)-extracellular signal-regulated kinase (ERK) [20], and the phosphoinositide 3-kinase (PI3K)–protein kinase B (Akt) signaling pathways [18] are known to mediate ion channel trafficking. Peptide Lv stimulates the activation/phosphorylation of ERK in photoreceptors and cardiomyocytes [1,2]. Thus, we postulated that peptide Lv promotes trafficking and membrane insertion of $K_{Ca}3.1$ in ECs in a similar manner. In this study, we employed patch-clamp recordings, biotinylation assays of plasma membrane-bound proteins, Western blots, and pharmacological tools to decipher the molecular signaling underlying the possibility that peptide Lv indeed promotes trafficking and plasma membrane insertion of $K_{Ca}3.1$ in cultured ECs.

2. Materials and Methods

2.1. Chemicals

Peptide Lv was custom-made from Peptide 2.0 Inc. (Chantilly, VA, USA). The polyclone antibody specifically against peptide Lv, anti-Lv, was obtained from Biomatik (Cambridge, ON, Canada). Peptide Lv and anti-Lv were made using the murine amino acid sequence DSLLAVRWFFAPDGSQEALMVKMTKLRIIQYYGNFSRTANQQRLRLLEE [1–3]. Peptide Lv and anti-Lv tested negative for endotoxin.

Other inhibitors and chemicals used in this study were TRAM-34 ($K_{Ca}3.1$ inhibitor; #AAJ60019-MB, Thermo Fisher Scientific, Waltham, MA, USA), β-escin (#E1378, Sigma-Aldrich, St. Louis, MO, USA), FR180402 (#SML0320, Sigma-Aldrich), PD98059 (#513000, Sigma-Aldrich), Akti (#A6730, Sigma-Aldrich), and LY294002 (#440202, Sigma-Aldrich).

2.2. Mice

C57BL/6J mice were originally purchased from the Jackson Laboratory (Bar Harbor, ME, USA), then bred and maintained at Texas A&M University. Mice were housed under temperature- and humidity-controlled conditions with 12:12 h light–dark cycles. Food and water were given ad libitum. All animal experiments were approved (AUP# 2020-0286) by the Institutional Animal Care and Use Committee of Texas A&M University.

2.3. Retinal Vasculature Immunofluorescent Staining

The fixation and trypsinization of the mouse retinas were processed as we previously described [21]. In brief, mouse eyes were collected and fixed with 4% paraformaldehyde at 4 °C overnight (for 20 h). Whole retinas were dissected and kept in double-deionized water with gentle rocking overnight at room temperature. The retinas were then incubated with 3% trypsin (#215250, BD Biosciences, Franklin Lakes, NJ, USA) for 1.5 h at 37 °C, for trypsinization followed by washing with double-deionized water by gentle pipetting to remove neural tissue. The remaining network of retinal vasculature was transferred to a glass slide and flattened (whole-mounted), blocked with 10% goat serum in phosphate-buffered saline (PBS) for 2 h at room temperature, and then incubated with primary antibodies overnight at 4 °C. After washing with PBS, the whole-mounted retinal vasculatures were incubated with secondary antibodies for 2 h at room temperature and mounted with Pro-Long Gold antifade with DAPI (#P36931, Invitrogen, Waltham, MA, USA). Images were obtained using a Zeiss Axiovert 200M microscope (Carl Zeiss AG, Jena, Germany) [21–23].

The following primary antibodies were used: anti-Lv (1:1000 dilution) and anti-CD31 (#3528, Cell Signaling Technology, Danvers, MA, USA). The following secondary antibodies were used: Fluor 488 goat anti-rabbit immunoglobulin G (IgG; 1:150 dilution; Thermo Fisher Scientific) and Cy5 goat anti-mouse IgG (1:150 dilution; Abcam, Cambridge, MA, USA).

2.4. Cell Culture

Cell cultures were maintained as we previously described [4]. Human umbilical vein ECs (HUVECs; #200-05n, Cell Applications Inc., San Diego, CA, USA) and human retinal microvascular ECs (HRMECs; #ACBRI 181, Cell Systems, Kirkland, WA, USA) were cultured in EGM-2 MV Microvascular Endothelial Cell Growth Medium-2 BulletKit (EGM; #CC-3202, Lonza, Walkersville, MD, USA) at 37 °C and 5% CO_2.

2.5. Patch-Clamp Electrophysiology

The β-escin-based perforated patch method [24] was used with the whole-cell patch-clamp configuration on cultured HUVECs as we previously described [1,2]. The methods and parameters for recording endothelial membrane potentials (current-clamp) and the $K_{Ca}3.1$ current (voltage-clamp) were based on our previous publications [1,4]. The external solution contained (in mM): 160 NaCl, 4.5 KCl, 1 $MgCl_2$, 2 $CaCl_2$, 10 HEPES, and 1 glucose, pH 7.5 adjusted with NaOH. The pipette solution contained (in mM): 120 KCl, 1.75 $MgCl_2$, 1 Na_2ATP, 10 EGTA, 4.1 $CaCl_2$, and 10 HEPES, pH 7.2 adjusted with KOH. The free calcium concentration in the pipette solution was calculated to be 100 nM using an online calcium chelator calculator [15]. β-escin was freshly prepared as a 35 mM stock solution in double-deionized water, kept on ice, and then added to the pipette solution to yield a final concentration of 35 μM.

HUVECs were seeded onto 12 mm acid-washed glass coverslips and placed in the incubator for 48 h to allow for adhesion. HUVECs were treated with PBS (vehicle; control) or peptide Lv (500 ng/mL) and maintained in the incubator at 37 °C and 5% CO_2 for 3 h prior to the patch-clamp recordings. There was no statistical difference in recorded amplitudes (either membrane potentials or currents) between the controls (either treated with PBS or without treatment), so their data were combined as a single control group.

Recordings were performed as we described previously [4]. In brief, all recordings were performed at room temperature (23 °C) using an A-M 2400 amplifier (A-M Systems Inc., Carlsborg, WA, USA). Signals were low-pass filtered at 1 kHz and digitized at 5 kHz with a Digidata 1550A interface (Axon Instruments/Molecular Devices, Union City, CA, USA), and pCLAMP 10.0 software (Molecular Devices) was used for data acquisition and analysis. Electrode capacitance was compensated after gigaohm (GΩ) seals were formed. The membrane capacitance, series resistance, and input resistance of the recorded HUVECs were measured by applying a +5 mV (100 ms) depolarizing voltage step from a holding potential of −60 mV. Any cell with an input resistance <1 GΩ was discarded. The membrane capacitance reading was used as the value for the whole-cell capacitance. The outward currents were elicited with a step command from a holding potential at −60 mV to 40 mV for 300 ms, as this current step command elicited the peak outward current [4 From the same cell, the total outward current containing $K_{Ca}3.1$ was first recorded, then TRAM-34 (10 μM; $K_{Ca}3.1$ inhibitor) was perfused into the recording chamber for 5 min and then a second current elicited and recorded in the presence of TRAM-34. The $K_{Ca}3.1$ current from a single cell was isolated by subtracting the current under TRAM-34 perfusion from the total outward current. The current density (pA/pF) was obtained by dividing the $K_{Ca}3.1$ current amplitude (measured at 200 ms, the tau point) by the whole-cell capacitance The membrane potentials were recorded under the current-clamp mode by injecting a 20 pA current for 750 ms.

2.6. Immunoblotting

Western blots were performed as we previously described [4]. HUVECs and HRMECs were seeded onto 60 mm culture plates and grown to 100% confluency. Cultured HUVECs and HMRECs were then treated with PBS (vehicle control) or peptide Lv (500 ng/mL for various times up to 3 h. Lysates were then collected for immunoblot analysis as we described previously [1,2,4]. In brief, cells were collected and lysed with an ice-cold RIPA lysis buffer, and proteins were denatured by mixing the lysate with 2X Lamelli sample buffer and heating for 5 min at 95 °C. Samples were separated using a 10% SDS-polyacrylamide gel and then transferred to a nitrocellulose membrane. Membranes were incubated with the primary antibodies $K_{Ca}3.1$ (1:400; #LS-C171766-100; LSBio, Seattle, WA, USA), phospho-p44/42 MAPK (1:1000; Erk1/2, Thr202/Tyr204; #4370, Cell Signaling Technology), p44/42 (1:1000; pan-Erk1/2; #4695, Cell Signaling Technology), phospho-Akt (1:800; ser473; #4060 Cell Signaling Technology), pan-Akt (1:1000; #4691, Cell Signaling Technology), and β-actin (1:1000; #4970S, Cell Signaling Technologies) overnight at 4 °C. The membranes were then washed with a tris-base saline (TBS)–tween solution and incubated with anti-rabbit IgG, HRP-linked secondary antibody (1:1000; #7074S, Cell Signaling Technologies) for 1 h at room temperature (23 °C). The membranes were visualized using Super Signal West Pico or Femto chemiluminescent substrate kit (#34078 or #34096, Pierce Biotechnology Inc., Rockford, IL, USA) with an immunoblot scanner (LI-COR Biosciences, Lincoln, NE USA). Band intensities were quantified using ImageJ software (https://imagej.nih.gov/ij/ National Institutes of Health; NIH, Bethesda, MA, USA). For $K_{Ca}3.1$ analysis, the band intensities were first normalized to the internal control, β-actin, and the relative changes were quantified according to the method described by Janes [25]. The pERK and pAkt were normalized to total ERK and total Akt, respectively.

2.7. Cell-Surface Biotinylation Assay

Cultured HUVECs were treated with PBS (vehicle control) or peptide Lv (500 ng/mL in the presence/absence of various inhibitors for 3 h. Biotinylation assays were performed as we described previously [26] and according to the company's protocol. In short, cultures were incubated with EZ-Link™ Sulfo-NHS-LC-Biotin (#21335; Pierce Biotechnology) for 30 min at room temperature while being rocked and for 30 min at 4 °C while being rocked The reaction was quenched with 100 mM glycine. Cells were lysed with a RIPA lysis buffer A portion of the lysate was collected to test for total protein concentration. The remaining sample was incubated with streptavidin agarose beads (#20353; Pierce Biotechnology) for 1 h at 4 °C under rotation. The supernatant was collected and tested for cytoplasmic protein concentration using immunoblot assays as described above. Lamelli buffer was added to the beads, and the resulting mixture was heated at 95 °C for 5 min to dissociate the protein from the beads.

2.8. Statistical Analysis

All data are presented as mean ± standard error of the mean (SEM). The comparisons between two groups were analyzed using Student's t test. Differences between multiple groups were analyzed by one-way ANOVA and Tukey *post hoc* tests. Origin 8.6 (OriginLab, Northampton, MA, USA) was used for statistical analyses. Throughout, $p < 0.05$ was considered significant.

3. Results

3.1. Peptide Lv Augments $K_{Ca}3.1$ Current Density and Endothelial Hyperpolarization through ERK Activation

Previously, we showed that peptide Lv is expressed in the neural retina [3]. To confirm its expression in retinal vasculature, mice retinas were digested with trypsin to isolate retinal vasculature from the neural tissue followed by co-immunostaining with an antibody against peptide Lv (anti-Lv) and CD31, an endothelial marker. Retinal vascular ECs expressed peptide Lv, as peptide Lv was localized in retinal microvasculature that was labelled with CD31 (Figure 1).

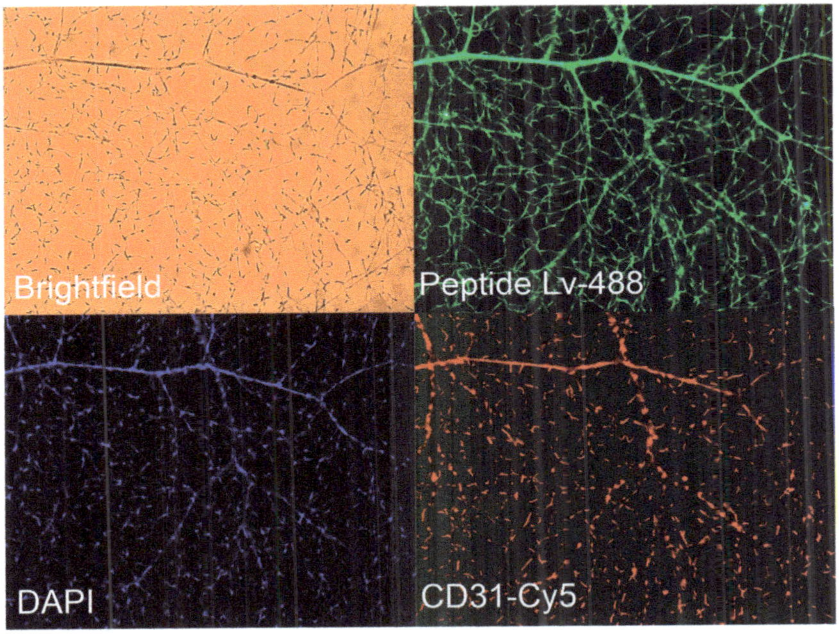

Figure 1. Peptide Lv is expressed in retinal vasculature. After trypsin digestion, the remaining retinal vasculature was stained with an antibody specific to peptide Lv (anti-Lv; green), endothelial marker CD31 (red), and DAPI (blue). Photos were taken under a 20× objective lens.

Since peptide Lv elicits ERK activation/phosphorylation in photoreceptors [1], we next determined whether exogenous peptide Lv also activated ERK in vascular ECs. Cultured HRMECs and HUVECs were treated with peptide Lv (500 ng/mL) for various periods of time and then harvested for Western blots. Treatment with peptide Lv for 120 min significantly elicited ERK phosphorylation in both HRMECs and HUVECs, but peptide Lv did not affect the total level of ERK in ECs (Figure 2). Activation of ERK signaling might be important in mediating the bioactivities of peptide Lv.

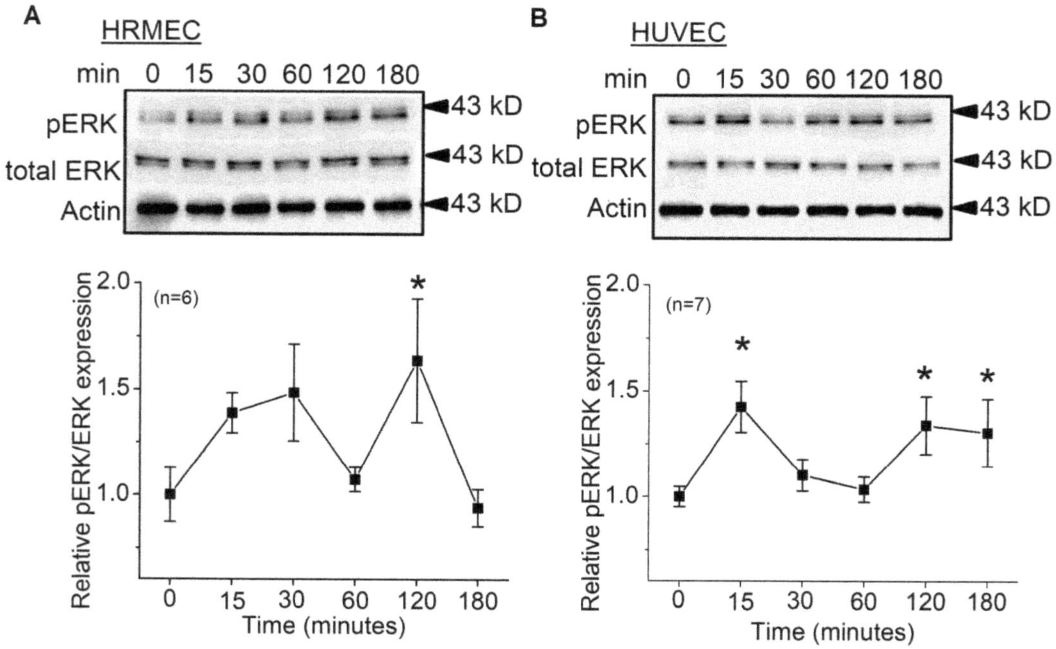

Figure 2. Peptide Lv elicits ERK phosphorylation in ECs. Cultured HRMECs and HUVECs were treated with peptide Lv for various durations (0, 15, 30, 60, 120, and 180 min). The Western blots show phosphorylated ERK (pERK) and total ERK in (**A**) HRMECs and (**B**) HUVECs. One-way ANOVA followed with Tukey *post hoc* tests were used for statistical analyses; * $p < 0.05$.

Previously, we showed that peptide Lv hyperpolarizes cultured ECs by increasing $K_{Ca}3.1$ current densities [4]. To determine whether the activation of ERK was needed for augmentation of $K_{Ca}3.1$ and endothelial hyperpolarization by peptide Lv, we performed patch-clamp electrophysiological recordings on cultured HUVECs after 3 h of treatment of PBS (vehicle control) or peptide Lv (500 ng/mL) in the presence or absence of FR180204 (10 μM; an ERK inhibitor; Figure 3). Representative traces (Figure 3A) showed that, after an outward current (black) was recorded, TRAM-34 ($K_{Ca}3.1$ inhibitor) was perfused followed by a second current recording (gray). The subtraction between these two currents indicates the isolated $K_{Ca}3.1$ current. Peptide Lv significantly increased the $K_{Ca}3.1$ current densities (2.32 ± 0.40 pA/pF) compared to the control (0.28 ± 0.07 pA/pF) or FR180204 treatment alone (0.69 ± 0.12 pA/pF; * $p < 0.05$; Figure 3A,B). Peptide Lv-elicited increases in $K_{Ca}3.1$ were attenuated by ERK inhibition (peptide Lv + FR180204: 0.90 ± 0.15 pA/pF, # $p < 0.05$ compared to the peptide Lv group; Figure 3A,B). In addition, peptide Lv-induced hyperpolarization of HUVECs (−79.23 ± 0.56 mV compared to the control (−73.14 ± 0.69 mV) or HUVECs treated with FR180204 alone (−74.07 ± 0.74 mV); * $p < 0.05$; Figure 3C) was attenuated when ERK activation was blocked with FR180204 (−74.68 ± 0.72 mV; # $p < 0.05$ compared to the peptide Lv group; Figure 3C). Thus, peptide Lv-elicited augmentation of $K_{Ca}3.1$ current densities and EC hyperpolarization are in part through ERK activation.

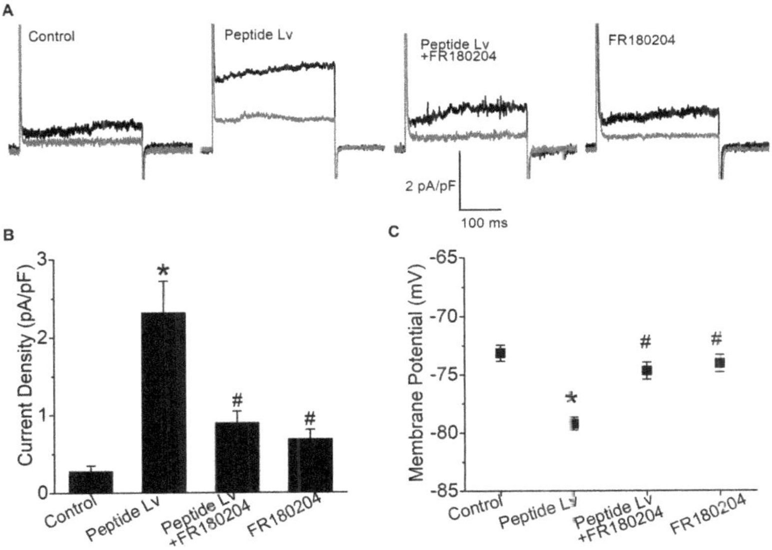

Figure 3. Blocking ERK activation attenuates peptide Lv-mediated increase in $K_{Ca}3.1$ current densities and endothelial hyperpolarization. HUVECs were seeded onto glass coverslips and kept in an incubator for 48 h to allow the cells to adhere. The cultures were then treated with PBS (vehicle control), peptide Lv (500 ng/mL), FR180204 (ERK inhibitor; 10 μM), or peptide Lv and FR180204 for 3 h prior to whole-cell electrophysiological recordings. (**A**) Representative traces are displayed with the total outward current (black) and the current after perfusion with TRAM-34 to isolate $K_{Ca}3.1$ currents (gray). (**B**) The current density (pA/pF) was obtained by dividing the $K_{Ca}3.1$ current amplitude (measured at 200 ms; the tau point) by the whole-cell capacitance. Peptide Lv significantly increases the $K_{Ca}3.1$ current densities, which was attenuated by FR180204. (**C**) Membrane potentials are significantly hyperpolarized in HUVECs treated with peptide Lv compared to the PBS-treated control, while treatment with FR180204 attenuates peptide Lv-elicited EC hyperpolarization. One-way ANOVA followed with Tukey *post hoc* tests were used for statistical analyses; n = 12–16 for each group; * $p < 0.05$. *: significantly different compared to the control; #: significantly different from the peptide Lv group.

3.2. Peptide Lv Promotes $K_{Ca}3.1$ Channel Trafficking and Membrane Insertion through the MEK1–ERK Signaling Pathway

After ion channel proteins are expressed, they need to be transported into the plasma membrane to be functional. The ERK signaling is known to promote ion channel trafficking and plasma membrane insertion in neurons [20]. As peptide Lv can elicit ERK activation/phosphorylation and increase $K_{Ca}3.1$ current densities, we postulated that activated ERK mediates peptide Lv-elicited increases in $K_{Ca}3.1$ trafficking in ECs. We performed biotinylation assays after HUVECs were treated for 3 h with peptide Lv in the absence or presence of PD98059 (50 μM; a MEK1 inhibitor) or FR180204 (10 μM; an ERK inhibitor) to determine if blocking MEK1–ERK signaling would prevent peptide Lv-stimulated $K_{Ca}3.1$ trafficking and plasma membrane insertion. We previously showed that treatment with peptide Lv for 3 h increases both mRNA and protein expression in cultured RMECs and HUVECs [4]. We found that peptide Lv not only increased the total protein expression of $K_{Ca}3.1$ (* $p < 0.05$), but also indeed promoted $K_{Ca}3.1$ insertion into the plasma membrane, as the level of membrane-bound $K_{Ca}3.1$ was significantly higher in peptide Lv-treated ECs compared to the control (* $p < 0.05$; Figure 4). Blocking the activation of MEK1 (Figure 4A,B) or ERK (Figure 4C,D) significantly attenuated peptide Lv-elicited increases in membrane-bound $K_{Ca}3.1$ (* $p < 0.05$; Figure 4). Interestingly, inhibition of MEK1–ERK signaling did not affect peptide Lv-elicited increases in the total protein level of $K_{Ca}3.1$ (* $p < 0.05$). These data indicate that peptide Lv-stimulated increases in $K_{Ca}3.1$ current densities (Figure 3) are in part through MEK1–ERK signaling-mediated protein trafficking and plasma membrane insertion of $K_{Ca}3.1$ in ECs.

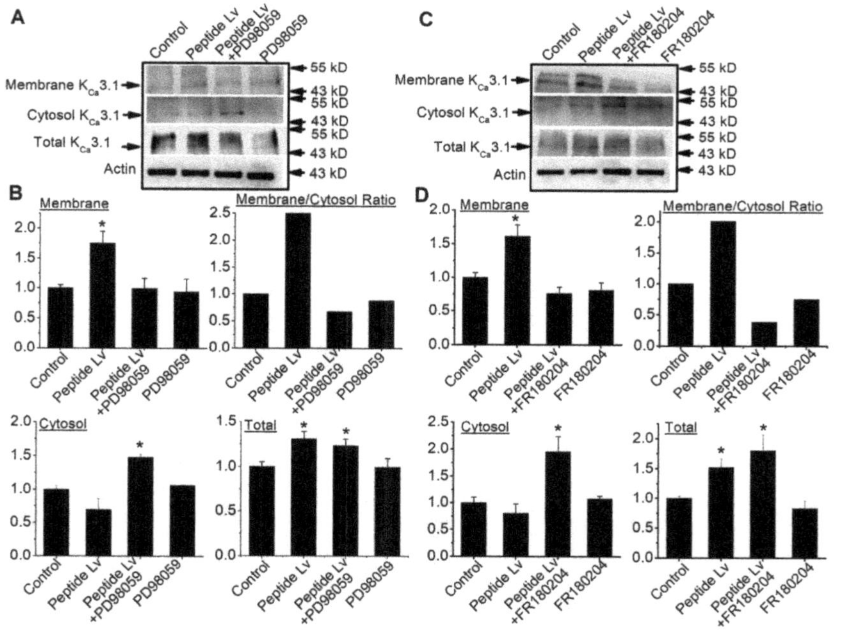

Figure 4. Blocking ERK activation decreases peptide Lv-elicited increase in membrane-bound $K_{Ca}3.1$ without affecting the total protein expression of $K_{Ca}3.1$. HUVEC cultures were treated with PBS (vehicle control) or peptide Lv (500 ng/mL) in the presence/absence of PD98059 (MEK1 inhibitor; 10 μM; (**A**,**B**)) or FR180204 (ERK inhibitor; 10 μM; (**C**,**D**)) for 3 h. Cell-surface biotinylation assays were performed on the cultures to separate the membrane-bound proteins from the cytoplasmic proteins. An aliquot of each sample was used for total protein analysis, and the remainder was used to separate the membrane-bound from cytosolic proteins. Actin was used as the loading control. The Y-axes of the "Membrane" (membrane-bound), "Cytosol" (cytosolic), and "Total" (total proteins) are presented as "Relative $K_{Ca}3.1$/Actin". (**A**,**B**) Peptide Lv significantly increases the total expression of $K_{Ca}3.1$ as well as the membrane-bound $K_{Ca}3.1$. PD98059 does not affect the total expression of $K_{Ca}3.1$ but significantly decreases peptide Lv-elicited increase in membrane-bound $K_{Ca}3.1$, which is also reflected as a significantly higher amount of cytosolic $K_{Ca}3.1$. (**C**,**D**) Similarly, FR180204 does not affect the total expression of $K_{Ca}3.1$ but significantly decreases peptide Lv-elicited increase in membrane-bound $K_{Ca}3.1$. One-way ANOVA followed with Tukey *post hoc* tests were used for statistical analyses; $n = 5$ for each group; * $p < 0.05$.

3.3. *Peptide Lv Augments $K_{Ca}3.1$ Current Density and Endothelial Hyperpolarization through Akt Activation*

The PI3K–Akt signaling pathway is another pathway known to promote ion channel trafficking and membrane insertion [18]. We found that HRMECs (Figure 5A) or HUVECs (Figure 5B) treated with peptide Lv (500 ng/mL) for 3 h had a significant increase in Akt phosphorylation at serine 473 (ser473), without affecting the total level of Akt.

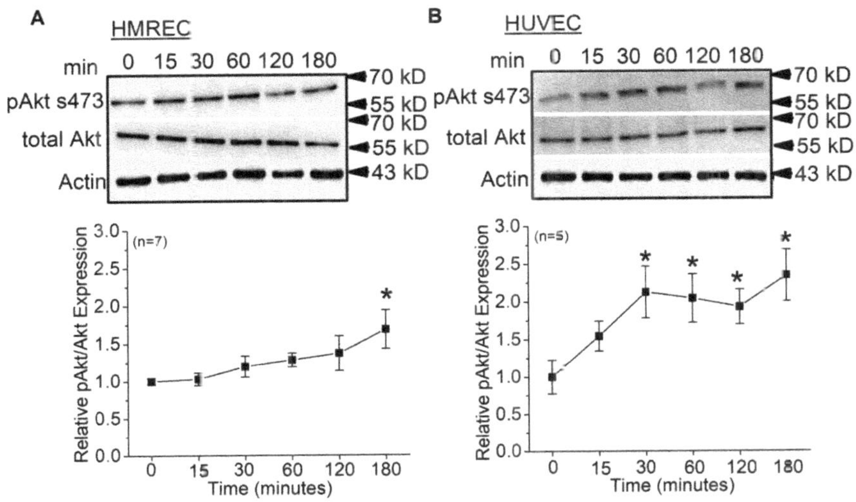

Figure 5. Peptide Lv elicits Akt phosphorylation/activation in ECs. Cultured HRMECs and HUVECs were treated with peptide Lv for various durations (0, 15, 30, 60, 120, and 180 min). The Western blots show phosphorylated Akt at ser473 and total Akt in (**A**) HMRECs and (**B**) HUVECs. One-way ANOVA followed with Tukey *post hoc* tests were used for statistical analyses; * $p < 0.05$.

To determine if the activation of PI3K–Akt signaling was needed for augmentation of $K_{Ca}3.1$ and endothelial hyperpolarization by peptide Lv, we performed patch-clamp electrophysiological recordings on cultured HUVECs after 3 h of treatment of PBS (vehicle control) or peptide Lv (500 ng/mL) in the presence or absence of LY294002 (50 μM; a PI3K inhibitor; Figure 6). Peptide Lv significantly increased the $K_{Ca}3.1$ current densities (2.28 ± 0.37 pA/pF) compared to the control (0.31 ± 0.08 pA/pF) or LY294002 treatment alone (0.63 ± 0.14 pA/pF; * $p < 0.05$; Figure 6A,B). Peptide Lv-elicited increases in $K_{Ca}3.1$ were attenuated by PI3K inhibition (peptide Lv + LY294002: 1.12 ± 0.22 pA/pF, # $p < 0.05$ compared to the peptide Lv group; Figure 6A,B). Peptide Lv-induced hyperpolarization of HUVECs (−79.86 = 0.90 mV compared to the control (−72.90 ± 0.87 mV) or HUVECs treated with LY204992 alone (−71.61 ± 0.93 mV); * $p < 0.05$; Figure 6C), was attenuated when PI3K–Akt activation was blocked by LY294002 (−72.49 ± 0.96 mV; # $p < 0.05$ compared to the peptide Lv group; Figure 6C). Thus, peptide Lv-elicited augmentation of $K_{Ca}3.1$ current densities and EC hyperpolarization are also mediated by PI3K–Akt signaling.

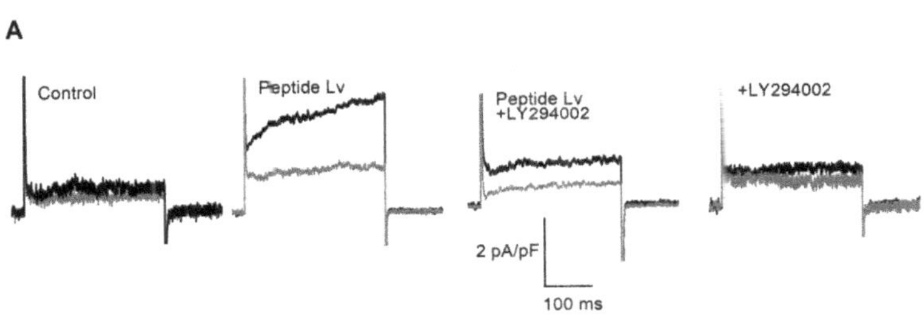

Figure 6. *Cont.*

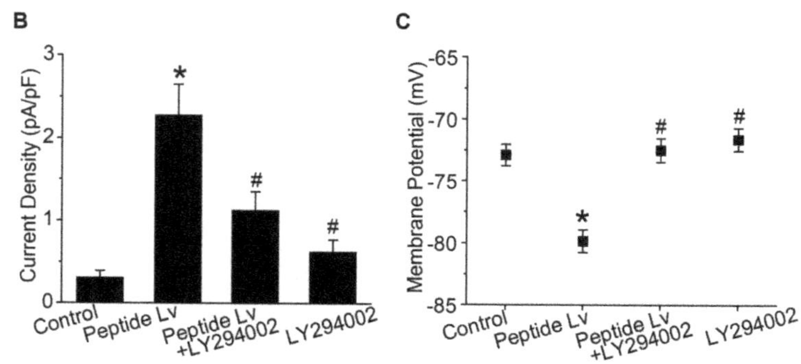

Figure 6. Blocking Akt activation attenuates peptide Lv-mediated increase in $K_{Ca}3.1$ current densities and endothelial hyperpolarization. HUVECs were seeded onto glass coverslips and kept in an incubator for 48 h to allow the cells to adhere. The cultures were then treated with PBS (vehicle control), peptide Lv (500 ng/mL), LY294002 (PI3K inhibitor, 10 μM), or peptide Lv and LY294002 for 3 h prior to whole-cell electrophysiological recordings. (**A**) Representative traces are displayed with the total outward current (black) and the current after perfusion with TRAM-34 to isolate $K_{Ca}3.1$ currents (grey). (**B**) The current density (pA/pF) was obtained by dividing the $K_{Ca}3.1$ current amplitude (measured at 200 ms; the tau point) by the whole-cell capacitance. Peptide Lv significantly increases the $K_{Ca}3.1$ current densities, which were attenuated by LY294002. (**C**) Membrane potentials are significantly hyperpolarized in HUVECs treated with peptide Lv compared to the PBS-treated control, while treatment with LY294002 significantly dampens peptide Lv-elicited EC hyperpolarization. One-way ANOVA followed with Tukey *post hoc* tests were used for statistical analyses; $n = 12–16$ for each group; * $p < 0.05$. *: significantly different compared to the control; #: significantly different from the peptide Lv group.

3.4. Peptide Lv Promotes $K_{Ca}3.1$ Channel Trafficking and Membrane Insertion through the PI3K–Akt Signaling Pathway

We next employed biotinylation assays to determine whether PI3K–Akt signaling also mediates peptide Lv-elicited channel trafficking and plasma membrane insertion of $K_{Ca}3.1$. Cultured HUVECs were treated for 3 h with PBS (vehicle control) or peptide Lv (500 ng/mL) with or without the presence of LY294002 (50 μM) or Akti (10 μM; an Akt inhibitor). We found that blocking the activation of PI3K (Figure 7A,B) or Akt (Figure 7C,D) significantly attenuated peptide Lv-elicited increases in membrane-bound $K_{Ca}3.1$ (* $p < 0.05$; Figure 7). In addition, inhibition of PI3K–Akt signaling did not affect peptide Lv-elicited increases in $K_{Ca}3.1$ total protein (* $p < 0.05$). These data indicate that peptide Lv-stimulated increases in $K_{Ca}3.1$ current densities (Figure 6) are in part through PI3K–Akt signaling-mediated protein trafficking and plasma membrane insertion of $K_{Ca}3.1$ in ECs.

3.5. Peptide Lv Concurrently Activates the MEK1–ERK and PI3K–Akt Signaling Pathways

Since both MEK1–ERK and PI3K–Akt signaling pathways mediated peptide Lv-elicited increases in $K_{Ca}3.1$ current densities in ECs, we next determined the possible interactions between these two signaling pathways and whether one was upstream/downstream to the other. We treated HUVECs for 3 h with PBS (vehicle control) or peptide Lv (500 ng/mL) in the presence or absence of an inhibitor of MEK1 (PD98059; 50 μM), ERK (FR180204; 10 μM), PI3K (LY294002; 50 μM), or Akt (Akti; 10 μM) to determine the levels of phosphorylated ERK and Akt. As peptide Lv activated/phosphorylated ERK (Figure 8A,B) and Akt (Figure 8C,D), inhibition of MEK1 prevented peptide Lv-stimulated activation of ERK without affecting Akt phosphorylation at Ser473. Similarly, inhibition of PI3K prevented peptide Lv-stimulated phosphorylation of Akt without affecting ERK phosphorylation. These findings suggest that peptide Lv can independently and concurrently activate MEK1–ERK and PI3K–Akt to mediate peptide Lv-elicited increases in $K_{Ca}3.1$ and its trafficking.

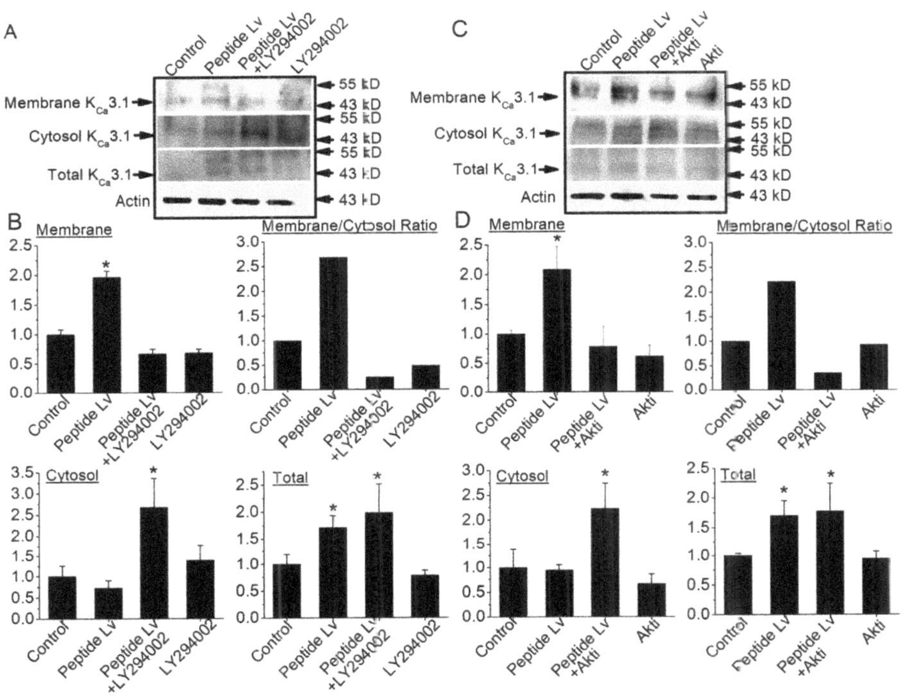

Figure 7. Blocking Akt activation decreases peptide Lv-elicited increase in membrane-bound $K_{Ca}3.1$ without affecting the total protein expression of $K_{Ca}3.1$. HUVEC cultures were treated with PBS (vehicle control) or peptide Lv (500 ng/mL) in the presence/absence of LY294002 (PI3K inhibitor; 10 μM; (**A**,**B**)) or Akti (Akt inhibitor; 10 μM (**C**,**D**)) for 3 h. Cell-surface biotinylation assays were performed on the cultures to separate the membrane-bound proteins from the cytoplasmic proteins. After sample collection, an aliquot of each was used for total protein analysis, and the remainder was used to separate the membrane-bound from cytosolic proteins. Actin was used as the loading control. The Y-axes of the "Membrane" (membrane-bound), "Cytosol" (cytosolic), and "Total" (total proteins) are presented as "Relative $K_{Ca}3.1$/Actin". (**A**,**B**) Peptide Lv significantly increases the total expression of $K_{Ca}3.1$ as well as the membrane-bound $K_{Ca}3.1$. LY294002 does not affect the total expression of $K_{Ca}3.1$ but significantly decreases peptide Lv-elicited increase in membrane-bound $K_{Ca}3.1$, which is also reflected as a significantly higher amount of cytosolic $K_{Ca}3.1$. (**C**,**D**) Similarly, Akti does not affect the total expression of $K_{Ca}3.1$ but significantly decreases peptide Lv-elicited increase in membrane-bound $K_{Ca}3.1$. One-way ANOVA followed with Tukey *post hoc* tests were used for statistical analyses; $n = 5$ for each group; * $p < 0.05$.

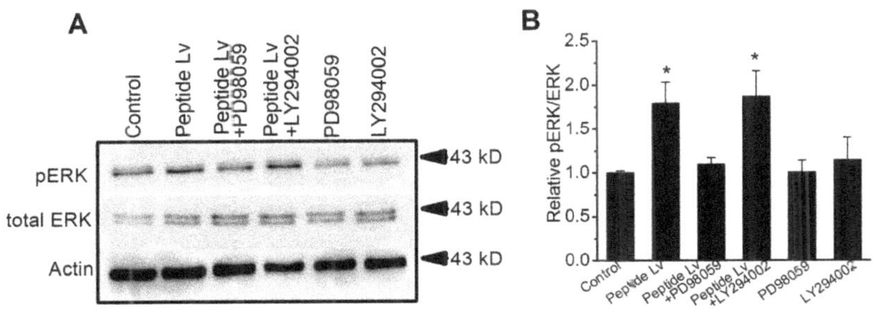

Figure 8. *Cont.*

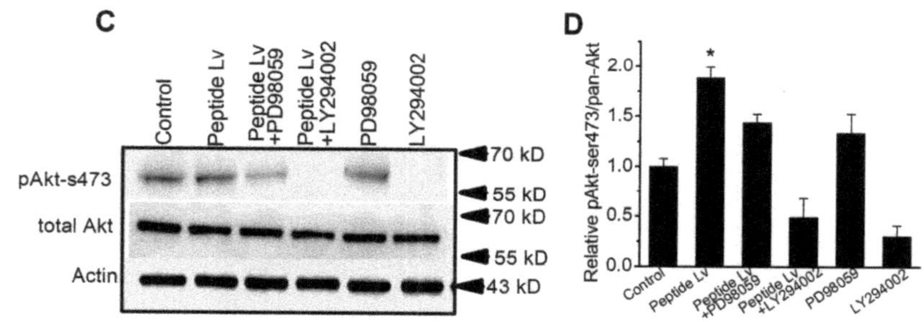

Figure 8. Peptide Lv concurrently activates the MEK1–ERK and PI3K–Akt signaling pathways. Immunoblot assays were performed on HUVEC cultures treated with PBS (vehicle control) or peptide Lv (500 ng/mL) in the presence/absence of FR180204 (ERK inhibitor) or LY294002 (PI3K inhibitor). (**A**,**B**) Peptide Lv increases phosphorylated ERK (pERK), which is attenuated by FR180204 but not LY294002. (**C**,**D**) Peptide Lv increases phosphorylated Akt at ser473, which is attenuated by LY294002 but not FR18204. One-way ANOVA followed with Tukey *post hoc* tests were used for statistical analyses; $n = 7$ for each group; * $p < 0.05$.

4. Discussion

Previously, we showed that peptide Lv enhances $K_{Ca}3.1$ current densities in vascular ECs to promote angiogenesis [4], but the mechanism by which these currents are enhanced was unclear. The current densities of ion channels can be augmented through various mechanisms, including posttranslational modifications [11,27,28], stimulating trafficking and insertion into the plasma membrane [18,20], or preventing internalization and degradation of the channel proteins [19,29,30].

Peptide Lv increases the protein expression and the current densities of endothelial $K_{Ca}3.1$ channels after ECs were treated with peptide Lv for 3 h [4], so it is likely that peptide Lv promotes the trafficking and insertion of $K_{Ca}3.1$ into the plasma membrane. The MEK1–ERK and PI3K–Akt signaling pathways are known to promote the trafficking and insertion of ion channels into the plasma membrane in neurons [18,20]. We found that peptide Lv stimulated both pathways in two different types of ECs, HRMECs and HUVECs (Figures 2 and 5), but the timing of peptide Lv-elicited activation of ERK and Akt were different in these two cell types: HUVECs appeared to have earlier activations of peptide Lv-elicited phosphorylation of ERK and Akt compared to that of HRMECs. This could be due to the different properties of ECs in large vessels (HUVECs) versus in microvasculature (HRMECs), and ERK and Akt were activated after these ECs treated with peptide Lv for less than 3 h. This indicates that peptide Lv elicited ERK and Akt activations first, and these two pathways further stimulated $K_{Ca}3.1$ trafficking onto the plasma membrane, so the timing of ERK and Akt activation was associated with the timing of peptide Lv-elicited increase in $K_{Ca}3.1$ current densities. Blocking these signaling pathways prevented peptide Lv-elicited increase in $K_{Ca}3.1$ current densities (Figures 3 and 6). While pharmacological inhibitors can have nonspecific effects, we employed a strategy of using two different inhibitors with different mechanisms of action in a single signaling pathway to confirm these results in our study. Interestingly, blocking these pathways only prevented peptide Lv-mediated increase in membrane-bound $K_{Ca}3.1$ but not the overall increase in $K_{Ca}3.1$ expression (Figures 4 and 7). These findings suggest that the MEK1–ERK and PI3K–Akt signaling pathways are responsible for peptide Lv-elicited increases in $K_{Ca}3.1$ current densities by promoting trafficking and membrane insertion of the channels, but they do not affect the overall protein expression. Since we performed these experiments in monolayer cultured ECs, these data might not be exactly the same as ECs in the three-dimensional environment in vivo, which is a limitation of this study. As we showed that peptide Lv elicits neovascularization in vivo [3], and endothelial $K_{Ca}3.1$ is required for EC-dependent neovascularization [14], nonetheless, these data in part reflect the molecular mechanism as to how peptide Lv elicits EC-dependent neovascularization through promoting $K_{Ca}3.1$ trafficking onto the plasma membrane.

These findings do not eliminate the possibility that peptide Lv can enhance $K_{Ca}3.1$ current densities through other mechanisms including posttranslational modulation. For example, phosphorylation of K_{Ca} channels can enhance its currents [20]. Previously, we showed that peptide Lv can elicit vasodilation in isolated arterioles after the vessels were bathed with a solution containing peptide Lv (1–10 μg/mL) for 2 or 3 min [3]. This finding suggests that peptide Lv may acutely elicit posttranslational modifications through activation of kinases, which then phosphorylates and opens $K_{Ca}3.1$ channels and leads to hyperpolarization of vascular ECs and vasodilation. As we stated earlier that peptide Lv-elicited vasodilation is partially NO-independent, further studies are needed to determine how peptide Lv might elicit posttranslational modulation of $K_{Ca}3.1$.

Previously, we found that peptide Lv increases cyclic adenosine monophosphate (cAMP) levels in cultured photoreceptors [1]. Cyclic AMP signaling is known to activate protein kinase A (PKA) [31], and the c-terminal phosphorylation of $K_{Ca}3.1$ by cAMP/PKA increases the channel activity [32,33], which could be a potential signaling pathway mediating the acute action of peptide Lv on vasodilation. Peptide Lv does not have any kinase activity, making it unlikely that peptide Lv directly activates the MEK1–ERK or PI3K–Akt signaling pathways. Further studies are needed to elucidate the upstream signaling of MEK1–ERK and PI3K–Akt activated by peptide Lv, whether acutely or chronically.

In addition to ion channel trafficking, the MEK1–ERK and PI3K–Akt signaling pathways are associated with angiogenesis through various processes [34–37]. Blocking ERK signaling impaired EC proliferation, migration, and sprouting [38,39]. EC migration and vessel tube formation is PI3K–Akt-dependent [40,41]. Knockout of Akt decreased EC proliferation, migration, and sprouting [42,43]. Both pathways are also downstream signaling of VEGF receptor 2 (VEGFR2) [44–46]. As we previously showed that peptide Lv is able to bind VEGFR2 [2], it is also possible that peptide Lv concurrently activates ERK and Akt signaling in part through binding VEGFR2, which can lead to EC proliferation, migration, and sprouting, and neovascularization [3].

5. Conclusions

This study provides further insights into the molecular mechanisms of peptide Lv in augmenting EC functions. Blocking peptide Lv may become a novel therapy to combat diseases associated with pathological angiogenesis, as we previously showed that using an antibody against peptide Lv blocked pathological neovascularization [3]. By contrast, perhaps peptide Lv may be used in situations that require angiogenesis, such as wound healing and recovery from ischemia. Previously, we showed that peptide Lv works synergistically with VEGF to promote EC proliferation [3], which may explain why a single dose of anti-VEGF therapy cannot block recurring neovascularization. Delineating the molecular mechanism of bioactivity of peptide Lv is important to combat diseases with pathological neovascularization.

In addition to promoting angiogenesis and vasodilation, peptide Lv may have other unknown functions. Peptide Lv is a widely expressed secretory peptide that has been shown to be important in early photoreceptor development and cardiomyocyte function [1,2]. Peptide Lv may also play a role in immune responses, as it dampens the inflammatory response in macrophages [47]. Much of the function and mechanism of peptide Lv remains unknown and should be investigated.

Author Contributions: Conceptualization, G.Y.-P.K.; methodology, G.Y.-P.K.; validation, G.Y.-P.K.; formal analysis, G.Y.-P.K., D.L.P., A.N., R.B., G.L. and J.T.; investigation, G.Y.-P.K. and D.L.P.; resources, G.Y.-P.K.; data curation, D.L.P., A.N., R.B., G.L. and J.T.; writing—original draft preparation, D.L.P. and G.Y.-P.K.; writing—review and editing, G.Y.-P.K. and M.L.K.; supervision, M.L.K. and G.Y.-P.K.; project administration, M.L.K. and G.Y.-P.K.; funding acquisition, G.Y.-P.K. All authors have read and agreed to the published version of the manuscript.

Funding: This work was funded by the National Eye Institute of the National Institutes of Health (grant No. NIHR21EY031813-01A1), a philanthropist gift fund awarded to Gladys Ko, and a Texas A&M School of Veterinary Medicine and Biomedical Sciences Graduate Student Research Trainee Grant awarded to Dylan Pham. The funders had no role in study design, data collection and analysis, decision to publish, or preparation of the manuscript.

Institutional Review Board Statement: Not applicable.

Informed Consent Statement: Not applicable.

Data Availability Statement: All data related to this study are presented in this manuscript.

Acknowledgments: We thank Dana Gaddy, Kayla Bayless, and Greg Johnson for constructive suggestions. This work was supported by the National Eye Institute of the National Institutes of Health (NIHR21EY031813-01A1), a philanthropist gift fund awarded to GK, and a Texas A&M School of Veterinary Medicine and Biomedical Sciences Graduate Student Research Trainee Grant awarded to DP.

Conflicts of Interest: The authors declare no conflict of interest.

References

1. Shi, L.; Ko, M.L.; Abbott, L.C.; Ko, G.Y. Identification of Peptide lv, a novel putative neuropeptide that regulates the expression of L-type voltage-gated calcium channels in photoreceptors. *PLoS ONE* **2012**, *7*, e43091. [CrossRef] [PubMed]
2. Shi, L.; Ko, S.; Ko, M.L.; Kim, A.J.; Ko, G.Y. Peptide Lv augments L-type voltage-gated calcium channels through vascular endothelial growth factor receptor 2 (VEGFR2) signaling. *Biochim. Biophys. Acta* **2015**, *1853*, 1154–1164. [CrossRef]
3. Shi, L.; Zhao, M.; Abbey, C.A.; Tsai, S.H.; Xie, W.; Pham, D.; Chapman, S.; Bayless, K.J.; Hein, T.W.; Rosa, R.H., Jr.; et al. Newly Identified Peptide, Peptide Lv, Promotes Pathological Angiogenesis. *J. Am. Heart Assoc.* **2019**, *8*, e013673. [CrossRef]
4. Pham, D.L.; Niemi, A.; Ko, M.L.; Ko, G.Y.P. Peptide Lv augments intermediate-conductance calcium-dependent potassium channels (KCa3.1) in endothelial cells to promote angiogenesis. *PLoS ONE* **2022**, *17*, e0276744. [CrossRef] [PubMed]
5. Hein, T.W.; Rosa, R.H., Jr.; Ren, Y.; Xu, W.; Kuo, L. VEGF Receptor-2-Linked PI3K/Calpain/SIRT1 Activation Mediates Retinal Arteriolar Dilations to VEGF and Shear Stress. *Investig. Ophthalmol. Vis. Sci.* **2015**, *56*, 5381–5389. [CrossRef] [PubMed]
6. Dabisch, P.A.; Liles, J.T.; Taylor, J.T.; Sears, B.W.; Saenz, R.; Kadowitz, P.J. Role of potassium channels in the nitric oxide-independent vasodilator response to acetylcholine. *Pharmacol. Res.* **2004**, *49*, 207–215. [CrossRef] [PubMed]
7. Edwards, G.; Dora, K.A.; Gardener, M.J.; Garland, C.J.; Weston, A.H. K^+ is an endothelium-derived hyperpolarizing factor in rat arteries. *Nature* **1998**, *396*, 269–272. [CrossRef] [PubMed]
8. Grgic, I.; Kaistha, B.P.; Hoyer, J.; Kohler, R. Endothelial Ca^+-activated K^+ channels in normal and impaired EDHF-dilator responses--relevance to cardiovascular pathologies and drug discovery. *Br. J. Pharmacol.* **2009**, *157*, 509–526. [CrossRef]
9. Hein, T.W.; Kuo, L. cAMP-independent dilation of coronary arterioles to adenosine: Role of nitric oxide, G proteins, and K(ATP) channels. *Circ. Res.* **1999**, *85*, 634–642. [CrossRef]
10. Michaelis, U.R.; Fleming, I. From endothelium-derived hyperpolarizing factor (EDHF) to angiogenesis: Epoxyeicosatrienoic acids (EETs) and cell signaling. *Pharmacol. Ther.* **2006**, *111*, 584–595. [CrossRef]
11. Dora, K.A.; Gallagher, N.T.; McNeish, A.; Garland, C.J. Modulation of endothelial cell KCa3.1 channels during endothelium-derived hyperpolarizing factor signaling in mesenteric resistance arteries. *Circ. Res.* **2008**, *102*, 1247–1255. [CrossRef] [PubMed]
12. Ungvari, Z.; Csiszar, A.; Koller, A. Increases in endothelial Ca^{2+} activate K(Ca) channels and elicit EDHF-type arteriolar dilation via gap junctions. *Am. J. Physiol. Heart Circ. Physiol.* **2002**, *282*, H1760–H1767. [CrossRef] [PubMed]
13. Archer, S.L.; Huang, J.M.; Hampl, V.; Nelson, D.P.; Shultz, P.J.; Weir, E.K. Nitric oxide and cGMP cause vasorelaxation by activation of a charybdotoxin-sensitive K channel by cGMP-dependent protein kinase. *Proc. Natl. Acad. Sci. USA* **1994**, *91*, 7583–7587. [CrossRef] [PubMed]
14. Grgic, I.; Eichler, I.; Heinau, P.; Si, H.; Brakemeier, S.; Hoyer, J.; Kohler, R. Selective blockade of the intermediate-conductance Ca^{2+}-activated K^+ channel suppresses proliferation of microvascular and macrovascular endothelial cells and angiogenesis in vivo. *Arterioscler. Thromb. Vasc. Biol.* **2005**, *25*, 704–709. [CrossRef] [PubMed]
15. Adams, R.H.; Alitalo, K. Molecular regulation of angiogenesis and lymphangiogenesis. *Nat. Rev. Mol. Cell Biol.* **2007**, *8*, 464–478. [CrossRef]
16. Carmeliet, P. VEGF as a key mediator of angiogenesis in cancer. *Oncology* **2005**, *69* (Suppl. S3), 4–10. [CrossRef]
17. Conway, E.M.; Collen, D.; Carmeliet, P. Molecular mechanisms of blood vessel growth. *Cardiovasc. Res.* **2001**, *49*, 507–521. [CrossRef]
18. Chae, K.S.; Martin-Caraballo, M.; Anderson, M.; Dryer, S.E. Akt activation is necessary for growth factor-induced trafficking of functional K(Ca) channels in developing parasympathetic neurons. *J. Neurophysiol.* **2005**, *93*, 1174–1182. [CrossRef]
19. Bertuccio, C.A.; Lee, S.L.; Wu, G.; Butterworth, M.B.; Hamilton, K.L.; Devor, D.C. Anterograde trafficking of KCa3.1 in polarized epithelia is Rab1- and Rab8-dependent and recycling endosome-independent. *PLoS ONE* **2014**, *9*, e92013. [CrossRef]

120. Lhuillier, L.; Dryer, S.E. Developmental regulation of neuronal KCa channels by TGFbeta 1: Transcriptional and posttranscriptional effects mediated by Erk MAP kinase. *J. Neurosci.* **2000**, *20*, 5616–5622. [CrossRef]
121. Yu, F.; Chapman, S.; Pham, D.L.; Ko, M.L.; Zhou, B.; Ko, G.Y. Decreased miR-150 in obesity-associated type 2 diabetic mice increases intraocular inflammation and exacerbates retinal dysfunction. *BMJ Open. Diabetes Res. Care* **2020**, *8*, e001446. [CrossRef] [PubMed]
122. Yu, F.; Ko, M.L.; Ko, G.Y. Decreased MicroRNA-150 Exacerbates Neuronal Apoptosis in the Diabetic Retina. *Biomedicines* **2021**, *9*, 1135. [CrossRef] [PubMed]
123. Chang, J.Y.; Yu, F.; Shi, L.; Ko, M.L.; Ko, G.Y. Melatonin Affects Mitochondrial Fission/Fusion Dynamics in the Diabetic Retina. *J. Diabetes Res.* **2019**, *2019*, 8463125. [CrossRef] [PubMed]
124. Fan, J.S.; Palade, P. Perforated patch recording with beta-escin. *Pflugers Arch.* **1998**, *436*, 1021–1023. [CrossRef] [PubMed]
125. Janes, K.A. An analysis of critical factors for quantitative immunoblotting. *Sci. Signal.* **2015**, *8*, rs2. [CrossRef] [PubMed]
126. Huang, C.C.; Ko, M.L.; Ko, G.Y. A new functional role for mechanistic/mammalian target of rapamycin complex 1 (mTORC1) in the circadian regulation of L-type voltage-gated calcium channels in avian cone photoreceptors. *PLoS ONE* **2013**, *8*, e73315. [CrossRef]
127. Wong, R.; Schlichter, L.C. PKA reduces the rat and human KCa3.1 current, CaM binding, and Ca^{2+} signaling, which requires Ser332/334 in the CaM-binding C terminus. *J. Neurosci.* **2014**, *34*, 13371–13383. [CrossRef]
128. Robertson, B.E.; Schubert, R.; Hescheler, J.; Nelson, M.T. cGMP-dependent protein kinase activates Ca-activated K channels in cerebral artery smooth muscle cells. *Am. J. Physiol.* **1993**, *265*, C299–C303. [CrossRef]
129. Lee, B.S.; Devor, D.C.; Hamilton, K.L. Modulation of Retrograde Trafficking of KCa3.1 in a Polarized Epithelium. *Front. Physiol.* **2017**, *8*, 489. [CrossRef]
130. Estadella, I.; Pedros-Gamez, O.; Colomer-Molera, M.; Bosch, M.; Sorkin, A.; Felipe, A. Endocytosis: A Turnover Mechanism Controlling Ion Channel Function. *Cells* **2020**, *9*, 1833. [CrossRef]
131. Turnham, R.E.; Scott, J.D. Protein kinase A catalytic subunit isoform PRKACA., History, function and physiology. *Gene* **2016**, *577*, 101–108. [CrossRef] [PubMed]
132. Gerlach, A.C.; Gangopadhyay, N.N.; Devor, D.C. Kinase-dependent regulation of the intermediate conductance, calcium-dependent potassium channel, hIK1. *J. Biol. Chem.* **2000**, *275*, 585–598. [CrossRef] [PubMed]
133. Gerlach, A.C.; Syme, C.A.; Giltinan, L.; Adelman, J.P.; Devor, D.C. ATP-dependent activation of the intermediate conductance, Ca^{2+}-activated K^+ channel, hIK1, is conferred by a C-terminal domain. *J. Biol. Chem.* **2001**, *276*, 10963–10970. [CrossRef] [PubMed]
134. Shiojima, I.; Walsh, K. Role of Akt signaling in vascular homeostasis and angiogenesis. *Circ. Res.* **2002**, *90*, 1243–1250. [CrossRef]
135. Song, M.; Finley, S.D. ERK and Akt exhibit distinct signaling responses following stimulation by pro-angiogenic factors. *Cell Commun. Signal.* **2020**, *18*, 114. [CrossRef]
136. Somanath, P.R.; Razorenova, O.V.; Chen, J.; Byzova, T.V. Akt1 in endothelial cell and angiogenesis. *Cell Cycle* **2006**, *5*, 512–518. [CrossRef]
137. Okuda, K.S.; Keyser, M.S.; Gurevich, D.B.; Sturtzel, C.; Mason, E.A.; Paterson, S.; Chen, H.; Scott, M.; Condon, N.D.; Martin, P.; et al. Live-imaging of endothelial Erk activity reveals dynamic and sequential signalling events during regenerative angiogenesis. *eLife* **2021**, *10*, e62196. [CrossRef]
138. Srinivasan, R.; Zabuawala, T.; Huang, H.; Zhang, J.; Gulati, P.; Fernandez, S.; Karlo, J.C.; Landreth, G.E.; Leone, G.; Ostrowski, M.C. Erk1 and Erk2 regulate endothelial cell proliferation and migration during mouse embryonic angiogenesis. *PLoS ONE* **2009**, *4*, e8283. [CrossRef]
139. Costa, G.; Harrington, K.I.; Lovegrove, H.E.; Page, D.J.; Chakravartula, S.; Bentley, K.; Herbert, S.P. Asymmetric division coordinates collective cell migration in angiogenesis. *Nat. Cell Biol.* **2016**, *18*, 1292–1301. [CrossRef]
140. Morales-Ruiz, M.; Fulton, D.; Sowa, G.; Languino, L.R.; Fujio, Y.; Walsh, K.; Sessa, W.C. Vascular endothelial growth factor-stimulated actin reorganization and migration of endothelial cells is regulated via the serine/threonine kinase Akt. *Circ. Res.* **2000**, *86*, 892–896. [CrossRef]
141. Kureishi, Y.; Luo, Z.; Shiojima, I.; Bialik, A.; Fulton, D.; Lefer, D.J.; Sessa, W.C.; Walsh, K. The HMG-CoA reductase inhibitor simvastatin activates the protein kinase Akt and promotes angiogenesis in normocholesterolemic animals. *Nat. Med.* **2000**, *6*, 1004–1010. [CrossRef] [PubMed]
142. Chen, J.; Somanath, P.R.; Razorenova, O.; Chen, W.S.; Hay, N.; Bornstein, P.; Byzova, T.V. Akt1 regulates pathological angiogenesis, vascular maturation and permeability in vivo. *Nat. Med.* **2005**, *11*, 1188–1196. [CrossRef] [PubMed]
143. Ackah, E.; Yu, J.; Zoellner, S.; Iwakiri, Y.; Skurk, C.; Shibata, R.; Ouchi, N.; Easton, R.M.; Galasso, G.; Birnbaum, M.J.; et al. Akt1/protein kinase Balpha is critical for ischemic and VEGF-mediated angiogenesis. *J. Clin. Investig.* **2005**, *115*, 2119–2127. [CrossRef] [PubMed]
144. Olsson, A.K.; Dimberg, A.; Kreuger, J.; Claesson-Welsh, L. VEGF receptor signalling—In control of vascular function. *Nat. Rev. Mol. Cell Biol.* **2006**, *7*, 359–371. [CrossRef]
145. Gerber, H.P.; McMurtrey, A.; Kowalski, J.; Yan, M.; Keyt, B.A.; Dixit, V.; Ferrara, N. Vascular endothelial growth factor regulates endothelial cell survival through the phosphatidylinositol 3′-kinase/Akt signal transduction pathway. Requirement for Flk-1/KDR activation. *J. Biol. Chem.* **1998**, *273*, 30336–30343. [CrossRef]

46. Takahashi, T.; Yamaguchi, S.; Chida, K.; Shibuya, M. A single autophosphorylation site on KDR/Flk-1 is essential for VEGF-A dependent activation of PLC-gamma and DNA synthesis in vascular endothelial cells. *EMBO J.* **2001**, *20*, 2768–2778. [CrossRef
47. Mukai, M.; Uchida, K.; Okubo, T.; Takano, S.; Matsumoto, T.; Satoh, M.; Inoue, G.; Takaso, M. Regulation of Tumor Necrosi Factor-alpha by Peptide Lv in Bone Marrow Macrophages and Synovium. *Front. Med.* **2021**, *8*, 702126. [CrossRef]

cells

MDPI

Article

Disheveled-1 Interacts with Claudin-5 and Contributes to Norrin-Induced Endothelial Barrier Restoration

Mónica Díaz-Coránguez [1], Laura González-González [2], Amy Wang [2], Xuwen Liu [2] and David A. Antonetti [2,*]

1 Department of Pharmacobiology, Center for Research and Advanced Studies of the National Polytechnic Institute (CINVESTAV-IPN), Mexico City 07360, Mexico; modiazco@cinvestav.mx
2 Department of Ophthalmology and Visual Sciences, Kellogg Eye Center, University of Michigan, Ann Arbor, MI 48105, USA; gonzalla@med.umich.edu (L.G.-G.); amywang@umich.edu (A.W.); xuwen@med.umich.edu (X.L.)
* Correspondence: dantonet@med.umich.edu; Tel.: +1-734-232-8230

Abstract: Previous studies have revealed that norrin can reverse vascular endothelial-growth-factor (VEGF)-induced permeability in a β-catenin-dependent pathway. Here, we have explored the contribution of disheveled-1 (DVL1) in norrin-induced blood-retinal barrier (BRB) restoration. We provide evidence that in addition to canonical signaling, DVL1 promotes tight junction (TJ) stabilization through a novel, non-canonical signaling pathway involving direct claudin-5 (CLDN5) binding. Immunofluorescence staining of rat retinal cross-sections showed enriched expression of DVL1 and 3 at endothelial capillaries and co-localization with CLDN5 and ZO-1 at the TJ complex in primary bovine retinal endothelial cells (BRECs). Barrier properties of BRECs were determined via measurements of trans-endothelial electrical resistance (TEER) or permeability to 70 kDa RITC-dextran. These studies demonstrated that norrin restoration of barrier properties after VEGF treatment required DVL1 as an siRNA knockdown of Dvl1 but not Dvl2 or Dvl3, reduced basal barrier properties and ablated norrin-induced barrier restoration. However, loss of Dvl1 did not decrease β-catenin signaling activity as measured by *Axin2* mRNA expression, suggesting the contribution of a non-canonical pathway. DVL and TJ protein interactions were analyzed via co-immunoprecipitation of endogenous protein in BRECs, which demonstrated that DVL1 interacts with both CLDN5 and ZO-1, while DVL3 interacts only with ZO-1. These interactions were most abundant after inducing BRB restoration by treating BRECs with VEGF and norrin. DVL has previously been shown to form intramolecular bindings between the C-terminal PDZ-binding motif (PDZ-BM) with an internal PDZ domain. Co-transfection of HEK293 cells with DVL1 and CLDN5 or relevant mutants revealed that DVL1 interacts with CLDN5 through the DVL PDZ domain binding, CLDN5 PDZ-BM, in competition with DVL1 PDZ-BM, since DVL/CLDN5 interaction increases with deletion of the DVL1 PDZ-BM and decreases by co-expressing the C-terminal fragment of DVL1 containing the PDZ-BM or through deletion of CLDN5 PDZ-BM. In BREC cells, transfection of the C-terminal fragment of DVL1 down-regulates the expression of CLDN5 but does not affect the expression of other proteins of the TJs, including ZO-1, occludin, CLDN1 or VE-cadherin. Blocking DVL1/CLDN5 interaction increased basal permeability and prevented norrin induction of barrier properties after VEGF. Combined with previous data, these results demonstrate that norrin signals through both a canonical β-catenin pathway and a non-canonical signaling pathway by which DVL1 directly binds to CLDN5 to promote barrier properties.

Keywords: blood–retinal barrier; disheveled; claudin-5; tight junction; norrin; endothelium; permeability; barriergenesis; retina; Wnt signaling

Citation: Díaz-Coránguez, M.; González-González, L.; Wang, A.; Liu, X.; Antonetti, D.A. Disheveled-1 Interacts with Claudin-5 and Contributes to Norrin-Induced Endothelial Barrier Restoration. *Cells* **2023**, *12*, 2402. https://doi.org/10.3390/cells12192402

Academic Editor: Hossein Ameri

Received: 5 September 2023
Revised: 27 September 2023
Accepted: 29 September 2023
Published: 4 October 2023

1. Introduction

Formation of the blood–retinal barrier (BRB), or barriergenesis, requires norrin, as evidenced by gene deletion studies of norrin [1–4] or any of the components of the norrin receptor complex including Frizzled 4 (FZD4) [5–7], low-density lipoprotein receptor-related protein 5 or 6 (LRP5/6) [8,9] and tetraspanin 12 (TSPAN12) [10,11]. Müller cells [12–14] and endothelial cells [15] of the developing retina express norrin that contributes to proper angiogenesis and the formation of BRB [1]. Norrin binds to the N-terminal, the extracellular, cysteine-rich domain of the FZD4 receptor and the β-propeller domains of the LRP 5/6 co-receptor [16] activating the β-catenin, canonical Wnt signaling pathway. In addition, TSPAN12 stabilizes FZD4 receptor at the cell membrane and enhances norrin-induced, but not Wnt-induced, β-catenin signaling [10].

The canonical pathway involves β-catenin stabilization and transcriptional gene regulation. In the absence of norrin or Wnt signaling, the β-catenin destruction complex formed by adenomatosis polyposis coli (APC) protein, axin, protein phosphatase 2a (PP2A), casein kinase 1α (CK1α) and glycogen synthase kinase 3 (GSK3), phosphorylates and targets β-catenin for ubiquitination and proteasomal degradation. This norrin-binding FZD4 receptor complex inactivates the APC degradation complex and inhibits GSK3 kinase, stabilizing β-catenin. Upon FZD4 receptor activation, canonical signaling is initiated through disheveled (DVL).

In mammals, there are three DVL proteins (DVL1, 2 and 3), which are between 78 and 90 kDa. DVL proteins contain a DIX domain (Dvl and axin), a PDZ domain (post-synaptic density protein-95, disc large tumor suppressor, zonula occludens-1) and a DEP domain (DVL, EGL-10, and Pleckstrin) and a C-terminal PDZ-binding motif (PDZ-BM) [17]. The activation of the canonical pathway releases DVL from FZD4, leading to DEP domain swapping between DVL molecules, initiating a head-to-tail polymerization of DVL through DIX domains [17,18]. This polymerized DVL inhibits the APC degradation complex and prevents β-catenin phosphorylation by GSK3, promoting protein stabilization. Recent studies suggest that upon ligand binding, the FZD4 receptor complex, phosphatidyl inositol [4,5] biphosphate (PIP2), increases at the plasma membrane, promoting DVL recruitment. This enhances DVL polymerization and DEP domain interaction with FZD4, followed by the inhibition of both the β-catenin destruction complex and GSK3, promoting β-catenin stabilization [19].

β-catenin stabilization and subsequent translocation to the nucleus allows interaction with the T-cell factor/lymphoid-enhancer-binding factor (TCF/LEF) complex to promote transcription (reviewed in [20]). This β-catenin-driven gene transcription is indeed required for proper vascular formation as gene deletion of β-catenin leads to fragile vessels, including malformed lumen and hemorrhagic vessels [21], and inhibition of β-catenin also leads to the loss of blood–brain barrier (BBB) properties in the remaining vessels, with CLDN3 downregulation and increased plasmalemma-vesicle-associated protein (PLVAP) expression [22]. Importantly, the genetic introduction of active β-catenin in mice with norrin loss can primarily restore barrier properties to the BBB and BRB [1]. Collectively, these studies demonstrate a requirement for β-catenin signaling in BBB and BRB formation and reveal norrin as specifically regulating BRB formation.

Dvl gene deletion studies, both individually and in combination, clearly demonstrate a requirement of DVL signaling in vascular development with some redundancy [23]. Mice with *Dvl1* gene deletion have structurally normal brains but display social interaction abnormalities in behaviors including whisker trimming, sleeping patterns, nest building, sensorimotor gating, huddling and recognition of social hierarchy and dominance [24,25]. However, deletion of either *Dvl2* or *Dvl3* leads to multiple phenotypes, including cardiac outflow tract abnormalities, with about half of $Dvl2^{-/-}$ mice dying in the perinatal period and about 87% of $Dvl3^{-/-}$ mice dying shortly after birth. Combined *Dvl1* and *Dvl3* deletion led to lethality by E15.5, while *Dvl2* heterozygotes plus *Dvl3* deletion was lethal by E9.5. Conversely, transgene expression of *Dvl1* or *2* can rescue *Dvl3* gene deletion phenotypes and the same effect of transgene rescue was observed for *Dvl1* or *2* gene deletion, revealing redundancy in DVL signaling [26].

DVL interacts with well over 20 proteins directly or in a complex [23]. These interactions both promote canonical signaling to β-catenin and non-canonical signaling, including planar cell polarity (PCP). Recent evidence indicates that the DVL PDZ-BM at the C-terminus can loop back and bind to the PDZ domain of DVL and contributes to the regulation of canonical versus non-canonical signaling. Solution nuclear magnetic resonance (NMR) spectroscopy and competition-binding studies have revealed that the C-terminal PDZ-BM binds with high affinity to the PDZ domain of DVL, forming a closed-loop structure that effectively activates β-catenin better than a C-terminal mutant [27]. Furthermore, in *Xenopus* embryos, the expression of Dsh (the *Xenopus* equivalent of DVL) with the deletion of the C-terminal 8 amino acids or deletion of the PDZ domain both led to a higher degree of two established PCP outcomes: convergent extension phenotype and greater JNK activation when compared to wild-type Dsh. These studies provide compelling evidence that the open structure of DVL/Dsh may promote PCP, while the closed-loop structure promotes canonical signaling.

Recently, we demonstrated that in addition to barriergenesis, norrin can reverse VEGF-induced endothelial permeability in culture and in the retinal vasculature in vivo and reverse diabetes-induced permeability [28]. VEGF was shown to induce endothelial permeability but also to promote TSPAN12 migration to the plasma membrane promoting norrin signaling and barrier restoration if norrin was provided. A role for canonical signaling through β-catenin was clearly established in this model. However, inhibition of GSK3 was not sufficient to promote barrier formation after VEGF-induced permeability even though β-catenin stabilization and signaling were observed, suggesting additional signaling pathways may contribute to norrin-induced barrier induction. Here, we provide evidence that DVL controls BRB function through canonical and non-canonical signaling pathways. Using knockdown strategies, we demonstrate that norrin signaling specifically requires DVL1 to stimulate barrier properties after VEGF-induced permeability. Further, we provide evidence for a novel, non-canonical signaling role of DVL1 in regulating barrier properties through direct binding of DVL1 PDZ domain to CLDN5 PDZ-BM. Collectively, the studies suggest that an open confirmation of DVL allows CLDN5 binding and, together with β-catenin signaling, promotes barrier formation in a novel, non-canonical pathway.

2. Materials and Methods

2.1. Primary Bovine Retinal Endothelial Cell Culture (BREC)

For in vitro studies, primary BRECs were isolated as previously described [29] Passages 2–8 were grown until confluence at 37 °C in dishes coated with 1 μg/cm^2 of fibronectin bovine plasma (Sigma-Aldrich, ST. Louis, MI, USA). Then, MCDB-131 media with 1% FBS and 100 nM of hydrocortisone (HC) (Sigma-Aldrich) was added to the monolayers to improve trans-endothelial electrical resistance (TEER) to over 3500 ohms. Then, cells were stimulated with the recombinant human proteins, VEGF$_{165}$ (R&D Systems) diluted in 0.1% BSA/PES and/or norrin (R&D Systems, Minneapolis, MN, USA) diluted in 4 mM HCl. All control monolayers were given the volume equivalent to drug or protein diluents (vehicle) used in each experiment.

Barrier properties in BRECs were determined via measurements of TEER using the electrical-cell-substrate-impedance-sensing (ECIS) Z-theta system in 8-well chamber slides (Applied Bio Physics Inc., Troy, NY, USA) [30] or through the quantification of the flux of 10 μM of 70 kDa rhodamine B isothiocyanate-(RITC) dextran (Sigma-Aldrich; R9379) across BREC monolayers over 3.5 h. Diffusive permeability (Po) (cm/sec) was calculated as previously described [31] using the equation

$$P_o = [(F_L/\Delta t)V_L]/(F_A A)$$

where P_o is in cm/s, F_L is basolateral fluorescence, F_A is apical fluorescence, Δt is the change in time, A is the surface area of the filter and V_L is the volume of the basolateral chamber.

2.2. BREC Transfection with siRNAs or DVL Mutants

Primary BRECs were transfected using Amaxa Nucleofector System (Amaxa, Koeln, Germany). Next, 0.5 × 10^6 cells were electroporated with 200 ng/mL of scramble (Scr) or siDvl sequences. Complete media were used to resuspend the cells and seed them at 0.125 × 10^6 cells/cm^2 on ECIS chambers, transwells, coverslips or plates coated with fibronectin. Media were changed after 24 h to MCDB-131 with 1% FBS and 100 nM of HC for two additional days before adding VEGF and/or norrin. The siRNA sequences used were GCAGAGUGAAGGAAGCAAAUU (siDvl1a), ACGUCAACUUCGAGAACAUUU (siDvl1b), CGACAUGAACUUUGAGAAUUU (siDvl2) (Dharmacon, Lafayette, CO, USA) and siDvl3 (Santa Cruz, Biotechnology, Santa Cruz, CA, USA). BRECs co-transfected with plasmids encoding DVL mutants and CLDN5 or ZO-1 were resuspended in complete media and seeded at confluency (0.25 × 10^6 cells/cm^2). The attached cells were washed after 3 h, and the media were replaced. Then, after 3 more hours, the media were changed to MCDB-131 with 1% FBS and 100 mM of HC to develop barrier properties. The next day more HC was added to the wells, and 24 h after transfection, cells were collected, fixed, or stimulated with $VEGF_{165}$ for 30 min, then norrin for 30 min. Human DVL1-WT or human DVL3-WT and its mutants were cloned in a pcDNA3.1(+)-N-HA vector, and FLAG-CLDN5 was cloned in pcDNA3.1(+)-N-DYK vector, all of them by Genscript. GFP-ZO-1 was a gift from Dr. Anuska Andejelkovic (University of Michigan). pcDNA3.1(+) empty vector (EV) was used as control. All constructs were amplified in DH5α competent cells (Thermo Fisher, Scientific, Asheville, NC, USA), purified using Plasmid Plus Maxi kit (QIAGEN, Hilden, Germany) and confirmed via sequencing.

2.3. HEK-293 Transfection

HEK-293 cells were grown in DMEM high glucose supplemented with 10% FBS and antibiotics. Cells were plated at 80% confluence on 6-well plates for transfection with DVL mutants. Transfection was performed with Lipofectamine 3000 (Thermo Fisher Scientific) following the standard protocol in the reduced serum medium Opti-MEM (Thermo Fisher Scientific). Next, 2 µg of plasmid was used for each DVL mutant, 4 µg of FLAG-CLDN5 or GFP-ZO-1 plasmids and 8 µg of FLAG-CLDN5-ΔPDZBM. Then, 24 or 48 h after transfection, cells were collected for co-immunoprecipitation.

2.4. Co-Immunoprecipitation

BREC or HEK-293 monolayers were collected in co-immunoprecipitation (co-IP) buffer containing 2% of non-ionic detergent NP-40. Lysates were incubated at 4 degrees for 20 min, collected via scraping then centrifuged at 13,000× *g* for 13 min. Supernatants were collected for protein quantification using BioRad protein assay. Then, 400 µg of protein was separated for co-IP using HA, FLAG or GFP tag antibodies or matching control at 1 µg of antibody per sample. Primary antibodies were incubated overnight at 4 degrees under rocking, then 60 µL of protein G agarose beads were added to each lysate for antibody capture. After 2 h, samples were centrifugated, the supernatants were discarded and the beads were washed with co-IP buffer four times. Protein was eluted from beads using 60 µL of low pH (2.8) glycine buffer for 5 min at room temperature, then NuPage sample buffer (LDS/DTT mix) was added and incubated at 70 degrees for 10 min. Supernatants were collected to be processed for Western blot analysis.

2.5. Western Blot (WB)

Cells were harvested in a Triton-X-100-deoxycolate-SDS-based, or 2% of non-ionic detergent NP-40 lysis buffer, and Western blotting was carried out in NuPAGE™ (Invitrogen™, Thermo Fisher Scientific, Asheville, NC, USA) system as described previously [32] (81). The primary antibodies used were monoclonal mouse α-Pan-DVL (B-4) (Santa Cruz, Biotechnology, Santa Cruz, CA, USA), monoclonal rabbit α-DVL2 (30D2) (Cell Signaling Technology, Danvers, MA, USA), polyclonal rabbit α-DVL3 (Sigma Aldrich, St. Louis, MO, USA), monoclonal rat α-HA high affinity (3F10) (Roahaha Roche, Indianapolis, IN, USA), mouse monoclonal α-FLAG® M2 (Sigma Aldrich), mouse monoclonal α-GFP [9F9.F9] (Abcam, Boston, MA, USA), polyclonal rabbit α-CLDN5 (Invitrogen™, Thermo Fisher Scientific), monoclonal rat α-ZO-1 clone R40.76 (Millipore, Sigma-Aldrich, Burlington, MA, USA) or polyclonal rabbit α-ZO-1 (Thermo Fisher Scientific). Horseradish peroxidase-conjugated secondary antibodies, α-mouse, α-rabbit, or α-rat (GE Healthcare, Logan UT, USA), were used to detect primary antibodies via chemiluminescence. Results were analyzed using AlphaView software FluorChem™ systems.

2.6. Immunofluorescence Staining

Transfected BRECs were plated on Thermanox™ round plastic coverslips of 13 mm diameter (Thermo Fisher Scientific). Immunofluorescence staining was performed in confluent monolayers fixed with 1% paraformaldehyde (Electron Microscopy Sciences, Hatfield, PA, USA), permeabilized 10 min with 0.2% Triton X-100, and blocked 1 h with 2% normal goat serum (Life Technologies, Thermo Fisher Scientific; 50062Z, Carlsbad, CA, USA) and 0.1% Triton X-100. The primary antibodies used were monoclonal mouse α-Pan-DVL (B-4) (Santa Cruz, Biotechnology), monoclonal rabbit α-DVL2 (30D2) (Cell Signaling Technology), polyclonal rabbit α-DVL3 (Sigma Aldrich), monoclonal rat α-HA high affinity (3F10) (Roahaha Roche); polyclonal rabbit α-CLDN5 (Invitrogen™, Thermo Fisher Scientific), monoclonal rat α-ZO-1 clone R40.76 (Millipore, Sigma-Aldrich), polyclonal rabbit α-ZO-1 (Thermo Fisher Scientific) or isolectin GS-IB4 Alexa Fluor 647 (Themo Fisher Scientific) for 2 days at 4 °C. These antibodies were detected with the secondary fluorescent antibodies, goat α-mouse, α-rabbit, or α-rat Alexa Fluor 488; goat α-rabbit or α-rat Alexa Fluor 594; goat α-rabbit or α-rat Alexa Fluor 647 (Life Technologies, Thermo Fisher Scientific); and Hoechst dye (Invitrogen™, Thermo Fisher Scientific) for nuclear staining overnight at 4 °C. Samples were imaged using confocal microscopy. Co-localization between DVL, CLDN5 and ZO-1 was calculated in *xz* planes from single stacks using Mander's correlation coefficient in ImageJ software. CLDN5 or ZO-1 cell border staining was analyzed via a semi-quantitative ranking score system on a scale of 5 categories indicating the percentage of loss in cell border staining (0–100%). Three independent observers were asked to assign a ranking score to four images per condition in a masked fashion. The results of three independent experiments were summed, and the frequency of each ranking score was calculated to determine differences between conditions.

Whole mount retinas or cryostat sections were obtained from adult Long Evans rats. Eyes were enucleated, fixed in 4% paraformaldehyde for 30 min, and washed with PBS three times. Retinas for whole mounts were dissected under a stereotaxic microscope for their staining following the same protocol as BRECs, with Pan-DVL and CLDN5 or ZO-1 antibodies. After the staining, retinas were flat mounted into microscope slices for their analysis via confocal microscopy. For cryostat sections, eyes were dehydrated in 30% sucrose after fixation and embedded in Tissue-Tek® OCT for their sectioning. Then, 10 µm slices were used for staining with IB4 and Pan-DVL, DVL2 or DVL3 antibodies. All animal experiments were conducted under the Association for Research in Vision and Ophthalmology Statement for the Use of Animals in Ophthalmic and Vision Research and the guidelines of the University Committee on Use and Care of Animals at the University of Michigan.

2.7. qRT-PCR

RNA was extracted using RNeasy Plus mini kit (Qiagen Inc., Hilden, Germany; 74134). cDNA was obtained from 400–1000 ng of RNA per sample processed with Omniscript Reverse Transcription kit (Qiagen Inc.). Then, specific TaqMan™ gene expression assays (Thermo Fisher Scientific) were used to detect *Dvl1*, *Dvl2* or *Axin2*. *Dvl3* was detected via non-quantitative PCR using TCTCCCAGTCCTCTCCCAAG (Fwd) and GATGGGGAC ATATGCCGGAG (Rev) primers and SybrSafe (Invitrogen) for quantification. Results were normalized to β-actin mRNA and expressed as a relative change using ΔΔCt method.

2.8. Statistical Analysis

GraphPad Prism software was used to analyze data from at least three independent experiments. Graphs represent the mean ± S.D. p values were calculated via t-test or one-way or two-way ANOVA, as indicated in each figure, and $p \leq 0.05$ was considered significant.

3. Results

3.1. DVL1 and 3 Are Highly Expressed in Retinal Capillaries, Co-Localizing with the Tight Junction Proteins, CLDN5 and ZO-1

Norrin restores barrier properties after VEGF-induced permeability in vivo and a primary culture of bovine retinal endothelial cells (BRECs). However, β-catenin activation by GSK3 inhibition failed to increase BREC barrier properties after VEGF despite β-catenin activation [28]. Therefore, we explored the hypothesis that DVL may signal through a non-canonical pathway to promote barrier properties in addition to β-catenin signaling. Immunostaining of cross-sections of rat retinas with Pan-DVL, DVL2 or DVL3 antibodies, along with isolectin GS-IB4 to identify retinal vessels, revealed DVL1 and 3 in the retinal capillaries (Figure 1a). Both Pan-DVL and DVL3 antibodies detected enriched expression in retinal vessels. This co-localization could be observed partially in larger arterioles but most clearly in the capillaries of the deep plexus. In contrast, DVL2 was broadly expressed across retinal tissue but with minimal vessel co-localization. Whole-mount images of retinas stained for Pan-DVL revealed localization at the cell border and co-staining with claudin-5 (CLDN5) and ZO-1 (Figure 1b).

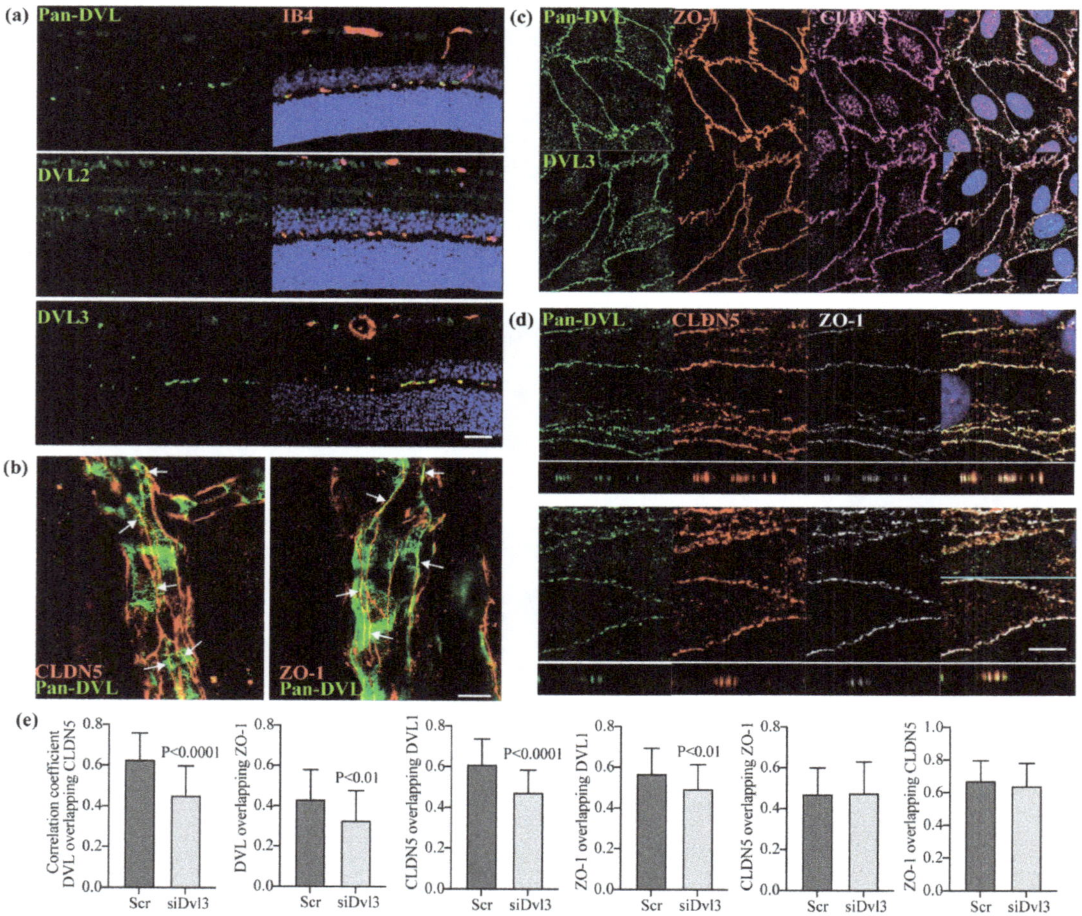

Figure 1. DVL1 and 3 are highly expressed in retinal capillaries, co-localizing with the TJ proteins CLDN5 and ZO-1. (**a**) Cross-sections of rat retinas immunostained with specific antibodies against DVL2 and DVL3 or with Pan-DVL antibody detecting both, DVL1 and DVL3 (green). Isolectin B4 (IB4)

was used as a marker of blood-vessels (red); scale bar = 50 µM. (**b**) Co-localization of DVL (green) with the TJ proteins CLDN5 or ZO-1 (red) in rat retina whole mounts. Arrows point to the sites of co-localization in the TJs. (**c**) Co-immunofluorescence staining of DVL proteins (green), ZO-1 (red) and CLDN5 (magenta) in BREC monolayers; scale bar = 10 µM. (**d**) BRECs transfected with siDvl3 or scramble (Scr) before the IF staining with Pan-DVL (green), CLDN5 (red) and ZO-1 (white) antibodies; scale bar = 10 µM. (**e**) Protein co-localization, analyzed in xz stacks using Mander's correlation coefficient, shows a high overlap of DVL1 with the TJ proteins in siDvl3 monolayers. *p* values were calculated via *t*-test analysis. Error bars, S.D.

We have confirmed the specificity of these antibodies by targeting DVL expression with siRNA in BREC cultures. Cells were grown from isolated bovine retinal capillaries as previously described [29,32]. Specific siRNA sequences were used to downregulate each DVL protein, and BREC monolayers were harvested to analyze protein expression. A representative Western blot is shown in supplementary Figure S1. Using siRNA to *Dvl2* and *Dvl3* demonstrates specific knockdown for DVL2 and DVL3 via Western blot, respectively. Pan-DVL antibody revealed a reduction of protein in the presence of siDvl1 or siDvl3, suggesting the antibody detected both forms of DVL. Unfortunately, we could not find a specific antibody for the DVL1 protein.

We further determined the localization of DVL1 and 3 in primary retinal endothelial cells. BREC monolayers were processed for the immunofluorescence staining of DVL, CLDN5 and ZO-1 proteins. Figure 1c shows enriched DVL staining at the cell contacts, co-localizing with the TJ proteins, CLDN5 and ZO-1. Because we could not find a specific antibody for DVL1 to corroborate that this protein is also expressed in retinal endothelial cells at the TJ, we knocked down *Dvl3* gene expression in BRECs using a specific siRNA. Pan-DVL antibody staining was analyzed through confocal microscopy in single z-stacks and *xz* planes. Figure 1d shows that the Pan-DVL antibody detects DVL protein enriched at the endothelial cell contacts in siDvl3 samples despite 70% *Dvl3* gene knockdown (Figure 2a). As expected, the staining was decreased in siDvl3 monolayers compared with scramble (Scr) controls due to the reduction in DVL3 content. The co-localization of this staining with CLDN5 and ZO-1 was analyzed in *xz* planes and quantified the Mander´s correlation coefficient of the pixel intensity between the two signals. Graphs in Figure 1e show over 40% correlation between Pan-DVL and TJ proteins, which was slightly decreased in the presence of siDvl3. Because siDvl3 reduced 70% of *Dvl3* gene expression (Figure 2a) and more than 80% of DVL3 protein (Figure S1), compared with Scr monolayers, we can conclude that most of the Pan-DVL staining in the siDvl3 samples indicated DVL1. Together, the data suggest that both DVL1 and 3 proteins are expressed in retinal vascular endothelial cells and co-localize at the TJ complexes.

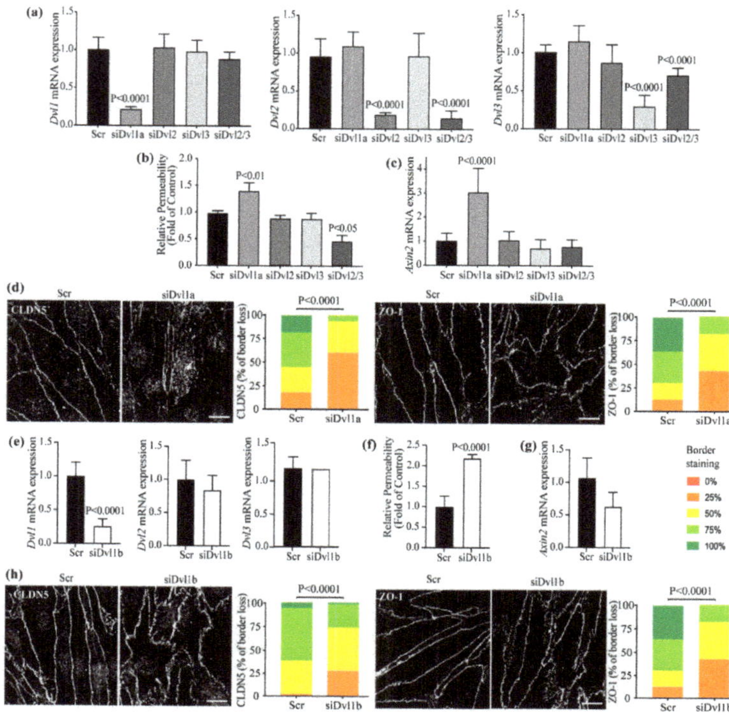

Figure 2. The knockdown of Dvl1 specifically, reduced basal barrier properties despite increased β-catenin signaling. (**a**) The specificity of siRNA sequences was analyzed via the qRT-PCR of Dvl1 or Dvl2 or Dvl3 PCR in BRECs transfected with siDvl1, siDvl2, siDvl3, the combined siDvl2/3 or scramble (Scr) control. Barrier properties were analyzed through the permeability to 70 kDa RITC-dextran molecule (**b**) and *Axin2* mRNA expression was measured, being accounted as β-catenin signaling activation (**c**). (**d**) Immunofluorescence staining of TJ proteins CLDN5 and ZO-1 in BREC monolayers after Dvl1 knockdown; scale bar = 10 μM. (**e–h**) Knockdown of Dvl1 with a second siRNA sequence. (**e**) qRT-PCR of Dvl1 or Dvl2, or Dvl3 PCR. (**f**) Solute flux. (**g**) qRT-PCR of *Axin2* (**h**) Immunofluorescence staining of the TJ proteins; scale bar = 10 μM. *p* values were calculated via one-way ANOVA followed by Sidak post hoc test (**a–c**), or via *t*-test analysis (**d–h**). Error bars, S.D.

3.2. *Dvl1 Knockdown Reduced Basal Barrier Properties despite Increased β-Catenin Signaling*

To evaluate the requirement of DVL proteins in the regulation of barrier properties, *Dvl* expression was reduced using targeted siRNA and the effect on endothelial permeability was determined. BRECs were transfected with siRNA sequences to target *Dvl 1*, *2* or *3* mRNA separately. The specificity and effectiveness in knockdown of these sequences was corroborated via qRT-PCR (Figure 2a), with each siRNA inducing a specific knockdown of at least 70%. To observe the effects of *Dvl* knockdown in basal barrier properties, we performed solute flux assays on BRECs transfected with Scr control, siDvl1a, siDvl2, siDvl3 separately or siDvl2 and 3 together (siDvl2/3). Figure 2b demonstrates a 50% increase in permeability only in BRECs transfected with siDvl1a, compared with Scr monolayers. Conversely, the basal permeability was significantly reduced when siDvl2 and 3 were transfected together. This result demonstrates that DVL1 specifically promotes the barrier of retinal endothelial cells.

Because DVL proteins can activate both canonical and non-canonical signaling, we measured the mRNA expression of *Axin2*, a downstream target of β-catenin, in the presence of *Dvl* siRNAs. Surprisingly, increased *Axin2* mRNA content was found with siDvl1a (Figure 2c). That β-catenin signaling did not decrease with siDvl1 suggests that additional signaling pathways may contribute to the control of endothelial barrier properties.

Given the co-localization of DVL with the TJs, we hypothesized that *Dvl1* knockdown might alter CLDN5 and ZO-1 localization at the endothelial cell contacts. BRECs transfected with Scr or siDvl1a were processed for the immunofluorescence staining of CLDN5 and ZO-1. Using masked scoring of confocal images, we determined the percent of border localization. As shown in Figure 2d, the knockdown of *Dvl1* led to the loss of border staining for both CLDN5 and ZO-1, supporting a role for DVL1 in TJ organization.

These results were further corroborated via a second siRNA, targeting *Dvl1* (siDvl1b), which reached a similar degree of *Dvl1* knockdown compared to the first sequence (Figures 2e and S2b). Consistent with our previous result, this sequence also increased basal permeability (Figure 2f) and disrupted CLDN5 and ZO-1 border localization (Figure 2h) but did not significantly alter *Axin2* mRNA expression (Figure 2g). Staining with Pan-DVL after siDvl1 knockdown revealed that the remaining DVL3 co-localized with ZO-1 both at the membrane and in the cytoplasm (S2C). Together, these results reveal that DVL1 regulates barrier properties through the regulation of TJ complex organization and suggests a non-canonical pathway that compliments β-catenin signaling.

3.3. Knockdown of Dvl1 but Not Dvl3, Ablated Norrin-Induced Blood–Retinal Barrier Restoration after VEGF

To determine the requirement of DVL signaling in norrin-induced BRB properties after VEGF, BRECs were transfected with Scr or siDvl1 and then grown to confluence on gold-plated chambers to measure TEER using the electrical cell-substrate impedance sensing (ECIS) Z-theta system at 4000 Hz once every hour (Figure 3a). Cells were switched to media with 1% FBS and 100 nM hydrocortisone and allowed to reach a stable TEER of over 3500 ohms. In Scr monolayers, VEGF addition (50 ng/mL) induced a loss of barrier properties, as observed by a reduction in the relative TEER measure to approximately 40% of control by 24 h. The stimulation of the monolayers with norrin (40 ng/mL) alone did not affect TEER; however, co-stimulation with norrin and VEGF led first to a loss of TEER and then a complete restoration of barrier properties after 72 h as previously reported [28]. BRECs transfected with siDvl1a, however, did not reach 3500 ohms; their basal TEER stabilized at 30% lower than that of Scr monolayers. VEGF addition reduced TEER by another ~30%, and importantly, norrin failed to reverse VEGF-induced barrier disruption, revealing a required role for DVL1 in norrin-induced BRB properties.

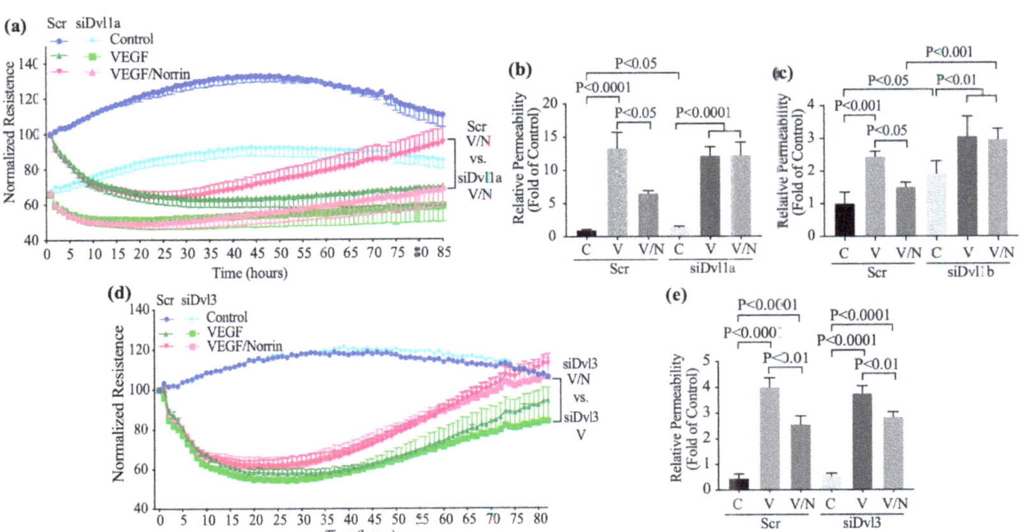

Figure 3. The knockdown of Dvl1 but not Dvl3-ablated norrin-induced blood–retinal barrier restoration after VEGF. (**a**,**d**) TEER or (**b**,**c**,**e**) permeability to a 70 kDa RITC-dextran molecule was measured

on BRECs transfected with siDvl1a (**a**,**b**), a second siDvl1b (**c**), siDvl3 (**d**,**e**) or a scramble (Scr) control sequence. Monolayers were stimulated with vehicle (control), VEGF 50 ng/mL or VEGF and norrin 40 ng/mL (VEGF/Norrin; V/N) 72 h after transfection. Solute flux was measured after 72 h of stimulation. *p* values were calculated via two-way ANOVA, followed by Tukey´s post hoc test. Error bars, S.D.

This effect was recapitulated by testing solute flux (Figure 3b,c). BRECs were transfected with siDvl1a or scramble and were grown on 12 mm diameter, Transwell® cell culture inserts, and diffusive permeability to 70 kDa dextran was determined (Po cm/sec) as previously described [31]. Scr monolayers (control, C) showed a basal permeability of 1.3×10^{-7} cm/s, increasing to 1.7×10^{-6} when the monolayers were stimulated with VEGF (Figure 3b). Co-stimulation with norrin and VEGF revealed a significant permeability reduction when compared with VEGF-stimulated monolayers. Consistent with measurements of basal TEER, siDvl1a monolayers had higher dextran permeability than Scr monolayers. VEGF increased the permeability of siDvl1a monolayers; however, norrin failed to reverse this effect after *Dvl1* knockdown. Similar results were found with a second small interfering sequence, siDvl1b (Figure 3c); in these experiments, the basal permeability started in 1.14×10^{-6} cm/s.

To determine the role of DVL3 in norrin-induced BRB properties, TEER and solute flux were also measured in BRECs transfected with Scr or siDvl3. As shown in Figure 3d,e, a similar VEGF and norrin response was observed with Scr control monolayers. Furthermore, in the presence of siDvl3, norrin is still able to recover TEER (Figure 3d) and to reduce VEGF-induced dextran permeability (Figure 3e) on solute flux assays, suggesting that DVL3 is not required for norrin-induced BRB properties. Together, these results suggest the requirement of DVL1 but not DVL3 in norrin-induced BRB properties after VEGF.

3.4. DVL1 Interacts with CLDN5 and ZO-1, While DVL3 Interacts with ZO-1

DVL proteins comprise DIX, PDZ and DEP domains and the relatively short PDZ-BM at the C-terminus. CLDN5 also possesses a PDZ-BM, while the ZO-1 structure contains three PDZ domains. PDZ domains are protein interaction modules that interact with other PDZ domains or PDZ-BM. Since our results demonstrate that DVL1 and 3 proteins co-localize with CLDN5 and ZO-1 in endothelial monolayers and that DVL1 is required for TJ organization and barrier properties, we tested whether DVL proteins interact directly with the TJ proteins to promote barrier function. To test this possibility, BREC monolayers were stimulated with norrin, VEGF or both for 72 h. At this time point, norrin had already begun to restore barrier properties (Figure 3). The DVL proteins were immunoprecipitated (IP), and the amount of CLDN5 and ZO-1 protein that co-IP'd with DVL was determined. As shown in Figure 4a,c, both Pan-DVL and DVL3 antibodies detected at least two bands in the total cell lysates. Based on previous literature, these are likely DVL phosphorylation sites [33] confirmed by treating with alkaline phosphatase, which collapsed all bands to one lower molecular weight (Figure S3). Norrin stimulated increased phosphorylation, as indicated via an increase in the upper molecular weight band. However, because mass spectrometry analyses have identified at least 50 serine and threonine sites of phosphorylation in DVL [34], we considered that the analysis of specific DVL phosphosites activated by norrin was beyond the scope of this manuscript. Importantly, pull-down samples denoted similar changes in DVL molecular weight, suggesting that DVL capture with these antibodies includes the phosphorylated protein.

Our results demonstrate that both CLDN5 and ZO-1 co-IP with DVL when the protein is pulled down with the Pan-DVL antibody (Figure 4a). CLDN5 had a basal interaction in non-stimulated monolayers (control, C), which was increased with the addition of VEGF (V), and even higher in the presence of both VEGF and norrin (V/N) (Figure 4b). ZO-1 had a low basal interaction with DVL; however, this was promoted with the addition of VEGF and again, the most robust interaction was induced with VEGF and norrin together. Similar changes were found by analyzing the interaction between ZO-1 and DVL3 (Figure 4c,d). However, in contrast to co-IP samples with Pan-DVL antibody, CLDN5 was not found in the co-IP of DVL3, suggesting that CLDN5 is bound to DVL1 in Pan-DVL IP samples. Together, these results indicate that ZO-1 interacts with both DVL1 and DVL3, while CLDN5 only interacts with DVL1, and these interactions are promoted by VEGF and norrin co-stimulation during the induction of BRB restoration.

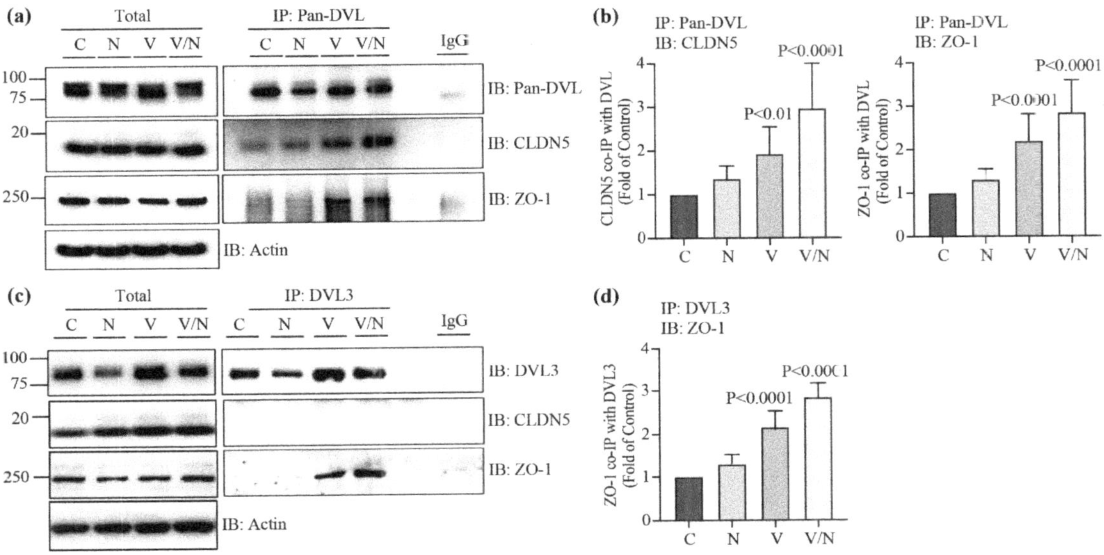

Figure 4. DVL1 interacts with CLDN5 and ZO-1, while DVL3 interacts with ZO-1, especially with the addition of VEGF and norrin. BREC monolayers were stimulated with vehicle (control, C), norrin 40 ng/mL (N), VEGF 50 ng/mL (V) or both (V/N). After 72 h of stimulation, lysates were collected for total protein or the immunoprecipitation (IP) of DVL using Pan-DVL (**a**,**b**) or DVL3 (**c**,**d**) specific antibodies, followed by the immunoblot (IB) with Pan-DVL, DVL3, CLDN5 or ZO-1 antibodies. Actin was used as a loading control of total protein. (**a**,**c**) Representative Western blots of total and IP protein. (**b**,**d**) Densitometry of CLDN5 or ZO-1 proteins relative to IP protein. *p* values were calculated by one-way ANOVA, followed by Sidak post hoc test. Error bars, S.D.

Since DVL1 co-IP'd with CLDN5 and because the siRNA studies revealed DVL1 as specifically required for TJ organization at the border and norrin-induced barrier restoration, we further analyzed the structural components in DVL1 that regulate its binding to CLDN5. CLDN5 interacts with ZO-1 through a C-terminal PDZ-BM that binds to the first PDZ domain of ZO-1 [35]. Furthermore, DVL possesses both PDZ and PDZ-BM that demonstrate intramolecular binding creating a loop, as demonstrated in *Xenopus* [27,36] and by fluorescence resonance energy transfer experiments in DVL3 [37]. Thus, we tested the hypothesis that the CLDN5 PDZ-BM competes with the DVL PDZ-BM for the DVL PDZ domain. We designed a DVL1 mutant composed only of the last 169 amino acids of the C-terminus of the protein, which contains the nuclear export signals and the PDZ-BM (HA-DVL1-CT). If CLDN5 binds to the DVL1 PDZ domain, the DVL1-CT fragment would be expected to compete for this interaction. Furthermore, we designed a mutant of DVL1 lacking the PDZ-BM lacking the 7 C-terminal amino acids (HA-DVL1-ΔC7), which we hypothesized would demonstrate increased binding to CLDN5 as the PDZ domain becomes exposed.

To evaluate the possibility that CLDN5 interacts with the DVL1 PDZ domain, HEK-293 cells were co-transfected with FLAG-tagged human full-length CLDN5 (FLAG-CLDN5) and human full-length DVL1 (HA-DVL1-WT) or its mutants. Then, we performed capture experiments of the transfected mutants using either HA or FLAG tags to determine the interaction between these proteins. Transfection of FLAG-CLDN5 alone (first line in all blots) or incubation of lysates with IgG (last line in capture blots) were used as negative controls. The Western blot of the whole-cell lysates (total) showed the efficiency of protein co-transfection with similar protein content between conditions (Figure 5a). Likewise, the IP of HA-DVL (IP: HA-DVL1) or FLAG-CLDN5 (IP: FLAG-CLDN5) denoted effective pull-down of both proteins, again with similar content between samples. Pull down of HA-DVL1-WT captured FLAG-CLDN5, and this binding was reduced when co-transfected with DVL1-CT (Figure 5a,b). Further, deletion of the PDZ-BM (HA-DVL1-ΔC7) increased the interaction with FLAG-CLDN5 compared to HA-DVL1-WT. This suggests that the PDZ domain of DVL1 is required for its interaction with CLDN5 and that CLDN5 competes with the intramolecular binding from DVL-PDZ-BM. Figure 5b shows the quantification of HA-DVL1-WT/FLAG-CLDN5 co-IP blots with co-transfection of 2 μg of DVL1-CT plasmid with a significant decrease in DVL1/CLDN5 interaction when the HA tag was used for IP. However, no significant difference was observed when the FLAG tag was used for IP. To confirm that DVL1-CT competes with the CLDN5 for its interaction with DVL1-WT, HEK-293 cells were then co-transfected with FLAG-CLDN5, HA-DVL1-WT and increasing doses of HA-DVL1-CT followed by FLAG-CLDN5 IP. As shown in Figure 5c,d, 4 μg of HA-DVL1-CT plasmid was sufficient to reduce DVL1 and CLDN5 interaction significantly and this was reduced even more as HA-DVL1-CT expression increased, thus corroborating that the C-terminus fragment of DVL1 competes with CLDN5 for the interaction to DVL1 PDZ domain.

In order to determine whether the CLDN5-PDZ-BM binds the DVL1-PDZ domain, we designed a CLDN5 lacking the carboxyl-terminal PDZBM, corresponding to the last four amino acids (FLAG-CLDN5-ΔPDZBM). FLAG-CLDN5 or FLAG-CLDN5-ΔPDZBM plasmids were transfected in HEK293 cells and co-transfected with HA-DVL1 or HA-DVL1-ΔC7 followed by FLAG-CLDN5 IP and blotting for FLAG-CLDN5, HA-DVL1 or endogenous ZO-1. Figure 5e,f shows that the IP of FLAG-CLDN5 captured HA-DVL1 and ZO-1 and that HA-DVL1-ΔC7 had increased interaction with FLAG-CLDN5. Furthermore, as expected, the deletion of the CLDN5-PDZBM impaired the binding of CLDN5 with ZO-1. The deletion of CLDN5-PDZBM reduced the interaction with HA-DVL1-WT and completely blocked binding to the HA-DVL1-ΔC7.

We also explored DVL1 or DVL3 interaction with ZO-1 and surprisingly found that each interacted with ZO-1 through distinct domains. Plasmids encoding HA-DVL1 or 3 or mutants were transfected into HEK-293 cells and co-transfected with GFP-ZO-1. Figure S4a,b shows that as with CLDN5, DVL1-WT interaction with ZO-1 was significantly reduced in the presence of DVL1-CT and increased with the deletion of DVL1 PDZ-BM, strongly implying binding to the DVL1 PDZ domain. DVL3-WT also interacted with ZO-1 (Figure S4c,d), but in contrast with DVL1, DVL3 utilizes its PDZ-BM for its interaction with ZO-1, as DVL3-ΔC7 mutant showed a shallow interaction with ZO-1. These results strongly suggest that ZO-1 can interact with DVL1 PDZ domain and DVL3 PDZ-BM.

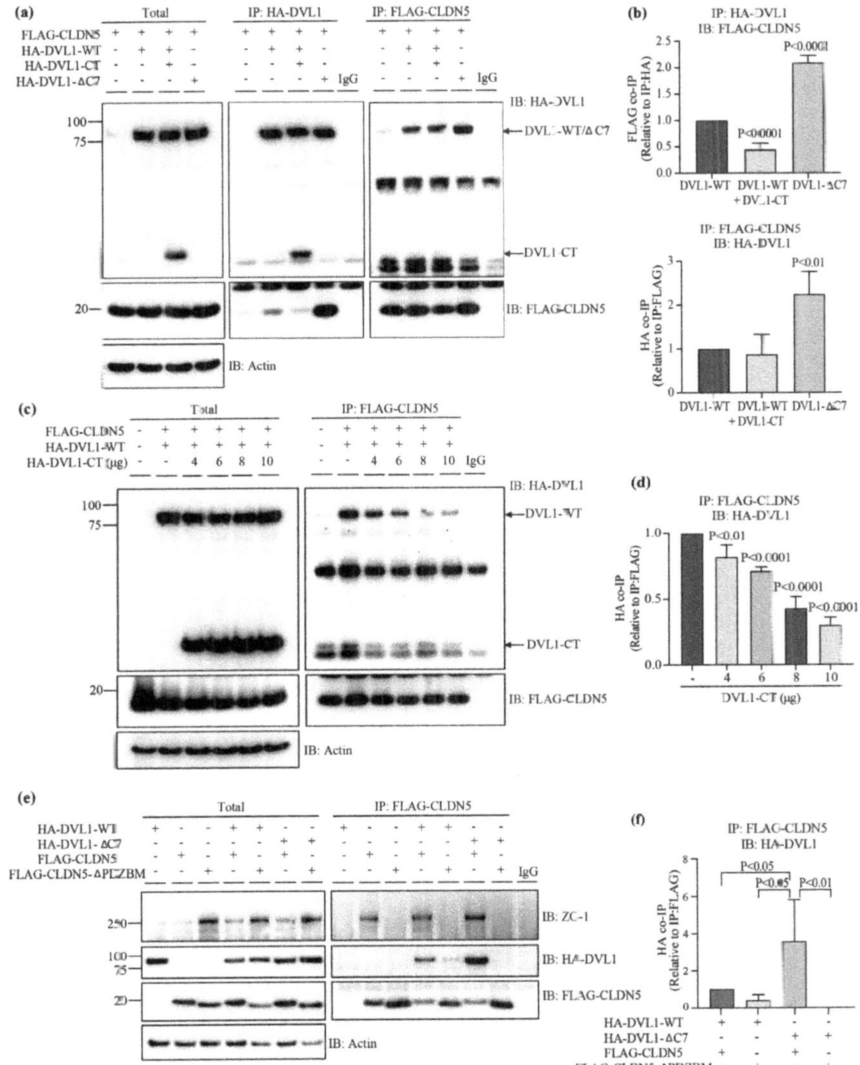

Figure 5. DVL1 interacts with CLDN5 through its PDZ domain. HEK-293 cells co-transfected with full-length DVL1 (HA-DVL1-WT), HA-DVL1-WT and DVL1 C-terminus (CT) 169aa fragment (HA-DVL1-CT) or Dvl1 deleted in the last 7aa which includes the PDZ-BM (HA-DVL1-ΔC7), together with FLAG-CLDN5 or FLAG-CLDN5Δ-PDZBM. Cell lysates were collected for total protein or the immunoprecipitation (IP) of DVL mutants or CLDN5, using specific antibodies against HA or FLAG tags, respectively. (**a**,**c**,**e**) Representative immunoblot (IB) of total and IP protein, using HA or FLAG antibodies. Actin was used as a loading control of total protein. (**b**,**d**,**f**) Densitometry of IP proteins. *p* values were calculated via one-way ANOVA, followed by Sidak post hoc test. Error bars, S.D.

3.5. DVL1 and CLDN5 Interaction Is Required for Basal and Norrin-Induced Endothelial Barrier Properties

To determine the role of DVL1/CLDN5 interaction in regulating endothelial barrier properties, BRECs were transfected with DVL1-CT to compete with the interaction between endogenous DVL1 and CLDN5. Measures of CLDN5 protein content when cells were transfected with the DVL1-CT revealed a dramatic loss of CLDN5 protein content (Figure 6a,b). This loss of CLDN5 was specific as CLDN1, ZO-1, occludin and VE-cadherin were not affected. Further, the immunoprecipitation of DVL1 with Pan-DVL antibody and measures of CLDN5 co-IP (Figure 6c,d) showed a dramatic loss of interaction, as expected with the loss of CLDN5 expression.

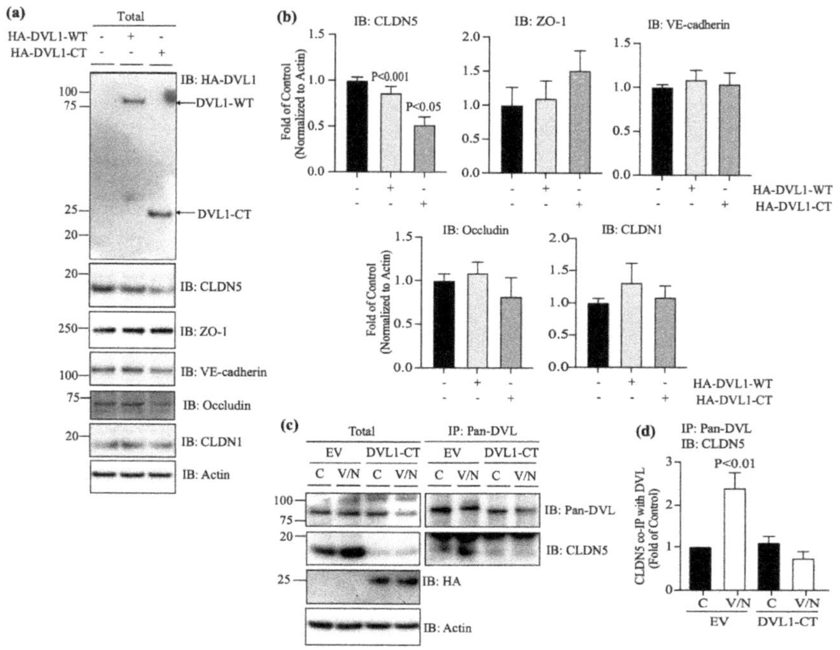

Figure 6. DVL1-CT fragment reduces CLDN5 content and DVL1/CLDN5 interaction in endothelial cells. BRECs were transfected with empty vector (EV), DVL1 WT (HA-DVL1) or DVL1 C-terminus (CT) 169aa fragment (HA-DVL1-CT). Some monolayers were stimulated with vehicle (control, C) or VEGF 50 ng/mL for 30 min, followed by Norrin 40 ng/mL (V/N) for additional 30 min. Cell lysates were collected for total protein or the immunoprecipitation (IP) of DVL using Pan-DVL antibody, followed by the immunoblot (IB) with Pan-DVL, CLDN5, ZO-1, VE-cadherin, occludin and CLDN1 antibodies. Actin was used as a loading control of total protein. (**a**,**c**) Representative Western blots. (**b**,**d**) Densitometry analysis. In IP experiments, CLDN5 protein was calculated relative to DVL IP protein. p values were calculated via one-way ANOVA, followed by Sidak post hoc test. Error bars, S.D.

Consistent with the loss of CLDN5, the basal permeability of BRECs transfected with DVL1-CT increased significantly in comparison with empty vector (EV)-transfected monolayers (Figure 7a). Measurements of the β-catenin-signaling target, *Axin2*, denoted no change in canonical Wnt signaling activation in the presence of DVL1-CT (Figure 7b). As expected, the immunofluorescence staining of cells expressing DVL1-CT demonstrates that CLDN5 and ZO-1 significantly reduced endothelial border staining (Figure 7c,d). Together, these data demonstrate that DVL1/CLDN5 interaction is required to maintain basal endothelial barrier properties. To evaluate the requirement of DVL1/CLDN5 interaction in the BRB restoration induced by norrin, BREC monolayers were transfected with DVL1-CT and stimulated with VEGF or VEGF and norrin, for measurements of endothelial permeability (Figure 7e). Consistent with our previous results, in BRECs transfected with EV, VEGF increased the permeability, and norrin restored this permeability. However, when cells were transfected with DVL1-CT, the basal (control, C) permeability of these monolayers was higher than EV control monolayers. VEGF increased the endothelial permeability, but norrin failed to restore barrier properties in the presence of DVL1-CT, indicating a requirement of DVL1/CLDN5 interaction for norrin to restore barrier properties after VEGF.

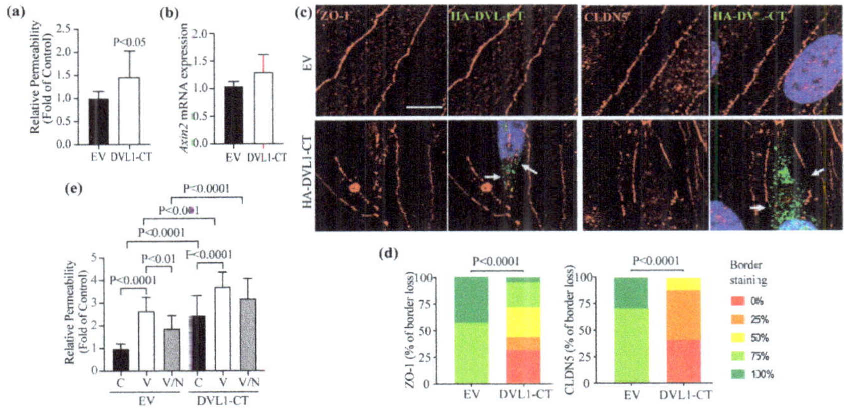

Figure 7. Effect of DVL1-CT in barrier properties. BRECs were transfected with empty vector (EV) or DVL1 C-terminus (CT) 169aa fragment (HA-DVL1-CT) for measurements of (**a**) basal permeability to a 70 kDa RITC-dextran molecule, (**b**) *Axin2* mRNA expression (β-catenin signaling activity) or (**c**) the IF staining of DVL1 mutants (HA-tag; green) and CLDN5 or ZO-1 TJ proteins (red); scale bar = 10 µM Arrows point to sites of TJ protein disruption in transfected cells. (**d**) Quantification of four images per experiment by masked scoring combined from three individuals. Data show the frequency of CLDN5 or ZO-1 loss at the cell contacts ranked in five categories, where 100% corresponds to complete loss (**e**) BREC monolayers transfected with EV or DVL1-CT fragment and stimulated with vehicle (control, C), VEGF (V) or VEGF/Norrin (V/N) for 24 h before measurements of their permeability to a 70 kDa RITC-dextran molecule. *p* values were calculated via *t*-test analysis (**a**,**b**,**d**) or via two-way ANOVA (**e**), followed by Tukey's post hoc test. Error bars, S.D.

Together, these results strongly suggest that norrin signals through DVL1 to promote BRB properties after VEGF-induced permeability. Furthermore, the studies suggest a novel non-canonical signaling pathway activated by norrin in which DVL1/CLDN5 interaction is required to organize endothelial junctional complex and barrier function in addition to canonical β-catenin signaling (Figure 8).

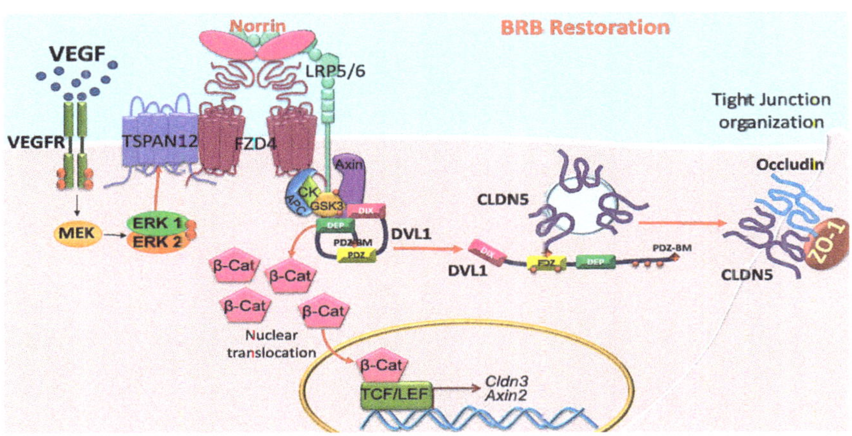

Figure 8. Norrin signals through DVL1 to stimulate barrier properties. Previous studies revealed that VEGF-induces endothelial permeability and simultaneously promotes norrin signaling by inducing

FZD4 co-receptor TSPAN12 migration to the plasma membrane. Norrin signals through DVL1 promoting the canonical pathway with β-catenin stabilization, likely through the closed confirmation of DVL1 involving intramolecular binding of DVL1 PDZ domain and DVL1 PDZ-BM. Data here reveal an additional non-canonical signaling pathway necessary for tight junction organization and induction of barrier properties involving the open conformation of DVL1 with its PDZ domain binding to claudin 5 PDZ-BM.

4. Discussion

Norrin signaling through Wnt/β-catenin is required for blood–retinal barrier (BRB) development and maintenance [1]. Moreover, our previous research demonstrates that norrin can restore BRB properties after VEGF-induced permeability and that canonical signaling through β-catenin is required for this norrin effect [28]. However, data from that study also suggest that additional, non-canonical signaling may be necessary. First, *Axin2* mRNA, as a readout of β-catenin activity, did not strictly match barrier properties, as norrin alone promoted canonical signaling, and expression of *Axin2* mRNA did not affect barrier properties. In contrast, VEGF and norrin together led to *Axin2* mRNA content similar to control but promoted barrier restoration. Second, inhibition of GSK3 promoted canonical signaling but failed to promote barrier properties and, instead inhibited the ability of norrin to promote barrier properties after VEGF. Finally, the literature provides compelling evidence for the role of DVL in polarity complex formation and TJs in particular [38]. In the current study, we demonstrated that knockdown of DVL1 specifically, but not DVL2 or DVL3, reduced basal retinal endothelial barrier properties and prevented norrin restoration of barrier properties after VEGF-induced permeability without decreasing β-catenin-induced signaling. It is noteworthy that siDvl2 and 3 together actually reduced barrier properties consistent with the specific role of DVL1 in barrier promotion.

Further, inhibiting DVL1 interaction with CLDN5 by expressing the CT fragment of DVL1 containing the PDZ-BM led to the same loss of basal barrier properties and prevention of norrin restoration. The literature reveals that in development, constitutively active β-catenin can largely compensate for norrin deletion [1]. However, in vivo, polarity cues and non-canonical signaling may come from separate ligands, such as Wnts. Furthermore, the response after VEGF signaling may be different than in development. Finally, when β-catenin forced expression was used to promote the vascular barrier in regions of the brain and eye with low barrier properties, such as the circumventricular organ of the brain and choriocapillaris of the eye, the phenotypic conversion was incomplete, suggesting that additional pathways contribute to BBB and BRB formation [39].

DVL interacts with well over 20 proteins directly or in a complex [23]. These interactions both promote canonical signaling to β-catenin and non-canonical signaling, including planar cell polarity (PCP). PCP is the collective alignment of cell polarity across a tissue plain and allows for the directional pattern of tissues, for example, cilia on an apical side of an epithelial sheet [40]. PCP may be developmental or through Wnt signaling and requires FZD and DVL, among other proteins [40]. Examples of DVL signaling include the regulation of axon development in neurons with downregulation of DVL abrogating axon differentiation and overexpression of DVL, yielding multiple axons by directly binding to and activating atypical PKC (aPKC) [41]. A role for PCP in angiogenesis has been identified using pulmonary endothelial cells, where the expression of a point mutant of DVL2 that acts in a dominant negative manner to prevent PCP (K446M in the DEP domain) reduced basal cell proliferation and inhibited VEGF-induced migration. In contrast, the expression of wild-type DVL2 had no effect [42].

Furthermore, the same group demonstrated that zebrafish with a Wnt5 mutation led to vascular defects of the intersegmental vessels, dorsal aorta, posterior cardinal vein and the anterior and middle cerebral veins of the head. In this model, the Wnt5 ligand specifically activates PCP signaling and not canonical Wnt signaling. Finally, these defects could be recapitulated with a drug targeting PCP signaling. These studies reveal that DVL contributes an essential role to cell polarity, which is an essential step in TJ formation. However, none of these studies have evaluated the possibility that DVL might regulate the assembly of TJ complexes.

Here, we show for the first time a direct interaction of DVL1, specifically with CLDN5, to induce BRB assembly in both basal conditions and during norrin-induced BRB restoration. DVL1 specifically binds to CLDN5 as observed by co-IP experiments, and this interaction increases upon norrin restoration of barrier properties after VEGF treatment. Capture assays strongly suggest that the PDZ-BM of CLDN5 competes with the intramolecular binding of the PDZ-BM of DVL1 for the DVL1 PDZ domain and is consistent with previous reports suggesting the closed conformation of DVL promotes canonical signaling and the open conformation allows non-canonical signaling and in this case binding to CLDN5. Future studies must address whether this is a dynamic interaction and the relation to CLDN5 binding ZO-1. However, the current studies reveal that TJ organization requires DVL1 binding CLDN5, as both siRNA to *Dvl1* and DVL1 C-terminal fragment overexpression preventing DVL1/CLDN5 interaction disrupt TJ organization and barrier properties.

Although many reports suggest redundancy in the function of DVL proteins, here we demonstrate that in retinal endothelial cells, DVL1 has a unique contribution to BRB regulation. Specific antibodies demonstrate minimal DVL2 expression in retinal endothelial cells in the retinal cross-section, although DVL2 was observed in the cell culture. In contrast, DVL3 is enriched in retinal capillaries and shows co-localization with the TJ proteins. Nevertheless, when we knocked down *Dvl3* in endothelial cell culture with a specific siRNA, basal barrier properties did not change, and norrin could still induce BRB restoration after VEGF. However, because previous research in *Dvl3* knockout mice showed profound vascular defects, and our results also indicate that DVL3 can interact with ZO-1, particularly with VEGF and norrin stimulation in BRECs, a role for DVL3 in TJ protein regulation remains. Importantly, here we have provided for the first time, a mechanistic insight into a novel non-canonical role of DVL in the TJ biology, in which DVL1 binds directly to the TJ CLDN5 to promote endothelial barrier function. Although gene deletion of *Dvl1* in mice results in abnormal social behavior [24,25], the molecular mechanisms that lead to this phenotype and the specific role of DVL1 in BBB or BRB in vivo remain to be determined.

DVL may have additional non-canonical pathways affecting TJ. DVL binding partners, such as the PDZ-domain-containing ubiquitin ligase, PDZRN3, have been demonstrated to have a role in vascular development through PCP signaling activation. In an endothelial-specific manner, the overexpression of this ubiquitin ligase leads to the loss of cortical brain vessel barrier properties and embryonic lethality [43]. Conversely, endothelial-specific gene deletion of *Pdrzn3* protects of TJ loss and brain edema in middle-cerebral-artery-occlusion-induced stroke [43]. Cell culture studies have revealed that PDZRN3 targets the polarity protein, MUPP1, to regulate the junctional complex. PDZRN3 appears to have a central role in regulating canonical versus non-canonical signaling in endothelial cells. Using c-Jun as a marker for PCP signaling and active β-catenin as a marker for canonical signaling, endothelial-specific loss of *Pdzrn3* gene was observed to promote canonical signaling in mice [44]. This was further supported in cell culture in which depletion of *Pdzrn3* with siRNA promoted canonical signaling and inhibited c-Jun-driven non-canonical signaling. In contrast, over-expression of PDZRN3 promoted non-canonical signaling and inhibited canonical signaling, as measured via transcription reporter assays. This canonical to non-canonical signaling may be through DVL, as PDZRN3 targets DVL3 for ubiquitination [44].

DVL possesses a PDZ domain and a PDZ-BM, while ZO-1 possesses 3 PDZ domains, and these PDZ domains can bind to either C-terminal PDZ-BM, for example, allowing the binding to CLDNs through PDZ1, or can bind other PDZ domains, such as allowing ZO-ZO interaction through PDZ2 (reviewed in [45]). Here, we have found that DVL1 and DVL3 utilize different regions to interact with ZO-1; while DVL1 requires its PDZ domain, DVL3 uses its PDZ-BM. Future studies will elucidate which PDZ domain of ZO-1 is used for this interaction. It is plausible that DVL1 and 3 utilize different PDZ domains of ZO-1, and this might explain the differences in their interactions and functions.

5. Conclusions

Collectively, these data provide compelling evidence for a novel, non-canonical norrin signaling pathway in which DVL1 binding to claudin 5 coordinates with the canonical signaling pathway to promote the restoration of endothelial barrier properties after VEGF-induced permeability. Furthermore, this study has revealed new interactions between the PDZ-BM of CLDN5 and DVL1 necessary for tight junction organization in the vascular endothelium.

Supplementary Materials: The following supporting information can be downloaded at https://www.mdpi.com/article/10.3390/cells12192402/s1, Figure S1: Knockdown of *Dvl1*, *Dvl2*, *Dvl3* or *Dvl2/3* in BRECs using specific siRNA sequences. BREC monolayers were transfected with siRNAs targeting *Dvl1*, *2*, *3*, *2/3* or the scramble (Scr) control. Western blot analysis showing specificity of DVL2 and DVL3 antibodies and Pan-DVL antibody targeting both DVL1 and DVL3. Figure S2: Knockdown of *Dvl1* in BREC monolayers, was stable for at least 72 h. qRT-PCR of *Dvl1*, *Dvl2* or *Axin2*, or *Dvl3* PCR in BREC monolayers transfected with two specific *Dvl1* siRNA sequences: (a) first; (b) second. p values were calculated via t-test analysis. *Error bars*, S.D. (c) Immunofluorescence staining of Pan-DVL and ZO-1 in BREC monolayers after Dvl1 knockdown; scale bar = 10 μM. Figure S3: Norrin induces DVL phosphorylation. BREC monolayers were stimulated with vehicle (control, C), norrin 40 ng/mL (N), VEGF 50 ng/mL (V) or both (V/N). After 72 h of stimulation, lysates were collected. Phosphorylation was depleted using alkaline phosphatase, as shown with collapse of high-molecular-weight bands into one, confirming that DVL changes in molecular weight are due to phosphorylation. Figure S4: DVL3, but not DVL1, requires its PDZ-BM (BM) to interact with ZO-1. HEK-293 cells co-transfected with (a, b) full-length DVL1 (HA-DVL1-WT), HA-DVL1-WT and DVL1 C-terminus (CT), 169aa fragment (HA-DVL1-CT) or DVL1 deleted in the last 7aa, which includes the PDZ-BM (HA-DVL1-ΔC7), or (c, d) full-length DVL3 (HA-DVL3-WT) or DVL3 deleted in the last 7aa, which includes the PDZ-BM (HA-DVL3-ΔC7), together with GFP-ZO-1. Cell lysates were collected for total protein or the immunoprecipitation (IP) of ZO-1, using specific antibody against GFP tag. (a,c) Representative immunoblots (IB) of total and IP protein, using HA or GFP antibodies. Actin was used as a loading control of total protein. (b,d) Densitometry of IP proteins. p values were calculated via one-way ANOVA, followed by *Sidak* post hoc test. *Error bars*, S.D.

Author Contributions: Conceptualization, M.D.-C. and D.A.A.; methodology and technical assistance, M.D.-C., L.G.-G., A.W., X.L. and D.A.A.; data analysis, M.D.-C., L.G.-G. and D.A.A.; writing—review and editing, M.D.-C., L.G.-G. and D.A.A.; supervision, M.D.-C. and D.A.A.; funding acquisition M.D.-C., D.A.A. All authors have read and agreed to the published version of the manuscript.

Funding: This work was supported by the National Institutes of Health Grant EY012021 (to D.A.A.) and Core Grants P30EY007003 and DK020572, the Jules and Doris Stein Professorship from Research to Prevent Blindness (to D.A.A.), the American Diabetes Association Minority Postdoctoral Fellowship Award 4-16-PMF-003 (to M.D.-C.).

Institutional Review Board Statement: All animals were treated in accordance with the Association for Research in Vision and Ophthalmology (ARVO) Statement on the Use of Animals in Ophthalmic and Visual Research and the guidelines established by the University of Michigan Institutional Animal Care and Use Committee.

Data Availability Statement: Data are contained within the article or supplementary material.

Acknowledgments: We thank Cheng-Mao Lin (University of Michigan) for support in animal management, Alyssa Dreffs (University of Michigan) for technical support, and Anuska Andejelkovic (University of Michigan) for the donation of GFP-ZO-1 clone.

Conflicts of Interest: Antonetti is an advisor for Eyebiotech.

References

1. Zhou, Y.; Wang, Y.; Tischfield, M.; Williams, J.; Smallwood, P.M.; Rattner, A.; Taketo, M.M.; Nathans, J. Canonical WNT signaling components in vascular development and barrier formation. *J. Clin. Investig.* **2014**, *124*, 3825–3846. [CrossRef]
2. Schafer, N.F.; Luhmann, U.F.O.; Feil, S.; Berger, W. Differential gene expression in Ndph-knockout mice in retinal development. *Investig. Opthalmology Vis. Sci.* **2009**, *50*, 906–916. [CrossRef] [PubMed]
3. Zuercher, J.; Fritzsche, M.; Feil, S.; Mohn, L.; Berger, W. Norrin stimulates cell proliferation in the superficial retinal vascular plexus and is pivotal for the recruitment of mural cells. *Hum. Mol. Genet.* **2012**, *21*, 2619–2630. [CrossRef] [PubMed]
4. Beck, S.C.; Feng, Y.; Sothilingam, V.; Garrido, M.G.; Tanimoto, N.; Acar, N.; Shan, S.; Seebauer, B.; Berger, W.; Hammes, H.-P.; et al. Long-term consequences of developmental vascular defects on retinal vessel homeostasis and function in a mouse model of Norrie disease. *PLoS ONE* **2017**, *12*, e0178753. [CrossRef]
5. Ngo, M.H.; Borowska-Fielding, J.; Heathcote, G.; Nejat, S.; Kelly, M.E.; McMaster, C.R.; Robitaille, J.M. Fzd4 Haploinsufficiency Delays Retinal Revascularization in the Mouse Model of Oxygen Induced Retinopathy. *PLoS ONE* **2016**, *11*, e0158320. [CrossRef]
6. Xu, Q.; Wang, Y.; Dabdoub, A.; Smallwood, P.M.; Williams, J.; Woods, C.; Kelley, M.W.; Jiang, L.; Tasman, W.; Zhang, K.; et al. Vascular development in the retina and inner ear: Control by Norrin and Frizzled-4, a high-affinity ligand-receptor pair. *Cell* **2004**, *116*, 883–895. [CrossRef] [PubMed]
7. Paes, K.T.; Wang, E.; Henze, K.; Vogel, P.; Read, R.; Suwanichkul, A.; Kirkpatrick, L.L.; Potter, D.; Newhouse, M.M.; Rice, D.S. Frizzled 4 is required for retinal angiogenesis and maintenance of the blood-retina barrier. *Investig. Opthalmol. Vis. Sci.* **2011**, *52*, 6452–6461. [CrossRef]
8. Xia, C.-H.; Liu, H.; Cheung, D.; Wang, M.; Cheng, C.; Du, X.; Chang, B.; Beutler, B.; Gong, X. A model for familial exudative vitreoretinopathy caused by LPR5 mutations. *Hum. Mol. Genet.* **2008**, *17*, 1605–1612. [CrossRef]
9. Chen, J.; Stahl, A.; Krah, N.M.; Seaward, M.R.; Dennison, R.J.; Sapieha, P.; Hua, J.; Hatton, C.J.; Juan, A.M.; Aderman, C.M.; et al. Wnt signaling mediates pathological vascular growth in proliferative retinopathy. *Circulation* **2011**, *124*, 1871–1881. [CrossRef]
10. Junge, H.J.; Yang, S.; Burton, J.B.; Paes, K.; Shu, X.; French, D.M.; Costa, M.; Rice, D.S.; Ye, W. TSPAN12 regulates retinal vascular development by promoting Norrin- but not Wnt-induced FZD4/beta-catenin signaling. *Cell* **2009**, *139*, 299–311. [CrossRef]
11. Zhang, C.; Lai, M.B.; Pedler, M.G.; Johnson, V.; Adams, R.H.; Petrash, J.M.; Chen, Z.; Junge, H.J. Endothelial Cell-Specific Inactivation of TSPAN12 (Tetraspanin 12) Reveals Pathological Consequences of Barrier Defects in an Otherwise Intact Vasculature. *Arter. Thromb. Vasc. Biol.* **2018**, *38*, 2691–2705. [CrossRef] [PubMed]
12. Ye, X.; Wang, Y.; Cahill, H.; Yu, M.; Badea, T.C.; Smallwood, P.M.; Peachey, N.S.; Nathans, J. Norrin, frizzled-4, and Lrp5 signaling in endothelial cells controls a genetic program for retinal vascularization. *Cell* **2009**, *139*, 285–298. [CrossRef] [PubMed]
13. Ye, X.; Smallwood, P.; Nathans, J. Expression of the Norrie disease gene (Ndp) in developing and adult mouse eye, ear, and brain. *Gene Expr. Patterns* **2011**, *11*, 151–155. [CrossRef] [PubMed]
14. Berger, W.; van de Pol, D.; Bächner, D.; Oerlemans, F.; Winkens, H.; Hameister, H.; Wieringa, B.; Hendriks, W.; Ropers, H.-H. An animal model for Norrie disease (ND): Gene targeting of the mouse ND gene. *Hum. Mol. Genet.* **1996**, *5*, 51–59. [CrossRef]
15. Lee, H.; Jo, D.H.; Kim, J.H.; Kim, J.H. Norrin expression in endothelial cells in the developing mouse retina. *Acta Histochem.* **2013**, *115*, 447–451. [CrossRef]
16. Ke, J.; Harikumar, K.G.; Erice, C.; Chen, C.; Gu, X.; Wang, L.; Parker, N.; Cheng, Z.; Xu, W.; Williams, B.O.; et al. Structure and function of Norrin in assembly and activation of a Frizzled 4-Lrp5/6 complex. *Genes Dev.* **2013**, *27*, 2305–2319. [CrossRef]
17. DeBruine, Z.J.; Xu, H.E.; Melcher, K. Assembly and architecture of the Wnt/beta-catenin signalosome at the membrane. *Br. J. Pharmacol.* **2017**, *174*, 4564–4574. [CrossRef]
18. Gammons, M.V.; Renko, M.; Johnson, C.M.; Rutherford, T.J.; Bienz, M. Wnt Signalosome Assembly by DEP Domain Swapping of Dishevelled. *Mol. Cell* **2016**, *64*, 92–104. [CrossRef]
19. Mahoney, J.P.; Bruguera, E.S.; Vasishtha, M.; Killingsworth, L.B.; Kyaw, S.; Weis, W.I. PI(4,5)P(2)-stimulated positive feedback drives the recruitment of Dishevelled to Frizzled in Wnt-beta-catenin signaling. *Sci. Signal.* **2022**, *15*, eabo2820. [CrossRef]
20. Nusse, R.; Clevers, H. Wnt/beta-Catenin Signaling, Disease, and Emerging Therapeutic Modalities. *Cell* **2017**, *169*, 985–999. [CrossRef]
21. Cattelino, A.; Liebner, S.; Gallini, R.; Zanetti, A.; Balconi, G.; Corsi, A.; Bianco, P.; Wolburg, H.; Moore, R.; Oreda, B.; et al. The conditional inactivation of the beta-catenin gene in endothelial cells causes a defective vascular pattern and increased vascular fragility. *J. Cell Biol.* **2003**, *162*, 1111–1122. [CrossRef] [PubMed]
22. Liebner, S.; Corada, M.; Bangsow, T.; Babbage, J.; Taddei, A.; Czupalla, C.J.; Reis, M.; Felici, A.; Wolburg, H.; Fruttiger, M.; et al. Wnt/beta-catenin signaling controls development of the blood-brain barrier. *J. Cell Biol.* **2008**, *183*, 409–417. [CrossRef] [PubMed]
23. Gao, C.; Chen, Y.-G. Dishevelled: The hub of Wnt signaling. *Cell. Signal.* **2010**, *22*, 717–727. [CrossRef]
24. Lijam, N.; Paylor, R.; McDonald, M.P.; Crawley, J.N.; Deng, C.X.; Herrup, K.; Stevens, K.E.; Maccaferri, G.; McBain, C.J.; Sussman, D.J.; et al. Social interaction and sensorimotor gating abnormalities in mice lacking Dvl1. *Cell* **1997**, *90*, 895–905. [CrossRef] [PubMed]
25. Long, J.M.; LaPorte, P.; Paylor, R.; Wynshaw-Boris, A. Expanded characterization of the social interaction abnormalities in mice lacking Dvl1. *Genes Brain Behav.* **2004**, *3*, 51–62. [CrossRef] [PubMed]
26. Etheridge, S.L.; Ray, S.; Li, S.; Hamblet, N.S.; Lijam, N.; Tsang, M.; Greer, J.; Kardos, N.; Wang, J.; Sussman, D.J.; et al. Murine dishevelled 3 functions in redundant pathways with dishevelled 1 and 2 in normal cardiac outflow tract, cochlea, and neural tube development. *PLoS Genet.* **2008**, *4*, e1000259. [CrossRef]

27. Lee, H.J.; Shi, D.L.; Zheng, J.J. Conformational change of Dishevelled plays a key regulatory role in the Wnt signaling pathways. *eLife* **2015**, *4*, e08142. [CrossRef]
28. Diaz-Coranguez, M.; Lin, C.M.; Liebner, S.; Antonetti, D.A. Norrin restores blood-retinal barrier properties after vascular endothelial growth factor-induced permeability. *J. Biol. Chem.* **2020**, *295*, 4647–4660. [CrossRef]
29. Antonetti, D.A.; Wolpert, E.B. Isolation and characterization of retinal endothelial cells. *Methods Mol. Med.* **2003**, *89*, 365–374.
30. Tiruppathi, C.; Malik, A.B.; Del Vecchio, P.J.; Keese, C.R.; Giaever, I. Electrical method for detection of endothelial cell shape change in real time: Assessment of endothelial barrier function. *Proc. Natl. Acad. Sci. USA* **1992**, *89*, 7919–7923. [CrossRef]
31. Harhaj, N.S.; Barber, A.J.; Antonetti, D.A. Platelet-derived growth factor mediates si redistribution and increases permeability in MDCK cells. *J. Cell Physiol.* **2002**, *193*, 349–364. [CrossRef] [PubMed]
32. Aveleira, C.A.; Lin, C.M.; Abcouwer, S.F.; Ambrosio, A.F.; Antonetti, D.A. TNF-alpha signals through PKCzeta/NF-kappaB to alter the tight junction complex and increase retinal endothelial cell permeability. *Diabetes* **2010**, *59*, 2872–2882. [CrossRef] [PubMed]
33. Bernatík, O.; Šedová, K.; Schille, C.; Ganji, R.S.; Červenka, I.; Trantírek, L.; Schambony, A.; Zdráhal, Z.; Bryja, V. Functional analysis of dishevelled-3 phosphorylation identifies distinct mechanisms driven by casein kinase 1ϵ and frizzled5. *J. Biol. Chem.* **2014**, *289*, 23520–23533. [CrossRef] [PubMed]
34. Hanáková, K.; Bernatík, O.; Kravec, M.; Micka, M.; Kumar, J.; Harnoš, J.; Ovesná, P.; Paclíková, P.; Rádsetoulal, M.; Potěšil, D.; et al. Comparative phosphorylation map of Dishevelled 3 links phospho-signatures to biological outputs. *Cell Commun. Signal.* **2019**, *17*, 170. [CrossRef] [PubMed]
35. Itoh, M.; Furuse, M.; Morita, K.; Kubota, K.; Saitou, M.; Tsukita, S. Direct binding of three tight junction-associated MAGUKs, ZO-1, ZO-2, and ZO-3, with the COOH termini of claudins. *J. Cell Biol.* **1999**, *147*, 1351–1363. [CrossRef] [PubMed]
36. Qi, J.; Lee, H.J.; Saquet, A.; Cheng, X.N.; Shao, M.; Zheng, J.J.; Shi, D.L. Autoinhibition of Dishevelled protein regulated by its extreme C terminus plays a distinct role in Wnt/beta-catenin and Wnt/planar cell polarity (PCP) signaling pathways. *J. Biol. Chem.* **2017**, *292*, 5898–5908. [CrossRef] [PubMed]
37. Harnoš, J.; Cañizal, M.C.A.; Jurásek, M.; Kumar, J.; Holler, C.; Schambony, A.; Hanáková, K.; Bernatík, O.; Zdráhal, Z.; Gömöryová, K.; et al. Dishevelled-3 conformation dynamics analyzed by FRET-based biosensors reveals a key role of casein kinase 1. *Nat. Commun.* **2019**, *10*, 1804. [CrossRef]
38. Li, L.; Mao, B.; Yan, M.; Wu, S.; Ge, R.; Lian, Q.; Cheng, C.Y. Planar cell polarity protein Dishevelled 3 (Dvl3) regulates ectoplasmic specialization (ES) dynamics in the testis through changes in cytoskeletal organization. *Cell Death Dis.* **2019**, *10*, 194. [CrossRef]
39. Wang, Y.; Sabbagh, M.F.; Gu, X.; Rattner, A.; Williams, J.; Nathans, J. Beta-catenin signaling regulates barrier-specific gene expression in circumventricular organ and ocular vasculatures. *eLife* **2019**, *8*, e43257. [CrossRef]
40. Devenport, D. The cell biology of planar cell polarity. *J. Cell Biol.* **2014**, *207*, 171–179. [CrossRef]
41. Zhang, X.; Zhu, J.; Yang, G.-Y.; Wang, Q.-J.; Qian, L.; Chen, Y.-M.; Chen, F.; Tao, Y.; Hu, H.-S.; Wang, T.; et al. Dishevelled promotes axon differentiation by regulating atypical protein kinase C. *Nat. Cell Biol.* **2007**, *9*, 743–754. [CrossRef] [PubMed]
42. Cirone, P.; Lin, S.; Griesbach, H.L.; Zhang, Y.; Slusarski, D.C.; Crews, C.M. A role for planar cell polarity signaling in angiogenesis. *Angiogenesis* **2008**, *11*, 347–360. [CrossRef] [PubMed]
43. Sewduth, R.N.; Kovacic, H.; Jaspard-Vinassa, B.; Jecko, V.; Wavasseur, T.; Fritsch, N.; Pernot, M.; Jeaningros, S.; Roux, E.; Dufourcq, P.; et al. PDZRN3 destabilizes endothelial cell-cell junctions through a PKCzeta-containing polarity complex to increase vascular permeability. *Sci. Signal.* **2017**, *10*, eaag3209. [CrossRef] [PubMed]
44. Sewduth, R.N.; Jaspard-Vinassa, B.; Peghaire, C.; Guillabert, A.; Franzl, N.; Larrieu-Lahargue, F.; Moreau, C.; Fruttiger, M.; Dufourcq, P.; Couffinhal, T.; et al. The ubiquitin ligase PDZRN3 is required for vascular morphogenesis through Wnt/planar cell polarity signalling. *Nat. Commun.* **2014**, *5*, 4832. [CrossRef]
45. Fanning, A.S.; Anderson, J.M. Zonula occludens-1 and -2 are cytosolic scaffolds that regulate the assembly of cellular junctions. *Ann. N. Y. Acad. Sci.* **2009**, *1165*, 113–120. [CrossRef]

cells

MDPI

Article

Isolation and Characterization of the Primary Marmoset (*Callithrix jacchus*) Retinal Pigment Epithelial Cells

Ha Young Jang [1], Chang Sik Cho [1], Young Mi Shin [1], Jina Kwak [2,3], Young Hoon Sung [4,5,*], Byeong-Cheol Kang [2,3,*] and Jeong Hun Kim [1,6,7,*]

1 Fight Against Angiogenesis-Related Blindness (FARB) Laboratory, Clinical Research Institute, Seoul National University Hospital, Seoul 03082, Republic of Korea; hyjang020710@gmail.com (H.Y.J.); kansayulgi@gmail.com (C.S.C.); pgym7234@gmail.com (Y.M.S.)
2 Graduate School of Translational Medicine, Seoul National University College of Medicine, Seoul 03080, Republic of Korea; jinawak@snu.ac.kr
3 Department of Experimental Animal Research, Biomedical Research Institute, Seoul National University Hospital, Seoul 03080, Republic of Korea
4 Asan Institute for Life Sciences, Asan Medical Center, Seoul 05505, Republic of Korea
5 Department of Convergence Medicine, University of Ulsan College of Medicine, Seoul 05505, Republic of Korea
6 Department of Biomedical Sciences & Ophthalmology, Seoul National University College of Medicine, Seoul 03080, Republic of Korea
7 Institute of Reproductive Medicine and Population, Seoul National University College of Medicine, Seoul 03080, Republic of Korea
* Correspondence: yhsung@amc.seoul.kr (Y.H.S.); bckang@snu.ac.kr (B.-C.K.); steph25@snu.ac.kr (J.H.K.)

Abstract: Marmosets have emerged as a valuable primate model in ophthalmic research due to their similarity to the human visual system and their potential for generating transgenic models to advance the development of therapies. In this study, we isolated and cultured primary retinal pigment epithelium (RPE) cells from marmosets to investigate the mechanisms underlying RPE dysfunction in aging and age-related macular degeneration (AMD). We confirmed that our culture conditions and materials supported the formation of RPE monolayers with functional tight junctions that closely resembled the in vivo RPE. Since serum has been shown to induce epithelial–mesenchymal transition (EMT) in RPE cells, we compared the effects of fetal bovine serum (FBS) with serum-free supplements B27 on transepithelial electrical resistance (TER), cell proliferation and morphological characteristics. Additionally, we assessed the age-related morphological changes of in vivo and primary RPE cells. Our results indicate that primary marmoset RPE cells exhibit in vivo-like characteristics, while cells obtained from an older donor show evidence of aging, including a failure to form a polarized monolayer, low TER, and delayed cell cycle. In conclusion, our primary marmoset RPE cells provide a reliable in vitro model for developing novel therapeutics for visual-threatening disorders such as AMD, which can be used before animal experiments using marmosets.

Keywords: marmoset; retinal pigment epithelium; primary cells; age-related macular degeneration

Citation: Jang, H.Y.; Cho, C.S.; Shin, Y.M.; Kwak, J.; Sung, Y.H.; Kang, B.-C.; Kim, J.H. Isolation and Characterization of the Primary Marmoset (*Callithrix jacchus*) Retinal Pigment Epithelial Cells. *Cells* **2023**, *12*, 1644. https://doi.org/10.3390/cells12121644

Academic Editor: Hossein Ameri

Received: 9 May 2023
Revised: 13 June 2023
Accepted: 13 June 2023
Published: 16 June 2023

1. Introduction

The retinal pigment epithelium (RPE) is the pigmented cell layer, consisting of tightly interconnected regular polygonal cells that create a semi-permeable barrier between the outside choroid and inside photoreceptor cells [1]. Microvilli at the apical surface of the RPE participate in the phagocytic function, while the basolateral surface of the RPE is connected to the basal folds of Bruch's membrane by half desmosomes. The gap, adherens, and tight junction at the apical periphery of contacting cells maintain cell polarity and control the movement of substances as the choroid–blood–retinal barrier (Figure 1) [2]. RPE function and structure alteration can lead to various retinopathies, including age-related macular degeneration (AMD), the leading cause of blindness in developed countries [3]. In AMD, the degeneration of RPE cells appears to begin with impaired phagocytosis properties, resulting in the formation of abnormal deposits called drusen, which impair the function of the RPE cells. Eventually, accumulation of drusen leads to progressive degeneration and cell death of photoreceptors and RPE, resulting in loss of central vision [4].

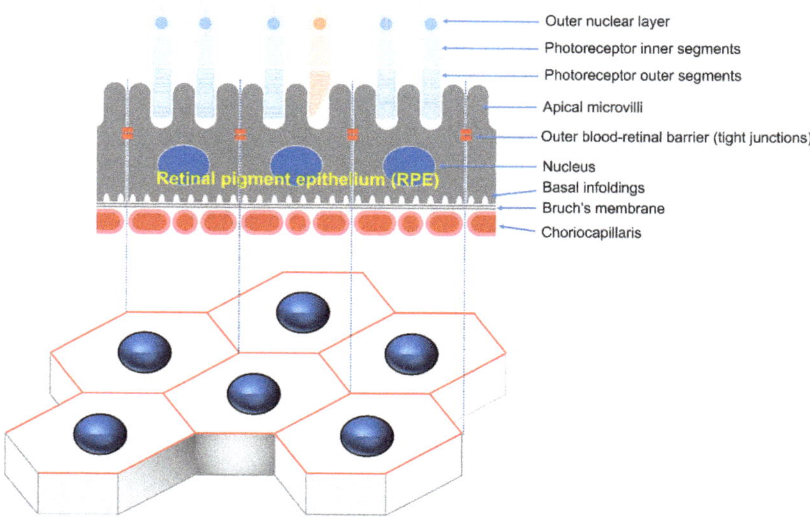

Figure 1. Schematic drawing of the RPE monolayer. RPE refers to a single layer of regular, polygonal RPE cells arranged at the outermost layer of the neural retina. The outside of the RPE is connected to Bruch's membrane and the choroid, while the inside is connected to the outside of photoreceptor cells. The tight junctions formed between the upper lateral membranes of RPE cells are responsible for a permeability barrier known as the outer blood–retinal barrier.

Although many animal models mimic several of the pathological features seen in AMD, they do not fully recapitulate the disease due to the unique features of the macula and complexities of AMD [5,6]. Cell culture models of RPE are valuable for understanding the RPE dysfunction and related pathophysiology in a controllable defined system. Various sources of human RPE cells are available, including primary prenatal and postnatal tissue, transformed cell lines, and stem cells, each with their extensibility, potential for differentiation, and the tendency to display RPE-like properties [7]. ARPE19 cells, a spontaneously arising human RPE-derived cell line, are not immortal and eventually exhibit replicative depletion and senescence, do not reliably express several RPE signature proteins, and do not achieve the desired differentiated phenotype [8]. Postnatal human RPE cells maintain relatively mature phenotypes in culture but have limited expansion in culture. Primary human prenatal cultures have the advantage of exhibiting many known characteristics of mature RPE and greater growth capacity than adult RPE, but they also include supply restrictions and ethical issues regarding their use [9]. Primary RPE cells have been widely utilized over the last few years to assess RPE dysfunction and basic cell biology as a more appropriate and reproducible research resource [10]. Although human cells are the most desirable for the pathophysiological study of RPE, ethical issues and the limited availability of human donors have led to studies of RPE isolation, culture, and characterization from non-human mammals [11–14]. Marmosets are a better alternative than other mammals considering the similarities with humans in ocular structure and visual function [15]. Unfortunately, to date, numerous studies have been reported on primary human RPE cells and non-human RPE cells, including those from mice and pigs, but none have been reported in marmosets [16–20].

The common marmoset (*Callithrix jacchus*), a small New World primate, provides a good model for human vision and can perform various visual identification and cognitive tasks by a human-like visual system. Marmosets mature quickly and reproduce easily in captivity, and they are suitable for genetic manipulation, which is vital in developing the first transgenic primate lines. Marmosets are ideal for genetic engineering models of ophthalmic diseases [15,21]. However, with increasing ethical considerations to reduce or replace the use of monkeys in biomedical research, the need for suitable in vitro models to complement animal models is also emerging. We aimed to utilize the primary marmoset RPE cells as a valuable in vitro model by establishing a simplified and reproducible environment for proliferation and differentiation with epithelial integrity. Moreover, we characterized age-related changes in primary marmoset RPE cells, which revealed variations depending on the donor's age. Therefore, primary marmoset RPE cells could be an efficient screening in vitro model with scalability to cross in vivo and in vitro studies.

2. Materials and Methods

2.1. Animals

Marmoset eyes were obtained from the Marmoset Model Network Center in Seoul National University Hospital. All marmoset experiment protocols were followed as set by the Institutional Animal Care and Use Committee of Seoul National University Hospital (IACUC No. 22-0069), an accredited research institute of the Association for Assessment and Accreditation of Laboratory Animal Care International (AAALAC) approved facility.

2.2. Preparation of the Growth and Maintenance Medium

For the RPE growth medium, 5 mL of N1 supplement (Sigma-Aldrich, St. Louis, MO, USA), 5 mL of Penicillin-Streptomycin (Gibco, Waltham, MA, USA), 5 mL of nonessential amino acids (Sigma-Aldrich, St. Louis, MO, USA), 125 mg of Taurine (Sigma-Aldrich, St. Louis, MO, USA), 10 µg of Hydrocortisone (Sigma-Aldrich, St. Louis, MO, USA), 0.0065 µg of Triiodo-thyronin (Sigma-Aldrich, St. Louis, MO, USA), and 10% of heat-inactivated FBS (Gibco, Waltham, MA, USA) were added to 500 mL of DMEM. This medium can be stored at 4 °C for up to 1 month [7,18,19]. The serum-free maintenance medium contained a 7:3 mixture of high glucose DMEM (Gibco, Waltham, MA, USA) and Ham's F12 (Gibco, Waltham, MA, USA) with 1% of Penicillin-Streptomycin. A total of 2% of 50× solution of B27 was added to the serum-free medium [7,17].

2.3. Isolation of the Primary RPE Cells

After enucleation, each eye was dipped in 70% ethanol and then placed into a 50 mL centrifuge tube containing 15 mL of medium on ice and moved within 1 h. The steps to isolate primary RPE cells were performed in a laminar flow hood under sterile conditions using sterilized instruments. The globe was placed in a 60 mm sterile culture dish after briefly soaking in 70% ethanol and washing 3 times with sterile PBS. The connective tissue and muscle attached to the sclera was carefully removed using Dumont #5 tweezers and curved iris scissors under a dissecting stereomicroscope (Nikon SMZ745T, Nikon Corporation, Tokyo, Japan). The medium was changed regularly to prevent contamination by connective tissue or blood. A step incision was performed using a sterilized #11 scalpel blade at a distance of 5 mm caudally from the limbus (Figure 2a). The anterior segment of the eye containing the corneal–iris complex was carefully removed by circumference cutting 5 mm posterior to the limbus using curved iris scissors (Figure 2b). The globe was cut into quadrants as close as possible to the optic disc for the eye cup to lay flat. It was cut enough to flatten the eyecup while keeping the petals connected (Figure 2c). The vitreous was grasped with forceps and gently pulled to remove it. A pair of tweezers was used to gently separate the retina being careful to avoid damage to the underlying RPE with the retina cut at the optic nerve attachment to facilitate separation. The neural retina was removed by gently pulling it from the edges with tweezers while holding the RPE–choroid–sclera (RCS) complex with other tweezers (Figure 2d). The RPE–choroid layer was gently peeled off from the sclera with a pair of tweezers taking care not to tear through the blunt dissection. The RPE–choroid sheet was transferred to a 60 mm sterile dish containing 3 mL of 2% (wt/vol) dispase solution (Sigma-Aldrich, St. Louis, MO, USA). It was incubated for 30 min at 37 °C in 5% CO_2 (Figure 2e). The RPE–choroid sheet was transferred to a 60 mm sterile culture dish containing 1 mL of medium following washing the sheet 3 times using the pre-warmed medium to stop the dispase activity. RPE cells were harvested by repeated gentle aspiration from the RPE–choroid sheet with a P200 micropipette while avoiding choroidal aspiration. The RPE cells were washed 3 times with the pre-warmed medium. After each washing step, the cells were carefully collected, avoiding other tissues, such as choroid debris. The RPE cells were pelleted by centrifuge at 340× *g* for 2 min at RT. The supernatant was discarded. The cells were carefully resuspended in 1 mL of medium. The cells were then placed into a T75 flask with an additional 10 mL of the growth medium volume and incubated at 37 °C, in 5% CO_2. The cells were cultured for 72 h without the change of medium. After 3 days, the medium was changed twice per week. After 1 week the cells became approximately more than 90% confluent.

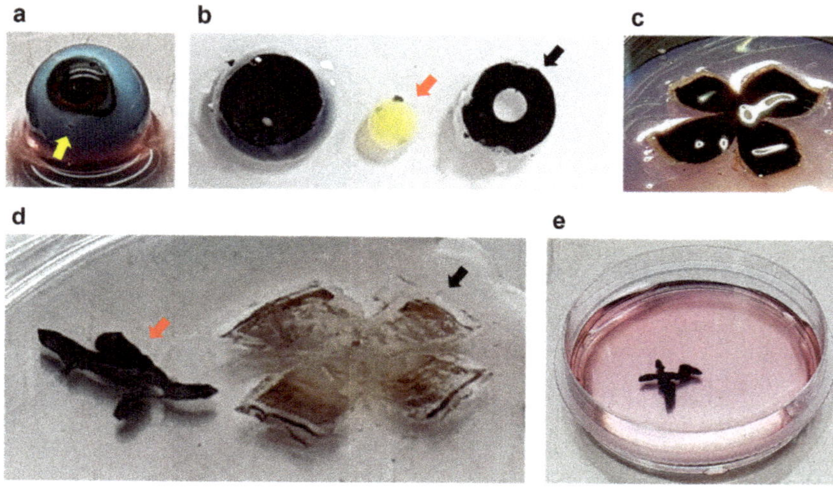

Figure 2. Dissection of the marmoset eyes. Dissection processing for the isolation of the primary RPE cells in marmoset. (**a**), Marmoset eye with the first step incision at a distance of 5 mm caudally from the limbus (yellow arrow). (**b**), The eye split into cornea–iris (black arrow), lens (red arrow), and posterior eye cup. The anterior segment is removed at once and is easily separated into the lens and cornea–iris complex. (**c**), The posterior eye cup is cut into quadrants to lay flat. (**d**) The retina (black arrow) and RPE–choroid sheet (red arrow) were gently separated. (**e**) RPE–choroid sheet incubating with 2% (wt/vol) dispase solution for 30 min at 37 °C in 5% CO_2.

2.4. Culture of Polarized Monolayer RPE Cells

The Transwell membranes with 0.4 µm pores (Corning Costar, Cambridge, MA, USA) were prepared by coating the upper compartment membrane with fibronectin (5 µg/mL^{-1}) dissolved in 100 µL of ddH_2O for 2 h. The fibronectin was then removed by aspiration and the membrane rinsed with PBS. The coated membranes were air-dried overnight at RT in the laminar flow hood [18]. After the first week of isolation, the medium was removed, and the cells in the T75 flask were washed with 10 mL of PBS. The PBS was then removed and 2 mL of 1× EDTA-trypsin was added. Trypsinization was usually complete within 10 min. The cells were resuspended in medium to stop the trypsin reaction, centrifuged at 340× *g* for 2 min at RT, and the supernatant was discarded. The cells were gently pipetted with the serum-free medium and the RPE suspension plated at a density of 1 × 10^5 cells per well (200 µL) drop-wise into the center of the prepared Transwell inserts. The cells were checked to ensure they were evenly distributed under the microscope. A volume of 700 μL of the serum-free medium was introduced into the lower compartment. The Transwells were placed in an incubator at 37 °C, 5% CO_2. The medium was completely changed every 3–4 days. Within up to 6 weeks, polarized RPE monolayers were characterized by measuring TER and immunohistochemistry. In addition, to assess whether the composition of the medium was associated with the generation of polarized RPE monolayers, we cultured cells under the same condition in medium containing 10% FBS.

2.5. Transepithelial Resistance (TER) Measurement

TER measurements were taken using an EVOM3 epithelial volt-ohmmeter and STX4 electrode (World Precision Instruments, Sarasota, FL, USA) according to the manufacturer's instruction. The electrodes were sterilized with 70% ethanol, rinsed in ddH_2O. and equilibrated in a pre-warmed culture medium. Measurement was performed in the hood within 10 min after the cells were taken out from the incubator. Net TERs (Ω cm^2) were calculated by subtracting the value of a blank insert, fibronectin-coated membrane without cells, from the experimental value and multiplying it by the area of the insert membrane.

2.6. Immunohistochemistry of RPE Cells

The cells grown on the Transwell insert membrane were washed twice with PBS and fixed with 4% paraformaldehyde for 15 min at 37 °C. A volume of 500 μL was added to the top and 1000 μL to the bottom chamber. After PBS washing, the membranes were removed from the inserts with a sterile #11 scalpel blade. The cells were then permeabilized with 0.25% Triton X-100 for 5 min and blocked with 1% BSA in PBS for 30 min at RT. Antibodies were diluted in 1% BSA/PBS and incubated with the cells overnight at 4 °C (as described in Supplementary Table S1). After washing with PBS, the cells were incubated for 1 h at RT in the dark with the Alexa Fluor-conjugated secondary antibody (Invitrogen, Carlsbad, CA, USA); in this case the Alexa Fluor-conjugated primary antibody was not used. The cells were washed with PBS and counterstained with 1 μg/mL 4′,6′-diamidino-2-phenylindole (DAPI) for 5 min. They were placed on a glass slide with the cell side up, covered with a coverslip, and fluorescence images were obtained using a laser scanning confocal microscope (Leica TCS STED, Leica Microsystem Ltd., Wetzlar, Germany). We analyzed four cellular morphometric parameters, including cell area, perimeter, aspect ratio, and circularity, from images obtained via confocal microscopy, according to the Image J user guidelines. We excluded the cells at the image edges and those that could not be fully identified. The spatial scale of the images was defined so that the measurement result could be displayed in calibrated units in μm. We calculated the area (μm^2) and length (μm) of the selected region by selecting the outside boundary of the f-actin or ZO-1-stained each cell. Aspect ratio and circularity were calculated using the shape descriptor of Image J in the selected cell area. Aspect ratio (major axis/minor axis) is calculated as the ratio between the major and minor axes of an ellipse, and this parameter is higher in elongated cells. Circularity ($[4\pi(area/perimeter)^2]$) is a shape descriptor that mathematically indicates the degree of similarity to a perfect circle, with a value close to 1.0 designating a perfect circle and a value close to 0.0 indicating a less circular shape.

2.7. Immunohistochemistry of RCS Complex

RCS complexes were carefully separated from the contralateral eye of the donor and fixed with 4% paraformaldehyde using a dissecting microscope (Nikon SM2745T, Nikon Corporation, Tokyo, Japan). The RCS complexes were permeabilized by incubating them with Perm/Block solution (0.2% Triton-X 100 and 0.3% BSA in PBS) for 2 h at RT. They were then incubated with antibodies diluted with Perm/Block solution (0.2% Triton-X 100 and 0.3% BSA in PBS) overnight at 4 °C, using the antibodies described in Supplementary Table S1. The RCS complexes were washed with PBS and counterstained with 1 μg/mL 4′,6′-diamidino-2-phenylindole (DAPI) for 15 min. After washing with PBS, the RCS complexes were placed on a glass microscope slide with the RPE layer facing up. After coverslip, fluorescence images were obtained using a laser scanning confocal microscope (Leica TCS STED, Leica Microsystem Ltd., Wetzlar, Germany). Based on the research finding that the location within the eye can affect cell growth, we classified the RPE whole mount into 3 regions: central, equatorial, and peripheral. Six images were randomly acquired from each region, and the cell morphology was analyzed using the same method [22,23].

2.8. Crystal Violet Staining Assay

RPE cells were seeded at 1×10^4 cells per wells of a 12-well plate with medium containing 10%, 5%, and 2.5% FBS or in the serum-free medium. After 3, 5, or 7 days of culture, each well was washed with PBS and fixed with 4% paraformaldehyde. Cells were stained with 0.5% crystal violet solution (Sigma-Aldrich, St. Louis, MO, USA) for 30 min. The plate was washed 5 times with PBS and air-dried. Crystal violet was eluted from the cells by incubation with 1% SDS solution for at least 1 h and the absorbance was read at 570 nm under a microplate reader (Infinite® 200 pro, Tecan, Switzerland). ARPE19 cells cultured in medium containing 10% FBS were used as a control group for quantification of primary marmoset RPE cells.

2.9. Cell Cycle Analysis

RPE cells were harvested and fixed with 70% ethanol for 1 h at 4 °C. The cells were washed twice with PBS, and stained with Propidium iodide/RNAses solution (Invitrogen, Carlsbad, CA, USA) for 30 min before being examined using a flow cytometer (Accuri C6 Plus, BD Biosciences, Franklin Lakes, NJ, USA). The data were analyzed using a BD Accuri C6 software (Version 1.0.264.21, BD Biosciences, Franklin Lakes, NJ, USA).

2.10. Western Blot Analysis

The protein extracts were prepared using RIPA buffer (1% Triton X-100). The protein contents were measured using the BCA protein assay kit (Pierce, Rockford, IL, USA). Proteins were separated using SDS-PAGE and transferred onto polyvinylidene fluoride (PVDF) membranes (Merck Millipore Ltd., Tullagreen, Ireland). The membrane was incubated in primary antibody diluted with 1% non-fat milk overnight at 4 °C with gentle rocking. The membrane was washed 3 times and incubated with the appropriate diluted secondary antibody for 1 h at RT. The antibodies used for immunoblotting are summarized in Supplementary Table S1. Immunoreactive bands were visualized using ImageQuant™ LAS4000 (GE Healthcare, Chicago, IL, USA).

2.11. Statistical Analysis

The results are presented as the mean ± SD. Differences between two groups were assessed with the two-tailed Student's unpaired *t*-test. The one-way ANOVA with Tukey's multiple comparison test was used to assess differences between more than two groups as indicated in figure legends. Statistical analysis was performed with GraphPad Prism (San Diego, CA, USA) version 7.0 software. Data were considered statically significant at $p < 0.05$ (* $p < 0.05$, ** $p < 0.01$).

3. Results

3.1. Characterization of RPE Cells

After 24 h of initial incubation using the RPE medium containing 10% FBS, cells of a marmoset aged 6 days old were attached to T75 flasks with high pigmentation (Figure 3a). The cells with high pigmentation proliferated rapidly (Figure 3b), exhibiting packed round shape colonies (Figure 3c). After 7 days, the typical morphology of RPE monolayer resembling RPE in vivo was visible (Figure 3d). Proteins abundantly expressed in the RPE, including RPE65, were identified through Western blot (Figure 3e), similar to ARPE19 cells (Figure 3f).

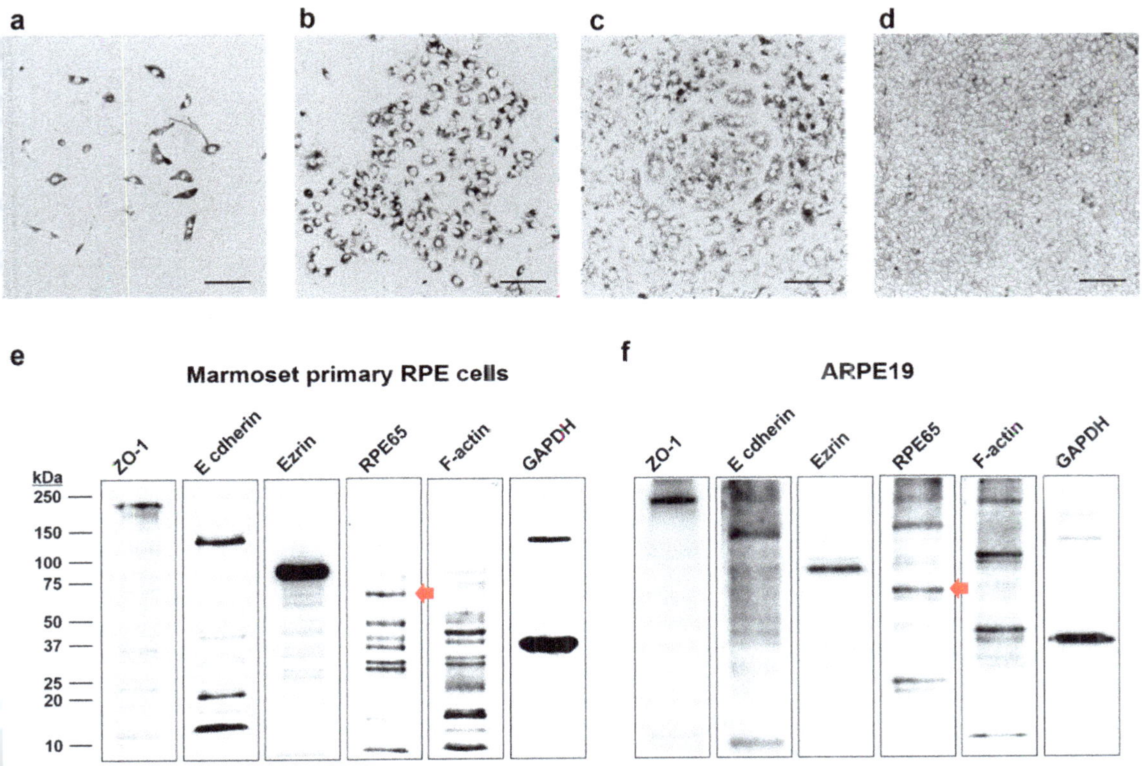

Figure 3. Light microscopic images and Western blot analysis of primary marmoset RPE cells. Light microscopic images show the time-dependent growth pattern of RPE cells in a T75 flask. (**a**) Day 1 cells were attached, and the initial growth of the culture with high pigmentation. (**b**) Day 3 cells were grown rapidly while maintaining high pigmentation. (**c**) Day 5 cells continued to undergo rapid proliferation while forming round-shaped colonies. (**d**) Day 7 cells in an early passage reached confluency and appeared with a regular polygonal shape on a T75 flask. (**e**,**f**) Western blot analysis of ZO-1 (molecular weight: 220 kDa), E-cadherin (molecular weight: 135 kDa), ezrin (molecular weight: 81 kDa), f-actin (molecular weight: 42 kDa), GAPDH (molecular weight: 37 kDa), and RPE65 (molecular weight: 61 kDa, red arrow), the RPE-specific protein, in the primary marmoset RPE cells (**e**) and ARPE19 cells (**f**). Scale bar, 20 μm.

The cultured monolayer's similarity to natural RPE was assessed with TER measurement and immunohistochemistry. In Figure 4a, it is shown that RPE cells were able to adhere and form highly differentiated monolayers with a high TER for over six weeks using B27 alone. Notably, a high TER of 163.65 ± 4.16 Ω cm^2 was achieved after only 2 weeks of Transwell culture, which significantly increased to 229.55 ± 7.27 Ω cm^2 at 5 weeks. The generated RPE monolayer showed regular polygonal geometrical features with ZO-1 and f-actin overlapping in apical-lateral contact with neighboring cells, similar to those observed in the contralateral eye of the same donor (Figure 4b).

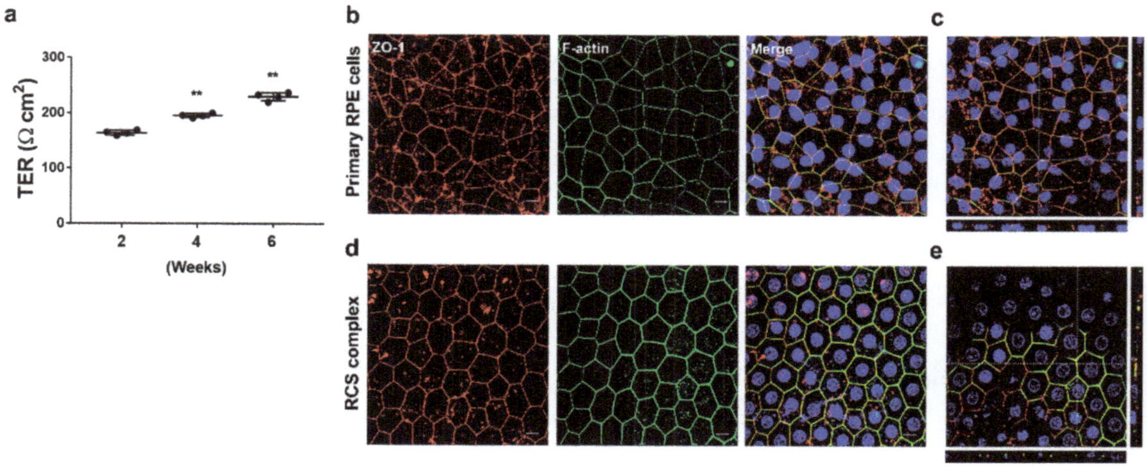

Figure 4. Characterization of RPE monolayers in comparison to the RPE in vivo. Monolayers with functional tight junctions were generated by culturing for 6 weeks in the serum-free medium alone in a fibronectin-coated Transwell insert and comparing the morphological features of RPE in the contralateral eye of the same donor. (**a**) Time-dependent increase in TER from multiple Transwells containing primary marmoset RPE cells from a single donor. (**b**) Merged images of primary marmoset RPE cells for immunofluorescence patterns for tight junction protein ZO-1 and f-actin visualized by confocal microscopy. (**c**) Upper panel: En-face view of a polarized RPE layer through the z-axis. Bottom and right panel: Cross-section through the z-plane of multiple optical slices at the location indicated by the white reference line in the corresponding upper panel. (**d**) Merged images of RCS complex for immunofluorescence in vivo patterns for ZO-1 and f-actin. (**e**) Upper panel: En-face view of the RCS complex through the z-axis. Bottom and right panel: Cross-section through the z-plane of multiple optical slices at the location indicated by the white reference line in the corresponding upper panel. Scale bar, 10 μm. Data are presented as the mean ± SD and analyzed by ANOVA. Tukey post hoc test: ** $p < 0.01$.

3.2. Changes in Cellular Morphology Depending on the Culture Medium

To assess the effect of FBS on the formation of monolayers with functional tight junctions during the initial RPE culture, the cells were cultured in a medium containing 10% FBS or a serum-free medium for 14 days after reaching confluency. Although the cells achieved a high TER of 181.19 ± 10.99 Ω cm^2 in the 10% FBS-containing medium, the tight junction function was significantly impaired compared to the serum-free medium (Figure 5a). The presence of FBS in the medium resulted in an overall reduction in the area of cell-to-cell contact, and it also induced morphological changes such as an increase in cell size and elongated spindle-like shapes that resemble those observed during an epithelial–mesenchymal transition (EMT) (Figure 5b,c). We reduced the time required to form polarized monolayers with a high TER (204.99 ± 8.64 Ω cm^2) to 14 days by changing to the serum-free medium after cells reached confluency. In addition, it was suggested that the serum-free medium is more suitable for forming highly differentiated monolayers with high TER than the FBS-containing medium in primary marmoset RPE cell culture.

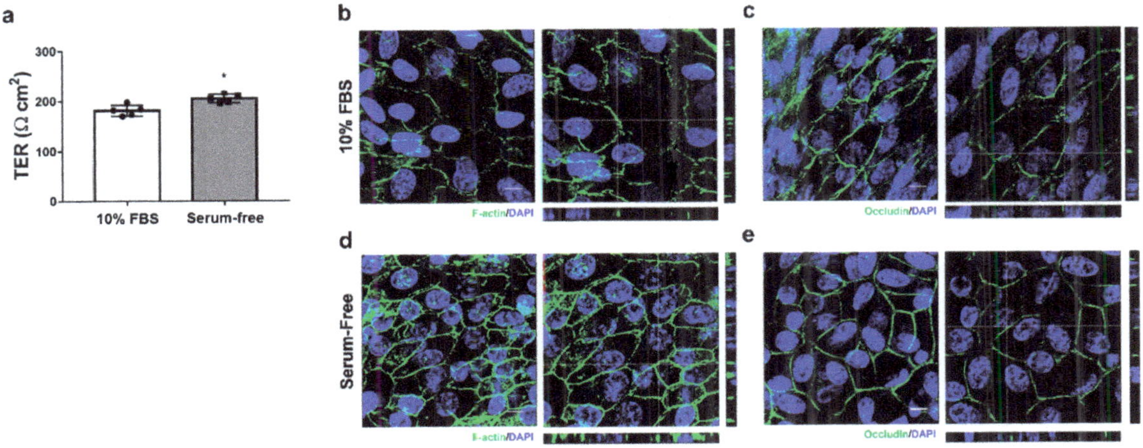

Figure 5. Comparison of RPE monolayer formation ability. The generation of RPE monolayers in a fibronectin-coated Transwell insert was compared between a serum-free medium and a medium containing 10% FBS. (**a**) TER increased with culture in the serum-free medium. (**b**,**c**) Immunofluorescence patterns for ZO-1 and occludin visualized by confocal microscopy in the RPE cells cultured with the growth medium containing 10% FBS for 13 days. Cell size increased and morphological features changed to an elongated-spindle shape. (**d**,**e**) Immunofluorescence patterns for ZO-1 and occludin visualized by confocal microscopy in the RPE cells cultured for 10 days with the serum-free medium after 3 days incubation with the growth medium containing 10% FBS. Scale bar, 10 µm. Data are presented as the mean ± SD and analyzed by two-tailed Student's unpaired *t*-test: * $p < 0.05$.

3.3. Changes in Cellular Morphology and TER according to Donor Age

Primary RPE cells from a 7-year-old and a 6-day-old marmoset were isolated and cultured in Transwell inserts for six weeks using the serum-free medium. The RPE cells isolated from the 7-year-old marmoset exhibited significantly low TER (100.69 ± 2.58 Ω cm^2) than cells isolated from the 6-day-old marmoset (214.29 ± 4.42 Ω cm^2) (Figure 6a). In RPE cells isolated from the 6-day-old marmoset, we observed that ZO-1 and F-actin were highly expressed at the periphery of the cells and colocalized with each other, forming a clear line at the lateral boundaries between adjacent cells. The polygonal cells formed a regular geometric pattern by contacting their neighbor cells, facilitating morphological analysis (Figure 6c). The cells had an area of 279.25 ± 27.77 µm^2, a perimeter of 65.16 ± 3.54 µm, an aspect ratio of 1.47 ± 0.29, and a circularity of 0.83 ± 0.05 on average. The cells derived from the 7-year-old marmoset exhibited fragmented ZO-1 expression at the cell borders, making it difficult to identify individual cells for morphological analysis. F-actin was distributed not only along the cell border but also inside the cell with stress fibers and hardly colocalized with ZO-1 (Figure 6b). The cells grew into multiple layers without contact with adjacent cells after they reached confluency. The morphological analysis could not be carried out because identifying the boundaries of individual cells was challenging.

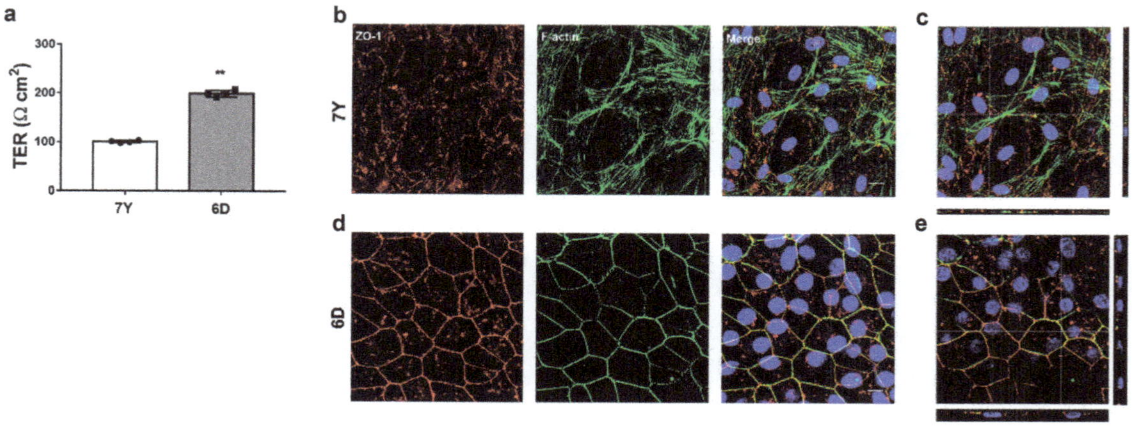

Figure 6. Comparison of morphological characteristics of primary RPE cells depending on donor's age. RPE cells isolated from a 7-year-old and a 6-day-old marmoset were cultured for 6 weeks in the serum-free medium alone in a fibronectin-coated Transwell insert. (**a**) TER significantly decreased in RPE cells isolated from the 7-year-old marmoset. (**b**) Immunofluorescence patterns for tight junction protein ZO-1 and f-actin visualized by confocal microscopy in RPE cells isolated from the 7-year-old marmoset. RPE cells failed to form monolayers with functional tight junctions in the features of overlapping ZO-1 and f-actin at the periphery of each cell. (**c**) Upper panel: En-face view of a polarized RPE layer through the *z*-axis. Bottom and right panel: Cross-section through the *z*-plane of multiple optical slices at the location indicated by the white reference line in the corresponding upper panel. (**d**) Immunofluorescence patterns for tight junction protein ZO-1 and f-actin visualized by confocal microscopy in RPE cells isolated from the 6-day-old marmoset. (**e**) Upper panel: En-face view of the RCS complex through the *z*-axis. Bottom and right panel: Cross-section through the *z*-plane of multiple optical slices at the location indicated by the white reference line in the corresponding upper panel. Scale bar, 10 μm. Data are presented as the mean ± SD and analyzed by two-tailed Student's unpaired *t*-test: ** $p < 0.01$.

To assess whether isolated RPE cells formed a similar structure to in vivo RPE, the RCS complex of the opposite eye of each donor was stained with ZO-1 and f-actin for immunohistochemistry. The RPE was observed in three geographic locations of the RPE flat mounts (Figure 7a). In the central and equatorial RPE of the 7-year-old marmoset, ZO-1 and f-actin were not only distributed along the cell border but also scattered throughout the cell inside. At the cell border, ZO-1 was fragmented or not colocalized with f-actin, which made it challenging to identify the boundaries of individual cells. In the peripheral region, ZO-1 and f-actin were mainly localized and colocalized at cell borders, enabling the identification of cell borders (Figure 7b). In the 6-day-old marmoset, ZO-1 and F-actin were highly expressed at the cell border and colocalized throughout the eyeball, making it easier to identify the cell boundaries. The cells formed a regular geometric pattern with hexagonal shapes (Figure 7c). While aspect ratio and circularity showed no significant differences among the regions, cell area and perimeter were significantly higher in the central region than in the equatorial and peripheral regions in the RPE of the 6-day-old marmoset (Figure 7d). The cell area and perimeter of the polarized monolayer of primary RPE cells were similar to the RPE in the equatorial region. However, it may depend on the seeding cell density.

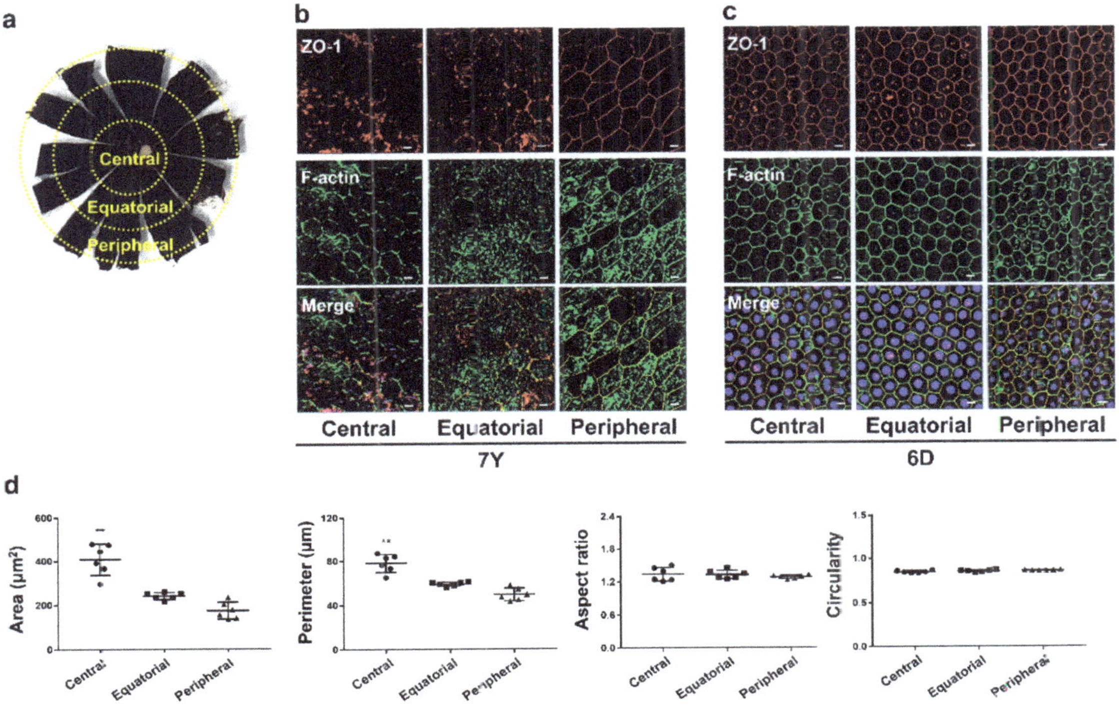

Figure 7. Characterization of age-related morphological changes in marmoset RPE. (**a**) RPE flat mount image showing the different geographic locations from which the images were obtained. (**b**) Immunofluorescence patterns for tight junction protein ZO-1 and f-actin visualized by confocal microscopy in RCS complex of the 7-year-old marmoset. Compared to the 6-day-old marmoset, morphological features are changed, including increased cell size, loss of uniform geometry, and inability to overlap ZO-1 and f-actin at the cell interface. (**c**) Immunofluorescence patterns for tight junction protein ZO-1 and f-actin visualized by confocal microscopy in RCS complex of the 6-day-old marmoset. (**d**) Morphological analysis of the 6-day-old marmoset was performed on six images, each from categorized locations within the eye, including cell area (μm^2), perimeter (μm), aspect ratio, and circularity. Scale bar, 10 μm. Data are presented as the mean ± SD and analyzed by ANOVA. Tukey post hoc test: ** $p < 0.01$.

3.4. Changes in Cell Proliferation with Donor Age and Growth Medium Composition

To evaluate the effects of donor age and fetal bovine serum (FBS) concentration in medium on the proliferation of primary RPE cells, we cultured cells isolated from a 7-year-old and a 6-day-old marmoset monkeys in media containing 0%, 2.5%, 5%, or 10% FBS for 7 days. We then stained the cells with crystal violet to assess proliferation (Figure 8a). The proliferation of RPE cells isolated from the 7-year-old marmoset in growth media containing 10% FBS significantly decreased compared to cells isolated from the 6-day-old marmoset. As a reference with ARPE cells, the most commonly used cells in RPE research, cell proliferation of cells isolated from the 6-day-old marmoset was 15.78% higher, while the cells isolated from the 7-year-old marmoset were around 12.64% lower in media containing 10% FBS (Figure 8b). The proliferation of RPE cells isolated from the 7-year-old (Figure 8c) and the 6-day-old marmoset (Figure 8d) significantly decreased as the concentration of FBS in the growth media decreased. Unlike ARPE19 cells, primary RPE cells did not show significant proliferation in the serum-free medium regardless of donor age. This suggests that culture medium may differ depending on the cellular proliferation and maturation of primary RPE cells as an in vitro model, emphasizing the importance of selecting appropriate media based on the experimental objectives.

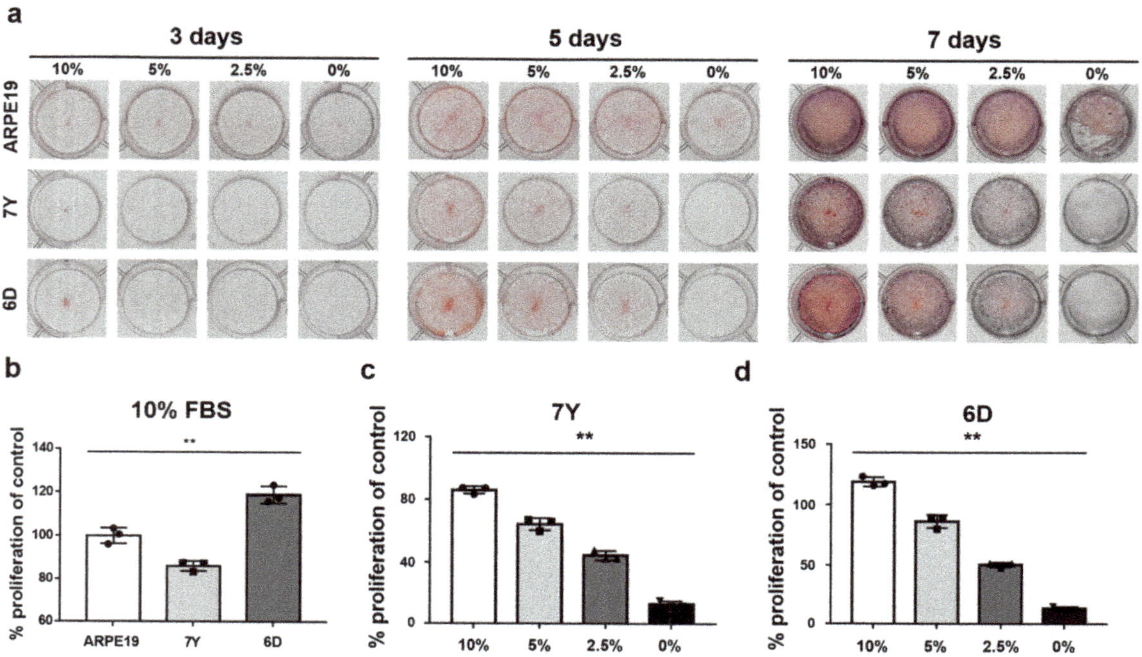

Figure 8. Proliferative potential of primary RPE cells changed by donor's age and FBS concentration in culture medium. Crystal violet staining was used to assess the proliferative changes of RPE cells isolated from a 7-year-old and a 6-day-old marmoset and ARPE19 cells depending on the various concentration of FBS in medium. ARPE19 cells cultured in medium containing 10% FBS were used as a control group for quantification of primary marmoset RPE cells. (**a**) Cells in 12-well plates stained with 0.5% crystal violet after culture for 3, 5, and 7 days in medium containing 10%, 5%, 2.5% FBS, and the serum-free medium. (**b**) RPE cells isolated from the 7-year-old marmoset significantly decrease proliferation even in 10% FBS-containing medium as measured by microplate reader by eluting crystal violet. (**c**,**d**) RPE proliferation significantly decreases as the FBS content decreases in the medium regardless of the donor's age. Data are represented as the mean ± SD and analyzed by ANOVA. Tukey post hoc test: ** $p < 0.015$.

3.5. Changes in Cell Cycle and Proteins Involved in Cell Cycle with Donor Age

To investigate whether the decrease in proliferative potential and morphological changes in RPE cells isolated from a 7-year-old marmoset was due to cellular senescence we analyzed cell cycle progression. RPE cells isolated from the 7-year-old marmoset exhibited an increased distribution in the G0/G1 phase, while the percentage of cells in the S phase decreased compared to cells isolated from the 6-day-old marmoset (Figure 9a,b) Additionally, the protein levels of cyclin D1, cyclin E, and CDK6 were significantly reduced whereas p53 and p21 were markedly upregulated in the cells isolated from the 7-year-old marmoset (Figure 9c,d). These results indicate that RPE cells isolated from the 7-year-old marmoset exhibited G0/G1 arrest.

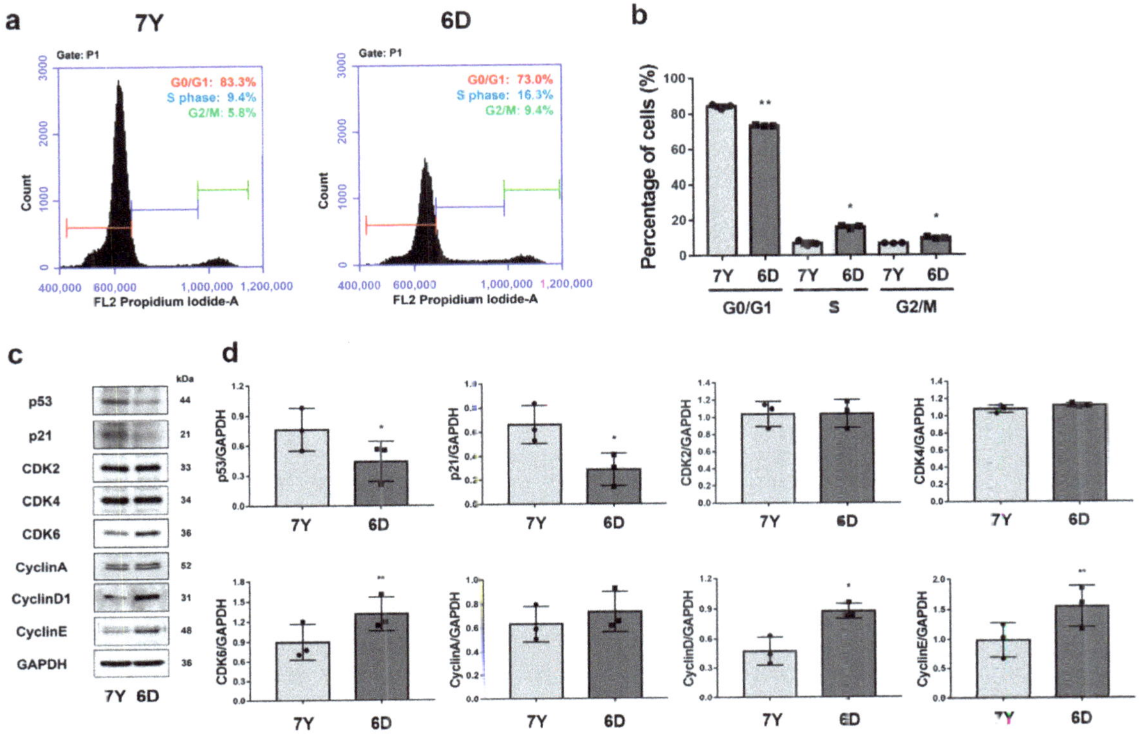

Figure 9. RPE cells isolated from a 7-year-old marmoset showed cell cycle delay in G0/G1 phase. Cell cycle analysis and Western blot assay for cell cycle protein expression perform to assess whether the cell proliferation was associated with an age-related cell cycle arrest in primary marmoset RPE cells. (**a**,**b**) In cell cycle analysis, RPE cells from the 7-year-old marmoset (7Y) are in cell cycle delay at G0/G1 and decrease in percentages of S and G2/M phases compared to RPE cells from the 6-day-old marmoset (6D). (**c**,**d**) RPE cells isolated from the 7-year-old marmosets show decreased cyclin D and cyclin E while elevated p53 and p21 expression levels, compared to RPE cells from the 6-day-old marmoset, suggesting the suppression of G1-to-S-phase progression. Data are represented as the mean ± SD and analyzed by ANOVA. Tukey post hoc test: ** $p < 0.01$ and * $p < 0.05$.

4. Discussion

We aimed to provide the potential of primary marmoset RPE cells as an in vitro model by developing a simple and defined method for their isolation, proliferation, and differentiation while maintaining epithelial integrity, with the characterization of cells from donors of different ages. There are several cell culture models to study RPE biology, including spontaneously formed cell lines (e.g., ARPE19), immortalized cell lines (e.g., hTERT-RPE1, RPE-J, and D407), primary human or animal RPE cells, and embryonic and induced pluripotent stem cell (iPSC) derived RPE [24]. ARPE-19 cells are widely used due to their ability to mimic many in vivo hallmarks when cultured under conditions that promote differentiation [25]. However, they have abnormal karyotypes and have been passaged over many years, reducing the ability to demonstrate the hallmarks of differentiated RPE. Additionally, cell phenotypes such as hexagonicity, pigmentation, junction resistance, and polarity are considered medium to low compared to primary RPE cells [18]. Primary RPE cultures were established from freshly harvested RPE obtained from mouse, pig, or human donors [24]. The time to maturity varied by species, but cultures formed well-differentiated monolayers with tight junctions and high TER. Unfortunately, while there has been a report of differentiating RPE from marmoset embryonic stem cells, no primary cultures of RPE have been reported In marmosets [16,26]. The common marmoset (*Callithrix jacchus*) possesses typical ocular features as a primate, and genetic modification is relatively more feasible to generate than other laboratory monkeys, making them valuable for research on congenital visual disorders [15]. The advantages of the in vitro culture models are potential reliability, consistency, and low cost compared to in vivo studies. Development and characterization of primary marmoset RPE cells as an in vitro model that recapitulates the RPE's morphology and function in vivo can be a reliable high-throughput platform for preclinical studies before using marmosets.

The cell culture model of RPE in marmosets can play an important role in better understanding RPE cell biology and RPE dysfunction related to AMD pathogenesis. Primary RPE cultures have been established from mouse, porcine, and human donors [16]. The use of marmosets to obtain primary RPE cells is not more widely accessible than rodents and is not supported by the reference provided, which details characterization of isolated primary mouse RPE [19]. Since many cells could be isolated from one eye, various molecular biological analyses, including proteomic analysis, were possible without cell pooling. In addition, primary mouse RPE could not maintain high TER, pigmentation, and similar morphology in vivo after division with trypsin [19]. In contrast, marmoset cells maintained morphological characteristics and high TER after trypsin division and cryopreservation. Most of the commercial antibodies in this study responded well to antigen–antibody reactions, even though there was no application information in monkeys. RPE cells isolated from newborn marmoset monkeys within one week of birth exhibit similar characteristics to those of humans, including high pigmentation, a polarized monolayer with high TER, and RPE-specific expression proteins, including ZO-1, E cadherin, ezrin, f-actin, and RPE65. These cells attach to fibronectin-coated Transwell membranes in the serum-free media and, over 6 weeks, achieve high TER. After reaching confluency in serum-containing media, they form similar geometric patterns of RPE within 2 weeks in the serum-free medium. ZO-1 and f-actin were mainly observed at the periphery of the cells and co-localized, facilitating the demarcation of cell boundaries, allowing morphological analysis, including cell area (μm^2), perimeter (μm), aspect ratio, and circularity of each cell. The morphological features of matured primary RPE cells were similar to the RPE in vivo from the same donor. Morphological analysis revealed that the cells exhibited a comparable cell area and perimeter to the equatorial region of the RPE whole mount. However, it may vary depending on the cell seeding density. As is known, FBS in the media inhibits the formation of monolayers in primary marmoset RPE cells and induces morphological changes, such as increased cell size and an EMT-like spindle shape. Although serum is necessary for the growth and initial adhesion of RPE cells in culture plates, its presence in culture medium can be problematic due to various unknown hormones, cytokines, and growth factors. Exposure to conventional serum levels retracts experimental reproducibility due to a variable and batch-dependent mixing of growth factors, resulting in morphological heterogeneity, dedifferentiation, and EMT-like phenotypes [27,28]. However, there was a notable difference in the proliferative potential of RPE cells between the serum-free medium with B27 supplements and media containing FBS. B27 supplement, a highly enhanced form of N2 initially designed to support the maintenance of hippocampal neurons and neuronal cell lines, can support the proliferation and maintenance of RPE cells even in the absence of N2 supplement [7,29,30]. Similarly, in the serum-free media supplemented with B27, cells primarily attached and reached confluency naturally over time, followed by adopting a compact, polygonal morphology and repigmentation over six weeks. The addition of serum shortened the time required to reach confluency, and subsequently the serum-free media supported the formation of a polarized monolayer with regular geometric features and high TER. Despite serum exposure during sample collection persisting as serum contaminants despite multiple washings and incubation in the serum-free medium [31], the characteristic morphological integrity of the RPE cells was exhibited after at least 14 days of the serum-free culture. TER measurement is commonly used to measure the establishment of functional tight junctions between RPE cells with high sensitivity and reliability. Primary marmoset RPE cells, which formed monolayers with functional tight junction using the serum-free media, achieved a high TER of over 200. It is higher than the low TER (<50 Ω cm^2) observed in cell lines such as ARPE19 and D407, and comparable to the typical TER value for human RPE monolayer in vivo (approximately 150 Ω cm^2) and cultured RPE monolayer (25–200 Ω cm^2) [18,32–34].

Primary RPE cells undergo dedifferentiation by losing their pigmentation and characteristic morphology and acquire a fibroblastic morphology during passages, which is considered useful for studying the epigenetic assessment of AMD progression [24]. The primary RPE cells in humans are ideal for research, but there are limitations, such as the availability of donor tissue, restrictions on the age of available donors, and the need to collect cells within a short post-mortem interval. Marmoset RPE culture has several advantages, including high visual and immune systems similarity with humans and a relatively feasible use of donor tissues with a wide range of donor ages. In addition, it was possible to compare single-cell and tissue levels by observing in vivo changes in the contralateral eye of the donor. Morphological changes were exhibited in the cells isolated from a 7-year-old marmoset, including increased cell size, loss of uniform geometry, and inability to overlap ZO-1 and f-actin at the cellular boundary. These morphological changes were also observed similarly in vivo, particularly prominent in the central and equatorial regions of the RPE whole mount. Isolated RPE cells also did not form a polarized monolayer with high TER. RPE cells form a monolayer network through cell–cell adhesion, including tight junctions with neighboring cells. Tight junctions are a type of cell–cell adhesion that functions as a paracellular by blocking the movement of plasma and toxic substances into the retina and controlling the flow of fluids and ions across the RPE from the choroid to the retina. ZO-1 is a membrane-associated tight junction adapter protein connecting transmembrane proteins to the cytoplasmic actin for initial formation and distinct organization of tight junctions [2]. Depletion of ZO-1 in MDCK cells results in the specific defects in the barrier for large solutes, accompanied by morphological changes and reorganization of apical actin and myosin [35]. Overexpression of ZONAB of knockdown of its cellular inhibitor ZO-1 leads to enhanced proliferation of RPE cells and induced structural changes an EMT-like phenotype [36]. The presence of rich stress fibers in the f-actin cytoskeleton was directly correlated with poor phagocytic activity in RPESC-RPE cells, while the lateral circumferential f-actin and the lack of stress fibers were directly associated with a high phagocytic activity [37]. RPE cells isolated from a 7-year-old marmoset exhibited an abundance of stress fibers and a lack of contiguous lateral circumferential f-actin, indicating a potential decline in RPE cell function, including phagocytic activity. The improper colocalization of ZO-1 and f-actin leads to compromised cell-to-cell junctions with neighboring cells, potentially resulting in decreased TER and impaired barrier function. Along with morphological changes, signs of cellular senescence, such as decreased cell proliferation and delayed G0/G1 phase, were identified, suggesting that RPE cells isolated from aged marmosets could be an in vitro model for age-related RPE dysfunction, including AMD.

We isolated RPE cells from marmosets for the first time and cultured native RPE monolayers that closely resembled those observed in the donor's RPE in vivo. We facilitated the formation of highly differentiated RPE monolayers with high TER by using B27 supplements instead of serum. We also aimed to characterize primary RPE cells isolated from a 7-year-old marmoset to identify the morphological changes that occur naturally in marmosets. The cells exhibited known age-related morphological changes, including cell size increase, uniform geometric loss, abundant stress fibers, and loss of ZO-1 and f-actin colocalization at cell boundaries with signs of cellular senescence. While further studies are required to determine whether primary marmoset RPE cells recapitulate the morphology and function of the RPE, primary marmoset RPE cultures can serve as an in vitro model to develop effective novel therapies in the future.

5. Conclusions

Primary marmoset RPE cells share similar features with RPE cells in vivo and primary human RPE cells, making them suitable for molecular studies. In addition, RPE cells isolated from the older marmoset exhibit age-related changes. Establishing an in vitro model based on the RPE of marmosets can promote the identification of epigenetic factors underlying RPE dysfunction, including AMD, and enable the development of cell-based model systems that mimic the morphological and functional abnormalities associated with AMD. This model provides a reliable and high-throughput platform for preclinical testing of potential therapies and pharmaceuticals, which can be combined with animal experiments using marmosets.

Supplementary Materials: The following supporting information can be downloaded at: https://www.mdpi.com/article/10.3390/cells12121644/s1, Table S1: List of used antibodies.

Author Contributions: J.H.K. and Y.H.S. conceptualized experiments. B.-C.K. conceptualized experiment using marmosets. H.Y.J. performed isolation, culture, and characterization of primary marmoset RPE cells. Y.M.S. and C.S.C. participated in cell cycle analysis. J.K. performed experiments using marmosets. All authors have read and agreed to the published version of the manuscript.

Funding: This research was supported by the Creative Materials Discovery Program through the National Research Foundation of Korea (NRF) funded by the Ministry of Science and ICT (2018M3D1A1058826 to J.H.K.), the National Research Foundation of Korea (NRF) Grants (NRF-2021M3H9A1097442 to B.C.K., NRF-2022M3A9E4082652 to Y.H.S., and NRF-2022M3A9E4017127 to J.H.K.), the Kun-hee Lee Child Cancer & Rare Disease Project, Republic of Korea (202200004004 to J.H.K.), and the Development of Platform Technology for Innovative Medical Measurement funded by Korea Research Institute of Standards and Science (KRISS-GP2022-0006 to J.H.K.).

Institutional Review Board Statement: Marmoset eyes were obtained from the Marmoset Model Network Center in Seoul National University Hospital. All marmoset experiment protocols were followed as set by the Institutional Animal Care and Use Committee of Seoul National University Hospital (IACUC No. 22-0069), an accredited research institute of the Association for Assessment and Accreditation of Laboratory Animal Care International (AAALAC) approved facility.

Informed Consent Statement: Not applicable.

Data Availability Statement: The data that support the findings of this study are available from the corresponding author upon reasonable request.

Conflicts of Interest: The authors declare no conflict of interest.

References

1. Hewitt, A.; Adler, R. The retinal pigment epithelium and interphotoreceptor matrix: Structure and function. In *Retina: Basic Science and Inherited Retinal Disease*, 2nd ed.; Ogden, T.E., Ryan, S.J., Eds.; The C.V. Mosby Company: St. Louis, MI, USA, 1994; Volume 1, pp. 57–64.
2. Naylor, A.; Hopkins, A.; Hudson, N.; Campbell, M. Tight junctions of the outer blood retina barrier. *Int. J. Mol. Sci.* **2019**, *21*, 211. [CrossRef] [PubMed]
3. Yang, S.; Zhou, J.; Li, D. Function and diseases of the retinal pigment epithelium. *Front. Pharmacol.* **2021**, *12*, 727870. [CrossRef] [PubMed]
4. Fleckenstein, M.; Keenan, T.D.L.; Guymer, R.H.; Chakravarthy, U.; Schmitz-Valckenberg, S.; Klaver, C.C.; Wong, W.T.; Chew, E.Y. Age-related macular degeneration. *Nat. Rev. Dis. Prim.* **2021**, *7*, 31. [CrossRef] [PubMed]
5. Pennesi, M.E.; Neuringer, M.; Courtney, R.J. Animal models of age related macular degeneration. *Mol. Asp. Med.* **2012**, *33*, 487–509. [CrossRef] [PubMed]
6. Forest, D.L.; Johson, L.V.; Clegg, D.O. Cellular models and therapies for age-related macular degeneration. *Dis. Models Mech.* **2015**, *8*, 421–427. [CrossRef]
7. Gamm, D.M.; Melvan, J.N.; Shearer, R.L.; Pinilla, I.; Sabat, G.; Svendsen, C.N.; Wright, L.S. A novel serum-free method for culturing human prenatal retinal pigment epithelial cells. *Investig. Ophthalmol. Vis. Sci.* **2008**, *49*, 788–799. [CrossRef]
8. Pfeffer, B.A.; Fliesler, S.J. Reassessing the suitability of APRE-19 cells as a valid model of native RPE biology. *Exp. Eye Res.* **2022**, *219*, 109046. [CrossRef]

9. Maminishkis, A.; Chen, S.; Jalickee, S.; Banzon, T.; Shi, G.; Wang, F.E.; Ehalt, T.; Hammer, J.A.; Miller, S.S. Confluent monolayers of cultured human fetal retinal pigment epithelium exhibit morphology and physiology of native tissue. *Investig. Ophthalmol. Vis Sci.* **2006**, *47*, 3612–3624. [CrossRef]
10. Shang, P.; Stepicheva, N.A.; Hose, S.; Zigler, J.S., Jr.; Sinha, D. Primary cell cultures from the mouse retinal pigment epithelium *J. Vis. Exp.* **2018**, *133*, 56997.
11. Julien, S.; Schraermeyer, U. Lipofuscin can be eliminated from the retinal pigment epithelium of monkeys. *Neurobiol. Aging* **2012**, *33*, 2390–2397. [CrossRef]
12. Uhl, P.B.; Szober, C.M.; Amann, B.; Alge-Priglinger, C.; Ueffing, M.; Hauck, S.M.; Deeg, C.A. In Situ surface proteomics reveals differentially expressed membrane proteins in retinal pigment epithelial cells during autoimmune uveitis. *J. Proteom.* **2014**, *109*, 50–62. [CrossRef]
13. Xia, H.; Krebs, M.P.; Kaushal, S.; Scott, E.W. Enhanced retinal pigment epithelium regeneration after injury in MRL/MpJ mice. *Exp. Eye Res.* **2011**, *93*, 862–872. [CrossRef] [PubMed]
14. Baek, D.S.; Liang, H.; Zhao, X.; Pankova, N.; Wang, H.; Boyd, S. Fundus autofluorescence (FAF) non-invasively identifies chorioretinal toxicity in a rat model of retinal pigment epithelium (RPE) damage. *J. Pharmacol. Toxicol. Methods* **2015**, *71*, 77–82. [CrossRef] [PubMed]
15. Mitchell, J.F.; Leopold, D.A. The marmoset monkey as a model for visual neuroscience. *Neurosci. Res.* **2015**, *93*, 20–46. [CrossRef] [PubMed]
16. Pfeffer, B.A. Improved methodology for cell culture of human and monkey retinal pigment epithelium. *Prog. Retin. Res.* **1991**, *10*, 251–291. [CrossRef]
17. Fronk, A.H.; Vargis, E. Methods for culturing retinal pigment epithelial cells: A review of current protocols and future recommendations. *J. Tissue Eng.* **2016**, *7*, 2041731416650838. [CrossRef]
18. Sonoda, S.; Spee, C.; Barron, E.; Ryan, S.J.; Kannan, R.; Hinton, D.R. A protocol for the culture and differentiation of highly polarized human retinal pigment epithelial cells. *Nat. Protoc.* **2009**, *4*, 662–673. [CrossRef]
19. Fernandez-Godino, R.; Garland, D.L.; Pierce, E.A. Isolation, culture and characterization of primary mouse RPE cells. *Nat. Protoc.* **2016**, *11*, 1206–1218. [CrossRef]
20. Toops, K.A.; Tan, L.X.; Lakkaraju, A. A detailed three-step protocol for live imaging of intracellular traffic in polarized primary porcine RPE monolayers. *Exp. Eye Res.* **2014**, *124*, 74–85. [CrossRef]
21. Sasaki, E.; Suemizu, H.; Shimada, A.; Hanazawa, K.; Oiwa, R.; Kamioka, M.; Tomioka, I.; Sotomaru, Y.; Hirakawa, R.; Eto, T.; et al. Generation of transgenic non-human primates with germline transmission. *Nature* **2009**, *459*, 523–527. [CrossRef]
22. Kim, Y.K.; Yu, H.; Summers, V.R.; Donaldson, K.J.; Ferdous, S.; Shelton, D.; Zhang, N.; Chrenek, M.A.; Jiang, Y.; Grossniklaus, H.E.; et al. Morphometric analysis of retinal pigment epithelial cells from C57BL/6J mice during aging. *Investig. Ophthalmol. Vis. Sci.* **2021**, *62*, 32. [CrossRef] [PubMed]
23. Chen, M.; Rajapakse, D.; Fraczek, M.; Luo, C.; Forrester, J.V.; Xu, H. Retinal pigment epithelial cell multinucleation in the agining eye—A mechanism to repair damage and maintain homoeostasis. *Aging Cell* **2016**, *15*, 436–445. [CrossRef] [PubMed]
24. Bharti, K.; Hollander, A.I.; Lakkaraju, A.; Sinha, D.; Williams, D.S.; Finnemann, S.C.; Bowes-Rickman, C.; Malek, G.; D'Amore, P. Cell culture models to study retinal pigment epithelium-related pathogenesis in age-related macular degeneration. *Exp. Eye Res.* **2022**, *222*, 109170. [CrossRef] [PubMed]
25. Samuel, W.; Jaworski, C.; Postnikova, O.A.; Kutty, R.K.; Duncan, T.; Tan, L.X.; Poliakov, E.; Lakkaraju, A.; Redmond, T.M. Appropriately differentiated ARPE-19 cells regain phenotype and gene expression profiles similar to those of native RPE cells. *Mol. Vis.* **2017**, *23*, 60–89. [PubMed]
26. Torrez, L.B.; Perez, Y.; Yang, J.; Nieden, N.I.Z.; Klassen, H.; Liew, C.G. Derivation of neural progenitors and retinal pigment epithelium from common marmoset and human pluripotent stem cells. *Stem Cells Int.* **2012**, *2012*, 417865. [CrossRef]
27. Tian, H.; Xu, J.; Tian, Y.; Cao, Y.; Lian, C.; Ou, Q.; Wu, B.; Jin, C.; Gao, F.; Wang, J.; et al. A cell culture condition that induces the mesenchymal-epithelial transition of dedifferentiated porcine retinal pigment epithelial cells. *Exp. Eye Res.* **2018**, *177*, 160–172. [CrossRef]
28. Chen, Z.; Shao, Y.; Li, X. The roles of signaling pathways in epithelial-to-mesenchymal transition of PVR. *Mol. Vis.* **2015**, *21*, 706–710.
29. Hu, J.; Bok, D. A cell culture medium that supports the differentiation of human retinal pigment epithelium into functionally polarized monolayers. *Mol. Vis.* **2001**, *7*, 14–19.
30. Tezel, T.H.; Del Priore, L.V. Serum-free media for culturing and serial-passaging of adult human retinal pigment epithelium. *Exp. Eye Res.* **1998**, *66*, 807–815. [CrossRef]
31. An, E.; Lu, X.; Flippin, J.; Devaney, J.M.; Halligan, B.; Hoffman, E.P.; Strunnikova, N.; Csaky, K.; Hathout, Y. Secreted proteome profiling in human RPe cell cultures derived from donors with age related macular degeneration and age matched healthy donors. *J. Proteome Res.* **2006**, *5*, 2599–2610. [CrossRef]
32. Pequignot, M.O.; Provost, A.C.; Sallé, S.; Menasche, M.; Saule, S.; Jaïs, J.; Abitbol, M. The retinal pigment epithelium undergoes massive apoptosis during early differentiation and pigmentation of the optic cup. *Mol. Vis.* **2011**, *17*, 989–996.
33. Liu, Y.; Zhang, D.; Wu, Y.; Ji, B. Docosahexaenoic acid aggravates photooxidative damage in retinal pigment epithelial cells via lipid peroxidation. *J. Photochem. Photobiol. B* **2014**, *140*, 85–93. [CrossRef] [PubMed]
34. Rizzolo, L.J. Barrier properties of cultured retinal pigment epithelium. *Exp. Eye Res.* **2014**, *126*, 16–26. [CrossRef] [PubMed]

35. Van Itallie, C.M.; Fanning, A.S.; Bridges, A.; Anderson, J.M. Zo-1 stanilizes the tight junction solute barrier through coupling to the perijunctional cytoskeleton. *Mol. Biol. Cell* **2009**, *20*, 3930–3940. [CrossRef]
36. Georgiadis, A.; Marion Tschernutter, M.; Bainbridge, J.W.B.; Balaggan, K.S.; Mowat, F.; West, E.L.; Munro, P.M.G.; Thrasher, A.J.; Matter, K.; Balda, M.S.; et al. The tight junction associated signalling proteins ZO-1 and ZONAB regulate retinal pigment epithelium homeostasis in mice. *PLoS ONE* **2010**, *5*, e15730.
37. Müller, C.; Charniga, C.; Temple, S.; Finnemann, S.C. Quantified f-actin morphology is predictive of phagocytic capacity of stem cell-derived retinal pigment epithelium. *Stem Cell Rep.* **2018**, *10*, 1075–1087. [CrossRef] [PubMed]

cells MDPI

Article

Visual Function and Survival of Injured Retinal Ganglion Cells in Aged Rbfox1 Knockout Animals

Lei Gu [1], Jacky M. K. Kwong [1], Joseph Caprioli [1,2] and Natik Piri [1,2,*]

1 Stein Eye Institute, University of California, Los Angeles, CA 90095, USA
2 Brain Research Institute, University of California, Los Angeles, CA 90095, USA
* Correspondence: piri@jsei.ucla.edu; Tel.: +310-825-9850; Fax: +310-794-2144

Abstract: Rbfox1 is a multifunctional RNA binding protein that regulates various aspects of RNA metabolism important for neuronal differentiation and normal physiology. Rbfox1 has been associated with neurodevelopmental and neurological conditions as well as age-related neurodegenerative diseases such as Alzheimer's and Parkinson's. We have shown that in mammalian retinas Rbfox1 is expressed in retinal ganglion cells (RGCs) and in amacrine cells (ACs). This study investigates the effect of advanced age (22-month-old mice) on visual function, retinal morphology and survival of injured retinal ganglion cells (RGC) in Rbfox1 knockout (KO) animals. A visual cliff test, which was used to evaluate visual function, showed that 22-month old Rbfox1 KO mice have profound depth perception deficiency. Retinal gross morphology in these animals appeared to be normal. Optic nerve crush (ONC) induced axonal injury resulted in approximately 50% of RGC loss in both Rbfox1 KO and age-matched control animals: the average RGC densities in uninjured control and Rbfox1 KO animals were 6274 ± 1673 cells/mm^2 and 6004 ± 1531 cells/mm^2, respectively, whereas 1 week after ONC, RGC numbers in the retinas of control and Rbfox1 KO mice were reduced to 2998 ± 858 cells/mm^2 and 3036 ± 857 cells/mm^2, respectively (Rbfox1 KO vs. Rbfox1 KO + ONC, $p < 0.0001$ and control vs. control + ONC, $p < 0.0001$). No significant difference between RGC numbers in Rbfox1 KO – ONC and age-matched control + ONC animals was observed, suggesting that Rbfox1 has no effect on the survival of injured RGCs. Interestingly, however, contrary to a commonly accepted view that the number of RGCs in old (18 month of age) compared to young animals is reduced by approximately 40%, the RGC densities in 22-month-old mice in this study were similar to those of 4-month-old counterparts.

Keywords: retina; ganglion cells; amacrine cells; optic nerve crush; depth perception

Citation: Gu, L.; Kwong, J.M.K.; Caprioli, J.; Piri, N. Visual Function and Survival of Injured Retinal Ganglion Cells in Aged Rbfox1 Knockout Animals. *Cells* **2022**, *11*, 3401. https://doi.org/10.3390/cells11213401

Academic Editor: Hossein Ameri

Received: 23 August 2022
Accepted: 26 October 2022
Published: 27 October 2022

1. Introduction

Rbfox1 is a one of three members of the Rbfox family of multifunctional RNA binding proteins that are involved in regulation of RNA metabolism, including alternative splicing, transcription, mRNA stability and translation [1,2]. The diverse functions of these protein are supported by spatiotemporal expression of different isoforms generated by alternative splicing or from alternative promoters. Rbfox proteins, Rbfox1, Rbfox2 and Rbfox3, are expressed in neurons, heart and skeletal muscle, and disruption of their normal functions have been associated with neurodevelopmental and neuropsychiatric disorders, including autism spectrum disorder (ASD), intellectual disability, epilepsy, ADHD, bipolar disorder, schizoaffective disorder and schizophrenia, sleep latency and heart disease [3–9]. The expression of Rbfox genes overlap in most areas of the brain, but their spatial pattern of expression in the cerebellar cortex, for instance, is quite different: granule cells express Rbfox1 and Rbfox3, whereas Purkinje cells express Rbfox1 and Rbfox2 [10,11]. The Rbfox proteins also exhibit different subcellular localization. Rbfox1 expression is observed in both the cytoplasm and nucleus of Purkinje cells, whereas Rbfox2 is restricted to the nucleus. The Rbfox genes have distinct patterns of expression also during cerebellar development.

In the retina, expression of Rbfox1, Rbfox2 and Rbfox3 overlaps in retinal ganglion cells (RGCs) and amacrine cells (ACs) [12–15]. However, there are notable differences: Rbfox2 and Rbfox3, but not Rbfox1, are expressed in horizontal cells (HCs) and Rbfox2 is expressed in a wider population of ACs and displaced ACs (dACs) subtypes within the inner nuclear layer (INL) and ganglion cell layer (GCL), respectively, than Rbfox1. Furthermore, although both Rbfox1 and Rbfox2 are expressed throughout retinal development; Rbfox1 expression shifts from cytoplasmic to predominantly nuclear at around P0 and remains so in mature retinas, whereas Rbfox2 localization is predominantly nuclear during retinogenesis and in adult retinas [15]. These differences in spatial and temporal expression suggest specific roles of Rbfox paralogs during retinal neurogenesis and in maintaining normal physiology of differentiated neurons. Transcriptome analysis identified multiple genes involved in synaptic functions that are regulated by Rbfox1 and Rbfox2 as well genes associated with circadian rhythm/entrainment pathways regulated by Rbfox2. Interestingly, a single gene deletion of Rbfox1, Rbfox2 or Rbfox3 in adult animals had no effect on gross retinal morphology, although both Rbfox1$^{-/-}$ and Rbfox2$^{-/-}$ mice experienced depth perception deficiency, which indicates the role of these proteins in regulation of gene networks associated with the retino-geniculo-cortical pathway [12,14,15]. Since all three Rbfox paralogs recognize the same (U)GCAUG element within their target genes, the absence of cellular phenotype in single Rbfox gene knockout (KO) or knockdown models can be explained by functional redundancy of Rbfox proteins in cells that have overlapping expression of these genes. On the other hand, individual Rbfox proteins have specific roles during neuronal development and in mature neuron physiology [10,11,16–22] and deficiencies in different Rbfox paralogs have been associated with the various neurodevelopmental and neurological conditions mentioned above. Recent studies also implicate Rbfox proteins, particularly Rbfox1, in the pathogenesis of age-related neurodegenerative diseases such as Alzheimer's and Parkinson's diseases [23–29]. Based on these observations, we can speculate that the effect of Rbfox1 functional deficiency on normal cellular physiology and integrity may be influenced by additional genetic/epigenetic factors, including those that are associated with age- and stress-related changes. Earlier, we showed that downregulation of Rbfox1 in young adult mice (4 months old) has no effect on the number of surviving RGCs after ONC injury [30]. The current study evaluates the effect of advanced age and optic nerve injury on retinal morphology, visual function and the survival of RGCs in Rbfox1 KO animals.

2. Experimental Procedures

2.1. Generation of Rbfox1 KO Animals

Experimental procedures with animals were approved by the Animal Research Committee of the University of California at Los Angeles and were performed in compliance with the National Institutes of Health Guide for the Care and Use of Animals and the ARVO (The Association for Research in Vision and Ophthalmology) Statement for the Use of Animals in Ophthalmic and Vision Research. Animals were housed in a 12 h light-dark cycle with food and water available ad libitum. Rbfox1 KO animals were generated by crossing homozygous transgenic Rbfox1$^{fl/fl}$ mice (loxP sites flanking Rbfox1 gene exons 11–12; kindly provided by Dr. Douglas Black, UCLA; [11]) with Tg(UBC-Cre/ERT2)1Ejb mice (Jackson Laboratory, Bar Harbor, ME). The UBC-Cre-ERT2 transgenic mouse line, in which the expression of tamoxifen-inducible Cre recombinase (Cre) gene is controlled by the human ubiquitin C (UBC) promoter, was used in this study since it provides a strong expression of Cre in RGCs (www.informatics.jax.org/recombinase/specificity?id=MGI 3707333&system=sensory+organs, accessed on 20 August 2022; [15,31]). The resulting heterozygous Rbfox1$^{fl/+}$/UBC-Cre$^{+/-}$ animals were bred with Rbfox1$^{fl/fl}$ mice to generate homozygous Rbfox1$^{fl/fl}$/UBC-Cre$^{+/-}$ mice. Cre activity in homozygous Rbfox1$^{fl/fl}$/UBC-Cre$^{+/-}$ animals was induced with tamoxifen (Sigma, St. Louis, MO, USA). Rbfox1$^{fl/fl}$/UBC-Cre$^{+/-}$ and age-matched heterozygous Rbfox1$^{fl/+}$ control mice were administered 5 doses of tamoxifen solution (200 mg/kg; once a day) by oral gavage. Rbfox1 KO animals were viable, had a normal growth rate and behaviorally had no apparent anomalies compared to control animals. Age-matched heterozygous Rbfox1$^{fl/+}$ mice were used as controls.

2.2. Visual Cliff Test

Visual function in Rbfox1 KO animals was examined by evaluation of depth perception with cliff test. The protocol described earlier was followed [14,15,32,33]. The test uses a glass-bottomed box placed on the edge of the table so its one half is resting on the table surface ("shallow" side) and the other half is suspended above the floor (the "cliff drop" or "deep" side; Figure 1A). The illusion of the cliff was created by covering the table and the floor under the test box with a black and white checkered tablecloth. The time the animal spent on the "shallow" and "deep" sides during 5 min of testing was recorded. The vibrissae of the tested mice were removed so the tactile placing responses do not interfere with evaluation of visual function. The glass surface of the test box was thoroughly cleaned after each test. Six Rbfox1 KO and six control animals were tested. Five independent experiments for each animal were performed.

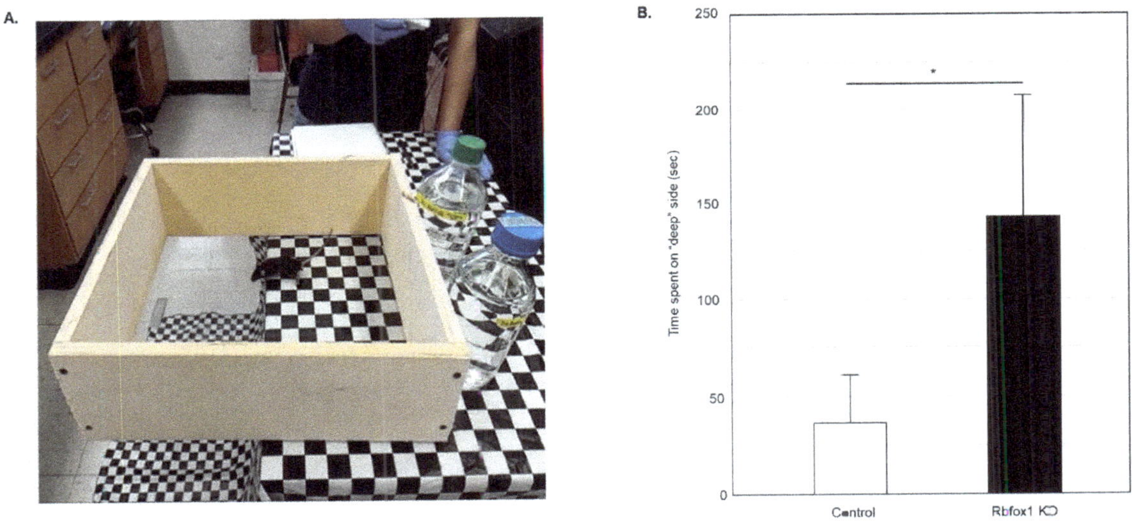

Figure 1. Evaluation of visual function in old Rbfox1 KO mice. (**A**) Visual cliff test setup. A visual cliff illusion was created by placing a glass-bottomed test box on the edge of the table so that one part sits on the table ("shallow" side) and the rest is suspended above the floor ("deep" side). The table and the floor under the test box were covered with black and white checkered table cloth. Animals tested for 5 min and time the spent on the "shallow" and "deep" sides was recorded. Each animal was tested 5 times. (**B**) Control mice spent most of their time on the "shallow" side of the field. The time they spent on the "deep" side was only 36.60 +/− 25.34 s (mean +/− SD; n = 5) out of 5 min of testing time. Rbfox1 KO mice showed no preference to either side of the test field. The time the Rbfox1 KO animals spent on the "deep" side was 141.75 +/− 64.63 s (mean +/− SD; n = 5). The mean difference in time spent on the "deep" side between two groups was statistically significant: 105.15 s (* $p = 4.55 \times 10^{-7}$).

2.3. Optic Nerve Crush (ONC) Injury

Prior to the ONC, external ocular examination was performed on all the anesthetized mice. Adnexal (the eyelids and conjunctiva), and anterior segment structures (cornea, iris and sclera) were examined by slit lamp biomicroscopy at ×16 magnification using broad-beam illumination. ONC was performed as described previously with minor modifications (Nadal-Nicolás et al., 2009). Briefly, the optic nerve was exposed and crushed with fine forceps (#5 Dumont, Fine Science Tools, Foster City, CA, USA) approximately 1 mm behind the posterior pole of the eye for 1 sec. Care was taken not to damage the blood vessels. The procedure was performed unilaterally.

2.4. Inclusion and Exclusion Criteria

Twenty-two-month-old Rbfox1$^{fl/fl}$/UBC-Cre$^{+/-}$ and age-matched heterozygous Rbfox1$^{fl/+}$ control mice were included in this study. Prior to the ONC procedure, exclusion criteria were visible corneal abrasion, opacity, inflammation or edema and cataract. A fundus examination was performed under an operating microscope immediately after the ONC procedure to exclude animals with blockage of retinal blood flow from the study. The exclusion criteria were also included: surgical complications including infection, bleeding, lens discoloration, or purulent drainage from the wound. No animal that entered the study was rejected due to the above-listed exclusion criteria.

2.5. Retinal Sections and Immunohistochemistry

Eyes were enucleated, fixed with ice-cold 4% paraformaldehyde and cryoprotected in 30% sucrose. 14-μm thick retinal sections were cut with cryostat. For immunohistochemistry, sections were incubated with blocking solution (20% fetal calf serum, 5% goat serum, 0.1% Triton X-100 in PBS) for 30 min and then with primary antibodies at 4 °C overnight. Primary antibodies used in this study: anti-Rbfox1 produced in mouse (1:200; Novus Biologicals, Littleton, CO, USA), anti-Rbpms produced in guinea pig (1:1000; PhosphoSolutions, Aurora, CO, USA; [34]) and anti-calbindin D-28K produced in rabbit (1:500; EMD Millipore, Billerica, MA). These primary antibodies have been well characterized and have high specificity for their target protein. After incubation with the primary antibodies, sections were washed with 0.1% Triton X-100 in PBS and incubated with secondary antibodies for 1 h at room temperature. Secondary antibodies used in this study: Alexa Fluor 488-conjugated goat anti-rabbit IgG, Alexa Fluor 568-conjugated goat anti-mouse IgG and Alexa Fluor 488-conjugated goat-anti-guinea pig IgG (all 1:500; Thermo Fisher Scientific, Canoga Park, CA, USA). Sections were mounted with mounting medium containing DAPI for nuclear counterstaining and imaged with a confocal laser scanning microscope Olympus FV3000 (Olympus FV3000, Tokyo, Japan).

2.6. Cell Quantification in Whole Mount Retinas

A standard protocol was followed for cell quantification in retinal flat mounts [34]. Briefly, the enucleated eyeballs were fixed in 4% paraformaldehyde in 0.1 M phosphate buffer, the retinas were dissected and incubated with 10% serum for 1 h to reduce non-specific staining and then with anti-Rbfox1 produced in mouse (1:200; Novus Biologicals) or anti-Rbpms produced in guinea pig (1:1000; PhosphoSolutions) primary antibodies overnight at 4 °C. Retinas were washed and then incubated with the corresponding secondary antibody overnight at 4 °C. Secondary antibodies used in this experiment: Alexa Fluor 568-conjugated donkey anti-mouse IgG and Alexa Fluor 488-conjugated goat-anti-guinea pig IgG (all 1:500; Thermo Fisher Scientific). The retinas were placed flat with the GCL facing upward, divided into superior, inferior, nasal and temporal quadrants with several radial cuts and mounted flat on the glass slide. Four sampling fields per retinal quadrant (0.24 × 0.24 mm each) were imaged at 0.5, 1, 1.5 and 2 mm from the center of the optic nerve disc with a confocal laser scanning microscopy (Olympus FV3000). Retinas from four animals per group were used in these experiments. Quantification was performed in a masked manner; the person who counted the cells had no knowledge about samples used for the quantitative analysis.

2.7. Statistical Analysis

An unpaired Student's *t*-test was used for analysis of the visual cliff test data and cell density data. $p < 0.05$ was considered statistically significant. Data are presented as mean ± SD.

3. Results

3.1. Depth Perception Deficiency in Old Rbfox1 KO Animals

To evaluate visual function in aged Rbfox1 null mice, a visual cliff test was used (Figure 1A). Animals with normal vision have the innate tendency to avoid the "cliff" field and stay on the "shallow" side for most of the testing time. Twenty-two-month-old Rbfox1 KO mice and age-matched control heterozygote Rbfox1$^{fl/+}$ animals were evaluated. Downregulation of Rbfox1 was completed one month prior to the test. Visual function evaluation in this test is based on the time the animals spend in the "deep" versus "shallow" side of the field. Control mice, as expected, showed a strong preference for the "shallow" side of the field. The mean (+/−SD) time that the control group (n = 5) spent on the "deep" side was 36.6 +/− 25.3 s during 300 s of testing time (Figure 1B). Rbfox1 KO animals, unlike their control counterparts, were indifferent to either side of the test field and spent almost as much time on the "deep" side as on the "shallow" side: the mean time that the Rbfox1 KO animals (n = 5) spent on the "deep" side was 141.8 +/− 64.6 s during 300 s of testing time (Figure 1B). The mean difference in time spent on the "deep" side between two groups was statistically significant: 105.2 s (p < 0.0001).

3.2. The Effect of Aging and ONC-Induced Injury on Retinal Morphology in Rbfox1 KO Animals

Four groups of aged (twenty-two-month-old) animals were evaluated: uninjured control heterozygote Rbfox1$^{fl/+}$ mice, control mice injured by ONC, Rbfox1 KO uninjured and Rbfox1 KO injured by ONC. Tamoxifen-induced downregulation of Rbfox1 was completed one month and ONC was performed 1 week before animals were euthanized for histological evaluation. In the uninjured control group, the expression of Rbfox1 was observed in cells located in the GCL and in cells within the innermost layer of the INL (Figure 2A). In the GCL, these Rbfox1-positive cells were also stained with the RGC marker, Rbpms, or with calbindin, a neuronal marker, which was used to label ACs (most ACs in the INL and dACs in the GCL express calbindin). Virtually all Rbpms positive cells (RGCs) and the vast majority of calbindin positive cells in the GCL (dACs and a small number of RGCs express calbindin) in these aged animals were Rbfox1-positive (Figure 2A,B). In the INL, most ACs adjacent to the inner plexiform layer (IPL) were Rbfox1-positive (Figure 2B). Calbindin-stained HCs (pointed with white arrows) in the outermost INL were Rbfox1-negative. ONC resulted in significant reduction in RGCs: very few Rbfox1/Rbpms-positive cells were present, but the majority of Rbfox1 immunostained cells in the GCL were calbindin-positive, suggesting that most of them are dACs (Figure 2C,D). ONC had no effect on intensity and pattern of Rbfox1 expression in the INL. Quantitative analysis of ONC-induced cell loss is presented in "The effect of aging on RGC numbers and their survival after axonal injury in Rbfox1 KO animals" section below.

In Rbfox1 KO animals, a dramatic reduction of Rbfox1 immunostaining intensity within the GCL and INL was observed (Figure 3A–D). Very few faintly stained for Rbfox1 RGCs and ACs were present. The immunostaining pattern or intensity for Rbpms and calbindin was not affected by Rbfox1 downregulation (Figure 3A,B). The retinal gross morphology in these animals was normal and similar to that of the control mice described above. An extensive loss of Rbpms-labeled RGCs was observed 1 week after ONC (Figure 3C). The staining intensity for Rbpms in remaining RGCs was much weaker compared to that of uninjured cells (Figure 3A,C). Interestingly, in both control (Figure 2B,D) and Rbfox1 KO (Figure 3B,D) animals, the intensity of calbindin immunostaining 1 week after ONC was also visibly reduced in the GCL, IPL, INL and especially in the outer plexiform layer (OPL) where photoreceptors make synaptic connections with bipolar cells and HCs.

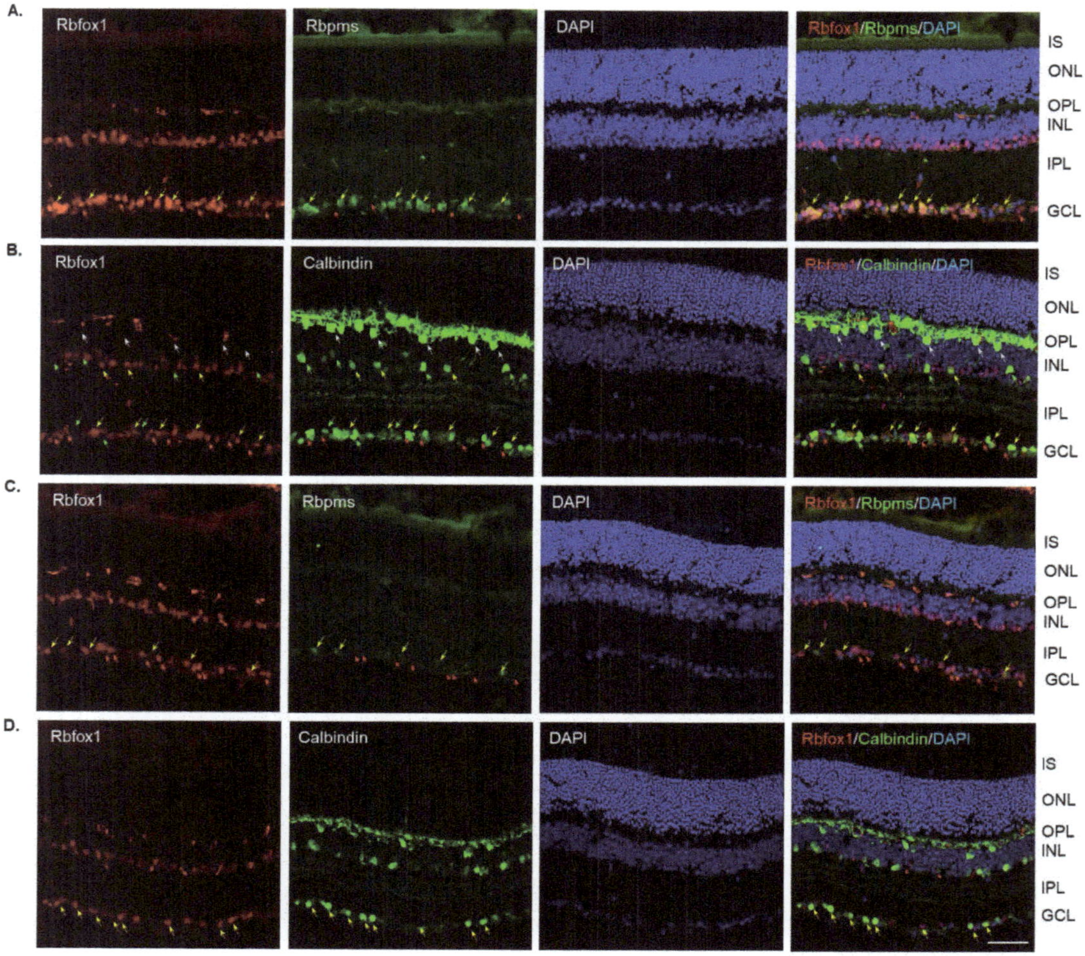

Figure 2. Expression of Rbfox1 in 22-month-old mouse retinas. As in young animals, Rbfox1 expression in the retinas of old mice was restricted to the cells within the GCL, which contains somas of RGCs and dACs in a ratio of approximately 1:1, and the innermost layer of INL, which is occupied by AC somas. (**A**) Virtually all Rbpms-labeled RGCs were also immunostained with Rbfox1 (pointed with yellow arrows). Some dACs in the GCL-Rbfox1-positive/Rbpms-negative cells–are pointed with red arrows. (**B**) Rbfox1 is expressed in the majority of calbindin-positive cells in the GCL (dACs and some RGCs express calbindin) and in the INL (ACs; yellow arrows). Rbfox1-negative dACs and ACs are pointed with green arrows. RGCs (Rbfox1 positive/calbindin negative) are indicated by red arrows. Antibodies against calbindin also stain HCs in the outmost layer of INL (adjacent to the OPL) are pointed with white arrows. (**C**) ONC resulted in significant loss of RGCs. Several remaining Rbpms/Rbfox1-positive cells are indicated by yellow arrows. Red arrows point at Rbfox1 positive dACs. (**D**) Most Rbfox1-positive cells in the GCL were also positive for calbindin, confirming that these cells are dACs (yellow arrows). A decrease in calbindin immunostaining intensity in retinas of ONC mice compared to that of uninjured animals was observed. IS, photoreceptor inner segments, ONL, outer nuclear layer; OPL, outer plexiform layer; INL, inner nuclear layer; IPL, inner plexiform layer; GCL, ganglion cell layer. Scale bar, 50 μm.

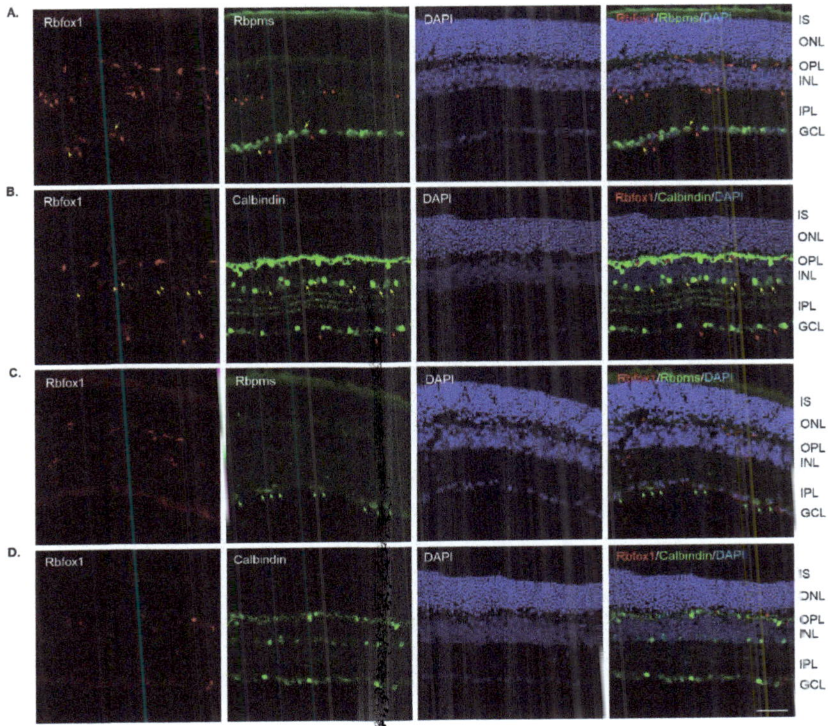

Figure 3. Downregulation of Rbfox1 expression in aged Rbfox1 KO animals. (**A–D**). The number of Rbfox1 positive cells and the immunostaining intensity in the few remaining Rbfox1-positive cells were dramatically decreased in retinas of both uninjured and ONC animals. Very few Rbfox1-positive RGCs or ACs were present (**A**,**B**), respectively; yellow arrows). Red arrows point at Rbfox1-immunostained ACs and dACs (**A**) and RGCs (**B**). The number of RGCs, dACs and ACs in Rbfox1 KO animals appeared to be normal. (**C**) Significant loss of RGCs 1 week after ONC was observed. Some of the remaining RGCs are pointed with green arrows. (**D**). The pattern of calbindin immunostaining is unaffected by ONC, although, the staining intensity appears to be lower than that in uninjured retinas. IS, photoreceptor inner segments; ONL, outer nuclear layer; OPL, outer plexiform layer, INL, inner nuclear layer; IPL, inner plexiform layer; GCL, ganglion cell layer. Scale bar, 50 µm.

3.3. The Effect of Aging on RGC Numbers and Their Survival after Axonal Injury in Rbfox1 KO Animals

To determine the effect of aging on the number of RGCs in Rbfox1 KO mice and the rate of their survival after ONC-induced axonal injury, Rbpms- and Rbpms/Rbfox1-labeled cells were counted in whole mounted retinas. As for immunohistochemical analysis, for groups of twenty-two-month-old animals were used: uninjured and ONC-injured control mice and uninjured and ONC-injured Rbfox1 KO mice. Representative images of retinas from each of these groups that were immunostained for Rbpms and Rbfox1 are shown in Figure 4. Images representing ONC retinas (Figure 4B,D) show a significant reduction in Rbpms labeled cells compared to corresponding uninjured controls (Figure 4A,C). Reduction of Rbfox1-labeld cells in retinas of control mice injured by ONC is due to the loss of RGCs (Figure 4B), whereas the loss of Rbfox1 immunoreactivity in Rbfox1 KO transgenes is the result of Rbfox1 downregulation (Figure 4C) or a combination of Rbfox1 downregulation and ONC-induced loss of RGCs (Figure 4D). Quantitative data for Rbpms- and Rbpms/Rbfox1-positive cells that were counted in superior, inferior, nasal and temporal retinal quadrants at 0.5, 1, 1.5 and 2 mm from the center of the optic nerve disc are presented in Figures 5 and 6 and Tables 1 and 2. The number of Rbpms-positive cells in the retinas of uninjured control and Rbfox1 KO mice were very similar when compared at corresponding locations (Figure 5A, Table 1). The average densities were: 6273 ± 1673 and 6004 ± 1531 Rbpms-positive cells/mm^2 for control and Rbfox1 KO mice, respectively (n = 4/group; Figure 5B). One week after ONC, the numbers of Rbpms-stained cells in retinas of control and Rbfox1 KO mice were decreased to 2998 ± 858 cells/mm^2 and 3036 ± 857 cells/mm^2, respectively (n = 4/group; Figure 5B). These numbers indicate approximately 50% RGC loss caused by ONC compared to uninjured animals (control vs. control + ONC, $p = 2.75 \times 10^{-27}$ and Rbfox1 KO vs. Rbfox1 KO + ONC, $p = 2.55 \times 10^{-26}$).

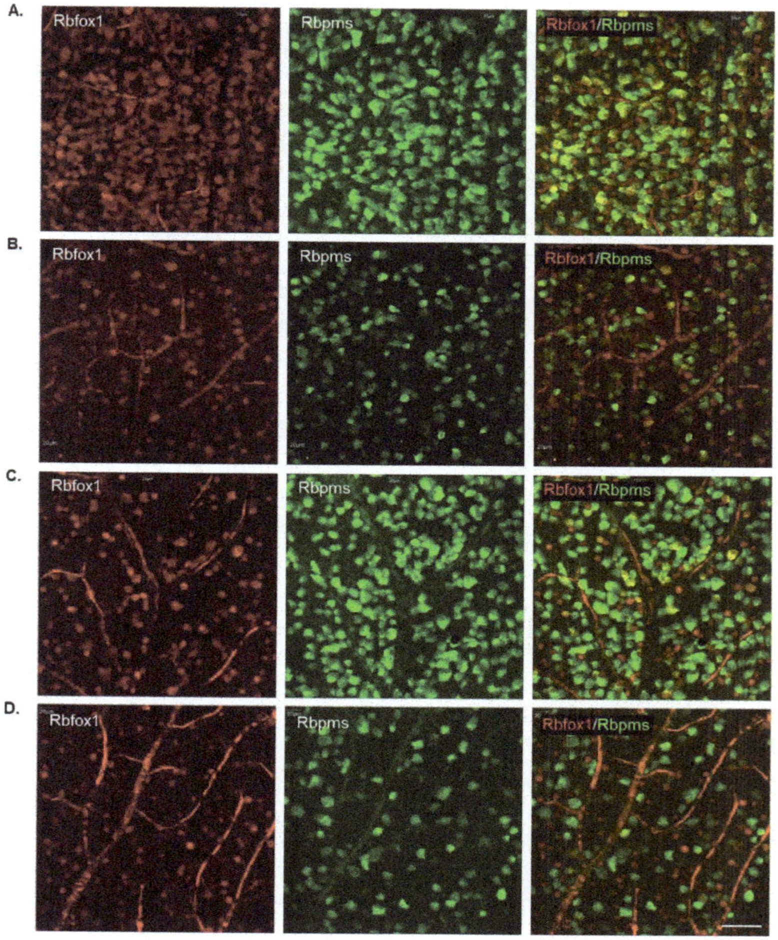

Figure 4. Representative images of immunostained whole-mounted retinas used for quantitative analysis of Rbpms and Rbpms/Rbfox1-positive cells in aged Rbfox1 KO animals with and without ONC injury. For quantitative analysis, Rbpms-positive and Rbpms/Rbfox1-positive cells were counted in the superior, inferior, nasal and temporal retinal quadrants at 0.5, 1, 1.5 and 2 mm from the center of the optic disc. Images represent Rbpms and Rbfox1 immunostained retinas at 1 mm from the center of the optic nerve head (**A–D**). A significant reduction in the numbers of Rbpms-labeled cells in ONC groups (**B**,**D**) compared to uninjured animals (**A**,**C**) is evident. Loss of Rbfox1-positive cells in control + ONC (**B**) is due to the loss of injured RGCs. (**A**) dramatic reduction of Rbfox1-positive cells was observed in retinas of Rbfox1 KO animals (**C**). Their numbers were further decreased in Rbfox1 KO + ONC (**D**) due to the loss of RGCs. Scale bar, 50 μm.

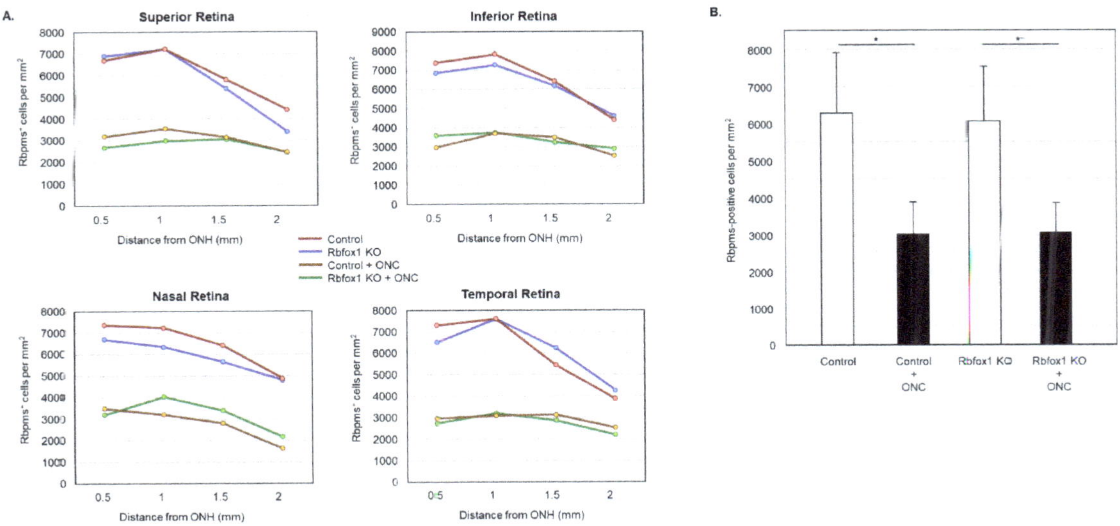

Figure 5. RGC quantification in aged Rbfox1 KO animals with and without ONC. (**A**) The density of Rbpms-positive cells in the superior, inferior, nasal and temporal retinal quadrants at 0.5, 1, and 1.5 mm from the center of the optic disc. (**B**) The number of Rbpms-positive cells in in retinas of uninjured control and Rbfox1 KO mice were very similar when compared at corresponding locations (Figure 5A, Table 1). The average RGC densities in retinas of control and Rbfox1 KO mice were 6273 ± 1673 cells/mm^2 and 6004 ± 1531 cells/mm^2, respectively (n = 4/group). The average RGC densities in retinas of control and Rbfox1 KO mice one week after ONC were reduced to 2998.32 ± 858.40 cells/mm^2 and 3036 ± 857 cells/mm^2, respectively (n = 4/group). Thus, ONC resulted in approximately 50% RGC loss compared to uninjured animals (control vs. control + ONC, * $p = 2.75 \times 10^{-27}$ and Rbfox1 KO vs. Rbfox1 KO + ONC, ** $p = 2.55 \times 10^{-26}$). No significant difference between RGC densities in control vs. Rbfox1 KO animals was observed. ONC, optic nerve crush; ONH, optic nerve head.

Downregulation of Rbfox1 in Rbfox1 KO mice retinas is not uniform and there are cells both ACs and RGCs, in which Rbfox1 expression is preserved (Figure 3A,B). This raises a question as to whether there is a difference in survival rate of Rbfox1-positive and Rbfox1-negative RGCs after ONC. To address this question, we counted the number of Rbfox1/Rbpms-immunostained cells in the control and Rbfox1 KO groups of animals with and without ONC injury (n = 4/group). The data indicate a significant reduction of Rbfox1/Rbpms-positive cells in all four retinal quadrants of Rbfox1 KO mice compared to the control (Figure 6A,B and Table 2). The average densities of Rbfox1/Rbpms-immunostained cells in control and Rbfox1 KO retinas were 5911 ± 1557 cells/mm^2 and 1062 ± 634 cells/mm^2, respectively, which translates to more than 80% of Rbfox1-negative RGCs in Rbfox1 KO animals ($p = 4.50 \times 10^{-47}$; Figure 6B). ONC injury resulted in the reduction of Rbfox1/Rbpms-positive cells by ~ 58% in control animals (5911 ± 1557 cells/mm^2 in control vs. 2215 ± 849 cells/mm^2 in control + ONC, $p = 1.14 \times 10^{-33}$) and by ~ 50% in Rbfox1 KO animals (1062 ± 634 cells/mm^2 in Rbfox1 KO vs. 489 ± 340 cells/mm^2 in Rbfox1 KO + ONC, $p = 3.19 \times 10^{-9}$; Figure 6B). Quantitative analysis of Rbfox1/Rbpms-positive Cells shows that Rbfox1 has no significant effect on survival of RGCs injured by ONC.

Table 1. Densities of Rbpms-positive cells in retinas of control and Rbfox1 KO animals with and without ONC injury (per mm^2).

Group \ Location	S 0.5	S 1.0	S 1.5	S 2.0	I 0.5	I 1.0	I 1.5	I 2.0	*n* 0.5	N 1.0	N 1.5	N 2.0	T 0.5	T 1.0	T 1.5	T 2.0
Control	6710 ± 1144	7235 ± 780	5824 ± 997	4431 ± 760	7382 ± 803	7826 ± 742	6419 ± 1769	4388 ± 2271	7369 ± 1121	7235 ± 282	6415 ± 1472	4879 ± 1364	7318 ± 1146	7617 ± 901	5447 ± 1635	3872 ± 1337
Control + ONC	3194 ± 605	3563 ± 1010	3168 ± 506	2474 ± 560	2973 ± 823	3702 ± 713	3498 ± 749	2526 ± 475	3490 ± 608	3212 ± 758	2799 ± 1619	1636 ± 322	2969 ± 302	3099 ± 771	3134 ± 956	2535 ± 931
Rbfox1 KO	6918 ± 930	7235 ± 672	5399 ± 1306	3433 ± 1040	6861 ± 509	7282 ± 965	6184 ± 1535	4605 ± 1475	6701 ± 1120	6345 ± 749	5651 ± 1846	4796 ± 1159	6519 ± 262	7604 ± 1170	6258 ± 1265	4266 ± 859
Rbfox1 KO + ONC	2687 ± 530	2995 ± 904	3069 ± 1473	2461 ± 980	3602 ± 975	3750 ± 520	3242 ± 737	2908 ± 530	3212 ± 739	4041 ± 594	3390 ± 887	2183 ± 983	2747 ± 543	3212 ± 698	2873 ± 280	2209 ± 467

Rbpms positive cells (RGCs) were counted in superior (S), inferior (I), nasal (N) and temporal (T) retinal quadrants at 0.5, 1, 1.5 and 2 mm from the center of the optic disc.

Table 2. Densities of Rbfox1/Rbpms-positive cells in retinas of control and Rbfox1 KO animals with and without ONC (per mm^2).

Group \ Location	S 0.5	S 1.0	S 1.5	S 2.0	I 0.5	I 1.0	I 1.5	I 2.0	N 0.5	N 1.0	N 1.5	N 2.0	T 0.5	T 1.0	T 1.5	T 2.0
Control	6155 ± 1077	6992 ± 688	5577 ± 983	4193 ± 742	6784 ± 428	7465 ± 736	6150 ± 1806	4236 ± 2194	6641 ± 931	6762 ± 278	6137 ± 1445	4635 ± 1361	6641 ± 1108	7231 ± 903	5286 ± 1602	3689 ± 1293
Control + ONC	2633 ± 897	2940 ± 978	2714 ± 797	1979 ± 522	2315 ± 775	3096 ± 1023	2789 ± 959	2228 ± 708	2911 ± 1002	2824 ± 1037	2488 ± 1582	1285 ± 254	2361 ± 437	2436 ± 384	2830 ± 1048	2083 ± 838
Rbfox1 KO	990 ± 580	1246 ± 707	1094 ± 805	690 ± 399	1089 ± 564	1385 ± 669	1185 ± 656	755 ± 386	1194 ± 711	1202 ± 631	1215 ± 834	794 ± 537	1076 ± 629	1328 ± 1103	1155 ± 822	586 ± 395
Rbfox1 KO + ONC	438 ± 186	525 ± 326	725 ± 535	543 ± 420	490 ± 215	690 ± 520	638 ± 486	608 ± 270	490 ± 183	499 ± 429	473 ± 348	239 ± 173	265 ± 142	395 ± 359	369 ± 405	430 ± 388

Rbpms/Rbfox1 positive cells were counted in superior (S), inferior (I), nasal (N) and temporal (T) retinal quadrants at 0.5, 1, 1.5 and 2 mm from the center of the optic disc.

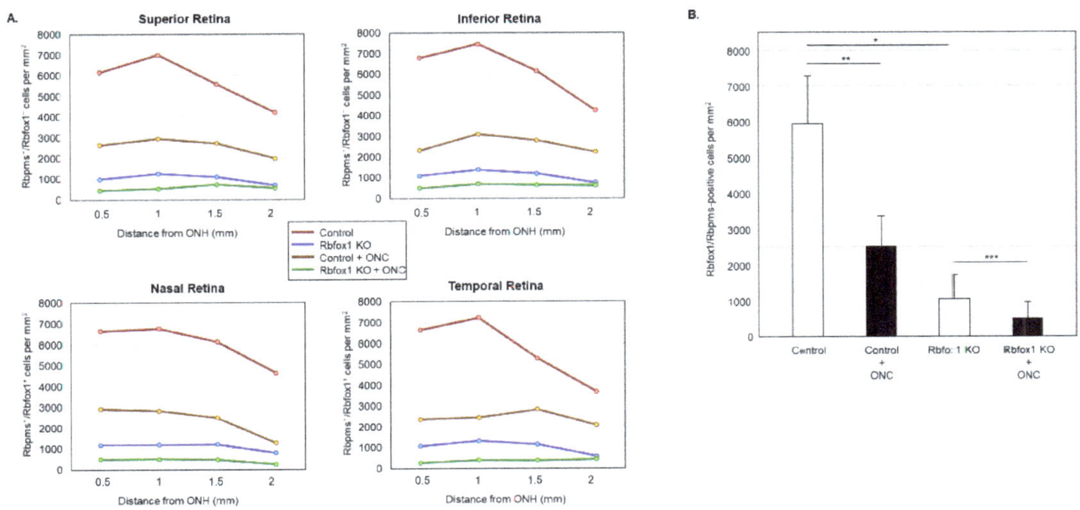

Figure 6. Downregulation of Rbfox1 and its effect on the survival of injured RGCs. For the quantitative assessment of Rbfox1 downregulation within RGC population, the numbers of Rbfox1/Rbpms-immunostained cells in control and Rbfox1 KO animals with and without ONC injury were counted (n = 4/group). (**A**) A significant reduction of Rbfox1/Rbpms-positive cells in all four retinal quadrants of Rbfox1 KO mice compared to the control was observed. (**B**) The average densities of Rbfox1/Rbpms-positive cells in control and Rbfox1 KO retinas were 5911 ± 1557 cells/mm^2 and 1062 ± 634 cells/mm^2, respectively. This indicates that more than 80% of RGCs in Rbfox1 KO animals are negative for Rbfox1 (* $p = 4.50 \times 10^{-47}$). ONC-induced loss of Rbfox1/Rbpms-positive cells was approximately 58% in control animals (5911 ± 1557 cells/mm^2 in control vs. 2215 ± 849 cells/mm^2 in control + ONC, ** $p = 1.14 \times 10^{-33}$) and ~50% in Rbfox1 KO animals (1062 ± 634 cells/mm^2 in Rbfox1 KO vs. 489 ± 340 cells/mm^2 in Rbfox1 KO + ONC, *** $p = 3.19 \times 10^{-9}$). ONC, optic nerve crush; ONH, optic nerve head.

4. Discussion

The present study evaluates the effect of aging on visual function, retinal morphology and survival of RGCs after axonal injury in Rbfox1 KO animals. Twenty-two-month-old Rbfox1 KO and age-matched control mice were used as "old" mice. Mice between 18–24 months of age meet the definition of "old," which is characterized by the presence of senescent changes in almost all biomarkers in all animals [35]. Significant impairment of cognitive function is present at 22 months of age [36].

We first evaluated the effect of aging on visual function in Rbfox1 KO animals. The visual cliff test that assesses the integrity and function of retino-geniculo-cortical pathway was used. In our earlier work, we have shown that downregulation of Rbfox1 in young adult mice (4 months old) results in deficient depth perception [15]. These animals spent more time on the "deep" side of the test field than on the "shallow" side. The aged Rbfox1 KO animals were also depth perception "blind" and spent almost as much time on the "deep" side as on the "shallow" side. Both young and aged control animals as expected, had a clear preference for the "shallow" side and avoided the illusionary cliff. The Rbfox1 KO model used in this study produced a robust downregulation of Rbfox1 in RGCs. However, downregulation of Rbfox1 in this model is not restricted to RGCs and can take place in other neurons, including those involved in the retino-geniculo-cortical pathway within the lateral geniculate nucleus or visual cortical regions. Therefore, the observed deficiency in depth perception may be associated with altered retinal function or with neuronal dysfunction in brain regions that process this visual information. Even though we cannot pinpoint the exact location at which retino-geniculo-cortical pathway is affected at this time, the results of this test show the importance of Rbfox1 in regulation of genes associated with this visual function.

Retinal gross morphology in aged Rbfox1 KO animals, in which more than 80% of RGCs were Rbfox1 negative, was unremarkable compared to age-matched controls. With respect to the effect of aging on the survival of injured neurons in Rbfox1 KO animals, we anticipated to see: (a) an increased rate of RGC degeneration in aged Rbfox1 KO animals compared to younger animals under similar conditions or (b) an increased loss of RGCs in Rbfox1 KO animals compared to the age-matched control with normal level of Rbfox1 expression. The first assumption is based on a decreased capacity of aged cells to oppose the destabilizing effects of metabolic stressors. Normal neuronal aging can be viewed as a metabolic state characterized by a decreased homeostatic reserve. Cellular homeostasis under normal conditions and in the presence of various stressors is maintained with the help of stress-response signaling pathways, such as the heat shock response (HSR) and the unfolded protein responses of the mitochondria (UPR^{MT}) and endoplasmic reticulum (UPR^{ER}), all of which are known to deteriorate during aging [37]. Therefore, acute surges of metabolic activity (induced by injury or by a loss of a protein, such as Rbfox1, which regulates the expression of genes important for normal neuronal functions) can reach the limits of homeostatic reserve defenses that are incompatible with the normal levels of functional load and result in an increased age-dependent vulnerability and extensive neuronal death [38]. The second assumption is based on several studies implicating Rbfox1 in the stress-induced regulation of several mechanisms that stimulate cell survival, including miR-132/Rbfox1-mediated mechanisms that promote neuronal survival against amyloid β-peptide (Aβ) and glutamate excitotoxicity in the brain of the mouse model for Alzheimer's disease [26,39]

The effect of Rbfox1 downregulation on survival of injured RGCs was evaluated 1 week after ONC, the time point at which the progressive RGC loss is well underway (~50–60% RGC loss). Quantitative analysis of RGC numbers in aged ONC-injured and uninjured Rbfox1 KO animals showed that downregulation of Rbfox1 has no effect on the RGC density compared to age-matched controls. No significant changes were observed in any of four retinal quadrants; the average densities were 6273 cells/mm^2 and 6004 cells/mm^2 for control and Rbfox1 KO mice, respectively and 2998 cells/mm^2 and 3036 cells/mm^2 for control + ONC and Rbfox1 KO + ONC, respectively. Recall that, in young mice, downregulation of Rbfox1 also had no detectable effect on the number of uninjured and ONC-injured RGCs [15,30]. What is interesting though, is that the number of RGCs in young (4-month-old) and old (22-month-old) control animals were very similar: 6438 cells/mm^2 in young [30] vs. 6273 cells/mm^2 in old animals. Progressive decline of RGCs with age has been documented in studies on rodents, primates and humans. RGC quantification indicates that the average rate of RGC loss in mice is 2.3% per month and in rats is 1.5% per month with a total loss of 41% in mice during 18 months of lifespan and 36% in rats during 24 months of lifespan [40,41]. In monkeys and humans RGC soma and axon quantification estimate 1.7% per year and 0.5% per year RGC loss, respectively resulting in the total loss of 44% during 26 years of a monkey lifespan and 38% during 76 years of a human lifespan [40,42–46]. However, there are studies that question the linearity of RGC loss progression during a lifetime and some even report no significant age-related loss of RGCs or their axons in the optic nerve. For instance, RGC loss in C57BL6 mice was found to start between 12 and 15 months of age with a total of 46% reduction by 18 months of age [47]. On the other hand, a comparison of all major retinal neuronal types, including RGCs, in young adult (3–5 months old) and aged (24–28 months old) mice showed no significant change in their number with age [48]. In Brown Norway rats, no axonal loss was observed in animals up to 24 months of age and only approximately 12% loss was reported by 30 months of age [49]. No age-dependent loss of RGC axons was detected in rhesus monkeys [50]. These discrepancies may be explained by a number of considerations used in the experimental design that can ultimately affect the outcomes of the study. These include the size of each age group, sampling techniques (i.e., a part or an entire optic nerve or retina was analyzed) and RGC labeling technique (immunohistochemical or retrograde; retrograde labeling may fail to label RGC somas if the axoplasmic transport is affected by age). Therefore, a standardized approach would be beneficial for a comprehensive evaluation of RGC loss during aging.

Considering the role of Rbfox1 in the regulation of splicing and transcriptional networks important for neurogenesis and neuronal function, and that the disruption of its functions have been associated with several neurodevelopmental, neuropsychiatric and neurodegenerative diseases, it was expected that deletion of this gene in RGCs would have a dramatic effect on their phenotype and survival in response to injury. The absence of cellular phenotype in this model, as well as other single Rbfox gene KO or knockdown models, can be explained by functional redundancy of Rbfox proteins in cells that have overlapping expression of these genes: all three members of the family, Rbfox 1, Rbfox2 and Rbfox3, are expressed in RGCs, have the same recognition site within their target genes, and therefore may substitute each other's function.

In conclusion, aged Rbfox1 KO mice have deficient depth perception, which suggest the involvement of Rbfox1 in the regulation of gene(s) responsible for normal function of the retino-geniculo-cortical pathway. Retinal morphology in old Rbfox1 KO animals was normal. Axonal injury resulted in approximately 50% RGC loss in both control and Rbfox1 KO animals 1 week after ONC suggesting that Rbfox1 has no significant effect on vulnerability or survival of injured RGCs.

Author Contributions: N.P. designed research; L.G. and J.M.K.K performed research; L.G., J.M.K.K., J.C. and N.P. analyzed data; N.P. wrote the paper. All authors have read and agreed to the published version of the manuscript.

Funding: This work was supported by an Unrestricted Grant from Research to Prevent Blindness, Inc. to the Department of Ophthalmology at UCLA.

Institutional Review Board Statement: The study was conducted according to the guidelines of the Declaration of Helsinki and approved by the UCLA institutional animal care and use committee (protocol # ARC-2017-099; PI Natik Piri; approval date 10/25/2021; expiration date 10/24/2024).

Informed Consent Statement: Not applicable.

Data Availability Statement: The raw data supporting the conclusions of this article will be made available by the authors, without undue reservation.

Conflicts of Interest: The authors declare no conflict of interest.

References

1. Conboy, J.G. Developmental Regulation of RNA Processing by Rbfox Proteins. *Wiley Interdiscip. Rev. RNA* **2017**, *8*, e1398. [CrossRef]
2. Kuroyanagi, H. Fox-1 Family of RNA-Binding Proteins. *Cell. Mol. Life Sci.* **2009**, *66*, 3895–3907. [CrossRef]
3. Misra, C.; Bangru, S.; Lin, F.; Lam, K.; Koenig, S.N.; Lubbers, E.R.; Hedhli, J.; Murphy, N.P. Parker, D.J.; Dobrucki, L.W.; et al. Aberrant Expression of a Non-Muscle RBFOX2 Isoform Triggers Cardiac Conduction Defects in Myotonic Dystrophy. *Dev. Cell* **2020**, *52*, 748–763.e6. [CrossRef]
4. McKean, D.M.; Homsy, J.; Wakimoto, H.; Patel, N.; Gorham, J.; DePalma, S.R.; Ware, J.S.; Zaidi, S.; Ma, W.; Patel, N.; et al. Loss of RNA Expression and Allele-Specific Expression Associated with Congenital Heart Disease. *Nat. Commun.* **2016**, *7*, 12824. [CrossRef]
5. Amin, N.; Allebrandt, K.V.; Van Der Spek, A.; Müller-Myhsok, B.; Hek, K.; Teder-Laving, M.; Hayward, C.; Esko, T.; Van Mill, J.G.; Mbarek, H.; et al. Genetic Variants in RBFOX3 Are Associated with Sleep Latency. *Eur. J. Hum. Genet.* **2016**, *24*, 1488–1495. [CrossRef]
6. Sebat, J.; Lakshmi, B.; Malhotra, D.; Troge, J.; Lese-Martin, C.; Walsh, T.; Yamrom, B.; Yoon, S.; Krasnitz, A.; Kendall, J.; et al. Strong Association of de Novo Copy Number Mutations with Autism. *Science* **2007**, *316*, 445–449. [CrossRef]
7. Martin, C.L.; Duvall, J.A.; Ilkin, Y.; Simon, J.S.; Arreaza, M.G.; Wilkes, K.; Alvarez-Retuerto, A.; Whichello, A.; Powell, C.M.; Rao, K.; et al. Cytogenetic and Molecular Characterization of A2BP1/FOX1 as a Candidate Gene for Autism. *Am. J. Med. Genet. B Neuropsychiatr. Genet.* **2007**, *144B*, 869–876. [CrossRef]
8. Lal, D.; Reinthaler, E.M.; Altmüller, J.; Toliat, M.R.; Thiele, H.; Nürnberg, P.; Lerche, H.; Hahn, A.; Møller, R.S.; Muhle, H.; et al. RBFOX1 and RBFOX3 Mutations in Rolandic Epilepsy. *PLoS ONE* **2013**, *8*, e73323. [CrossRef]

9. Bhalla, K.; Phillips, H.A.; Crawford, J.; McKenzie, O.L.D.; Mulley, J.C.; Eyre, H.; Gardner, A.E.; Kremmidiotis, G.; Callen, D.F. The de Novo Chromosome 16 Translocations of Two Patients with Abnormal Phenotypes (Mental Retardation and Epilepsy) Disrupt the A2BP1 Gene. *J. Hum. Genet.* **2004**, *49*, 308–311. [CrossRef]
10. Gehman, L.T.; Meera, P.; Stoilov, P.; Shiue, L.; O'Brien, J.E.; Meisler, M.H.; Ares, M.; Otis, T.S.; Black, D.L. The Splicing Regulator Rbfox2 Is Required for Both Cerebellar Development and Mature Motor Function. *Genes Dev.* **2012**, *26*, 445–460. [CrossRef]
11. Gehman, L.T.; Stoilov, P.; Maguire, J.; Damianov, A.; Lin, C.-H.; Shiue, L.; Ares, M.; Mody, I.; Black, D.L. The Splicing Regulator Rbfox1 (A2BP1) Controls Neuronal Excitation in the Mammalian Brain. *Nat. Genet.* **2011**, *43*, 706–711. [CrossRef]
12. Lin, Y.-S.; Kuo, K.-T.; Chen, S.-K.; Huang, H.-S. RBFOX3/NeuN Is Dispensable for Visual Function. *PLoS ONE* **2018**, *13*, e0192355. [CrossRef]
13. Gu, L.; Caprioli, J.; Piri, N. Rbfox1 Expression in Amacrine Cells Is Restricted to GABAergic and VGlut3 Glycinergic Cells. *Biosci. Rep.* **2022**, *42*, BSR20220497. [CrossRef]
14. Gu, L.; Kawaguchi, R.; Caprioli, J.; Piri, N. The Effect of Rbfox2 Modulation on Retinal Transcriptome and Visual Function. *Sci. Rep.* **2020**, *10*, 19683. [CrossRef]
15. Gu, L.; Bok, D.; Yu, F.; Caprioli, J.; Piri, N. Downregulation of Splicing Regulator RBFOX1 Compromises Visual Depth Perception. *PLoS ONE* **2018**, *13*, e0200417. [CrossRef]
16. Vuong, C.K.; Wei, W.; Lee, J.-A.; Lin, C.-H.; Damianov, A.; de la Torre-Ubieta, L.; Halabi, R.; Otis, K.O.; Martin, K.C.; O'Dell, T.J.; et al. Rbfox1 Regulates Synaptic Transmission through the Inhibitory Neuron-Specific VSNARE Vamp1. *Neuron* **2018**, *98*, 127–141.e7. [CrossRef]
17. Jacko, M.; Weyn-Vanhentenryck, S.M.; Smerdon, J.W.; Yan, R.; Feng, H.; Williams, D.J.; Pai, J.; Xu, K.; Wichterle, H.; Zhang, C. Rbfox Splicing Factors Promote Neuronal Maturation and Axon Initial Segment Assembly. *Neuron* **2018**, *97*, 853–868.e6. [CrossRef]
18. Lee, J.-A.; Damianov, A.; Lin, C.-H.; Fontes, M.; Parikshak, N.N.; Anderson, E.S.; Geschwind, D.H.; Black, D.L.; Martin, K.C. Cytoplasmic Rbfox1 Regulates the Expression of Synaptic and Autism-Related Genes. *Neuron* **2016**, *89*, 113–128. [CrossRef]
19. Hamada, N.; Ito, H.; Nishijo, T.; Iwamoto, I.; Morishita, R.; Tabata, H.; Momiyama, T.; Nagata, K.-I. Essential Role of the Nuclear Isoform of RBFOX1, a Candidate Gene for Autism Spectrum Disorders, in the Brain Development. *Sci. Rep.* **2016**, *6*, 30805. [CrossRef]
20. Hamada, N.; Ito, H.; Iwamoto, I.; Morishita, R.; Tabata, H.; Nagata, K.-I. Role of the Cytoplasmic Isoform of RBFOX1/A2BP1 in Establishing the Architecture of the Developing Cerebral Cortex. *Mol. Autism* **2015**, *6*, 56. [CrossRef]
21. Fogel, B.L.; Wexler, E.; Wahnich, A.; Friedrich, T.; Vijayendran, C.; Gao, F.; Parikshak, N.; Konopka, G.; Geschwind, D.H. RBFOX1 Regulates Both Splicing and Transcriptional Networks in Human Neuronal Development. *Hum. Mol. Genet.* **2012**, *21*, 4171–4186. [CrossRef]
22. Lee, J.-A.; Tang, Z.-Z.; Black, D.L. An Inducible Change in Fox-1/A2BP1 Splicing Modulates the Alternative Splicing of Downstream Neuronal Target Exons. *Genes Dev.* **2009**, *23*, 2284–2293. [CrossRef]
23. Kasap, M.; Dwyer, D.S. Na^+ Leak-Current Channel (NALCN) at the Junction of Motor and Neuropsychiatric Symptoms in Parkinson's Disease. *J. Neural Transm.* **2021**, *128*, 749–762. [CrossRef]
24. Lin, L.; Göke, J.; Cukuroglu, E.; Dranias, M.R.; VanDongen, A.M.J.; Stanton, L.W. Molecular Features Underlying Neurodegeneration Identified through In Vitro Modeling of Genetically Diverse Parkinson's Disease Patients. *Cell Rep.* **2016**, *15*, 2411–2426. [CrossRef]
25. Kunkle, B.W.; Schmidt, M.; Klein, H.-U.; Naj, A.C.; Hamilton-Nelson, K.L.; Larson, E.B.; Evans, D.A.; de Jager, P.L.; Crane, P.K.; Buxbaum, J.D.; et al. Novel Alzheimer Disease Risk Loci and Pathways in African American Individuals Using the African Genome Resources Panel. *JAMA Neurol.* **2021**, *78*, 102. [CrossRef]
26. el Fatimy, R.; Li, S.; Chen, Z.; Mushannen, T.; Gongala, S.; Wei, Z.; Balu, D.T.; Rabinovsky, R.; Cantlon, A.; Elkhal, A.; et al. MicroRNA-132 Provides Neuroprotection for Tauopathies via Multiple Signaling Pathways. *Acta Neuropathol.* **2018**, *136*, 537–555. [CrossRef]
27. Alkallas, R.; Fish, L.; Goodarzi, H.; Najafabadi, H.S. Inference of RNA Decay Rate from Transcriptional Profiling Highlights the Regulatory Programs of Alzheimer's Disease. *Nat. Commun.* **2017**, *8*, 909. [CrossRef]
28. Raghavan, N.S.; Dumitrescu, L.; Mormino, E.; Mahoney, E.R.; Lee, A.J.; Gao, Y.; Bilgel, M.; Goldstein, D.; Harrison, T.; Engelman, C.D.; et al. Association Between Common Variants in RBFOX1, an RNA-Binding Protein, and Brain Amyloidosis in Early and Preclinical Alzheimer Disease. *JAMA Neurol.* **2020**, *77*, 1288. [CrossRef]
29. Meng, X.; Wei, Q.; Meng, L.; Liu, J.; Wu, Y.; Liu, W. Feature Fusion and Detection in Alzheimer's Disease Using a Novel Genetic Multi-Kernel SVM Based on MRI Imaging and Gene Data. *Genes* **2022**, *13*, 837. [CrossRef]
30. Gu, L.; Kwong, J.M.; Caprioli, J.; Piri, N. Loss of Rbfox1 Does Not Affect Survival of Retinal Ganglion Cells Injured by Optic Nerve Crush. *Front. Neurosci.* **2021**, *15*, 687690. [CrossRef]
31. Ruzankina, Y.; Pinzon-Guzman, C.; Asare, A.; Ong, T.; Pontano, L.; Cotsarelis, G.; Zediak, V.P.; Velez, M.; Bhandoola, A.; Brown, E.J. Deletion of the Developmentally Essential Gene ATR in Adult Mice Leads to Age-Related Phenotypes and Stem Cell Loss. *Cell Stem Cell* **2007**, *1*, 113–126. [CrossRef]
32. Fox, M.W. The Visual Cliff Test for the Study of Visual Depth Perception in the Mouse. *Anim. Behav.* **1965**, *13*, 232–233. [CrossRef]
33. Gibson, E.J.; Walk, R.D. The "Visual Cliff". *Sci. Am.* **1960**, *202*, 64–71. [CrossRef]
34. Kwong, J.M.K.; Caprioli, J.; Piri, N. RNA Binding Protein with Multiple Splicing: A New Marker for Retinal Ganglion Cells. *Investig. Ophthalmol. Vis. Sci.* **2010**, *51*, 1052–1058. [CrossRef]

35. Flurkey, K.; Mcurrer, J.; Harrison, D. Mouse Models in Aging Research. In *The Mouse in Biomedical Research*; Elsevier: Amsterdam, The Netherlands, 2007; pp. 637–672.
36. Yanai, S.; Endo, S. Functional Aging in Male C57BL/6J Mice Across the Life-Span: A Systematic Behavioral Analysis of Motor, Emotional, and Memory Function to Define an Aging Phenotype. *Front. Aging Neurosci.* **2021**, *13*. [CrossRef]
37. Dutta, N.; Garcia, G.; Higuchi-Sanabria, R. Hijacking Cellular Stress Responses to Promote Lifespan. *Front. Aging* **2022**, *3*. [CrossRef]
38. Toescu, E.C. Normal Brain Ageing: Models and Mechanisms. *Philos. Trans. R. Soc. B Biol. Sci.* **2005**, *360*, 2347–2354. [CrossRef]
39. Kucherenko, M.M.; Shcherbata, H.R. Stress-Dependent MiR-980 Regulation of Rbfox1/A2bp1 Promotes Ribonucleoprotein Granule Formation and Cell Survival. *Nat. Commun.* **2018**, *9*, 312. [CrossRef]
40. Neufeld, A.; Gachie, E. The Inherent, Age-Dependent Loss of Retinal Ganglion Cells Is Related to the Lifespan of the Species. *Neurobiol. Aging* **2003**, *24*, 167–172. [CrossRef]
41. Ricci, A.; Bronzetti, E.; Amenta, F. Effect of Ageing on the Nerve Fibre Population of Rat Optic Nerve. *Gerontology* **1988**, *34*, 231–235. [CrossRef]
42. Harman, A.; Abrahams, B.; Moore, S.; Hoskins, R. Neuronal Density in the Human Retinal Ganglion Cell Layer from 16-77 Years. *Anat. Rec.* **2000**, *260*, 124–131. [CrossRef]
43. Jonas, J.B.; Müller-Bergh, J.A.; Schlötzer-Schrehardt, U.M.; Naumann, G.O. Histomorphometry of the Human Optic Nerve. *Investig. Ophthalmol. Vis. Sci.* **1990**, *31*, 736–744.
44. Kerrigan-Baumrind, L.A.; Quigley, H.A.; Pease, M.E.; Kerrigan, D.F.; Mitchell, R.S. Number of Ganglion Cells in Glaucoma Eyes Compared with Threshold Visual Field Tests in the Same Persons. *Investig. Ophthalmol. Vis. Sci.* **2000**, *41*, 741–748.
45. Morrison, J.C.; Cork, L.C.; Dunkelberger, G.R.; Brown, A.; Quigley, H.A. Aging Changes of the Rhesus Monkey Optic Nerve. *Investig. Ophthalmol. Vis. Sci.* **1990**, *31*, 1623–1627.
46. Balazsi, A.G.; Rootman, J.; Drance, S.M.; Schulzer, M.; Douglas, G.R. The Effect of Age on the Nerve Fiber Population of the Human Optic Nerve. *Am. J. Ophthalmol.* **1984**, *97*, 760–766. [CrossRef]
47. Danias, J.; Lee, K.C.; Zamora, M.-F.; Chen, B.; Shen, F.; Filippopoulos, T.; Su, Y.; Goldblum, D.; Podos, S.M.; Mittag, T. Quantitative Analysis of Retinal Ganglion Cell (RGC) Loss in Aging DBA/2NNia Glaucomatous Mice: Comparison with RGC Loss in Aging C57/BL6 Mice. *Investig. Ophthalmol. Vis. Sci.* **2003**, *44*, 5151. [CrossRef]
48. Samuel, M.A.; Zhang, Y.; Meister, M.; Sanes, J.R. Age-Related Alterations in Neurons of the Mouse Retina. *J. Neurosci.* **2011**, *31*, 16033–16044. [CrossRef]
49. Cepurna, W.O.; Kayton, R.J.; Johnson, E.C.; Morrison, J.C. Age Related Optic Nerve Axonal Loss in Adult Brown Norway Rats. *Exp. Eye Res.* **2005**, *80*, 877–884. [CrossRef]
50. Fortune, B.; Reynaud, J.; Cull, G.; Burgoyne, C.F.; Wang, L. The Effect of Age on Optic Nerve Axon Counts, SDOCT Scan Quality, and Peripapillary Retinal Nerve Fiber Layer Thickness Measurements in Rhesus Monkeys. *Transl. Vis. Sci. Technol.* **2014**, *3*, 2. [CrossRef]

cells

MDPI

Article

In Vivo Efficacy and Safety Evaluations of Therapeutic Splicing Correction Using U1 snRNA in the Mouse Retina

Sebastian Swirski [1,†], Oliver May [1], Malte Ahlers [2], Bernd Wissinger [3], Martin Greschner [3,4], Christoph Jüschke [1] and John Neidhardt [1,4,*]

1 Human Genetics, Department of Human Medicine, Faculty of Medicine and Health Sciences, University of Oldenburg, Carl-von-Ossietzky-Straße 9-11, 26129 Oldenburg, Germany
2 Visual Neuroscience, Department of Neuroscience, Faculty of Medicine and Health Sciences, University of Oldenburg, Carl-von-Ossietzky-Straße 9-11, 26129 Oldenburg, Germany
3 Institute for Ophthalmic Research, Centre for Ophthalmology, University of Tübingen, Elfriede-Aulhorn-Straße 7, 72076 Tübingen, Germany
4 Research Center Neurosensory Science, University of Oldenburg, Carl-von-Ossietzky-Straße 9-11, 26129 Oldenburg, Germany
* Correspondence: john.neidhardt@uni-oldenburg.de; Tel.: +49-441-7983810
† Current address: Research Data Service Group, Leibniz Centre for Tropical Marine Research, 28359 Bremen, Germany.

Abstract: Efficacy and safety considerations constitute essential steps during development of in vivo gene therapies. Herein, we evaluated efficacy and safety of splice factor-based treatments to correct mutation-induced splice defects in an *Opa1* mutant mouse line. We applied adeno-associated viruses to the retina. The viruses transduced retinal cells with an engineered U1 snRNA splice factor designed to correct the *Opa1* splice defect. We found the treatment to be efficient in increasing wild-type *Opa1* transcripts. Correspondingly, Opa1 protein levels increased significantly in treated eyes. Measurements of retinal morphology and function did not reveal therapy-related side-effects supporting the short-term safety of the treatment. Alterations of potential off-target genes were not detected. Our data suggest that treatments of splice defects applying engineered U1 snRNAs represent a promising in vivo therapeutic approach. The therapy increased wild-type *Opa1* transcripts and protein levels without detectable morphological, functional or genetic side-effects in the mouse eye. The U1 snRNA-based therapy can be tailored to specific disease gene mutations, hence, raising the possibility of a wider applicability of this promising technology towards treatment of different inherited retinal diseases.

Keywords: subretinal injection; gene therapy; U1 snRNA; U6 snRNA; AAV; splice correction; Opa1; ADOA; optic atrophy; therapeutic treatment

Citation: Swirski, S.; May, O.; Ahlers, M.; Wissinger, B.; Greschner, M.; Jüschke, C.; Neidhardt, J. In Vivo Efficacy and Safety Evaluations of Therapeutic Splicing Correction Using U1 snRNA in the Mouse Retina. *Cells* **2023**, *12*, 955. https://doi.org/10.3390/cells12060955

Academic Editor: Hossein Ameri

Received: 20 January 2023
Revised: 14 March 2023
Accepted: 14 March 2023
Published: 21 March 2023

1. Introduction

Splicing of nuclear pre-mRNAs is defined by the excision of introns and the re-ligation of exons to generate mature transcripts. It depends on the identification of exonic sequences within the pre-mRNA as well as the recognition of splice sites at the borders between exons and introns [1]. Splice factors have an essential role in this process, including the U1 small nuclear ribonucleoproteins (U1). They are composed of proteins and a small nuclear RNA (snRNA), and mediate splicing by dynamic and orchestrated interactions between pre-mRNAs and splicing relevant proteins [2]. Spliceosome formation is initiated by recruitment of U1 to the splice donor site (SDS), a process that involves Watson-Crick base-pairing of U1 with the first six nucleotides of the 5′ end of the intron (+1 to +6) and the last three nucleotides of the 3′ end of the exon (−3 to −1) [3,4]. In a later step, U6 is recruited to the SDS, as part of a heterotrimeric complex containing U4-U5-U6. While U1 is released from the spliceosome, U6 generates base-pairs with the nucleotides at positions +4 to +6 of the SDS [5,6]. With the completion of splicing, the splice factors are released from the mature transcript and recycled [3].

Splicing defects are the cause of many human diseases. About 10–20% of human mutations affect canonical splice sites and lead to pathogenic splice defects (Human Gene Mutation Database [HGMD], http://www.hgmd.cf.ac.uk/ac/index.php, accessed on 20 March 2022) [7]. It has been demonstrated that aberrant splicing can be caused by mutations that reduce the interaction between the SDS and U1 [8]. To compensate for mutation-induced defects in pre-mRNA splicing, we and others previously engineered U1 to increase complementarity with the mutated SDS. This gene therapeutic approach on transcript level has been employed successfully in vitro and in cell cultures [9–18]. Furthermore, we previously demonstrated that a combination of engineered U1 and U6 can lead to improved correction of mutation-induced splice defects [12]. Additional studies suggested that U1 can be efficient in other mouse tissues than the retina [19–21].

Mutations in the human Optic Atrophy 1 gene (*OPA1*) lead to a progressive degeneration mainly of retinal ganglion cells (RGC) and their axons which project through the retina and into the optic nerve [22]. *OPA1* mutations are associated with autosomal dominant optic atrophy (ADOA; OMIM: 165500) [23,24], and are the major cause of ADAO [25]. Symptoms of ADOA usually involve slow bilateral vision loss with highly variable severity [26–28]. In about 20% of cases, *OPA1* mutations may also lead to syndromic ADOAplus characterized by additional extraocular clinical features including myopathy, peripheral neuropathy, ataxia, encephalopathy, and sensorineural hearing loss [29] These multisystemic manifestations indicate that other cell types in addition to RGCs may be affected by *OPA1* mutations. Effective therapeutic interventions to attenuate or reverse *OPA1*-associated ADOA do not yet exist.

Opa1 mouse models provided valuable tools to analyze the pathomechanisms of *Opa1* mutations and to develop potential treatment options [30–33]. In this study, we used the *Opa1*:c.1065+5G>A (*Opa1*$^{enu/+}$) mouse line as a model to evaluate treatments of pathogenic splice donor site defects. The splice defect in the mouse line *Opa1*$^{enu/+}$, carrying the splice donor site mutation *Opa1*:c.1065+5G>A, is well characterized [30,34–36]. Notably, *Opa1*$^{enu/+}$ is the only published Opa1 mouse model affected by a SDS mutation. The phenotype in the mouse model is a late onset partial RGC degeneration with slow progression detectable in approximate 1.5-year-old heterozygous mice. The extent of RGC degeneration is variable among heterozygous animals. Interestingly, the same splice donor site mutation (*OPA1*:c.1065+5G>A) occurred in human ADOA patients and caused pathogenic transcript alterations that are highly comparable between the mouse retina and patient-derived cells [37].

Herein, we present a proof-of-principle study that aimed to evaluate short-term efficacy and safety of U1-mediated treatments in vivo in the mouse eye. Our study is the first to demonstrate U1-based splice correction and safety in the retina of an in vivo model, an essential prerequisite for further therapy development.

2. Materials and Methods

2.1. Generation and Cloning of Mutation-Adapted U1/U6 Cassettes

Wild-type human U1 including its endogenous promotor (U1.wt, 614 bp) was amplified from genomic DNA introducing restriction sites for *BcuI* (BcuI_BshTI_U1_fwd: 5′-AGCTACTAGTACCGGTGTTAGCGTACAGTCTACTTTTGA-3′) and *XbaI* (XbaI_U1_rev: 5′-CGCGTCTAGAGTAAGGACCAGCTTCTTTGG-3′). Site-directed mutagenesis was performed on U1.wt template in three steps using Phusion Polymerase (Thermo Fisher, Darmstadt, Germany). Fragment 1 was mutagenized using BcuI_BshTI_U1_fwd and mutagenesis primer 1 (U1_opa1_rev: 5′-ggcccaagatctcatatttacatcgcaggggagatacc-3′). Fragment 2 was mutagenized using mutagenesis primer 2 (U1_opa1_fwd: 5′- ggtatctcccctgcgatgtaaatatgagatcttgggcc-3′) and XbaI_U1_rev. Both fragments were fused by PCR with primers BcuI_BshTI_U1_fwd and XbaI_U1_rev to yield mutation-adapted U1 cassette (U1.ad, 614 bp).

Wild-type human U6 including its endogenous promotor (U6.wt, 936 bp) was amplified from genomic DNA introducing restriction sites for *XbaI* (XbaI_U6_fwd: 5′- CGCGTCTAGACTCGCTAGGGTCACGTCTCTC-3′ and XbaI_U6_rev: 5′- CGCGTCTAGACAGAGGCAAGATGGGAAAGATC-3′). Site-directed mutagenesis and PCR were performed on U6.wt template as described above using primers XbaI_U6_fwd and XbaI_U6_rev with two mutagenesis primers (U6_opa1_fwd: 5′-ctaaaattggaacgatattgagaagattagcatggcccctg-3′ and U6_opa1_rev: 5′- caggggccatgctaatcttctcaatatcgttccaattttag-3′) to yield mutation-adapted U6 cassette (U6.ad, 936 bp).

The target vector pscAAV2.1_EF1a_GFP_SV40pA was linearized by restriction digestion with *BcuI* and *XbaI* excising its EF-1a, eGFP and poly(A) cassettes in the process. U1.wt and U1.ad cassettes were digested with *BcuI* and *XbaI* and ligated to the linearized pscAAV2 vector using T4 Ligase (Thermo Fisher, Darmstadt, Germany) yielding plasmids pscAAV2_U1.wt and pscAAV2_U1.ad.

U6.wt and U6.ad cassettes were digested with *XbaI* and ligated to the *XbaI* linearized pscAAV2_U1.wt and pscAAV2_U1.ad plasmids yielding four plasmids: pscAAV2_U1.wt_U6.wt (U1/U6.wt), pscAAV2_U1.ad_U6.wt (U1.ad), pscAAV2_U1.wt_U6.ad (U6.ad, not used in this study) and pscAAV2_U1.ad_U6.ad (U1/U6.ad). All plasmids were verified by sequencing.

2.2. Generation and Purification of AAV2/8

Generation of AAV2/8 was performed for each of the three pscAAV2_U1_U6 plasmids (pscAAV2_U1.wt_U6.wt (U1/U6.wt), pscAAV2_U1.ad_U6.wt (U1.ad), and pscAAV2_U1.ad_U6.ad (U1/U6.ad)) as previously described [38]. We used molar ratios of 1:2:1 of the capsid plasmid pLT-RC08, the pHGTI-Adeno1 helper plasmid, and the pscAAV2 plasmid and yielded a total of approx. 500 µL viral suspension each. Virus titer was quantified by qPCR using the linearized pscAAV2 vector with primers binding to the U1 cassette (U1_fwd: 5′-CACATTTGGGGAAATCGCAGG-3′ and U1_rev 5′-CCGTGTGTGTAAAGAGTGAGG-3′). The AAV2/8 carrying the GFP-expressing plasmid (pssAAV-CMV-eGFP) was generated correspondingly.

2.3. Mice and Subretinal Injection

All animal experiments adhered to the ARVO Statement for the Use of Animals in Ophthalmic and Vision Research and were approved by the Niedersächsisches Landesamt für Verbraucherschutz und Lebensmittelsicherheit (LAVES, ref. no. 15-1829).

C3HeB/FeJ-OPA1c.1065+5G>A mice were bred to generate heterozygous (*Opa1*$^{enu/+}$) and wild-typic (*Opa1*$^{+/+}$) offspring. In a randomized fashion, young adult mice (between 1 and 4 months old) from both genotypes were injected with AAV2/8 that carried either U1/U6.wt, U1.ad or U1/U6.ad cassettes mixed 5:1 with AAV2/8 carrying GFP-expressing plasmid (pssAAV-CMV-eGFP) in one eye and PBS or AAV2/8-GFP in the other. A volume of 1.5 µL at a titer of 1.8×10^{11} viral particles/mL was injected per eye.

Mice were anaesthetized with a mixture of Fentanyl (0.05 mg/kgbw), Medetomidin (0.5 mg/kgbw) and Midazolam (5 mg/kgbw) intraperitoneally and pupil-dilating eye-drops were applied (5% Neosynephrin and 1% Tropicamid) for 2 minutes. Fundus imaging was carried out prior to injection on a Phoenix Micron III high-resolution retinal imaging microscope (Phoenix Technology Group, Bend, OR, USA). Subretinal injections were performed by the transcorneal approach [39]: The cornea was punctured at the nasal side close to or at the limbus with a 30 G cannula, and a 33 G needle attached to a Hamilton microliter syringe (Hamilton, Reno, NV, USA) was inserted through the hole. While the iris may be punctured by this method, touching the lens can be avoided in a free-hand injection as the angle of the injection needle is steep enough to pass it. When the tip of the needle penetrated the retina a volume of 1.5 µL of viral solution was applied. The needle was removed and optical gel (Visc Ophthal, Berlin, Germany) was applied. Animals were placed on a heating mat and remained anaesthetized for one hour after injection; at that time, the antidote mixture (50 μL of Naloxon [0.12 mg/kgbw], Flumazenil [0.2 mg/kgbw] and Atipamezol [0.75 mg/kgbw]) was applied subcutaneously.

Mice were anesthetized after 3–4 weeks of therapy and fundus images including GFP-fluorescence were acquired as described above. To verify the therapeutic duration, RNA analyses (see below) were additionally performed after 8 weeks of therapy. Afterward the treatment, cervical dislocation was performed and eyecups were harvested for further analyses. Eyes displaying GFP signals in less than approximately 50% of the fundus area were excluded from any further analyses, except for non-injected or PBS-injected samples.

2.4. Analysis of Transcript Levels with Semi-Quantitative RT-PCR

We extracted RNA from full eyecups preparations using the NucleoSpin RNA kit (Macherey-Nagel, Düren, Germany) according to the manufacturer's manual. Reverse transcription was carried out using SuperScript III reverse transcriptase (Thermo Fisher, Darmstadt, Germany) and 200-500 ng of RNA per sample. A control reaction was performed without reverse transcriptase (-RT).

RT-PCR reactions were performed using HotFire Taq (Solis Biodyne, Tartu, Estonia) with 1 μL of cDNA as template and *Opa1*-specific primers (OPA1_1011_fwd: 5′-GACGACAAAGGC-ATCCACCA-3′ and OPA1_1510_rev: 5′-GTTTCCTTTGTCTCGGGAGC-3′) binding to exons 7 and 13, respectively, resulting in PCR fragments of 500 bp from the wild-type allele and 419 bp from the mutated allele (exon 10 skipping during splicing resulted in shorter fragments). Relative band intensities were quantified with Bio-Rad ImageLab v6.0 (Bio-Rad, Feldkirchen, Germany) using the "Auto-detect Bands" function and manual alignment. Values were compared between the eye receiving one of the U1-AAV2/8 and the PBS- or mock-injected control eye from the same animal. All RT-PCRs were technically replicated (two separate cDNA syntheses and two PCR analyses each) at least three times and the mean value was used for statistical analyses. We compared values from eyes that received the same therapeutic construct with control eyes resulting in the following group sizes for semi-quantitative RT-PCR analyses: PBS/GFP controls (n = 25), U1.wt_U6.wt (n = 9), U1.ad_U6.wt (n = 13) and U1.ad_U6.ad (n = 9).

2.5. Sanger Sequencing of Vectors and Opa1 Transcripts

PCR fragments were excised from agarose gels and purified using Macherey-Nagel's NucleoSpin Gel and PCR Clean-up kit (Macherey-Nagel, Düren, Germany) according to the manufacturer's manual. DNA concentrations were measured with an Eppendorf BioSpectrometer (Eppendorf, Hamburg, Germany) and 10-20 ng of each sample were used for sequencing. Cycle sequencing was performed using BigDye Terminator v3.1 (Thermo Fisher, Darmstadt, Germany) with one of the primers applied in RT-PCR reactions. Alignment and verification with the reference sequence were performed in SnapGene v5.1.3.1 (GSL Biotech, San Diego, CA, USA).

Vectors were sequence verified either with primer BcuI_BshTI_U1_fwd, XbaI_U1_rev, XbaI_U6_fwd, or XbaI_U6_rev. Two vector-specific primers (pscAAV2_3prime_seq_R: 5′-gcaacaggaaaaacgctcatgg-3′ and pscAAV2_ ori_seq_fwd: 5′-gcgtcgatttttgtgatgctcg-3′) were also applied for Sanger sequencing.

2.6. Western-Blot Analysis

Retina and RPE were isolated and lysed in RIPA buffer supplemented with protease inhibitors (S8830, Merck, Taufkirchen, Germany). After 5 min incubation on ice, the supernatant was cleared from insoluble debris by centrifugation (10 min, 6000× *g*, 4 °C). 15 μg of protein lysate was loaded per lane, separated on a 10% PAGE, and blotted onto a PVDF-membrane. After blocking with 5% BSA/TEST for 1 h at room temperature, the membrane was incubated over night at 4 °C with an OPA1-specific antibody (1:1000 diluted, mouse anti-OPA1 clone 18, cat. 612607, BD Biosciences, Heidelberg, Germany) in blocking buffer. As loading control, a mouse anti-GAPDH antibody (Merck, Chemicon, MAB374) was used. A peroxidase-conjugated goat anti-mouse antibody (NB7539, Novus) was used as the secondary antibody, followed by ECL detection. Relative protein levels were calculated based on band intensity quantifications using ImageLab software (Bio-Rad, Feldkirchen, Germany).

2.7. Histological Analyses of Retinal Slices

Eyecups were prepared after euthanizing the animals. The cornea was pierced with a lancet at the border between cornea and sclera and the whole eye was incubated in 4% paraformaldehyde for 5 minutes. Afterwards, the cornea was cut off along the equator between cornea and sclera and eyecups were incubated for 2 hours in 4% PFA. Post fixation, eyecups were washed with PBS and incubated in PBS with 30% sucrose overnight at 4°C, followed by embedding in Tissue-Tek O.C.T. compound (Sakura Finetek, Torrance, CA, USA) and storage at −20°C until cutting (similar as previously described [38]).

Histological slices of 20 μm were cut from an area around the optic nerve and the middle third of the eyecup on a Thermo Scientific CryoStar NX70 cryotome (Thermo Fisher, Darmstadt, Germany). GFP and DAPI fluorescence was analyzed on an AxioScan Z1 microscope (Zeiss, Oberkochen, Germany) at 5–10× magnification. Adjacent slices from the region including the optic nerve head were used for further analyses.

Outer nuclear layer thickness was measured at two regions, proximal and distal to the optic nerve head, using FIJI software [40]. Analyzed groups included: wild-type/non-injected (n = 6), wild-type/U1.wt-injected (n = 5), wild-type/U1.ad-injected (n = 9), heterozygous/non-injected (n = 5), heterozygous/GFP-injected (n = 9), heterozygous/U1.wt-injected (n = 9) and heterozygous/U1.ad-injected (n = 16). For each n, retinal thickness was measured in 2-3 adjacent slices at both regions and the means were used for comparison between groups.

2.8. Electroretinogram (ERG) Measurements

Eyes were grouped by their fundus image into three categories of injection quality. The lowest quality category was excluded from ERG analyses. High and medium categories were analyzed by ERGs. Animals were dark-adapted overnight (>12 h) before ERG recordings. The animals were handled under dim red light (660 nm). Mice were anaesthetized with a mixture of Fentanyl (0.05 mg/kgbw), Medetomidin (0.5 mg/kgbw) and Midazolam (5 mg/kgbw) intraperitoneally and pupil-dilating eye-drops were applied (5% Neosynephrin and 1% Tropicamid) for 2 minutes. Eyes were moistened with a thin layer of ViscOphthal gel (Dr. Winzer Pharma GmbH, Berlin, Germany). Animals of either sex were used and their ages ranged from 3 to 8 months. For the ERG recordings, the experimenter was blinded to the treatment conditions of the eyes. Full-field ERGs were recorded binocularly. Two platinum coil electrodes were placed on the corneal surface of the left and right eye, respectively. A platinum needle reference electrode was inserted subcutaneously on the forehead. Another platinum needle grounding electrode was inserted at the base of the tail. Mice were placed on a custom-built temperature-regulated stage during the recording. Corneal potentials were amplified (500×) and bandpass filtered (0.1–300 Hz) using a differential amplifier (Model 3000, A-M Systems, WA, USA) and digitized at 10 kHz (PowerLab, ADInstruments). A calibrated Ganzfeld ERG stimulation arena was used for light stimulation (Q450, Roland Consult). ISCEV standard intensities were used [41]. For scotopic ERGs, white-light flashes with intensities of 0.0003, 0.0009, 0.003, 0.0095, 0.03, 0.095, 0.3, 0.95, 3, and 9.5 cd·s/m^2 were used. Each intensity was repeated 10 times. Inter-stimulus intervals were 2 s for intensities 0.0003–0.0095 cd·s/m^2, 5 s for intensities 0.03–0.3 cd·s/m^2, 10 s for intensities 0.95–3 cd·s/m^2, and 20 s for intensity 9.5 cd·s/m^2, respectively. For photopic flash stimulation, the animals were light-adapted at a background intensity of 25 cd/m^2 for 600 s. This intensity was also used as a background during the photopic stimulation, which consisted of 10 repetitions each, with intensities of 0.03, 0.095, 0.3, 0.95, 3, and 9.5 cd·s/m^2. The a-wave amplitude was defined as the voltage minimum after stimulus onset (0–300 ms) relative to the average voltage before stimulus onset (−100–0 ms). The a-wave implicit time was defined as the time of the voltage minimum relative to the stimulus onset. The b-wave peak was defined as the maximum of the low-pass filtered (cutoff frequency: 75 Hz) response voltage in a time window of 0–300 ms after stimulus onset. The b-wave amplitude was measured from the trough of the a-wave to the b-wave peak. The b-wave implicit time was defined as the time of the voltage maximum relative to the stimulus onset. The a- and b-wave amplitudes and implicit times were visually inspected and manually corrected if necessary. In addition, a photopic flicker ERG was measured over a range of frequencies (2, 4, 8, 16, 32 Hz) at an intensity of 3 cd·s/m^2. For a given intensity and frequency, responses were averaged across at least 10 stimulus periods and the amplitude was defined as the difference between the maximum and the minimum mean response.

2.9. Off-Target Gene Analyses

Potential off-target sites for mutation-adapted U1 were identified by bioinformatic searches for full binding sequence (GATGTAAAT) in coding regions of the mouse genome (mm39) and filtering by the following criteria: target sequence located between 50 and 200 bp downstream of the beginning of a coding exon (potentially shortening the exon if recognized as a bona fide donor site) or between 1 and 60 bp downstream of the beginning of an intron (potential for partial intron retention). Genes containing sites fitting these criteria were checked for expression in the visual system and association with retinal phenotypes on Mouse Gene Expression Database (http://www.informatics.jax.org/expression.shtml, accessed on 18 February 2021). For all of these genes with potential off-target effects, primers spanning the exon/intron region of the target sites were designed (Supplementary Table S1). RT-PCR and subsequent Sanger sequencing were performed on cDNA comparing injected and control eyes (n = 5 animals).

2.10. Statistical Analysis

Experiments were replicated at least three times unless otherwise specified. Data are presented as mean ± standard deviation (SD). Error bars indicate the SD. Statistical significance was calculated using Mann-Whitney U test of the R software package (version 3.6) unless otherwise specified (***: $p < 0.001$; **: $p < 0.01$; *: $p < 0.05$; n.s.: not significant, $p > 0.05$).

3. Results

3.1. Engineered U1 Treatment Rescues Opa1 Splice Defect

The *Opa1*$^{enu/+}$ mouse model carries an *Opa1*:c.1065+5G>A SDS mutation that causes a splice defect leading to exon 10 skipping and thus an in-frame deletion of 81 nucleotides (Figure 1A,B) [30]. The mutation interferes with the base-pairing between the *Opa1* SDS and the wild-type splice factors U1 and U6 (U1.wt and U6.wt; Figure 1A,B). To restore correct splicing of *Opa1* exon 10, we generated engineered U1 and U6 that show full base pair complementarity to the mutated SDS of exon 10 (U1.ad and U6.ad; Figure 1C). We produced AAV2/8 vectors expressing different combinations of the engineered or wild-type U1 and U6: U1.wt_U6.wt, U1.ad_U6.wt, and U1.ad_U6.ad.

After treatment, we extracted total RNAs from whole eyecups and quantified *Opa1* transcripts that either included or excluded exon 10 (Figure 1D,E). Sanger sequencing of RT-PCR products confirmed the identity of PCR fragments (Supplementary Figure S1). The ratio between wild-type *Opa1* and mutant *Opa1* transcripts skipping exon 10 was not altered in U1.wt_U6.wt injected eyes compared to contralateral mock injected eyes (Figure 1D). In contrast, injections of U1.ad_U6.wt or U1.ad_U6.ad showed therapeutic efficacy and increased the fraction of wild-type *Opa1* transcripts at the expense of mutated *Opa1* (Figure 1E and Supplementary Figure S2, respectively). Quantification of the relative RT-PCR band intensities from treated eyecups confirmed that U1.ad_U6.wt injections significantly increased the wild-type *Opa1* transcripts, while reducing mutated *Opa1* transcripts (U1.ad_U6.wt: 72.0 ± 3.4% (n = 13); U1.wt_U6.wt: 59.6 ± 1.7% (n = 9); PBS: 60.6 ± 1.9% (n = 25)) (Figure 1F). Similarly, a significant increase in *Opa1* wild-type transcripts was detected in eyes treated with U1.ad_U6.ad (71.2 ± 2.4% (n = 9)) (Figure 1F). Differences in *Opa1* splicing between U1.wt_U6.wt and mock injected eyes were not detected (Figure 1F).

Notably, no significant differences between the two treatments, either U1.ad_U6.wt or U1.ad_U6.ad, were revealed (Figure 1F), indicating that the co-expression of engineered U1 and U6 did not improve the efficacy of *Opa1* exon 10 inclusion to a detectable level. Consequently, we decided to continue our studies with U1.ad_U6.wt treatments only.

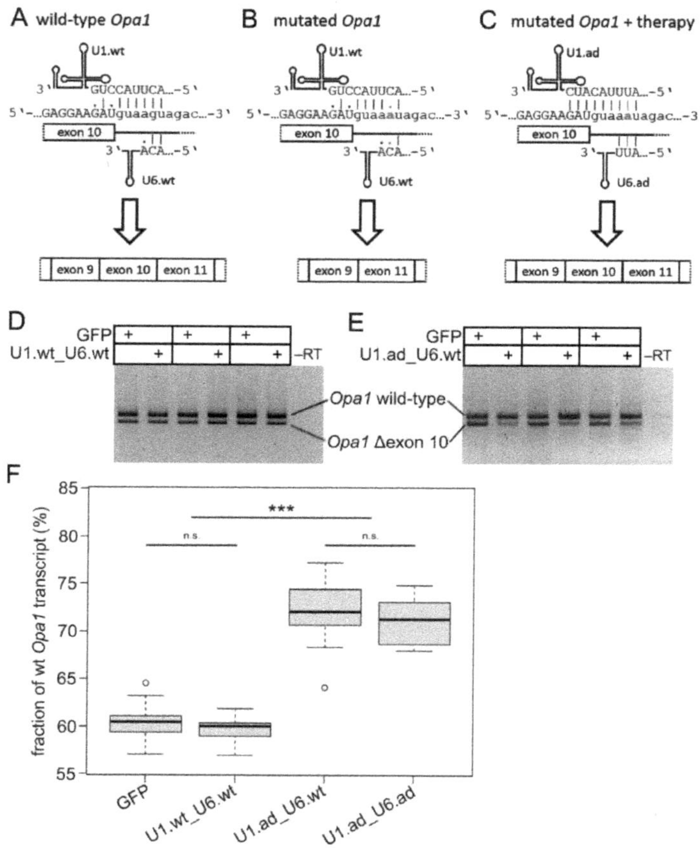

Figure 1. Correcting *Opa1* splice defects with engineered U1 splice factors. (**A**) Wild-type U1 (U1.wt) binds with seven Watson-Crick base-pairings to the *Opa1* wild-type splice donor site of exon 10. Wild-type U6 (U6.wt) binds with only two base-pairings to the *Opa1* pre-mRNA. (**B**) The mutation c.1065+5G>A abolishes Watson-Crick base-pairing at the highly-conserved +5 position in the intron (highlighted in red), leading to skipping of exon 10 during splicing. (**C**) Our treatment strategy uses engineered U1 (U1.ad) to overcome the effect of the mutation c.1065+5G>A and to bind to all nine base pairings of the mutated splice donor site of exon 10 in *Opa1*. Engineered U6 (U6.ad) can bind with all three possible base-pairings. (**D**,**E**) RT-PCR analyses of *Opa1* splicing after treatment with engineered U1 in three $Opa1^{enu/+}$ mice each. The larger RT-PCR fragment corresponds to the correctly spliced wild-type *Opa1*, the shorter corresponds to *Opa1* skipping exon 10. (**D**) The U1.wt_U6.wt treatment did not change *Opa1* splicing in $Opa1^{enu/+}$ eyes compared to the contralateral mock (GFP) controls. (**E**) The fraction of correctly spliced wild-type *Opa1* transcripts was increased upon treatment with U1.ad_U6.wt compared to the contralateral mock control. (**F**) Quantification of *Opa1* wild-type and exon10 skipping fragments revealed a significant increase in relative wild-type band intensity (with respect to total *Opa1* transcripts) upon treatment with U1.ad. Statistical significances were calculated by Mann-Whitney U test. ***: $p < 0.001$; n.s.: not significant, $p > 0.05$. –RT: cDNA reaction without reverse transcriptase.

Taken together, our results show a significant increase of wild-type *Opa1* transcript expression in the eye upon AAV2/8-mediated retinal treatment with engineered U1 splice factors. We did not observe unwanted splice-altering effects in *Opa1* transcripts after treatment with either wild-type or engineered U1/U6.

3.2. *Opa1 Protein Levels Are Increased upon U1 Therapy*

Next, we compared Opa1 protein expression in total homogenates of untreated heterozygous mutant ($Opa1^{enu/+}$) and wild-type ($Opa1^{+/+}$) mouse eyes. Untreated $Opa1^{enu/+}$ mice exhibited clearly reduced Opa1 levels compared to $Opa1^{+/+}$ mice (Figure 2A). Left and right eyes from the same mouse showed highly similar Opa1 protein levels. Our findings confirmed previous observations: Protein levels of Opa1 were reduced by appr. 50% in mouse models and human fibroblasts carrying heterozygous *Opa1* mutations [30,37].

To further evaluate the therapeutic potential of engineered U1, we asked whether the increase in wild-type *Opa1* transcripts after U1 therapy translates into elevated Opa1 protein expression. To exclude inter-individual variability, we performed different injections into each of the two eyes of the same animals. $Opa1^{enu/+}$ and $Opa1^{+/+}$ mice were injected with U1.ad in one eye, whereas the contralateral eye received U1.wt as control. We compared the treatment effects by measuring Opa1 protein expression in Western blots. While there was no detectable difference between the injected eyes in wild-type mice (Figure 2B, left), treatment with engineered U1 increased Opa1 protein expression in mutant $Opa1^{enu/+}$ mice (Figure 2B, right).

Quantification of Western blots was performed by comparing Opa1 expression of one eye (set to 100% as internal control) with the contralateral eye of the same animal. As expected, non-injected animals showed no significant difference in Opa1 protein levels between left and right eyes in both, $Opa1^{enu/+}$ and $Opa1^{+/+}$ mice (Figure 2A,C(left)). For injected animals, Opa1 levels in the U1.wt treated eye were set to 100% and compared to the contralateral U1.ad treated eye (see Figure 2B,C(right)). Our quantification showed that U1.ad treated $Opa1^{enu/+}$ eyes had significantly increased Opa1 protein levels (114.1 ± 5.4%, n = 5, normalized to the contralateral control-eyes) (Figure 2C), consistent with the increase of wild-type *Opa1* transcripts (Figure 1E,F). The effect was specific for $Opa1^{enu/+}$ eyes. Opa1 protein levels in wild-type eyes were not affected by injection of engineered or wild-type U1 (101.4 ± 4.2%, n = 4).

In summary, the in vivo treatment of the retina with engineered U1 in $Opa1^{enu/+}$ mutant eyes concomitantly increased *Opa1* transcript and protein levels. Our results indicate efficacy of the splice correction therapy on the molecular level.

3.3. *Off-Target Splicing Is Not Affected by the U1 Treatment*

To analyze the RNA safety profile of the U1 treatment, we verified whether splicing alterations in potential off-target genes occurred following AAV2/8-U1 transduction of the retina. We analyzed genes previously associated with diseases of the visual system and bioinformatically identified potential binding sites of the engineered U1 (GATG-TAAAT). This procedure predicted 57 or 37 binding sites with the potential to result in exon-truncation or intron-retention, respectively. From these two sets of binding sites, eleven U1 binding sites occurred in genes that have previously been associated with visual phenotypes (Supplementary Table S1). We analyzed the corresponding transcripts for splice alternating events and compared treated and control eyecups using RT-PCR (Supplementary Table S1). We did not detect alterations in either fragment sizes or intensities, and we did not find an association with the engineered U1 therapy (n = 5) (Figure 3). Sequencing the PCR products confirmed the wild-type sequences of the transcripts and did not identify sequence alterations in potential off-target transcripts (Figure 3). Despite several attempts, we were unable to establish RT-PCRs amplifying *Ankrd35*

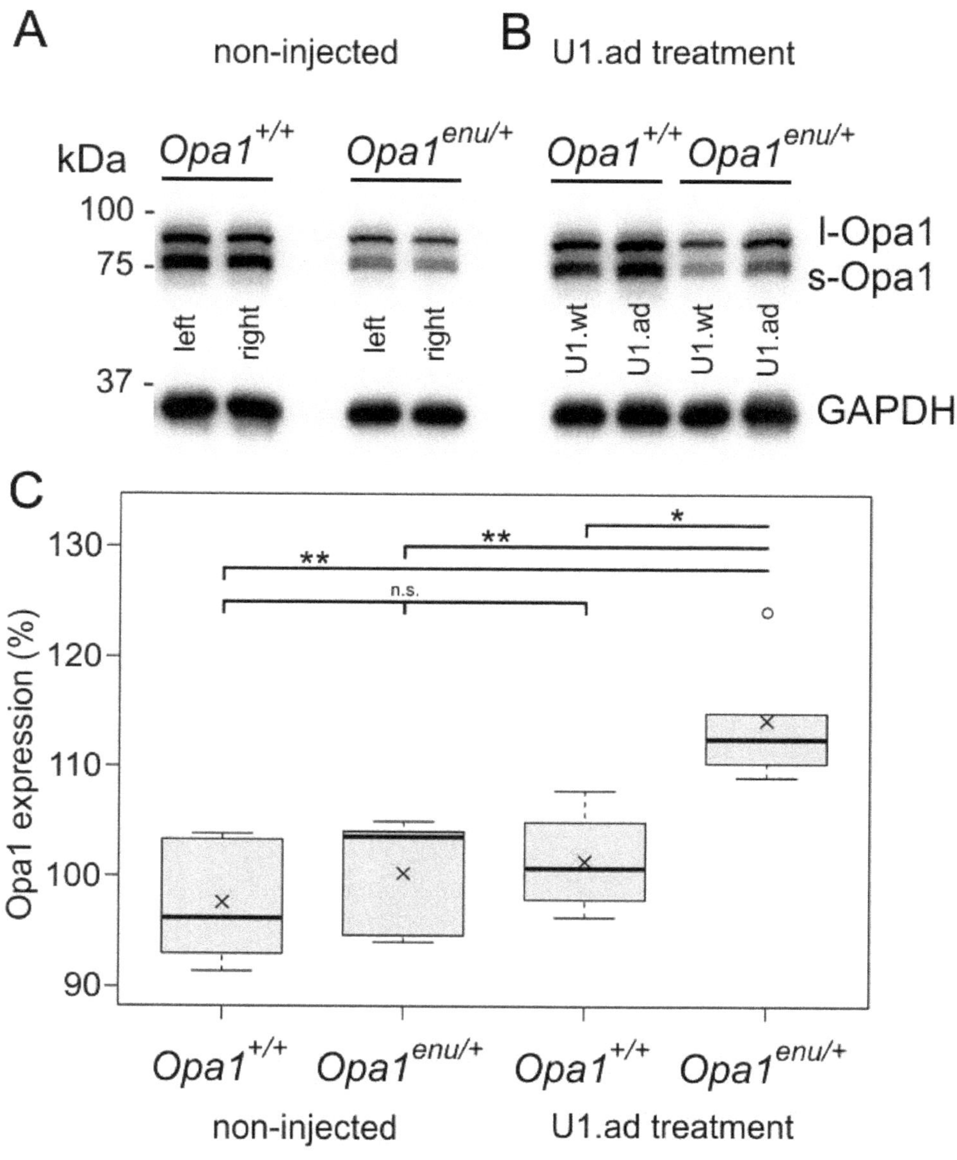

Figure 2. Treatment with engineered U1 specifically increased Opa1 protein expression in *Opa1*$^{enu/+}$ mutant mouse eyes. (**A**,**B**) Western blot analyses of Opa1 protein expression in total homogenates of retina and RPE. Left and right eyes of each animal are shown next to each other. GAPDH was used as a loading control. (**A**) In non-injected animals, Opa1 protein expression was reduced in *Opa1*$^{enu/+}$ eyes compared to wild-type eyes. (**B**) Opa1 protein levels were not changed in wild-type eyes upon U1 treatment (left). In *Opa1*$^{enu/+}$ eyes injected with engineered U1 (U1.ad), Opa1 protein levels were increased compared to the contralateral control eye injected with U1.wt (right). (**C**) *Opa1*$^{enu/+}$ eyes, but not wild-type eyes, treated with engineered U1.ad showed a significant increase of Opa1 protein relative to control (U1.wt) treated contralateral eyes. Quantification of Western blots was performed by comparing Opa1 protein expression of one eye (set to 100% as internal control) to the contralateral eye of the same animal. Statistical significances were calculated by Mann-Whitney U tests. ** $p < 0.01$; * $p < 0.05$; n.s., not significant, $p > 0.05$.

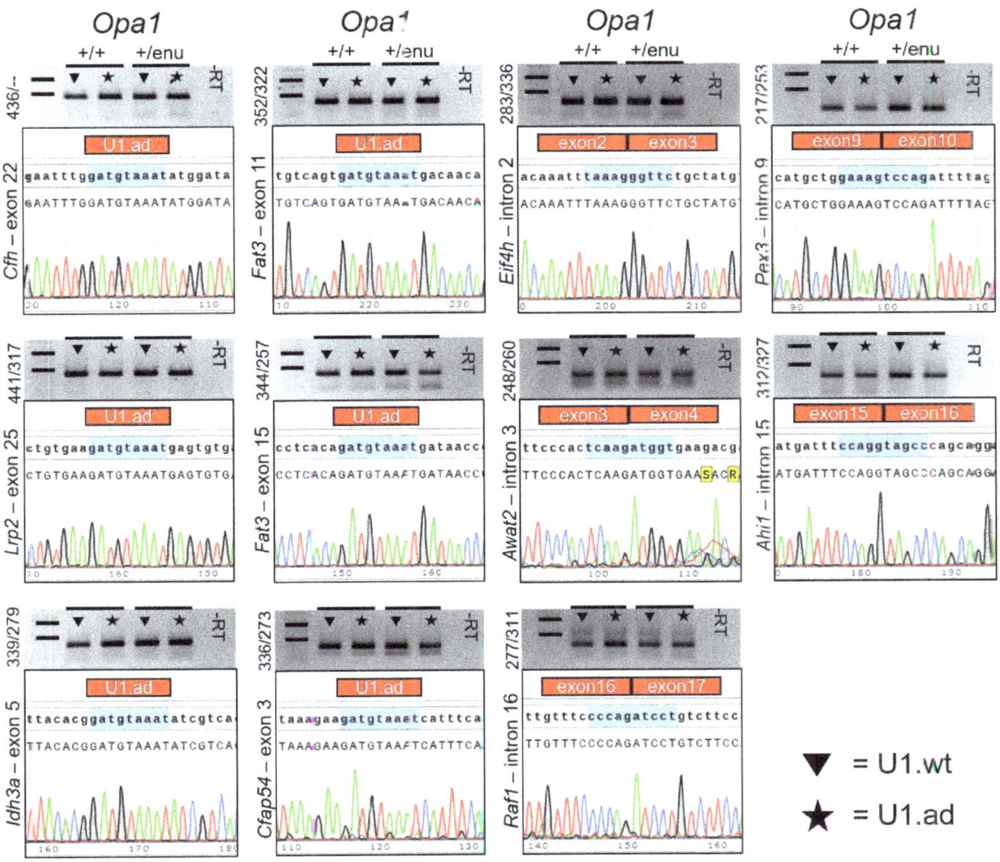

Figure 3. Off-target analyses of engineered U1 on pre-mRNA splicing. Potential off-target transcripts from 11 genes previously associated with retinal phenotypes were analyzed by RT-PCR using exon/intron-spanning primers. Off-target effects on splicing were not detected. Additional bands, indicating splice alterations, were not identified in any of the 11 transcripts upon U1-based treatments, neither in wild-type (*Opa1*$^{+/+}$, n = 5) nor in heterozygous (*Opa1*$^{enu/+}$, n = 5) mice. Sequencing of the RT-PCR fragments confirmed correct splicing of the potential off-target genes. U1.ad marks potential binding sites of engineered U1. Exon to exon borders are highlighted. Size marker represent 500 and 1000 base pairs (bp). Numbers to the left of each gel image indicate the expected wild-type fragment sizes (first number) as well as sizes of potential off-target splicing events (second number).

Overall, expression of engineered U1 did not induce detectable mis-splicing in visual system-associated transcripts. These results support the belief that splice defects in off-target transcripts are not an obvious side effect of U1 treatments.

3.4. Retinal Morphology Remains Inconspicuous after Subretinal AAV2/8 Injection

To monitor in vivo retinal morphology and transduction efficacy of AAV2/8 after subretinal therapy, we employed fundus imaging and collected full eyecups from sacrificed injected animals. Between three and eight weeks after injection, GFP signals were detectable across wide areas of the retina in fundus images of injected animals (Figure 4A–D). This observation was confirmed in retinal slices, mainly showing GFP signals in the retinal pigment epithelium (RPE) and photoreceptor layer (Figure 4A′–D′). The photoreceptor layer was widely transduced, but appeared to express less GFP than RPE cells (Figure 4C″–D″).

These observations are in line with previously published data of transduction efficacies of the AAV2/8 serotype [38,42,43]. Taken together, our results show a successful and wide-spread transduction of outer retinal layers following a single subretinal injection of AAV2/8 viral particles that expressed engineered U1.

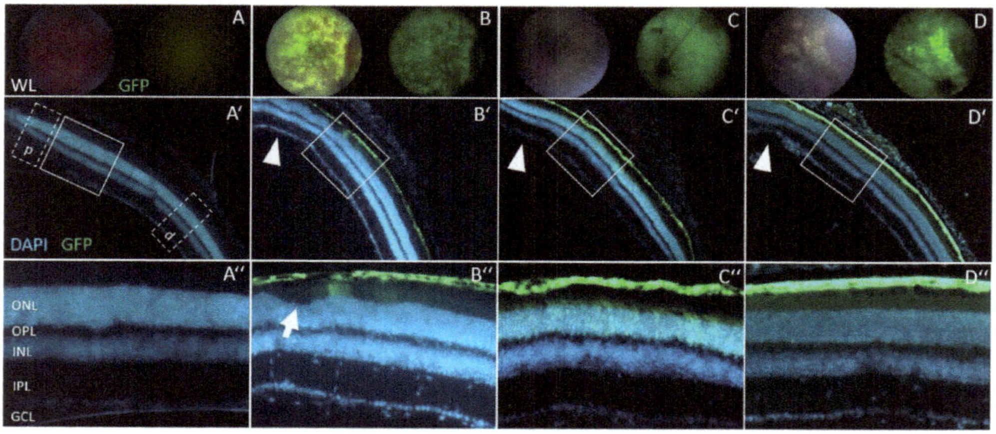

Figure 4. Fundus images and histological morphology of injected eyes. (**A–D**) Fundus images of injected eyes after white-light (WL, left) or UV-light with GFP channel detection (right). (**A**) PBS and non-injected animals only showed background signal. (**B**) GFP fluorescence appeared across the wide areas of the fundus in mice injected with GFP-expressing AAV2/8. (**C**) Wild-type U1 and (**D**) engineered U1 four weeks after subretinal injection. Fundus pictures documented wide-spread GFP-expression and successful transduction of retinal cells. (**A′–D″**) Retinal cryosections document AAV2/8-mediated transductions mainly to the retinal pigment epithelium and photoreceptor layers. ONL thickness was measured at regions marked by boxes (dotted lines) and labelled by p and d in **A′**. At and around the point of injection (arrowheads), mechanical damage to the retina can be detected (as in **D′**). (**A″–D″**) Magnification of boxes showed that the overall retinal morphology was similar between eyes regardless of the injected construct. Retinal detachments (arrow in **B″**) were observed occasionally. Panels A, C, D: *Opa1*$^{+/+}$; B: *Opa1*$^{enu/+}$. WL, white light; ONL, outer nuclear layer; OPL, outer plexiform layer; INL, inner nuclear layer; IPL, inner plexiform layer; GCL, ganglion cell layer.

To elucidate potential side-effects of treatment with engineered U1 in the retina, we analyzed the retinal layer morphology. Compared to non-injected eyes (*Opa1*$^{enu/+}$: n = 4, *Opa1*$^{+/+}$: n = 8), retinal slices from the area around the optic nerve head showed inconspicuous morphology after injection of AAV2/8 expressing engineered U1 (*Opa1*$^{enu/+}$: n = 14, *Opa1*$^{+/+}$: n = 12), or wild-type U1 (*Opa1*$^{enu/+}$: n = 8, *Opa1*$^{+/+}$: n = 10), or GFP/PBS (*Opa1*$^{enu/+}$: n = 19, *Opa1*$^{+/+}$: n = 12) (Figure 4A′–D″). At the site of injection, the retina was mechanically damaged by the injection needle (Figure 4D′). While physical disruption of the retinal tissue at the injection site can be considered a side-effect, the retinae tolerated detachment/re-attachment during the course of therapy and showed no detectable morphological differences caused by our therapeutic intervention.

To detect potential degeneration of retinal cells, we quantified retinal thickness near the site of injection (Table 1). Measurements of outer nuclear layer (ONL) thickness in retinal slices at two regions, proximal and distal to the optic nerve head (see boxes p and d in Figure 4A′), revealed no significant difference (Supplementary Table S2). Thus, alterations of the retinal morphology, e.g., a thinning or swelling of the retina, were undetectable between injected and non-injected *Opa1*$^{enu/+}$ animals (Table 1).

Table 1. ONL thickness of the central retina is not altered by the U1 treatments. The average outer nuclear layer (ONL) thickness in micro-meters (µm) is listed with standard deviations. Retinae from *Opa1*$^{+/+}$ (wild-type) and *Opa1*$^{enu/+}$ (heterozygous mutant) animals are compared. ONL thickness was measured at regions proximal and distal to the optic nerve head. Pairwise comparison of ONL thickness at proximal and distal regions between groups using Mann-Whitney U test with Bonferroni correction did not yield significant differences (Supplementary Table S2). The number of biological replicates is indicated in brackets (n = 5–16).

	Opa1$^{+/+}$ ONL Thickness (µm)		*Opa1*$^{enu/+}$ ONL Thickness (µm)	
	Proximal	**Distal**	**Proximal**	**Distal**
non-injected	56.05 ± 1.21 (n = 6)	52.71 ± 2.08 (n = 6)	57.99 ± 1.24 (n = 5)	56.75 ± 1.10 (n = 5)
GFP	-	-	56.08 ± 1.80 (n = 9)	53.97 ± 2.12 (n = 9)
U1.wt	56.77 ± 1.57 (n = 5)	52.84 ± 2.35 (n = 5)	56.28 ± 2.73 (n = 9)	53.75 ± 2.72 (n = 9)
U1.ad	54.77 ± 3.86 (n = 9)	53.25 ± 3.07 (n = 9)	56.08 ± 2.90 (n = 16)	53.24 ± 2.29 (n = 16)

Taken together, we did not detect any significant differences in ONL thickness, neither in proximal nor distal retinal areas of the optic nerve head. Comparing retinae between *Opa1*$^{+/+}$ and *Opa1*$^{enu/+}$ animals and different treatments resulted in highly similar values. These results indicated the retinal morphology was not affected by the AAV2/8-mediated gene therapy applying engineered U1 splice factors.

3.5. Electroretinogram Responses Are Unaltered after U1 Treatments

To verify retinal function and to explore potential side-effects of the U1 treatment, we performed Ganzfeld ERG measurements under photopic and scotopic conditions and compared U1-treated, mock-injected and control eyes. Following the ERG measurement, histological analyses of all analyzed eyes detected only minor mechanical damages caused by the injection procedures and confirmed wide-spread GFP signals across the retinae. Amplitude, waveform, and timing of the a-wave—reflecting the activity of the photoreceptors—were unaltered between treated and control eyes (Figure 5A–D). Furthermore, the b-wave, which reflects excitation and activity of bipolar cells in the inner retina, showed no significant differences across treatment conditions (Figure 5A–D). Groups tested were: injected with U1.ad_U6.wt (n = 7); mock-injected (n = 9); and non-injected (n = 24) (Figure 5B–D). The flicker ERG showed no significant difference between the therapeutic construct, control construct, and the non-injected groups, with overlapping confidence intervals (Supplementary Figure S3).

Overall, our data supported that AAV2/8-mediated U1 treatments of retinal diseases may be considered an efficient and safe gene therapeutic approach that lacks obvious side-effects towards retinal functionality, morphology and splicing.

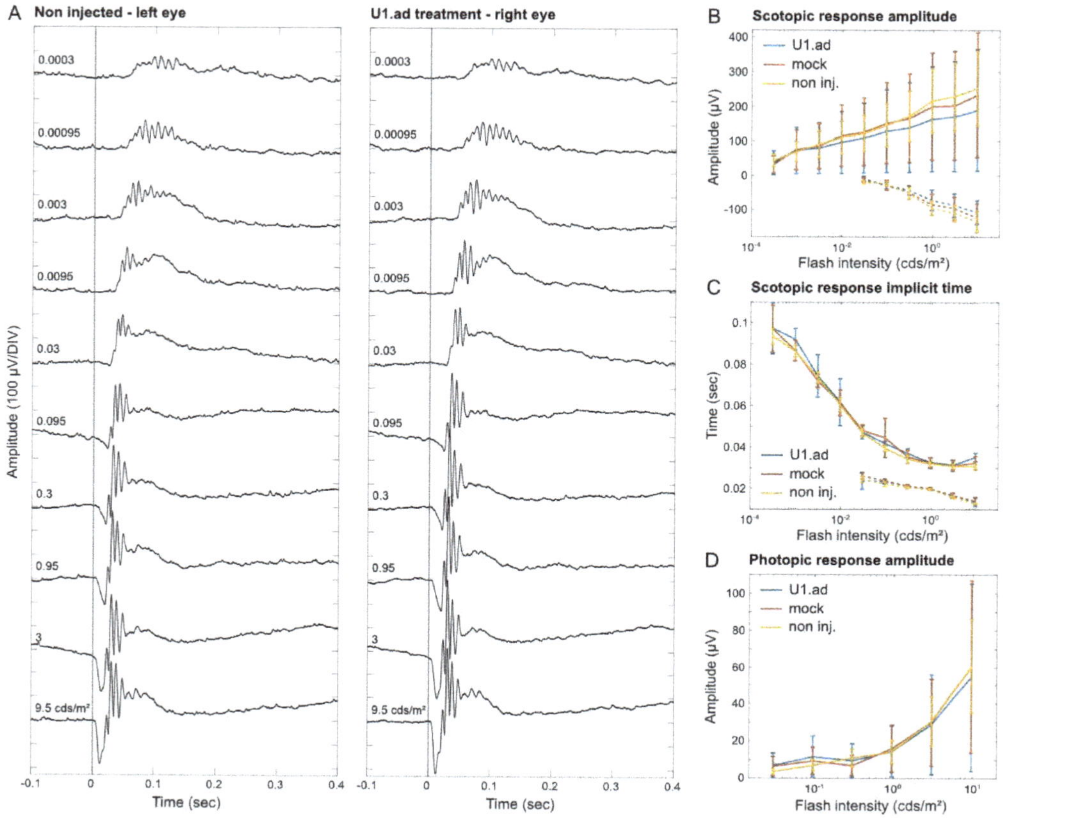

Figure 5. U1-mediated side effects on retinal function were not detected by electroretinography. (**A**) Representative scotopic ERG responses to light flashes of increasing intensity. The left eye was untreated. The right eye was injected with engineered U1 (U1.ad). The ERGs of both eyes were recorded simultaneously. (**B**) Mean scotopic ERG amplitudes. Error bars represent 95% confidence intervals (U1.ad: n = 7, mock: n = 9, non-injected: n = 24). A-wave dashed line, b-wave solid line. (**C**) Scotopic ERG implicit times corresponding to (**B**). (**D**) Photopic ERG amplitudes of light-adapted mice with stimulus flash of increasing intensities corresponding to (**B**).

4. Discussion

We evaluated the efficacy and safety of engineered U1 as a gene therapeutic agent in the retina. We made use of an ADOA mouse model carrying a splice-site mutation in the *Opa1* gene and show that subretinal delivery of engineered U1 via AAV2/8 leads to transduction of the RPE and photoreceptor layers. Following the U1 treatment, we detected increased wild-type *Opa1* expression on the transcript and protein level. Short-term side-effects of the U1-therapy were not detected, especially not on functional level (ERG). The results presented herein suggested that this therapeutic approach might be considered for further evaluation in RGCs, conceivably even in human clinical trials.

The *Opa1*$^{enu/+}$ mouse model is well-suited to evaluate treatments of splice defects, i.e., the *Opa1*$^{enu/+}$ mutation leads to exon skipping [30], one of the most frequently observed mutation-induced splice defects. However, the *Opa1*$^{enu/+}$-mouse model is less suitable to evaluate short-term safety and efficacy of RGC-specific treatments; variability and late onset as well as slow progression of RGC degeneration imply obstacles to short-term phenotypic evaluations [30]. Thus, we did not target RGCs, the most affected cell type in ADOA, but RPE and photoreceptor cells, the most abundant and highly vulnerable cell types in the retina. Their high abundance allowed us to analyze short-term efficacy and safety of the U1-mediated treatment in vivo, with the clear advantage of circumventing more emulated in vitro measurements with enriched and cultivated cells types from the retina. Safety and efficacy of the U1 treatment had not been analyzed in the in vivo retina before. The analyzed cell types enable well-established, highly-sensitive molecular (Figure 3), morphological (Figure 4 and Table 1), and physiological measurements (Figure 5) likely to detect potential short-term side-effects that might be associated with therapeutic interventions.

Efficient and sustained transduction of RPE and photoreceptor layers was previously achieved after subretinal AAV2/8 injections [42–45]. Therapeutic effects were transferred to the retina using subretinal injection of AAV2/8, e.g., inhibiting neovascularization [46]. Even more significant, subretinal injections were applied in approved gene therapies to treat retinal diseases in human, e.g., for RPE65 deficiencies [47–49]. Delivery routes of gene therapies were only accompanied by minor side effects [50], further supporting the enormous potential of genetic treatments in the retina.

Our present study was designed as a proof-of-concept at an early stage of retinal therapy. Hence, we did not investigate long-term therapeutic effects nor did we aim at attenuating or reversing the pathogenesis of ADOA. While we were able to show that the U1 therapy increases Opa1 expression on the molecular level—the precondition for any functional therapy—we did not address if the U1 treatment rescues mitochondrial functions or attenuates RGC degeneration on the cellular level; this will require future long-term studies. For isolated RGCs from *Opa1*^{Q285STOP} mice, mitochondrial functions were shown to be compromised [51]. It will be interesting to evaluate whether similar effects can be observed in *Opa1*$^{enu/+}$ derived RGCs and whether they can be rescued applying the U1 therapy.

Haploinsufficiency is the major disease mechanism underlying *OPA1* mutations, i.e., the disease is caused by the loss of gene dose from a single allele that carries a heterozygous mutation [52]. Engineered U1 resulted in a significant upregulation of correctly spliced *Opa1* transcripts from the mutated allele. Given the loss or very mild disease expression frequently observed in *OPA1*-linked ADOA [53,54], it is likely that the increased expression levels of Opa1 will attenuate or even stop the disease progression. Importantly, about 10–20% of all mutations affect splicing, an observation that is almost independent from the affected gene and thus, raises the chance to apply U1 therapies to many different genetic diseases [55].

We applied engineered U1 which showed increased binding capacities to the mutated SDS in order to recruit downstream splicing factors and to initiate normal splicing despite of the mutation [2,56,57]. The U1 therapy targets pre-mRNA transcripts derived specifically from the mutated allele and shows the capacity to restore expression levels and isoform ratios of the targeted gene within physiological ranges. In contrast, gene replacement/augmentation strategies typically overexpress a single protein isoform along with limited capacities in restoring physiological expression ratio among isoforms. Notably, viral vectors provide restricted insert sizes—a further limitation to gene augmentation therapies of large genes. In contrast, U1 expression cassettes are only a few hundred base pairs in size and can easily be applied using all viral vectors. The natural chromosomal context of the affected gene and hence regulation of gene expression is maintained, i.e., the endogenous promotor, pre-mRNA processing, and ratios of splice isoforms are not affected.

Opa1 is a ubiquitously expressed gene in which different isoforms are relevant to its multiple mitochondrial functions [29,58,59]. Mis-regulated expression of single isoforms or transcripts may thus counteract therapeutic effects. For the complete fusion of mitochondrial inner membranes both, the short and long isoform of Opa1 are required at stoichiometric levels [60–62], whereas an excess of the short isoform inhibits fusion [62]. These findings suggested that gene augmentation therapies for *OPA1* may have limited therapeutic efficacy. However, studies in a different *Opa1* mutant mouse line (*Opa1*^{delTTAG}) showed some promising therapeutic efficacy by overexpression of a single wild-type *Opa1* isoform including the restoration of visual function and preventing RGC loss [32]. In mice with rotenone induced mitochondrial dysfunction, gene augmentation using single *Opa1* isoforms protected spatial visual function [63]. However, a thorough safety study on gene augmentation therapy for *OPA1* has not been conducted. Regardless, U1-based therapies may provide attractive alternatives to gene augmentation approaches.

While subretinal injections mechanically disrupts the retina at the point of injection, reversible retinal detachment after application of the viral suspension was observed. The viral therapeutic agent did not lead to obvious side effects on retinal morphology. Our findings are in line with other studies showing ONL thickness in the range of approx. 54 µm to 57 µm in retinae of C57BL/6 mice [64,65]. In future studies, an intravitreal injection method and different AAV serotypes might be favorable to minimize mechanical stress to the retina and to specifically target RGCs [38,66]. However, the inner limiting membrane might reduce efficient transduction of retinal cells and transduction is mainly limited to retinal cells facing the vitreous [67]. Compared to subretinal injections, intravitreal injections exhibit a higher risk for intraocular inflammation and humoral immune response [68].

Mutations in the 5′ SDS binding region of U1 snRNA, i.e., r.3A>G and r.3A>C that pair with the +6 position of the splice donor site, have been associated with specific forms of cancer [69,70], highlighting the need for careful safety evaluation for therapies based on engineered U1. Importantly, the critical r.3A nucleotide has not been modified in our engineered U1 (see Figure 1A–C) and we did not observe any indication for cancerous cell alterations in the treated retinae. Furthermore, our data did not indicate off-target effects of the treatment on the molecular level, i.e., mis-splicing events caused by the U1 treatment. Apart from RNA and morphological analyses, we also tested for potential side-effects of the U1-based treatment on retinal functionality. ERG measurements of eyes injected with engineered U1 showed no significant differences in parameters representing outer retina function (including photoreceptors and bipolar cells). Although our molecular, morphological and functional analyses of the retina did not suggest splice defects in off-target transcripts, we cannot completely exclude this possibility; future studies might include transcriptome analyses to further address this issue.

Together, we showed for the first time that retinal applications of U1-based gene therapies lead to promising treatment profiles, providing support for further development towards splice donor site treatments of many different retinal diseases. Our findings strongly support that the U1 therapy, as presented herein, did not alter physiological function of the retina. Engineered U1 delivered via AAV2/8 to eyes of $Opa1^{enu/+}$ mice led to a significant increase in *Opa1* wild-type expression while showing no indications for short-term side-effects on splicing, retinal morphology, and retinal physiology. Future studies will verify these findings in RGCs, investigate long-term safety and efficacy of treatments, and will aim to further evaluate the potential of this promising therapeutic approach towards clinical trials.

Supplementary Materials: The following supporting information can be downloaded at: https://www.mdpi.com/article/10.3390/cells12060955/s1, Figure S1: Sequence verification of *Opa1* wild type and exon 10 skipping PCR fragments; Figure S2: RT-PCR analyses of *Opa1* splicing after treatment with engineered U1 and U6 in three $Opa1^{enu/+}$ mice; Figure S3: U1-mediated side effects on retinal function were not detected by flicker electroretinography; Table S1: Primer sequences for off-target analyses; Table S2: Comparison of ONL thickness in central retinal slices by injected construct.

Author Contributions: Conceptualization, J.N.; Methodology, S.S., O.M. and M.A.; Investigation, S.S., O.M. and M.A.; Resources, B.W., M.G. and J.N.; Writing—original draft, S.S., M.A. and C.J.; Writing—review & editing, M.G., C.J. and J.N.; Visualization, S.S., M.A. and C.J.; Supervision, M.G. and J.N.; Project administration, J.N.; Funding acquisition, J.N.. All authors have read and agreed to the published version of the manuscript.

Funding: J.N. acknowledges funding by a grant of the Deutsche Forschungsgemeinschaft (NE2118/2-1) as part of a joint research project 'TreatOPON' within the framework of the ERA-NET E-Rare 3 call.

Institutional Review Board Statement: The animal study protocol was approved by the Niedersächsisches Landesamt für Verbraucherschutz und Lebensmittelsicherheit (LAVES, ref. no. 15-1829).

Data Availability Statement: The data presented in this study are available on request from the corresponding author.

Acknowledgments: The authors thank Klaus Dieter Hübel for his help in breeding the animals at our facility and Sonja Zimmer for help in cloning and testing engineered U1/U5 constructs.

Conflicts of Interest: The authors declare no conflict of interest. The funders had no role in the design of the study; in the collection, analyses, or interpretation of data; in the writing of the manuscript; or in the decision to publish the results.

References

1. Wang, Z.; Burge, C.B. Splicing regulation: From a parts list of regulatory elements to an integrated splicing code. *RNA* **2008**, *14*, 802–813. [CrossRef] [PubMed]
2. Will, C.L.; Luhrmann, R. Spliceosome structure and function. *Cold Spring Harb. Perspect. Biol.* **2011**, *3*, a003707. [CrossRef] [PubMed]
3. Wahl, M.C.; Will, C.L.; Luhrmann, R. The spliceosome: Design principles of a dynamic RNP machine. *Cell* **2009**, *136*, 701–718. [CrossRef] [PubMed]
4. Lund, M.; Kjems, J. Defining a 5′ splice site by functional selection in the presence and absence of U1 snRNA 5′ end. *RNA* **2002**, *8*, 166–179. [CrossRef] [PubMed]
5. Kandels-Lewis, S.; Seraphin, B. Involvement of U6 snRNA in 5′ splice site selection. *Science* **1993**, *262*, 2035–2039. [CrossRef]
6. Lesser, C.F.; Guthrie, C. Mutations in U6 snRNA that alter splice site specificity: Implications for the active site. *Science* **1993**, *262*, 1982–1988. [CrossRef]
7. Krawczak, M.; Reiss, J.; Cooper, D.N. The mutational spectrum of single base-pair substitutions in mRNA splice junctions of human genes: Causes and consequences. *Hum. Genet.* **1992**, *90*, 41–54. [CrossRef]
8. Zhuang, Y.; Weiner, A.M. A compensatory base change in U1 snRNA suppresses a 5′ splice site mutation. *Cell* **1986**, *46*, 827–835. [CrossRef]
9. Breuel, S.; Vorm, M.; Brauer, A.U.; Owczarek-Lipska, M.; Neidhardt, J. Combining Engineered U1 snRNA and Antisense Oligonucleotides to Improve the Treatment of a BBS1 Splice Site Mutation. *Mol. Ther. Nucleic Acids* **2019**, *18*, 123–130. [CrossRef]
10. Balestra, D.; Giorgio, D.; Bizzotto, M.; Fazzari, M.; Ben Zeev, B.; Pinotti, M.; Landsberger, N.; Frasca, A. Splicing Mutations Impairing CDKL5 Expression and Activity Can be Efficiently Rescued by U1snRNA-Based Therapy. *Int. J. Mol. Sci.* **2019**, *20*, 4130. [CrossRef]
11. Matos, L.; Canals, I.; Dridi, L.; Choi, Y.; Prata, M.J.; Jordan, P.; Desviat, L.R.; Perez, B.; Pshezhetsky, A.V.; Grinberg, D.; et al. Therapeutic strategies based on modified U1 snRNAs and chaperones for Sanfilippo C splicing mutations. *Orphanet J. Rare Dis* **2014**, *9*, 180. [CrossRef] [PubMed]
12. Schmid, F.; Hiller, T.; Korner, G.; Glaus, E.; Berger, W.; Neidhardt, J. A gene therapeutic approach to correct splice defects with modified U1 and U6 snRNPs. *Hum. Gene Ther.* **2013**, *24*, 97–104. [CrossRef] [PubMed]
13. van der Woerd, W.L.; Mulder, J.; Pagani, F.; Beuers, U.; Houwen, R.H.; van de Graaf, S.F. Analysis of aberrant pre-messenger RNA splicing resulting from mutations in ATP8B1 and efficient in vitro rescue by adapted U1 small nuclear RNA. *Hepatology* **2015**, *61*, 1382–1391. [CrossRef]
14. Pinotti, M.; Rizzotto, L.; Balestra, D.; Lewandowska, M.A.; Cavallari, N.; Marchetti, G.; Bernardi, F.; Pagani, F. U1-snRNA-mediated rescue of mRNA processing in severe factor VII deficiency. *Blood* **2008**, *111*, 2681–2684. [CrossRef]
15. Tanner, G.; Glaus, E.; Barthelmes, D.; Ader, M.; Fleischhauer, J.; Pagani, F.; Berger, W.; Neidhardt, J. Therapeutic strategy to rescue mutation-induced exon skipping in rhodopsin by adaptation of U1 snRNA. *Hum. Mutat.* **2009**, *30*, 255–263. [CrossRef]
16. Glaus, E.; Schmid, F.; Da Costa, R.; Berger, W.; Neidhardt, J. Gene therapeutic approach using mutation-adapted U1 snRNA to correct a RPGR splice defect in patient-derived cells. *Mol. Ther. J. Am. Soc. Gene Ther.* **2011**, *19*, 936–941. [CrossRef]
17. Schmid, F.; Glaus, E.; Barthelmes, D.; Fliegauf, M.; Gaspar, H.; Nurnberg, G.; Nurnberg, P.; Omran, H.; Berger, W.; Neidhardt, J. U1 snRNA-mediated gene therapeutic correction of splice defects caused by an exceptionally mild BBS mutation. *Hum. Mutat* **2011**, *32*, 815–824. [CrossRef] [PubMed]
18. Boussaad, I.; Obermaier, C.D.; Hanss, Z.; Bobbili, D.R.; Bolognin, S.; Glaab, E.; Wolynska, K.; Weisschuh, N.; De Conti, L.; May, C.; et al. A patient-based model of RNA mis-splicing uncovers treatment targets in Parkinson's disease. *Sci. Transl. Med* **2020**, *12*, eaau3960. [CrossRef]
19. Balestra, D.; Faella, A.; Margaritis, P.; Cavallari, N.; Pagani, F.; Bernardi, F.; Arruda, V.R.; Pinotti, M. An engineered U1 small nuclear RNA rescues splicing defective coagulation F7 gene expression in mice. *J. Thromb. Haemost.* **2014**, *12*, 177–185. [CrossRef] [PubMed]
20. Balestra, D.; Scalet, D.; Ferrarese, M.; Lombardi, S.; Ziliotto, N.; C, C.C.; Petersen, N.; Bosma, P.; Riccardi, F.; Pagani, F.; et al. A Compensatory U1snRNA Partially Rescues FAH Splicing and Protein Expression in a Splicing-Defective Mouse Model of Tyrosinemia Type I. *Int. J. Mol. Sci.* **2020**, *21*, 2136. [CrossRef]
21. Lee, N.C.; Lee, Y.M.; Chen, P.W.; Byrne, B.J.; Hwu, W.L. Mutation-adapted U1 snRNA corrects a splicing error of the dopa decarboxylase gene. *Hum. Mol. Genet.* **2016**, *25*, 5142–5147. [CrossRef] [PubMed]

22. Kjer, P.; Jensen, O.A.; Klinken, L. Histopathology of eye, optic nerve and brain in a case of dominant optic atrophy. *Acta Ophthalmol.* **1983**, *61*, 300–312. [CrossRef] [PubMed]
23. Alexander, C.; Votruba, M.; Pesch, U.E.; Thiselton, D.L.; Mayer, S.; Moore, A.; Rodriguez, M.; Kellner, U.; Leo-Kottler, B.; Auburger, G.; et al. OPA1, encoding a dynamin-related GTPase, is mutated in autosomal dominant optic atrophy linked to chromosome 3q28. *Nat. Genet.* **2000**, *26*, 211–215. [CrossRef] [PubMed]
24. Delettre, C.; Lenaers, G.; Griffoin, J.M.; Gigarel, N.; Lorenzo, C.; Belenguer, P.; Pelloquin, L.; Grosgeorge, J.; Turc-Carel, C.; Perret, E.; et al. Nuclear gene OPA1, encoding a mitochondrial dynamin-related protein, is mutated in dominant optic atrophy. *Nat. Genet.* **2000**, *26*, 207–210. [CrossRef]
25. Le Roux, B.; Lenaers, G.; Zanlonghi, X.; Amati-Bonneau, P.; Chabrun, F.; Foulonneau, T.; Caignard, A.; Leruez, S.; Gohier, P.; Procaccio, V.; et al. OPA1: 516 unique variants and 831 patients registered in an updated centralized Variome database. *Orphanet J. Rare Dis.* **2019**, *14*, 214. [CrossRef]
26. Kjer, B.; Eiberg, H.; Kjer, P.; Rosenberg, T. Dominant optic atrophy mapped to chromosome 3q region. II. Clinical and epidemiological aspects. *Acta Ophthalmol. Scand.* **1996**, *74*, 3–7. [CrossRef]
27. Votruba, M.; Moore, A.T.; Bhattacharya, S.S. Clinical features, molecular genetics, and pathophysiology of dominant optic atrophy. *J. Med. Genet.* **1998**, *35*, 793–800. [CrossRef]
28. Votruba, M.; Fitzke, F.W.; Holder, G.E.; Carter, A.; Bhattacharya, S.S.; Moore, A.T. Clinical features in affected individuals from 21 pedigrees with dominant optic atrophy. *Arch. Ophthalmol.* **1998**, *116*, 351–358. [CrossRef]
29. Lenaers, G.; Neutzner, A.; Le Dantec, Y.; Juschke, C.; Xiao, T.; Decembrini, S.; Swirski, S.; Kieninger, S.; Agca, C.; Kim, U.S.; et al. Dominant optic atrophy: Culprit mitochondria in the optic nerve. *Prog. Retin. Eye Res.* **2021**, *83*, 100935. [CrossRef]
30. Alavi, M.V.; Bette, S.; Schimpf, S.; Schuettauf, F.; Schraermeyer, U.; Wehrl, H.F.; Ruttiger, L.; Beck, S.C.; Tonagel, F.; Pichler, B.J.; et al. A splice site mutation in the murine Opa1 gene features pathology of autosomal dominant optic atrophy. *Brain* **2007**, *130*, 1029–1042. [CrossRef]
31. Davies, V.J.; Hollins, A.J.; Piechota, M.J.; Yip, W.; Davies, J.R.; White, K.E.; Nicols, P.P.; Boulton, M.E.; Votruba, M. Opa1 deficiency in a mouse model of autosomal dominant optic atrophy impairs mitochondrial morphology, optic nerve structure and visual function. *Hum. Mol. Genet.* **2007**, *16*, 1307–1318. [CrossRef] [PubMed]
32. Sarzi, E.; Seveno, M.; Piro-Megy, C.; Elziere, L.; Quiles, M.; Pequignot, M.; Muller, A.; Hamel, C.P.; Lenaers, G.; Delettre, C. OPA1 gene therapy prevents retinal ganglion cell loss in a Dominant Optic Atrophy mouse model. *Sci. Rep.* **2018**, *8*, 2468. [CrossRef] [PubMed]
33. Zhang, Z.; Wakabayashi, N.; Wakabayashi, J.; Tamura, Y.; Song, W.J.; Sereda, S.; Clerc, P.; Polster, B.M.; Aja, S.M.; Pletnikov, M.V.; et al. The dynamin-related GTPase Opa1 is required for glucose-stimulated ATP production in pancreatic beta cells. *Mol. Biol. Cell* **2011**, *22*, 2235–2245. [CrossRef] [PubMed]
34. Alavi, M.V.; Fuhrmann, N.; Nguyen, H.P.; Yu-Wai-Man, P.; Heiduschka, P.; Chinnery, P.F.; Wissinger, B. Subtle neurological and metabolic abnormalities in an Opa1 mouse model of autosomal dominant optic atrophy. *Exp. Neurol.* **2009**, *220*, 404–409. [CrossRef]
35. Gonzalez-Menendez, I.; Reinhard, K.; Tolivia, J.; Wissinger, B.; Munch, T.A. Influence of Opa1 Mutation on Survival and Function of Retinal Ganglion Cells. *Investig. Ophthalmol. Vis. Sci.* **2015**, *56*, 4835–4845. [CrossRef]
36. Heiduschka, P.; Schnichels, S.; Fuhrmann, N.; Hofmeister, S.; Schraermeyer, U.; Wissinger, B.; Alavi, M.V. Electrophysiological and histologic assessment of retinal ganglion cell fate in a mouse model for OPA1-associated autosomal dominant optic atrophy. *Investig. Ophthalmol. Vis. Sci.* **2010**, *51*, 1424–1431. [CrossRef]
37. Juschke, C.; Klopstock, T.; Catarino, C.B.; Owczarek-Lipska, M.; Wissinger, B.; Neidhardt, J. Autosomal dominant optic atrophy: A novel treatment for OPA1 splice defects using U1 snRNA adaption. *Mol. Ther. Nucleic Acids* **2021**, *26*, 1186–1197. [CrossRef]
38. Da Costa, R.; Roger, C.; Segelken, J.; Barben, M.; Grimm, C.; Neidhardt, J. A Novel Method Combining Vitreous Aspiration and Intravitreal AAV2/8 Injection Results in Retina-Wide Transduction in Adult Mice. *Investig. Ophthalmol. Vis. Sci.* **2016**, *57*, 5326–5334. [CrossRef]
39. Irigoyen, C.; Amenabar Alonso, A.; Sanchez-Molina, J.; Rodriguez-Hidalgo, M.; Lara-Lopez, A.; Ruiz-Ederra, J. Subretinal Injection Techniques for Retinal Disease: A Review. *J. Clin. Med.* **2022**, *11*, 4717. [CrossRef]
40. Schindelin, J.; Arganda-Carreras, I.; Frise, E.; Kaynig, V.; Longair, M.; Pietzsch, T.; Preibisch, S.; Rueden, C.; Saalfeld, S.; Schmid, B.; et al. Fiji: An open-source platform for biological-image analysis. *Nat. Methods* **2012**, *9*, 676–682. [CrossRef]
41. McCulloch, D.L.; Marmor, M.F.; Brigell, M.G.; Hamilton, R.; Holder, G.E.; Tzekov, R.; Bach, M. ISCEV Standard for full-field clinical electroretinography (2015 update). *Doc. Ophthalmol.* **2015**, *130*, 1–12. [CrossRef] [PubMed]
42. Natkunarajah, M.; Trittibach, P.; McIntosh, J.; Duran, Y.; Barker, S.E.; Smith, A.J.; Nathwani, A.C.; Ali, R.R. Assessment of ocular transduction using single-stranded and self-complementary recombinant adeno-associated virus serotype 2/8. *Gene Ther.* **2008**, *15*, 463–467. [CrossRef] [PubMed]
43. Allocca, M.; Mussolino, C.; Garcia-Hoyos, M.; Sanges, D.; Iodice, C.; Petrillo, M.; Vandenberghe, L.H.; Wilson, J.M.; Marigo, V.; Surace, E.M.; et al. Novel adeno-associated virus serotypes efficiently transduce murine photoreceptors. *J. Virol.* **2007**, *81*, 11372–11380. [CrossRef] [PubMed]

44. Igarashi, T.; Miyake, K.; Asakawa, N.; Miyake, N.; Shimada, T.; Takahashi, H. Direct comparison of administration routes for AAV8-mediated ocular gene therapy. *Curr. Eye Res.* **2013**, *38*, 569–577. [CrossRef]
45. Lebherz, C.; Maguire, A.; Tang, W.; Bennett, J.; Wilson, J.M. Novel AAV serotypes for improved ocular gene transfer. *J. Gene Med.* **2008**, *10*, 375–382. [CrossRef] [PubMed]
46. Igarashi, T.; Miyake, K.; Masuda, I.; Takahashi, H.; Shimada, T. Adeno-associated vector (type 8)-mediated expression of soluble Flt-1 efficiently inhibits neovascularization in a murine choroidal neovascularization model. *Hum. Gene Ther.* **2010**, *21*, 631–637. [CrossRef] [PubMed]
47. Hauswirth, W.W.; Aleman, T.S.; Kaushal, S.; Cideciyan, A.V.; Schwartz, S.B.; Wang, L.; Conlon, T.J.; Boye, S.L.; Flotte, T.R.; Byrne, B.J.; et al. Treatment of leber congenital amaurosis due to RPE65 mutations by ocular subretinal injection of adeno-associated virus gene vector: Short-term results of a phase I trial. *Hum. Gene Ther.* **2008**, *19*, 979–990. [CrossRef]
48. Maguire, A.M.; Simonelli, F.; Pierce, E.A.; Pugh, E.N., Jr.; Mingozzi, F.; Bennicelli, J.; Banfi, S.; Marshall, K.A.; Testa, F.; Surace, E.M.; et al. Safety and efficacy of gene transfer for Leber's congenital amaurosis. *N. Engl. J. Med.* **2008**, *358*, 2240–2248. [CrossRef]
49. Russell, S.; Bennett, J.; Wellman, J.A.; Chung, D.C.; Yu, Z.F.; Tillman, A.; Wittes, J.; Pappas, J.; Elci, O.; McCague, S.; et al. Efficacy and safety of voretigene neparvovec (AAV2-hRPE65v2) in patients with RPE65-mediated inherited retinal dystrophy: A randomised, controlled, open-label, phase 3 trial. *Lancet* **2017**, *390*, 849–860. [CrossRef]
50. Nuzbrokh, Y.; Kassotis, A.S.; Ragi, S.D.; Jauregui, R.; Tsang, S.H. Treatment-Emergent Adverse Events in Gene Therapy Trials for Inherited Retinal Diseases: A Narrative Review. *Ophthalmol. Ther.* **2020**, *9*, 709–724. [CrossRef]
51. Sun, S.; Erchova, I.; Sengpiel, F.; Votruba, M. Opa1 Deficiency Leads to Diminished Mitochondrial Bioenergetics With Compensatory Increased Mitochondrial Motility. *Investig. Ophthalmol. Vis. Sci.* **2020**, *61*, 42. [CrossRef] [PubMed]
52. Pesch, U.E.; Leo-Kottler, B.; Mayer, S.; Jurklies, B.; Kellner, U.; Apfelstedt-Sylla, E.; Zrenner, E.; Alexander, C.; Wissinger, B. OPA1 mutations in patients with autosomal dominant optic atrophy and evidence for semi-dominant inheritance. *Hum. Mol. Genet.* **2001**, *10*, 1359–1368. [CrossRef] [PubMed]
53. Cohn, A.C.; Toomes, C.; Potter, C.; Towns, K.V.; Hewitt, A.W.; Inglehearn, C.F.; Craig, J.E.; Mackey, D.A. Autosomal dominant optic atrophy: Penetrance and expressivity in patients with OPA1 mutations. *Am. J. Ophthalmol.* **2007**, *143*, 656–662. [CrossRef]
54. Xu, X.; Wang, P.; Jia, X.; Sun, W.; Li, S.; Xiao, X.; Hejtmancik, J.F.; Zhang, Q. Pathogenicity evaluation and the genotype-phenotype analysis of OPA1 variants. *Mol. Genet. Genom.* **2021**, *296*, 845–862. [CrossRef] [PubMed]
55. Hwu, W.L.; Lee, Y.M.; Lee, N.C. Gene therapy with modified U1 small nuclear RNA. *Expert. Rev. Endocrinol. Metab.* **2017**, *12*, 171–175. [CrossRef]
56. Ast, G. How did alternative splicing evolve? *Nat. Rev. Genet.* **2004**, *5*, 773–782. [CrossRef]
57. Faustino, N.A.; Cooper, T.A. Pre-mRNA splicing and human disease. *Genes Dev.* **2003**, *17*, 419–437. [CrossRef]
58. Akepati, V.R.; Muller, E.C.; Otto, A.; Strauss, H.M.; Portwich, M.; Alexander, C. Characterization of OPA1 isoforms isolated from mouse tissues. *J. Neurochem.* **2008**, *106*, 372–383. [CrossRef]
59. Delettre, C.; Griffoin, J.M.; Kaplan, J.; Dollfus, H.; Lorenz, B.; Faivre, L.; Lenaers, G.; Belenguer, P.; Hamel, C.P. Mutation spectrum and splicing variants in the OPA1 gene. *Hum. Genet.* **2001**, *109*, 584–591. [CrossRef]
60. Del Dotto, V.; Mishra, P.; Vidoni, S.; Fogazza, M.; Maresca, A.; Caporali, L.; McCaffery, J.M.; Cappelletti, M.; Baruffini, E.; Lenaers, G.; et al. OPA1 Isoforms in the Hierarchical Organization of Mitochondrial Functions. *Cell Rep.* **2017**, *19*, 2557–2571. [CrossRef]
61. Song, Z.; Chen, H.; Fiket, M.; Alexander, C.; Chan, D.C. OPA1 processing controls mitochondrial fusion and is regulated by mRNA splicing, membrane potential, and Yme1L. *J. Cell Biol.* **2007**, *178*, 749–755. [CrossRef]
62. Ge, Y.; Shi, X.; Boopathy, S.; McDonald, J.; Smith, A.W.; Chao, L.H. Two forms of Opa1 cooperate to complete fusion of the mitochondrial inner-membrane. *eLife* **2020**, *9*, e50973. [CrossRef] [PubMed]
63. Maloney, D.M.; Chadderton, N.; Millington-Ward, S.; Palfi, A.; Shortall, C.; O'Byrne, J.J.; Cassidy, L.; Keegan, D.; Humphries, P.; Kenna, P.; et al. Optimized OPA1 Isoforms 1 and 7 Provide Therapeutic Benefit in Models of Mitochondrial Dysfunction. *Front. Neurosci.* **2020**, *14*, 571479. [CrossRef] [PubMed]
64. Berger, A.; Cavallero, S.; Dominguez, E.; Barbe, P.; Simonutti, M.; Sahel, J.A.; Sennlaub, F.; Raoul, W.; Paques, M.; Bemelmans, A.P. Spectral-domain optical coherence tomography of the rodent eye: Highlighting layers of the outer retina using signal averaging and comparison with histology. *PLoS ONE* **2014**, *9*, e96494. [CrossRef]
65. Dysli, C.; Enzmann, V.; Sznitman, R.; Zinkernagel, M.S. Quantitative Analysis of Mouse Retinal Layers Using Automated Segmentation of Spectral Domain Optical Coherence Tomography Images. *Transl. Vis. Sci. Technol.* **2015**, *4*, 9. [CrossRef] [PubMed]
66. Hellstrom, M.; Ruitenberg, M.J.; Pollett, M.A.; Ehlert, E.M.; Twisk, J.; Verhaagen, J.; Harvey, A.R. Cellular tropism and transduction properties of seven adeno-associated viral vector serotypes in adult retina after intravitreal injection. *Gene Ther.* **2009**, *16*, 521–532. [CrossRef] [PubMed]
67. Dalkara, D.; Kolstad, K.D.; Caporale, N.; Visel, M.; Klimczak, R.R.; Schaffer, D.V.; Flannery, J.G. Inner limiting membrane barriers to AAV-mediated retinal transduction from the vitreous. *Mol. Ther. J. Am. Soc. Gene Ther.* **2009**, *17*, 2096–2102. [CrossRef]
68. Reichel, F.F.; Peters, T.; Wilhelm, B.; Biel, M.; Ueffing, M.; Wissinger, B.; Bartz-Schmidt, K.U.; Klein, R.; Michalakis, S.; Fischer, M.D.; et al. Humoral Immune Response After Intravitreal But Not After Subretinal AAV8 in Primates and Patients. *Investig. Ophthalmol. Vis. Sci.* **2018**, *59*, 1910–1915. [CrossRef]

69. Shuai, S.; Suzuki, H.; Diaz-Navarro, A.; Nadeu, F.; Kumar, S.A.; Gutierrez-Fernandez, A.; Delgado, J.; Pinyol, M.; Lopez-Otin, C Puente, X.S.; et al. The U1 spliceosomal RNA is recurrently mutated in multiple cancers. *Nature* **2019**, *574*, 712–716. [CrossRef]
70. Suzuki, H.; Kumar, S.A.; Shuai, S.; Diaz-Navarro, A.; Gutierrez-Fernandez, A.; De Antonellis, P.; Cavalli, F.M.G.; Juraschka, K Farooq, H.; Shibahara, I.; et al. Recurrent noncoding U1 snRNA mutations drive cryptic splicing in SHH medulloblastoma. *Natur* **2019**, *574*, 707–711. [CrossRef]

cells MDPI

Article

Mitochondrial Dysfunction and Impaired Antioxidant Responses in Retinal Pigment Epithelial Cells Derived from a Patient with *RCBTB1*-Associated Retinopathy

Zhiqin Huang [1,2], Dan Zhang [2], Shang-Chih Chen [2], Di Huang [2], David Mackey [1,2], Fred K. Chen [1,2,3,4,*] and Samuel McLenachan [1,2,*]

1 Centre for Ophthalmology and Visual Science, The University of Western Australia, Crawley, WA 6009, Australia; zhiqin.huang@lei.org.au (Z.H.)
2 Lions Eye Institute, Nedlands, WA 6009, Australia
3 Department of Ophthalmology, Royal Perth Hospital, Perth, WA 6000, Australia
4 Ophthalmology, Department of Surgery, The University of Melbourne, Parkville, VIC 3010, Australia
* Correspondence: fred.chen@lei.org.au (F.K.C.); samuel.mclenachan@lei.org.au (S.M.)

Abstract: Mutations in the *RCBTB1* gene cause inherited retinal disease; however, the pathogenic mechanisms associated with RCBTB1 deficiency remain poorly understood. Here, we investigated the effect of RCBTB1 deficiency on mitochondria and oxidative stress responses in induced pluripotent stem cell (iPSC)-derived retinal pigment epithelial (RPE) cells from control subjects and a patient with *RCBTB1*-associated retinopathy. Oxidative stress was induced with tert-butyl hydroperoxide (tBHP). RPE cells were characterized by immunostaining, transmission electron microscopy (TEM), CellROX assay, MitoTracker assay, quantitative PCR and immunoprecipitation assay. Patient-derived RPE cells displayed abnormal mitochondrial ultrastructure and reduced MitoTracker fluorescence compared with controls. Patient RPE cells displayed increased levels of reactive oxygen species (ROS) and were more sensitive to tBHP-induced ROS generation than control RPE. Control RPE upregulated *RCBTB1* and *NFE2L2* expression in response to tBHP treatment; however, this response was highly attenuated in patient RPE. RCBTB1 was co-immunoprecipitated from control RPE protein lysates by antibodies for either UBE2E3 or CUL3. Together, these results demonstrate that *RCBTB1* deficiency in patient-derived RPE cells is associated with mitochondrial damage, increased oxidative stress and an attenuated oxidative stress response.

Keywords: RCBTB1; inherited retinal disease; retinal pigment epithelium; oxidative stress; mitochondria

Citation: Huang, Z.; Zhang, D.; Chen, S.-C.; Huang, D.; Mackey, D.; Chen, F.K.; McLenachan, S. Mitochondrial Dysfunction and Impaired Antioxidant Responses in Retinal Pigment Epithelial Cells Derived from a Patient with *RCBTB1*-Associated Retinopathy. *Cells* **2023**, *12*, 1358. https://doi.org/10.3390/cells12101358

Academic Editor: Hossein Ameri

Received: 30 March 2023
Revised: 1 May 2023
Accepted: 4 May 2023
Published: 10 May 2023

1. Introduction

Mutations in the *RCC1 and BTB domain-containing protein 1* (*RCBTB1*) gene have been reported to cause a spectrum of inherited retinal diseases (IRDs) [1–5]. To date, fifteen cases from eleven unrelated families have been reported to be associated with biallelic variants in the *RCBTB1* gene; however, the pathogenic mechanisms associated with RCBTB1 deficiencies remain poorly understood. We previously described the natural history of chorioretinal atrophy progression of a patient with compound heterozygous *RCBTB1* mutations (c.170delG and c.707delA) over 5 years [3]. Consistent with our previous optical coherence tomography (OCT) findings, which demonstrated diffuse attenuation and irregularity of the retinal pigment epithelium (RPE) layer and the adjacent ellipsoid zone in the proband, a multimodal imaging study reported by Catomeris et al. also demonstrated widespread RPE irregularity and progressive RPE atrophy in patients carrying biallelic *RCBTB1* variants [5]. Collectively, clinical observations in *RCBTB1*-associated retinopathy suggest that the primary lesion is most likely located within the RPE layer, followed by secondary photoreceptor cell apoptosis. Recent recognition of the phenotypic similarity between RCBTB1-associated retinopathy and the most severe forms of mitochondrial retinopathy [6] further suggests RPE mitochondria may play a role in disease pathogenesis.

To study the effects of RCBTB1 deficiency in human RPE cells, we previously generated induced pluripotent stem cell (iPSC) lines from a patient with *RCBTB1*-associated retinopathy caused by biallelic frameshifting mutations in the *RCBTB1* gene [7]. RPE generated from these iPSC lines displayed reduced expression of genes involved in the NFE2L2 oxidative stress pathway [8], supporting similar findings reported in lymphocytes isolated from patients with *RCBTB1*-associated retinopathy [2]. To further elucidate the effects of RCBTB1 deficiency in RPE cells, we characterized reactive oxygen species (ROS) generation and activation of *NFE2L2* expression in patient- and control-derived RPE cells under basal conditions and in response to an oxidative stress challenge with tert-butyl hydroperoxide (tBHP). Additionally, we characterized mitochondria in patient-derived RPE cells by electron microscopy and Mitotracker labelling.

2. Materials and Methods

2.1. Human-Induced Pluripotent Stem Cell Lines

The human iPSC lines used in this study were obtained with informed consent following protocols approved by the Human Research Ethics Committee (2001-053) at Sir Charles Gairdner Hospital (Nedlands, Western Australia, Australia) and the Human Ethics Office of Research Enterprise at The University of Western Australia (RA/4/1/7916). This study was performed in accordance with the Declaration of Helsinki. The protocols used in this study were approved by the Institutional Biosafety Committee of the Harry Perkins Institute of Medical Research, University of Western Australia (NLRD 02-2020). The clinical phenotype of the patient and generation of iPSC lines was reported previously [3,7].

2.2. RPE Differentiation and Cell Culture

RPE monolayers were generated from iPSC lines from a patient with RCBTB1-associated retinopathy and two healthy donors using our published methods [8]. Media were changed twice per week and RPE cells were passaged using TrypLE™ Express Enzyme (Thermo Fisher Scientific, Waltham, MA, USA) at a split ratio of 1:4. RPE cells were used at passage 3–4 for experiments 8 weeks after seeding. Expression of RPE markers in iPSC-derived RPE cells was characterized by immunofluorescence staining and quantitative PCR (Supplementary Figure S1), as previously described [8]. Primers and antibodies used for RPE characterisation are listed in Supplementary Tables S1 and S2, respectively.

2.3. Phagocytosis Assay

Bovine rod photoreceptor outer segment (POS) was purchased from InVision BioResources (Seattle, WA, USA). For POS labelling, fluorescein isothiocyanate (FITC) was added to the POS following the published procedure [9]. Phagocytosis assays were performed according to McLaren and Buskin et al.'s protocols [10,11]. The ratio of POS and iPSC-RPE cells was 10:1, and cells were incubated with POS for 6 h at 37 °C. For each RPE cell line, one well without POS was included as negative control, while total and internalized FITC-POS were assessed in triplicate wells. In all phagocytosis experiments, unbound POS were washed away with phosphate-buffered saline supplemented with 1 mM $MgCl_2$ and 0.2 mM $CaCl_2$ (PBS-CM) before fixation. To quantify internalized POS, cells were treated with 0.4% trypan blue and incubated at room temperature for 10 min to quench fluorescence of surface-bound FITC-labelled POS. For flow cytometry, cells were suspended in FACS buffer (PBS with 1% FBS and 1 mM EDTA) and transferred to FACS tubes. Samples were fixed and run on a BD LSRFortessa X-20 flow cytometer, and at least 5000 events were collected per sample. Untreated iPSC-RPE cells were used as negative control to set the live cell gate in each experiment. Data were analysed using FlowJo 10.4 software.

For the fluorescence microscopy method, cells were rinsed with PBS-CM following trypan blue incubation and fixed with 4% PFA for 10 min. Nuclei were stained with DAPI, and internalized phagocytosed POS were visualized using the Nikon A1 Confocal Laser microscope (Nikon, Tokyo, Japan). The number of internal POS per cell and per area was counted, and data analysis of differences was determined by unpaired student *t*-test.

2.4. Quantitative Real-Time PCR

RNA was harvested from RPE cells using Trizol (Thermo Fisher Scientific) following the manufacturer's instructions. RPE cDNA was synthesized using the RT^2 First Strand Kit (Qiagen, Hilden, Germany). Quantitative real-time PCR analysis (qRT-PCR) was conducted using the CFX Connect Real-Time System (BioRad, Hercules, CA, USA) with the RT^2 SYBR Green qPCR Mastermix (Qiagen). Expression levels were normalized to *GAPDH* and expressed as mean fold-changes compared with untreated control RPE cells ($n = 3$ independently derived iPSC lines). Significance testing was performed using the student *t*-test, with p values < 0.05 considered significant. Primers used for qPCR are listed in Supplementary Table S1.

2.5. Transmission Electron Microscopy

For transmission electron microscopy (TEM), RPE cells were seeded on Millicell hanging cell culture inserts (24 well, 0.4 µm, Merck, Darmstadt, Germany) coated with Geltrex (Thermo Fisher Scientific). After 8 weeks of culture, inserts were removed and fixed with 2.5% glutaraldehyde (Grade I, Sigma-Aldrich, St. Louis, MO, USA) in 0.1 M phosphate buffer (pH 7.4) for 30 min at room temperature. The cells were secondarily fixed with 1% osmium tetroxide. Following dehydration using ethanol and acetone, RPE monolayers were embedded in resin in silicone rubber flat embedding moulds and cured for 48 h at 65 °C. Glass knives were made and mounted to a Leica EM UC6 ultramicrotome for ultrathin sectioning (80–120 nm). Sections were then collected onto copper grids and double stained with uranyl acetate (UA) and lead citrate. These sections were then visualized using a JEOL JEM-F200 transmission electron microscope (Jeol, Tokyo, Japan). Mitochondrial analysis was performed on TEM images collected from three wells of patient RPE and three wells of control RPE cells. Each well of RPE cells was derived from an independent, clonal iPSC line. Mitochondria length and area were measured in 6–13 TEM images per well using ImageJ and expressed as the mean values obtained from >200 mitochondria, with error bars indicating standard error of the mean (SEM). Statistical comparisons were performed using the unpaired *t*-test, with p values < 0.05 considered as significant.

2.6. MitoTracker Assay

The MitoTracker Orange CMTMRos Reagent (100 nM, ThermoFisher) was added to RPE cultures 8 weeks after seeding, and cells were incubated for 30 min, with or without 10 µM carbonyl cyanide chlorophenylhydrazone (CCCP, ThermoFisher). Following staining and fixation, cell fluorescence imaging was performed using the Olympus BX60 upright fluorescence microscope. Fluorescence signals in images were measured using ImageJ 1.53i (National Institute of Health, USA) and expressed as the mean values obtained from RPE cells generated from three independent iPSC lines, with error bars representing standard deviation. Statistical comparisons were performed using the unpaired *t*-test, with p values < 0.05 considered as significant.

2.7. CellROX Assay

CellROX Deep Red Reagent (Invitrogen, C10491) was used for measuring intracellular reactive oxidative species (ROS) levels. Healthy control and patient-derived iPSC-RPE cells were seeded in 96 well plates and cultured for 8 weeks. RPE cells were treated with tBHP at various concentrations (0 µM, 100 µM, 200 µM) for one hour, then stained with 5 µM CellROX® Deep Red Reagent by adding the probe to the no phenol cell culture medium and incubating the cells for 30 min at 37 °C. Fluorescence signals were analysed on the CLARIOstar® Plus microplate reader (BMG, Labtech, Ortenberg, Germany) The CellRox assay was performed in technical triplicates, and fluorescence signal data are expressed as the mean from three independent clonal iPSC lines, with error bars indicating standard deviation. The statistical analysis of differences was determined by unpaired *t*-test, with p values < 0.05 considered significant.

2.8. Immunoprecipitation Assay

Immunoprecipitation was carried out using the Immunoprecipitation kit (Abcam, Cambridge, United Kingdom, ab206996) as per the manufacturer's instructions (Method B). For each reaction, 40 µL of protein A/G sepharose beads were washed twice with 1 mL wash buffer by slow-speed centrifugations and suspended as 50% slurry in 40 µL of wash buffer. Primary antibody was added into the slurry of protein A/G. Negative control beads without the addition of antibodies were also prepared. The antibody-bead mixture was incubated for 4 h at 4 °C on a shaker and was subsequently collected by slow-speed centrifugations. After cell lysis with ice-cold non-denaturing lysis buffer and protease inhibitor, protein quantification was performed using Quick Start™ Bradford 1× Dye Reagent (Bio-Rad). A total of 100 µg of cell lysate was added into the bead/antibody conjugate mixture and incubated under rotary agitation overnight at 4 °C. Beads were then pelleted and washed three times in 1× Wash Buffer. After the final wash step, beads were resuspended in 15 µL of 4× SDS-PAGE loading buffer without 2-mercaptoethanol and incubated at 50 °C for 5 min. Beads were pelleted, and the supernatant was collected and supplemented with 1 µL of 2-mercaptoethanol (elution 1). The bead pellet was then resuspended in 15 µL of 4X SDS-PAGE loading buffer with 2-mercaptoethanol and incubated at 50 °C for 5 min (elution 2). The eluted samples were boiled for 5 min and subjected to SDS-PAGE for electrophoresis and western blotting. Antibodies used are listed in Supplementary Table S2.

2.9. Western Blotting

Total protein was extracted using RIPA buffer (Sigma-Aldrich) and 1% Protease Inhibitor Cocktail (Sigma-Aldrich). Following centrifugation, protein lysates were collected and the concentration was measured using the Quick Start™ Bradford Protein Assay (Biorad). Protein samples were loaded on 4–12% NuPAGE™ Bis-Tris Mini Protein Gels (ThermoFisher), and electrophoresis was conducted the XCell SureLock Mini-Cell Electrophoresis System (ThermoFisher). Protein was subsequently transferred to an Immobilon-FL PVDF membrane (Merck) using the Mini Protean II Trans-Blot Cell (Bio-Rad). Following transfer, membranes were cut at 50 kDa and 90 kDa to generate three pieces for staining with different antibodies. Membranes were blocked with 5% bovine serum albumin (BSA) in tris-buffered saline (TBS) for one hour at room temperature. The membrane was incubated in primary antibodies solution overnight at 4 °C, followed by incubation with secondary antibodies for 1.5 h at room temperature. Membranes were washed three times in TBS buffer supplemented with 0.1% TWEEN-20 and scanned using the Li-Cor Odyssey 9120 Infrared Imaging System (LI-COR Biosciences, Lincoln, NE, USA). Original fluorescence images of Western blots are shown in Supplementary Figure S4. Antibodies used are listed in Supplementary Table S2.

3. Results

3.1. RPE Differentiation

Expression of RPE markers was confirmed using immunocytochemistry and qRT-PCR in three control and three patient iPSC-derived RPE cell lines. Six weeks after seeding, iPSC-RPE cell cultures expressed RPE markers involved in the retinoid cycle (RPE65 and CRALBP), tight junction (ZO1), chloride channels (BEST1), phagocytosis function (MerTK), Na^+/K^+ ATPase, melanin biosynthesis (Tyrosinase) and transcription factors (PAX6 and MITF) (Supplementary Figure S1).

3.2. Photoreceptor Outer Segment Uptake in RCBTB1-Deficient RPE

To assess the ability of patient RPE cells to phagocytize POS, iPSC-RPE monolayers were incubated with FITC-labelled POS for six hours at a ratio of ten POS per cell. Analysis of the mean percentage of FITC-positive cells demonstrated no significant difference in phagocytosis function between control RPE and patient RPE (Supplementary Figure S2A). These findings were corroborated using a fluorescence microscopy method

(Supplementary Figure S2B). Data analysis of the mean number of internal POS per cell and per area showed no significant changes between control and patient groups.

3.3. Ultrastructural Changes in RCBTB1-Deficient RPE

Transmission electron microscopic (TEM) images of the control-derived iPSC-RPE monolayers revealed features of native RPE tissue, showing apical microvilli, cytoplasmic melanosomes, nucleus, abundant endoplasmic reticulum and mitochondria. In contrast, patient RPE monolayers appeared more disorganized compared with the control RPE cells. Additionally, mitochondrial abnormalities, including disrupted cristae and a swollen appearance, were observed in patient RPE cells (Figure 1A–C). To further examine mitochondrial ultrastructure in patient RPE cells, mean mitochondrial length was determined by measuring > 200 individual mitochondria from TEM micrographs of RPE cells derived from control and patient iPSC lines. Mitochondria length (Figure 1D) and mitochondria area (Figure 1E) in patient RPE cells (length: 0.79 ± 0.34 μm; area: 0.21 ± 0.12 μm^2) were significantly decreased compared with control RPE cells (length: 1.04 ± 0.43 μm, $p < 0.0001$; area: 0.37 ± 0.2 μm^2, $p < 0.0001$). Furthermore, a comparison of the percentages of mitochondria with aberrant cristae revealed increased frequencies of degenerate mitochondria in patient RPE cells (Figure 1F,G). Frequency distribution plots for mitochondrial length and area demonstrated decreased mitochondrial size in patient RPE cells compared with control RPE cells (Figure 1H,I). Representative TEM images illustrating organelle structures in control and patient iPSC-RPE cells are shown in Supplementary Figure S3.

3.4. Mitochondrial Membrane Potential Is Reduced in RCBTB1-Deficient RPE

To further characterize mitochondrial function in iPSC-RPE cells, we incubated the patient and control iPSC-RPE with the MitoTracker Orange CMTMRos probe, which accumulates in mitochondria in a mitochondrial membrane potential-dependent manner. Untreated control iPSC-RPE cells were brightly labelled with the MitoTracker probe, while treatment of RPE cells with the mitochondrial membrane potential disruptor CCCP significantly reduced MitoTracker fluorescence ($p = 0.0353$). In contrast, MitoTracker fluorescence was significantly reduced in untreated patient iPSC-RPE compared with untreated control RPE ($p = 0.0214$) and was unaffected by CCCP treatment (Figure 2), suggesting mitochondrial membrane potential is compromised in RCBTB1-deficient iPSC-RPE.

3.5. Reactive Oxygen Species Are Increased in RCBTB1-Deficient RPE

To measure ROS production in iPSC-derived RPE cell cultures, we utilized the CellROX assay (Figure 3A). Untreated patient iPSC-derived RPE cells showed significantly increased levels of ROS compared with iPSC-RPE derived from healthy controls ($p = 0.015$). In control iPSC-RPE cells, treatment with 100 μM tBHP induced a small, non-significant increase in ROS compared with untreated controls ($p = 0.0544$), while treatment with 200 μM induced a larger, significant increase in ROS levels ($p = 0.0018$). In contrast, patient iPSC-RPE showed a large, significant increase in ROS levels in response to 100 μM tBHP compared with untreated patient cells ($p = 0.0463$) or control iPSC-RPE cells treated with 100 μM tBHP ($p = 0.0037$). Treatment of patient iPSC-RPE with 200 μM tBHP resulted in no further increase in ROS levels and were not significantly different from control iPSC-RPE treated at this dose ($p = 0.1378$). Together, these results indicate RCBTB1-deficient RPE cells generate high levels of ROS and, compared to control iPSC-RPE cells, are more sensitive to tBHP-induced oxidative stress.

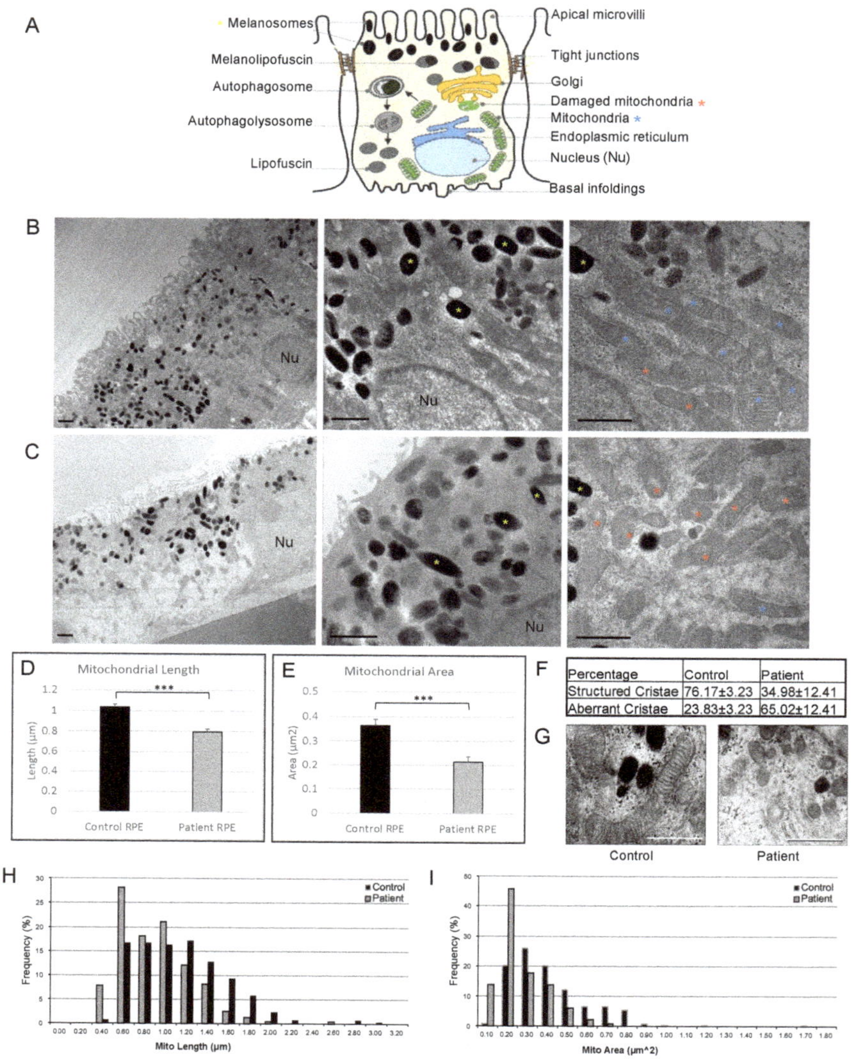

Figure 1. Electron microscopic analysis of cultured RPE cells showing morphology and structural features in patient-derived and control RPE cells. (**A**) A diagram showing the basic structure of RPE cell. (**B**,**C**) TEM images of RPE cells demonstrating the ultrastructure of control-(**B**) and patient-derived (**C**) RPE. Melanosomes (yellow asterisk) were observed in the apical RPE cytoplasm and nuclei (Nu) in the basal cytoplasm. Apical microvilli, tight junctions, endoplasmic reticulum and abundant mitochondria (blue asterisk) were observed. Note the disrupted cristae formation and architecture of damaged mitochondria (red asterisk) in patient-derived RPE. (**D**,**E**) Mean mitochondria length (**D**) and area (**E**) were significantly decreased inpatient RPE cells compared with control groups. (**F**,**G**) An increase in damaged mitochondria with aberrant cristae was observed in patient RPE cells, compared with control RPE cells. (**H**,**I**) Frequency distribution plots of mitochondria by size (**H**) and area (**I**). At least 200 mitochondria were measured in each group. Scale bars in (**B**,**C**,**G**) indicate 1 µm. *** $p < 0.0001$.

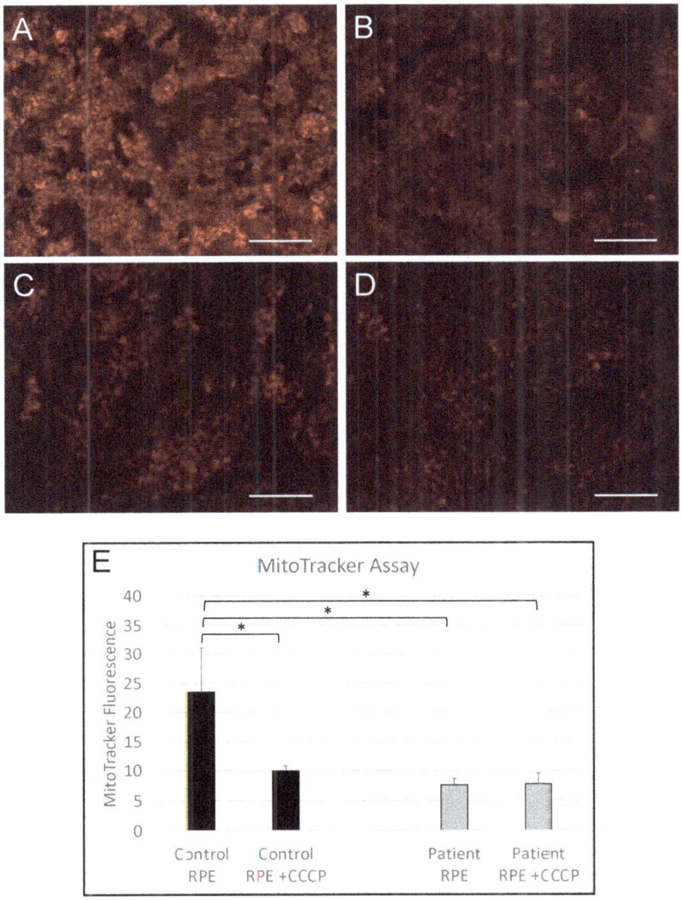

Figure 2. MitoTracker labelling of RPE mitochondria. (**A**–**D**) Fluorescence micrographs showing live control- (**A**,**B**) and patient-derived (**C**,**D**) RPE monolayers labelled with MitoTracker Orange CMTMRos reagent. Panels on the right (**B**,**D**) were treated with 10 µM CCCP. Scale bars indicate 100 µm. (**E**) Bar graph showing mean MitoTracker fluorescence intensities in control- and patient-derived RPE cells, with and without CCCP treatment. Each bar represents the mean Mitotracker signal from RPE monolayers derived from three independent iPSC lines. Error bars indicate standard deviation. * $p < 0.05$.

3.6. RCBTB1 Expression Is Induced by Oxidative Stress

Given that *RCBTB1* deficiency has previously been associated with decreased expression of the oxidative stress response gene *NFE2L2* [2], we hypothesized that *RCBTB1* might also be regulated by oxidative stress. To determine if oxidative stress can activate the expression of the *RCBTB1* gene in RPE cells, RNA was harvested from tBHP-treated iPSC-RPE cells and qPCR was performed to detect the expression level of *RCBTB1*. *RCBTB1* mRNA expression was significantly increased in control iPSC-RPE cells treated with either 100 µM ($p = 0.0142$) or 200 µM ($p = 0.0049$) tBHP, compared with untreated control cells. No significant differences in *RCBTB1* expression levels were detected in patient-derived iPSC-RPE cells treated with different concentrations of tBHP (Figure 3B). This result indicates *RCBTB1* expression is responsive to ROS levels in healthy RPE cells.

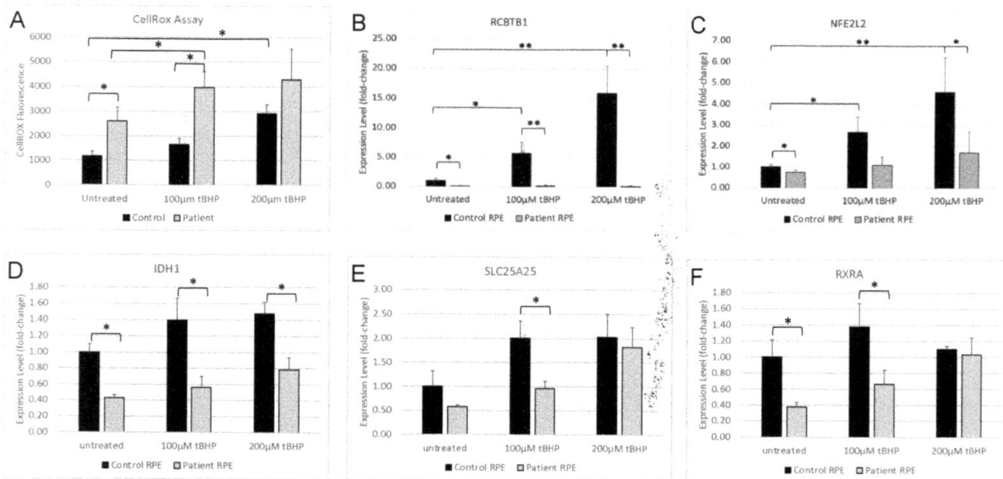

Figure 3. Response of RPE cells to oxidative stress induced by tBHP. Control- and patient-derived RPE monolayers were treated with 0, 100 or 200 µM of tBHP for 1 h. (**A**) ROS levels in live control- and patient-derived RPE monolayers were measured by CellROX assay. (**B–F**) Expression of *RCBTB1* (**B**), *NFE2L2* (**C**), *IDH1* (**D**), *SLC25A25* (**E**) and *RXRA* (**F**) was measured by qPCR. Each bar represents the mean values obtained from RPE cells generated from three independently derived iPSC lines. Error bars indicate standard deviation. * $p < 0.05$; ** $p < 0.01$.

3.7. Oxidative Stress-Induced Activation of NFE2L2 Is Attenuated in Patient RPE

To evaluate the NFE2L2-driven oxidative stress response in RPE cells, the expression level of *NFE2L2* was measured in tBHP-treated RPE cells by qPCR (Figure 3C). *NFE2L2* expression was significantly reduced in untreated patient iPSC-RPE cells compared with untreated control iPSC-RPE cells ($p = 0.0453$). In control iPSC-RPE cells, tBHP treatment significantly increased the expression of *NFE2L2* ($p = 0.0437$ at tBHP, $p = 0.0012$ at 200 µM). In contrast, patient-derived RPE cells showed no significant change in *NFE2L2* expression at either dose. Compared with control RPE, patient RPE cells showed a non-significant reduction in *NFE2L2* expression at 100 µM tBHP ($p = 0.0622$) and a significant reduction in *NFE2L2* expression following treatment with 200 µM tBHP ($p = 0.0161$). The expression of the NFE2L2 target genes *IDH1*, *SLC25A25* and *RXRA* was also reduced inpatient RPE cells compared with controls. Significantly reduced *IDH1* expression was observed in patient RPE cells at all tBHP doses tested ($p < 0.05$). Interestingly, *SLC25A25* and *RXRA* expression showed significant reductions in expression at 100 µM tBHP but were not significantly different from controls at 200 µM tBHP, suggesting NFE2L2-independent upregulation of these genes at high ROS levels. Together, these results demonstrate that oxidative stress-induced activation of the NFE2L2 response is attenuated in patient RPE, suggesting that RCBTB1 may play a role in the NFE2L2-mediated antioxidant responses that protect RPE cells against oxidative stress.

3.8. RCBTB1 Interacts with CUL3 and UBE2E3 in RPE Cells

To identify potential interactions between RCBTB1 and the CUL3 or UBE2E3 proteins, we performed immunoprecipitation (IP) assays using Sepharose beads conjugated to anti-CUL3 or anti-UBE2E3 antibodies, followed by western blot analysis (Figure 4). RPE cells derived from healthy control iPSC expressed all three proteins, with both UBE2E3 and RCBTB1 showing multiple immunoreactive bands (Figure 4, Input lanes). Following IP with anti-CUL3 beads, cullin-3, UBE2E3 and RCBTB1 proteins were detected by western blot (Figure 4, left side). Interestingly, anti-CUL3 IP pulled down 2/3 UBE2E3 isoforms and 2/4 RCBTB1 isoforms detected in total protein samples, suggesting selective inclusion of specific isoforms in cullin-3 complexes. Similar results were obtained using anti-UBE2E3 conjugated beads, which demonstrated coimmunoprecipitation of cullin-3, UBE2E3 and RCBTB1 proteins (Figure 4, right side). No proteins were detected following IP using unconjugated beads. Together, these results indicate that RCBTB1 is present in protein complexes containing cullin-3 and UBE2E3 in human RPE cells, consistent with its proposed function as a ubiquitin adapter protein [12].

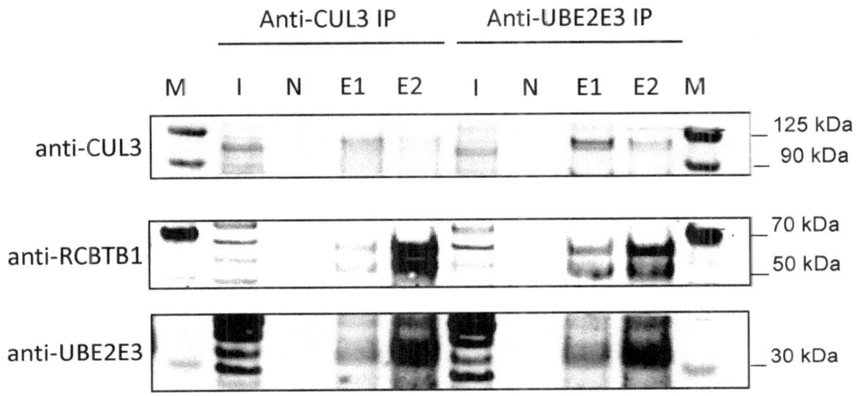

Figure 4. Western blots showing immunoprecipitation results from control iPSC-derived RPE. Protein lysates from control RPE cultures were immunoprecipitated using beads conjugated with either anti-CUL3 (left side) or anti-UBE2E3 (right side) antibodies and analysed by Western blotting using anti-CUL3 (top), anti-RCBTB1 (middle) or anti-UBE2E3 (bottom). Western blot analysis demonstrated coimmunoprecipitation of cullin-3, UBE2E3 and RCBTB1 proteins in RPE cells derived from healthy control iPSC lines using both antibodies. Lanes: M: Protein ladder; I: 10 µg input protein; N: control immunoprecipitation with beads alone (no conjugated antibody, elution 2); E1: Immunoprecipitation with antibody-conjugated beads, elution 1 (without beta-mercaptoethanol); E2: Immunoprecipitation with antibody-conjugated beads, elution 2 (with beta-mercaptoethanol).

4. Discussion

To date, phenotypic features of *RCBTB1*-associated retinopathy have been reported in fifteen cases from eleven families, among which eleven cases from nine unrelated families presented with progressive late-onset macular chorioretinal atrophy and peripheral retinal abnormalities in their 40s or 50s [2,3,5], while four cases manifested a typical retinitis pigmentosa phenotype in their 20s [2,4]. The clinical onset of retinal atrophy in the nine families ranged between 30 and 62 years of age. These patients mostly presented with gradually reduced visual acuity or visual distortion as a result of chorioretinal macular atrophy [3,5]. OCT and fundus autofluorescence (FAF) in patients with *RCBTB1*-associated retinopathy revealed attenuation of the RPE layer and the adjacent ellipsoid zone in the atrophic retinal lesions [3,5]. In light of the available multimodal retinal imaging data showing enlarging RPE atrophic lesions and widespread RPE irregularities, we speculated that the primary defect might occur in the central RPE with centrifugal expansion followed by severe RPE loss and secondary photoreceptor cell loss. To explore this hypothesis, we previously generated iPSC lines from a patient with *RCBTB1*-associated retinopathy and differentiated these into RPE cells [7]. Compared with healthy control RPE, patient RPE cells showed reduced primary cilia lengths, as well as reduced expression of *NFE2L2* and its target genes [8]. In the present study, we extend these observations to demonstrate that levels of ROS are significantly elevated in patient iPSC-derived RPE cells. We further showed that patient-derived RPE cells were more sensitive to tBHP-induced oxidative stress, generating significantly higher ROS than control RPE when treated with the lower dose of tBHP. The increased sensitivity of patient RPE cells to tBHP stimulation was associated with a failure to upregulate expression of the *NFE2L2* gene in response to the oxidative stress challenge. Robust upregulation of *NFE2L2* was observed in control RPE cells in response to tBHP stimulation but not in patient-derived RPE cells. Together, these results indicate that RCBTB1 deficiency is associated with increased oxidative stress and diminished antioxidant responses in RPE cells. In support of these findings, a recent abstract by Carron et al. reported that the knockdown of *RCBTB1* in ARPE19 cells caused changes in cellular responses to exogenous oxidative stress [13].

In a recent report describing three phenotypes of mitochondrial retinopathy, Birtel et al. commented on the resemblance of type 3 mitochondrial retinopathies to *RCBTB1*-associated retinopathy [6]. Here, we observed alterations in mitochondrial ultrastructure in patient-derived RPE by electron microscopy. Compared with healthy control RPE, mitochondria in patient-derived RPE were smaller and displayed disrupted cristae more frequently. Furthermore, mitochondria in patient-derived RPE were poorly labelled by the MitoTracker dye compared with control RPE, consistent with reduced mitochondrial membrane potential [14]. It remains unclear whether the mitochondrial dysfunction observed is a consequence or a cause of the increased ROS levels in RCBTB1-deficient RPE cells; however, this data indicates that damage to RPE mitochondria is a key pathogenic process in RCBTB1-associated retinopathy.

Current understanding of the cellular functions of the RCBTB1 protein is limited, although a number of studies have provided important insights. Using a yeast two-hybrid approach, Plafker et al. demonstrated that the RCBTB1 protein directly interacts with the ubiquitin-conjugating enzyme, UbcM2 (the mouse homologue of the human UBE2E3 protein), suggesting it functions as a CUL3 substrate adaptor in the ubiquitin proteosome system (UPS) [15]. These authors later showed that UbcM2 binds to and regulates the activity of the Nrf2 protein (the mouse homologue of human NFE2L2), a key transcription factor involved in the cellular response to oxidative stress [12]. Under homeostatic conditions, Nrf2 is constitutively ubiquitinated and degraded by the UPS; however, this process is interrupted in response to oxidative stress, leading to the stabilisation of Nrf2 protein and translocation to the nucleus. In the nucleus, Nrf2 binds to the antioxidant response element in the *Nrf2* gene promoter, amplifying its expression in a positive feedback loop. Additionally, Nrf2 activates the expression of network genes involved in the cellular antioxidant defence response [16–20]. Together, these studies suggest the intriguing possibility that RCBTB1 may interact with UBE2E3 to regulate the activation of cellular oxidative stress responses. In support of this hypothesis, our data showed *NFE2L2* expression was unresponsive to increasing ROS levels in patient RPE cells lacking RCBTB1. Furthermore, in healthy control, RPE cells, *RCBTB1* and *NFE2L2*, were both upregulated in a dose-dependent manner in response to increased ROS levels. Inspection of the *RCBTB1* genomic sequence revealed an NFE2L2 antioxidant response element (TGACCCGGC) located 25 base pairs upstream of the putative transcriptional start site in exon 1, suggesting the *RCBTB1* gene is a target for NFE2L2-mediated induction. Additionally, our data showing the coimmunoprecipitation of RCBTB1 and UBE2E3 in human RPE cells provides a direct link between RCBTB1 and NFE2L2 pathway activation.

5. Conclusions

In summary, our study provides new insights into the biological role of RCBTB1 in oxidative stress responses and highlights RPE mitochondria as a key target site in the pathogenesis of *RCBTB1*-associated retinopathy. Our results help to explain the similarity in disease phenotype between *RCBTB1*-associated retinopathy and the most severe form of mitochondrial retinopathy. The similarity in pathogenesis between these conditions suggests that therapeutic approaches aimed at alleviating mitochondrial dysfunction may be useful for treating *RCBTB1*-associated retinopathy.

Supplementary Materials: The following supporting information can be downloaded at: https://www.mdpi.com/article/10.3390/cells12101358/s1, Supplementary Figure S1: Characterization of iPSC-RPE Cells; Supplementary Figure S2: iPSC-RPE Phagocytosis Assays; Supplementary Figure S3: Electron microscopy of iPSC-RPE cells; Supplementary Figure S4: Original Western Blot Images; Supplementary Table S1: Primer List; Supplementary Table S2: Antibody Information.

Author Contributions: Conceptualization, Z.H., F.K.C. and S.M.; data curation, Z.H. and S.M.; formal analysis, Z.H. and S.M.; funding acquisition, F.K.C.; investigation, Z.H.; methodology, Z.H., D.Z., S.-C.C., D.H. and S.M.; project administration, F.K.C. and S.M.; resources, F.K.C. and S.M.; supervision, D.Z., D.M., F.K.C. and S.M.; validation, Z.H., D.Z., D.H. and S.M.; visualization, Z.H. and S.M.; writing—original draft, Z.H. and S.M.; writing—review and editing, Z.H., D.Z., D.M. and S.M. All authors have read and agreed to the published version of the manuscript.

Funding: This study was funded by the Australian National Health and Medical Research Council (GNT1116360, GNT1188694, GNT1054712 and MRF1142962), the Government of Western Australia Department of Health (WA Near-miss Fellowship Award 2021) and donations from the McCusker Charitable Foundation, the Saleeba family and the Lee and Low family.

Institutional Review Board Statement: The study was conducted in accordance with the Declaration of Helsinki and approved by the Human Ethics Office of Research Enterprise at THE UNIVERSITY OF WESTERN AUSTRALIA (2021/ET000151, 22 February 2021). The protocols used in this study were approved by the Institutional Biosafety Committee of the Harry Perkins Institute of Medical Research, THE UNIVERSITY OF WESTERN AUSTRALIA (NLRD 02-2020, 3 February 2020).

Informed Consent Statement: Informed consent was obtained from all subjects involved in the study.

Data Availability Statement: The data presented in this study are available in the Supplementary Materials.

Acknowledgments: We are thankful for the facilities and expertise provided by the Australian Microscopy and Microanalysis Research Facility at the Centre for Microscopy, Characterisation and Analysis (CMCA), funded by the University of Western Australia (UWA), the State of Western Australia and the Australian Commonwealth Government.

Conflicts of Interest: The authors declare no conflict of interest. The funders had no role in the design of the study; in the collection, analyses, or interpretation of data; in the writing of the manuscript; or in the decision to publish the results.

References

1. Wu, J.H.; Liu, J.H.; Ko, Y.C.; Wang, C.T.; Chung, Y.C.; Chu, K.C.; Liu, T.T.; Chao, H.M.; Jiang, Y.J.; Chen, S.J.; et al. Haploinsufficiency of RCBTB1 is associated with Coats disease and familial exudative vitreoretinopathy. *Hum. Mol. Genet.* **2016**, *25*, 1637–1647. [CrossRef] [PubMed]
2. Coppieters, F.; Ascari, G.; Dannhausen, K.; Nikopoulos, K.; Peelman, F.; Karlstetter, M.; Xu, M.; Brachet, C.; Meunier, I.; Tsilimbaris, M.K.; et al. Isolated and Syndromic Retinal Dystrophy Caused by Biallelic Mutations in RCBTB1, a Gene Implicated in Ubiquitination. *Am. J. Hum. Genet.* **2016**, *99*, 470–480. [CrossRef] [PubMed]
3. Huang, Z.; Zhang, D.; Thompson, J.A.; Jamuar, S.S.; Roshandel, D.; Jennings, L.; Mellough, C.; Charng, J.; Chen, S.C.; McLaren, T.L.; et al. Deep clinical phenotyping and gene expression analysis in a patient with RCBTB1-associated retinopathy. *Ophthalmic Genet.* **2021**, *42*, 266–275. [CrossRef] [PubMed]
4. Yang, J.; Xiao, X.; Sun, W.; Li, S.; Jia, X.; Zhang, Q. Variants in RCBTB1 are Associated with Autosomal Recessive Retinitis Pigmentosa but Not Autosomal Dominant FEVR. *Curr. Eye Res.* **2020**, *46*, 839–844. [CrossRef] [PubMed]
5. Catomeris, A.J.; Ballios, B.G.; Sangermano, R.; Wagner, N.E.; Comander, J.I.; Pierce, E.A.; Place, E.M.; Bujakowska, K.M.; Huckfeldt, R.M. Novel RCBTB1 variants causing later-onset non-syndromic retinal dystrophy with macular chorioretinal atrophy. *Ophthalmic Genet.* **2022**, *43*, 332–339. [CrossRef] [PubMed]
6. Birtel, J.; von Landenberg, C.; Gliem, M.; Gliem, C.; Reimann, J.; Kunz, W.S.; Herrmann, P.; Betz, C.; Caswell, R.; Nesbitt, V.; et al. Mitochondrial Retinopathy. *Ophthalmol. Retin.* **2022**, *6*, 65–79. [CrossRef] [PubMed]
7. Huang, Z.; Zhang, D.; Chen, S.C.; Thompson, J.A.; McLaren, T.; Lamey, T.; De Roach, J.N.; McLenachan, S.; Chen, F.K. Generation of three induced pluripotent stem cell lines from an isolated inherited retinal dystrophy patient with RCBTB1 frameshifting mutations. *Stem Cell Res.* **2019**, *40*, 101549. [CrossRef] [PubMed]
8. Huang, Z.; Zhang, D.; Chen, S.C.; Jennings, L.; Carvalho, L.S.; Fletcher, S.; Chen, F.K.; McLenachan, S. Gene replacement therapy restores RCBTB1 expression and cilium length in patient-derived retinal pigment epithelium. *J. Cell. Mol. Med.* **2021**, *25*, 10020–10027. [CrossRef] [PubMed]

9. Parinot, C.; Rieu, Q.; Chatagnon, J.; Finnemann, S.C.; Nandrot, E.F. Large-scale purification of porcine or bovine photorecepto outer segments for phagocytosis assays on retinal pigment epithelial cells. *J. Vis. Exp.* **2014**, *94*, e52100. [CrossRef]
10. McLaren, M.J.; Inana, G.; Li, C.Y. Double fluorescent vital assay of phagocytosis by cultured retinal pigment epithelial cell *Investig. Ophthalmol. Vis. Sci.* **1993**, *34*, 317–326.
11. Buskin, A.; Zhu, L.; Chichagova, V.; Basu, B.; Mozaffari-Jovin, S.; Dolan, D.; Droop, A.; Collin, J.; Bronstein, R.; Mehrotra, S.; et a Disrupted alternative splicing for genes implicated in splicing and ciliogenesis causes PRPF31 retinitis pigmentosa. *Nat. Commun* **2018**, *9*, 4234. [CrossRef] [PubMed]
12. Plafker, K.S.; Nguyen, L.; Barneche, M.; Mirza, S.; Crawford, D.; Plafker, S.M. The ubiquitin-conjugating enzyme UbcM2 ca regulate the stability and activity of the antioxidant transcription factor Nrf2. *J. Biol. Chem.* **2010**, *285*, 23064–23074. [CrossRef [PubMed]
13. Carron, M.; Naert, T.; Ascari, G.; Demuynck, S.; Van Nieuwenhuyzen, T.; Rosseel, T.; Priem, D.; Kremer, A.; Van Landeghem H.; Hoogstoel, S.; et al. Functional characterization of a Xenopus tropicalis knockout and a human cellular model of RCBTB1 associated inherited retinal disease shows involvement of RCBTB1 in the cellular response to oxidative stress. *Investig. Ophthalmo Vis. Sci.* **2020**, *61*, 1125.
14. Xiao, B.; Deng, X.; Zhou, W.; Tan, E.K. Flow Cytometry-Based Assessment of Mitophagy Using MitoTracker. *Front. Cell. Neurosc* **2016**, *10*, 76. [CrossRef] [PubMed]
15. Plafker, K.S.; Singer, J.D.; Plafker, S.M. The ubiquitin conjugating enzyme, UbcM2, engages in novel interactions with component of cullin-3 based E3 ligases. *Biochemistry* **2009**, *48*, 3527–3537. [CrossRef] [PubMed]
16. Kasai, S.; Shimizu, S.; Tatara, Y.; Mimura, J.; Itoh, K. Regulation of Nrf2 by Mitochondrial Reactive Oxygen Species in Physiolog and Pathology. *Biomolecules* **2020**, *10*, 320. [CrossRef] [PubMed]
17. Tonelli, C.; Chio, I.I.C.; Tuveson, D.A. Transcriptional Regulation by Nrf2. *Antioxid. Redox Signal.* **2018**, *29*, 1727–1745. [CrossRef [PubMed]
18. Sachdeva, M.M.; Cano, M.; Handa, J.T. Nrf2 signaling is impaired in the aging RPE given an oxidative insult. *Exp. Eye Res.* **2014** *119*, 111–114. [CrossRef] [PubMed]
19. Kosztelnik, M.; Kurucz, A.; Papp, D.; Jones, E.; Sigmond, T.; Barna, J.; Traka, M.H.; Lorincz, T.; Szarka, A.; Banhegyi, G.; et al Suppression of AMPK/aak-2 by NRF2/SKN-1 down-regulates autophagy during prolonged oxidative stress. *FASEB J.* **2019**, *33* 2372–2387. [CrossRef] [PubMed]
20. Kensler, T.W.; Wakabayashi, N.; Biswal, S. Cell survival responses to environmental stresses via the Keap1-Nrf2-ARE pathwa *Annu. Rev. Pharmacol. Toxicol.* **2007**, *47*, 89–116. [CrossRef] [PubMed]

cells

Article

Intraocular RGD-Engineered Exosomes and Active Targeting of Choroidal Neovascularization (CNV)

Dimitrios Pollalis [1], Dongin Kim [2], Gopa Kumar Gopinadhan Nair [3], Changsun Kang [2], Arjun V. Nanda [4] and Sun Young Lee [1,3,5,*]

1 USC Roski Eye Institute, USC Ginsburg Institute for Biomedical Therapeutics, Department of Ophthalmology, Keck School of Medicine, University of Southern California, Los Angeles, CA 90033, USA
2 Department of Pharmaceutical Sciences, College of Pharmacy, University of Oklahoma Health Science Center, Oklahoma City, OK 73104, USA
3 Department of Ophthalmology, Dean McGee Eye Institute, University of Oklahoma Health Science Center, Oklahoma City, OK 73104, USA
4 College of Medicine, University of Oklahoma Health Science Center, Oklahoma City, OK 73104, USA
5 Department of Physiology, University of Oklahoma Health Science Center, Oklahoma City, OK 73104, USA
* Correspondence: sunyoung.lee@med.usc.edu

Abstract: Purpose: To assess the transretinal penetration of intravitreally injected retinal multicell-derived exosomes and to develop exosome-based active targeting of choroidal neovascularization (CNV) by bioengineering with ASL, which is composed of a membrane Anchor (BODIPY), Spacer (PEG), and targeting Ligands (cyclic RGD peptide). Methods: Retinal multicell-derived exosomes were recovered from a whole mouse retina using differential ultracentrifugation. Their size, number, and morphology were characterized using nanoparticle tracking analysis (NTA) and transmission electron microscopy (TEM). Exosome markers were confirmed using an exosome detection antibody array. Intravitreal injection of fluorescent (PKH-26)-labeled or engineered ASL exosomes (1×10^6 exosomes/μL) were given to the wild-type mouse or laser-induced CNV mouse model. Retinal uptake of exosomes was assessed by in vivo retinal imaging microscopy and histological staining with DAPI, GSA, and anti-integrin α_v for retinal sections or choroid/RPE flat mounts. Active targeting of CNV was assessed by comparing retinal uptake between areas with and without CNV and by colocalization analysis of ASL exosomes with integrin α_v within CNV. Staining with anti-F4/80, anti-ICAM-1, and anti-GFAP antibodies on retinal sections were performed to identify intracellular uptake of exosomes and immediate reactive retinal gliosis after exosome treatment. Results: An average of 2.1×10^9 particles/mL with a peak size of 140 nm exosomes were recovered. Rapid retinal penetration of intravitreally injected exosomes was confirmed by retinal imaging microscopy at 3 and 24 h post-injection. Intravitreally delivered PKH-26-labeled exosomes reached inner and outer retinal layers including IPL, INL, OPL, and ONL at 1 and 7 days post-injection. Intravitreally injected ASL exosomes were predominantly delivered to the area of CNV including ONL, RPE, and choroid in laser-induced CNV mouse models with 89.5% of colocalization with integrin α_v. Part of exosomes was also taken intracellularly to vascular endothelial cells and macrophages. After intravitreal injection, neither naive exosomes nor ASL exosomes induced immediate reactive gliosis. Conclusions: Intravitreally delivered retinal multicell-derived exosomes have good retinal penetration, and ASL modification of exosomes actively targets CNV with no immediate reactive gliosis. ASL exosomes have a great potential to serve as an intraocular drug delivery vehicle, allowing an active targeting strategy.

Keywords: active targeting; choroidal neovascularization; exosome

Citation: Pollalis, D.; Kim, D.; Nair, G.K.G.; Kang, C.; Nanda, A.V.; Lee, S.Y. Intraocular RGD-Engineered Exosomes and Active Targeting of Choroidal Neovascularization (CNV). *Cells* **2022**, *11*, 2573. https://doi.org/10.3390/cells11162573

Academic Editor: Bruce A. Bunnell

Received: 10 July 2022
Accepted: 12 August 2022
Published: 18 August 2022

1. Introduction

Exosomes are naturally occurring, cell-secreted, and nano-sized (diameter ≈ 30–150 nm) extracellular vesicles (EVs) that steadily carry and transfer different biomolecules throughout the whole body. Exosomes physiologically carry various cargos, including microRNAs (miR), proteins, and lipids for cell-to-cell communication [1–4]. Based on these characteristics, there are several potential advantages for developing exosome-based intraocular therapy. Because exosomes are naturally produced by all cells, they may be superior to synthetic drug carriers as they provide higher biocompatibility, less toxicity, and better tissue penetration [5–7]. Intraocular treatment of the stem cell-derived exosomes that deliver their unique functional cargo has been investigated to reprogram the microenvironment in various ocular pathologies by stem cell-derived paracrine effects [8,9]. Exosomes are also highly engineerable [10,11]. Surface modification of exosomes can confer cell and tissue specificity. Successful exosomal cargo packing has also shown that exosomes can be an excellent natural drug delivery system.

Millions of intravitreal injections of vascular endothelial growth factor (VEGF) neutralizing proteins are performed each year in the United States as the current mainstay treatment for choroidal neovascularization (CNV) in neovascular age-related macular degeneration (NVAMD) [12,13]. However, some patients still experience suboptimal visual outcomes due to insufficient efficacy [14,15]. Several barriers contribute to the suboptimal efficacy of current anti-VEGF treatment. Local delivery of intravitreally injected anti-VEGF agents to the retina primarily depends on diffusion from the vitreous humor to the retina and choroid without active targeting of NV lesions [16]. Biological impedance within vitreoretinal interface-retina-RPE tight junction-choroid is a barrier that limits the local concentration of agents. Frequently, accompanying retinal fibrosis and gliosis secondary to NV are additional barriers to efficient penetration of intravitreally delivered drugs, and currently, there is no treatment for retinal fibrosis [17]. Anti-VEGF monotherapy does not target multiple signaling pathways involved in the pathogenesis of CNV. Clearance of intravitreally delivered anti-VEGF agents through the aqueous and vitreous humor shortens the half-life of these treatments that are resulting in the requirement of frequent intravitreal injection [16].

While exosome-based ocular therapy has been increasingly studied, supporting the potential benefits of their ophthalmological application, exosome-based active targeting strategies in ocular treatment have not been studied [8,9,18]. To employ an active targeting strategy for CNV, we engineered multiretinal cell-derived exosomes modified with ASL, composed of membrane Anchor (BODIPY), Spacer (PEG), and targeting Ligands (cyclic RGD peptide). We previously demonstrated that doxorubicin-loaded ASL exosomes substantially enhanced the anti-cancer efficacy of doxorubicin by active targeting of melanoma where integrin $\alpha_v\beta_5$ is overexpressed and by reduced systemic clearance of ASL exosomes [19]. RGD is a ligand for a subset of integrin, including integrin α_v [19–24]. Expression of integrin α_v is also increased in ocular tissues from NVAMD and proliferative DR patients [25].

In this study, we evaluated an ASL exosome-mediated active targeting strategy for CNV that has the functionality of delivering multi-molecular targets for posterior eye diseases. We also studied retinal penetration of the intravitreally delivered exosomes and subsequent reactive gliosis after intraocular exosome treatment to assess the potential translation of an exosome-assisted active targeting strategy to treat posterior eye disease.

2. Material and Methods

2.1. Animals

Wild-type C57BL/6 mice used for the experiments were treated according to the Association for Research in Vision and Ophthalmology guidelines on the care and use of animals in research. All protocols were approved by the Animal Care and Use Committee of the Oklahoma University Health Science Center (OUHSC) (20-055-HL, Approved 30 September 2021) and the University of Southern California (USC) (21415, Approved 26 May 2022).

2.2. Exosome Isolation and Characterization

The retinas from 4–6-week-old wild-type C57BL/6J mice were extracted and digested using the Papain Dissociation System (Worthington Biochemical Corporation, Lakewood, NJ, USA) according to the manufacturer's protocol. Exosomes were recovered using differential ultracentrifugation as previously described [26]. In brief, the supernatant received from the digestion process was mixed with cold HBSS and centrifuged at 25,000× RPM for 60 min at 4 °C in a Beckman Coulter Optima L80-XP ultracentrifuge with rotor SW60 Ti. The supernatant was again subjected to sequential ultra-centrifugation at 41,000× RPM for 60 min at 4 °C for 120 min. The final pellet (exosomes) was resuspended in 100 μL PBS and stored at −80 °C until further use (Figure 1A). The size, number, and morphology of exosomes were characterized by nanoparticle tracking analysis (NTA) (NanoSight, Malvern Instruments Ltd., Malvern, UK) and transmission electron microscopy (TEM), respectively (Figure 1B,D). Exosome markers were analyzed by an exosome detection antibody array (Exo-Check, System Biosciences, Palo Alto, CA, USA) according to the manufacturer's instructions (Figure 1C). Zeta potential of exosomes was measured using ZetaPALS (Brookhaven Instruments, Holtsville, NY, USA) (Figure 1E).

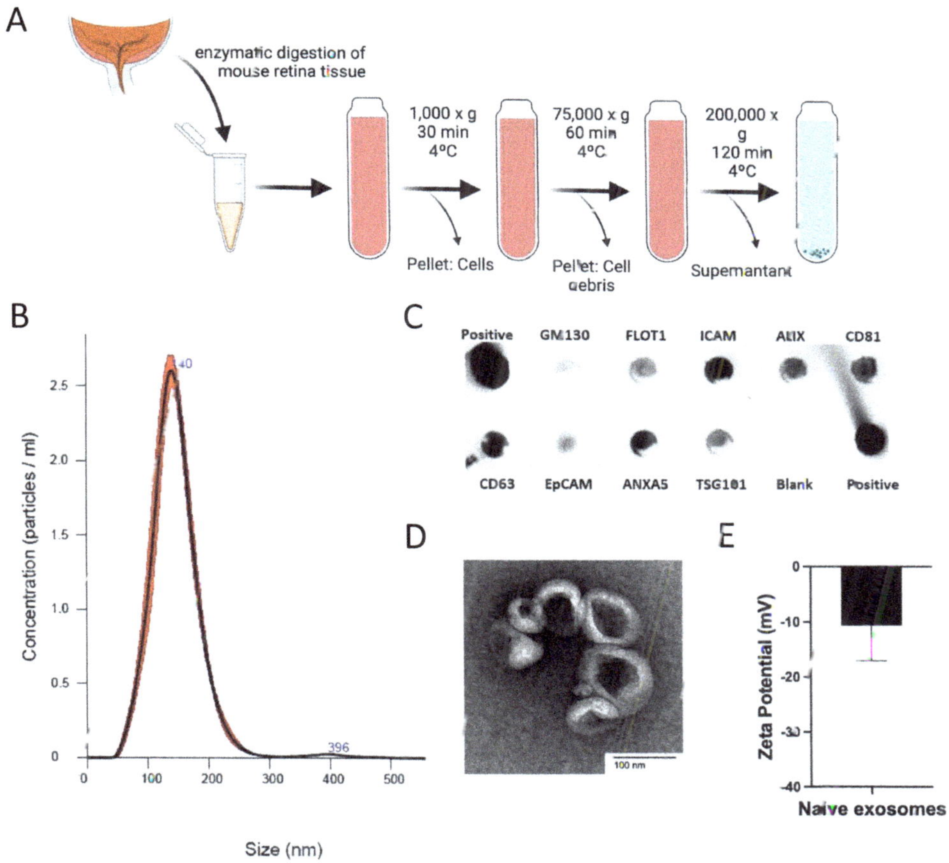

Figure 1. Recovery and characterization of retinal multicell-derived exosomes. Exosomes were recovered from the whole mouse retina by differential ultracentrifugation (**A**). The average peak size of exosome particles (140 nm) and an average number of particles of exosomes (2.1×10^9 particles/mL)

was determined by nanoparticle tracking analysis (NAT) (**B**). Exosome markers including FLOT1 ICAM, ALIX, CD81, CD63, ANXA5, and TSG101 were confirmed by exosome detection antibody array (**C**). The morphology of exosomes was confirmed by transmission electron microscopy (TEM (**D**) and Zeta potential of exosomes was measured (**E**).

2.3. Fluorescent Labeling of Exosomes

For fluorescent labeling of the exosomes, PKH-26 Red Cell Linker Mini Kit for Genera Cell Membrane Labeling (Sigma-Aldrich, St. Louis, MO, USA) was used according to the manufacturer's instructions. In brief, resuspended in 1 mL Diluent C exosome pellets and 1 mL Diluent C mixed with 4 μL PKH-26 were incubated for four minutes. Then, 10% BSA was used to stop the labeling reaction. Labeled exosomes were isolated and purified by sucrose density gradient ultracentrifugation (41,000× RPM for 120 min at 4 °C).

2.4. Bioengineering of ASL Exosomes

ASL was synthesized as previously described [19]. In brief, fluorescent lipophilic boron-dipyrromethene (BODIPY), TR-X NHS Ester (7.08 μM) (Thermo Fisher Scientific (Rockford, IL, USA), and TEA (7.08 μM) were completely dissolved in dichloromethane (DCM). Carboxy-PEG_{12}-Amine (5.66 μM) (Thermo Fisher Scientific (Rockford, IL, USA) in DCM was added dropwise, and the resulting mixture was stirred at room temperature for 6 h. The reaction mixture was precipitated with cold diethyl ether. Anchor-Spacer-COOH (AS) was obtained as a purple powder. A measure of 6.95 μM of AS, 30.30 μM of 1-Ethyl 3-(3-dimethylaminopropyl) carbodiimide, and 8.34 μM of N-hydroxysuccinimide were dissolved in dimethyl sulfoxide (DMSO), followed by the addition of TEA (8.34 μM). The mixture was stirred for 1 h in the dark. A measure of 10.43 μM of Cyclic Arg-Gly-Asp-D-Phe-Lys (RGD) (Peptide International (Louisville, KY, USA) was completely dissolved in DMSO and added to the mixture. The reaction was kept at room temperature overnight and the product was purified using the membrane dialysis tubing method (Cutoff MW 1000 Spectrum Laboratories, Inc., Rancho Dominguez, CA, USA). The product was obtained using a sephacryl S-100 HR column (Sigma-Aldrich, St. Louis, MO, USA) and lyophilized A measure of 7.9 mg of ASL (40 mg/mL) was added to 100 μL of exosomes (1.8×10^9/mL) The ASL exosomes were formulated using a sonication method, and the sample was incubated at 37 °C for 1 h to recover the integrity of the membrane [19].

2.5. In Vitro Exosome Uptake

Human retinal microvascular endothelial cells (HRMEC), donated by Dr. Jian-Xing Ma from the University of Oklahoma, were cultivated in an EBM endothelial cell growth medium (Lonza Bioscience, Basel, Switzerland) supplemented with an EGM endothelial cell growth medium SingleQuots kit, including hEGF, hydrocortisone, GA-1000, BBE, ascorbic acid, and FBS (Lonza Bioscience) in a humidified incubator at 37 °C and 5% CO_2. The cells were treated with either $CoCl_2$ (200 μM or 300 μM) or the complete routine medium once they reached confluency of 70–80%. Forty-eight hours later the cells were stained with anti-α_v for further assessment. Briefly, the cells were fixed in 4% PFA and blocked in 10% horse serum. Then, they were stained with anti-α_v integrin (SC9969, Santa Cruz CA, USA) primary antibodies. The secondary antibody used was anti-mouse Alexa Fluor 594 (A11005, Invitrogen, Waltham, MA, USA) and 4′6′-diamino-2-phenylindole (DAPI H-1500, Vector Laboratories, Newark, CA, USA) was used for nuclei visualization. Using HRMECs cultured as described above, ASL exosome cellular uptake was also evaluated After confluence reached 70–80%, either $CoCl_2$ or the regular routine medium was applied for 24 h. ASL exosomes were then added, and cells were incubated for a further 24 h. After a total 48-h incubation they were fixed and stained with anti-α_v integrin antibodies.

2.6. Laser-Induced Choroidal Neovascularization Mouse Model

Laser photocoagulation was performed on wild-type C57BL/6J 4–6-week-old mice [27] Briefly, pupils were dilated using 2.5% phenylephrine hydrochloride and 1% cyclopen tolate hydrochloride. Mice were anesthetized with ketamine (100 mg/mL) and xylazine (100 mg/mL) via intraperitoneal (IP) injection. Laser photocoagulation was performed using the Micron IV retinal imaging system (Phoenix Research Lab, Pleasanton CA, USA) with the Meridian Merilas 532 green laser (50 µm, 70 ms, 240 mW). Three lesions were induced located approximately one disk diameter from the optic nerve with respect to the large vessels. Laser-induced disruption of Bruch's membrane was confirmed by the appearance of a bubble at the site of photocoagulation. Induction of CNV was confirmed by fundus photography, optical coherence tomography (OCT), fluorescent angiography, and H&E stain of the retinal section seven days after laser photocoagulation.

2.7. Intravitreal Injection of Exosomes

C57BL/6J 4–6-week-old mice (n = 9) received intravitreal injection of PKH-26 labeled exosomes or ASL exosomes. Laser-induced CNV mouse models received intravitreal injection exosome treatment three days after laser photocoagulation. For intravitreal injection, mice were anesthetized with IP ketamine (100 mg/mL)/xylazine (100 mg/mL) and pupils were dilated with 2.5% phenylephrine/1% cyclopentolate. A small scleral incision was made posterior to the limbus using a 31-gauge (G) insulin needle. Then, a 33G blunt needle attached to a 10 µL NanoFil syringe (World Precision Instruments, Sarasota, FL, USA) was used to deliver 1 µL (1×10^6 exosome particles/1 µL) of solution in the vitreous cavity through the same incision.

2.8. Intravenous Injection of Exosomes

C57BL/6J 4–6-week-old mice (n = 2) received an intravenous injection of PKH-26-labeled exosomes through the tail vein (5×10^7 exosome particles/50 µL). Mice were euthanized three hours after injection, and their eyes were harvested to study the retinal uptake of systemically administered exosomes.

2.9. In Vivo Imaging Analysis

Fundus photography, in vivo retinal imaging microscopy, and fluorescein angiography (FA) were performed using the Micron IV retinal imaging system (Phoenix Research Lab, Pleasonton, CA, USA). Mice were anesthetized with IP ketamine (100 mg/mL)/xylazine (100 mg/mL) and pupils were dilated with 2.5% phenylephrine/1% cyclopentolate. For FA, mice received an intraperitoneal injection of 10% of fluorescein sodium (AK-Fluor, Akorn, Lake Forest, IL, USA). Optical coherence tomography (OCT) was performed using OCT (Bioptigen, Durham, NC, USA).

2.10. Histological Analysis

The mice were euthanized seven days after the intravitreal injection of the exosomes, and the eyes were harvested for either flat-mount histology or cryopreserved for retinal sections (8 µm). For flat mount immunostaining, the cornea, lens, vitreous, and retina were removed, and the RPE/choroid complex was fixed in 4% PFA for 3 h. Then, the tissue was blocked in 10% horse serum, stained with Rhodamine-conjugated GSA (RL-1102, Vector Laboratories) overnight, and flat-mounted. For cryosection immunohistochemical staining, eyes were fixed in 4% PFA, embedded in an optimal cutting temperature medium (Sakura, Japan), frozen with liquid nitrogen, and stored at −20 °C. Leica CM3050S was used to obtain 8 µm thick frozen sections. The sections were stained with anti-α_V integrin (SC9969, Santa Cruz, Dallas, TX, USA), anti-F4/80 (SC 377009, Santa Cruz, Dallas, TX, USA), and anti-ICAM-1 (AF796, R&D Systems, Minneapolis, MN, USA), with the primary antibodies and GSA-lectin rhodamine labeled (RL-1102, Vector Laboratories Burlingame, CA, USA). The secondary antibodies used were anti-mouse Alexa Fluor 594 (A11005, Invitrogen), anti-rabbit Alexa Fluor 594 (A11037, Invitrogen, Waltham, MA, USA), anti-rabbit Alexa Fluor 488 (A11034, Invitrogen, Waltham, MA, USA), anti-goat Alexa Fluor 555 (A21432, Invitrogen, Waltham, MA, USA), and 4′6′-diamino-2-phenylindole (DAPI, H-1500, Vector

Laboratories, Burlingame, CA, USA) for nuclei staining. Stained RPE/choroid flat mounts and frozen sections were examined using Olympus FluoView FV1200 confocal microscope.

2.11. Image Analysis

FIJI was used for image analysis. For integrin α_v expression, z-stacks of the RPE/choroid flat mounts were used. The acquired z-stacks were converted to binary stacks, and the threshold areas were used for volumetric analysis using the formula: $V = h \times \sum_{i=1}^{N} A_i$, V: total volume, h: interval between z-stack sections, A: threshold area. The JACoP plugin was applied for colocalization analysis, and Pearson's and Manders' coefficients were reported and evaluated.

2.12. Statistical Analysis

Unless otherwise described, all values were reported as mean $\pm$ standard deviation (SD). GraphPad Prism was used for the statistical analysis and graph plotting. Statistical differences were measured with Student's t-test to compare the two groups. JMP Pro 14 was used for graph plotting. p values < 0.005 were considered to be significant.

3. Results

3.1. Exosome Recovery and Characterization

Exosomes were recovered from the whole retina of the 4–6-week-old wild-type (C57BL/6) mice with 2.1×10^9 particles/mL with a peak size of 140 nm. Exosome markers were positive, including FLOT1, ICAM, ALIX, CD81, CD63, ANXA5, and TSG101, and the morphology of exosomes was confirmed by TEM (Figure 1).

3.2. Retinal Uptake of Intravitreally Delivered Exosomes and BRB crossing of Systemically Delivered Exosomes

In vivo retinal imaging microscopy confirmed that PKH26-labeled fluorescent exosomes started to be taken from the vitreous humor to the retina at 3 h post intravitreal injection, and mostly, they were brought into the retina in a scattered fashion 24 h post intravitreal injection (Figure 2A). Confocal microscopy of flat-mounted retinas (B) and retinal sections (C) obtained one and seven days after the intravitreal injection of exosomes showed that exosomes were taken into both the inner and outer retina in a scattered fashion (Figure 2B,C). Systemically administered PKH-26 labeled exosomes via the tail vein were observed within the retina 3 h after intravenous injections of exosomes, suggesting that exosomes bypassed the blood-retinal barrier (BRB) (Figure 2D).

3.3. Engineered Anchor, Spacer, and RGD Ligand Modified ASL Exosomes

Multiple retinal cell-derived exosomes were modified with membrane Anchor (BODIPY), Spacer (PEG), and targeting Ligands, and cyclic Arg-Gly-Asp (RGD) peptide (Figure 3A). A peak size of 125 nm of ASL exosomes was generated and confirmed by NTA (Figure 3B). The morphology of ASL exosomes was also confirmed by TEM (Figure 3C) and the Zeta potential of ASL exosomes was measured (Figure 3D). Fluorescent (green) BODIPY allowed for the study of the biodistribution of intravitreally delivered ASL exosomes.

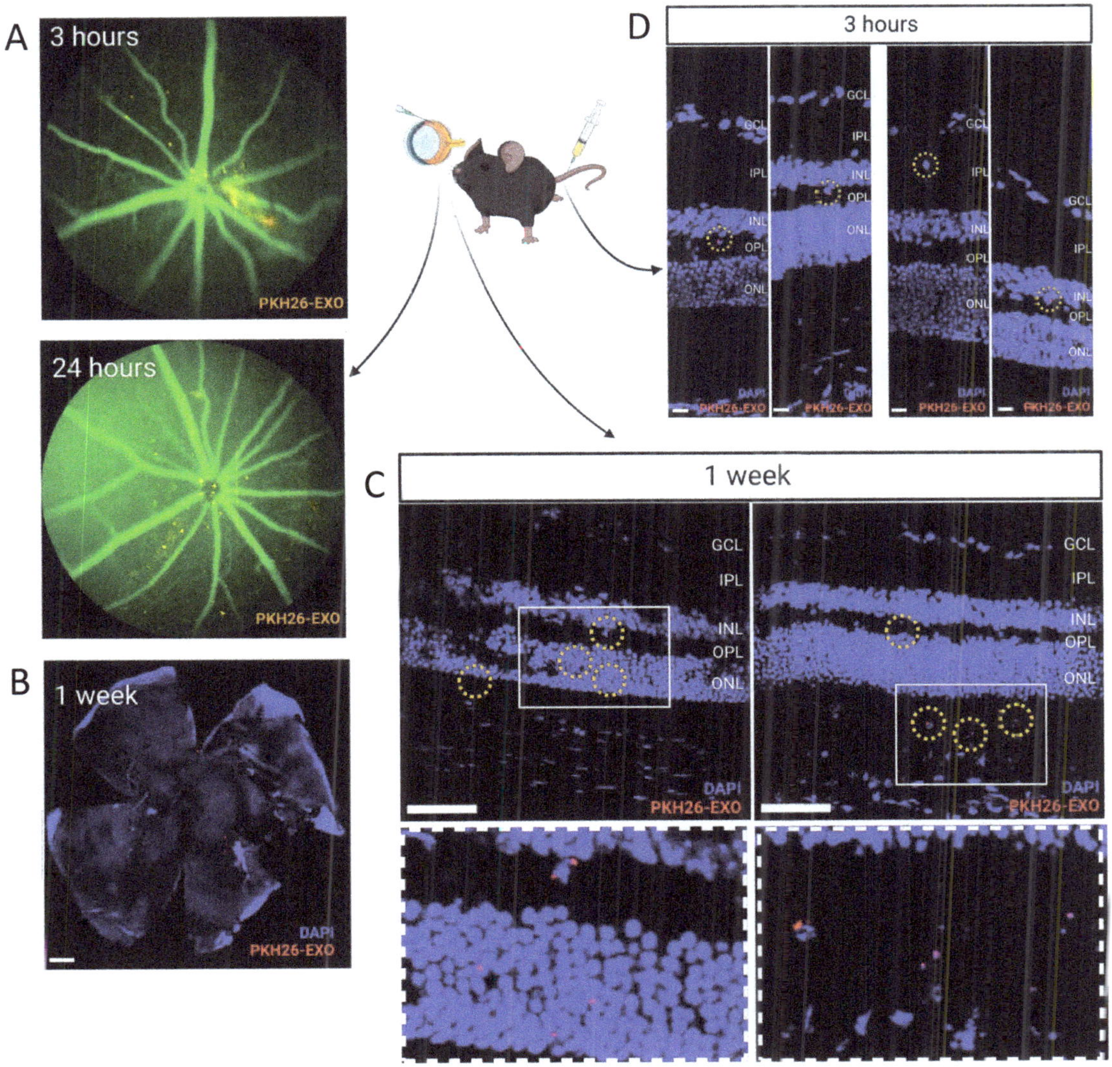

Figure 2. Retinal uptake of intravitreally delivered exosomes and BRB crossing of systemically delivered exosomes. In vivo retinal imaging microscopy performed at 3 and 24 h and retinal flat-mounts obtained 1 day and 1 week after intravitreal injection of PKH-26-labeled exosomes (orange-red) show immediate retinal penetration and wide distribution of intravitreally delivered exosomes (**A**,**B**). Retinal sections (8 μm) obtained 24 h after intravitreal injection of PKH-26-labeled exosomes (ret dots with yellow circles) shows that exosomes were taken up by both the inner and outer retina (Scale bar, 500 μm) (**C**). Retinal sections (8 μm) obtained 3 h after systemic injection of PKH-26-labeled exosomes (ret dots with yellow circles) via tail vein injection shows that exosomes were taken up by the retina, suggesting that exosomes can cross the blood-retina barrier (BRB). (Scale bar, 10 μm) (**D**).

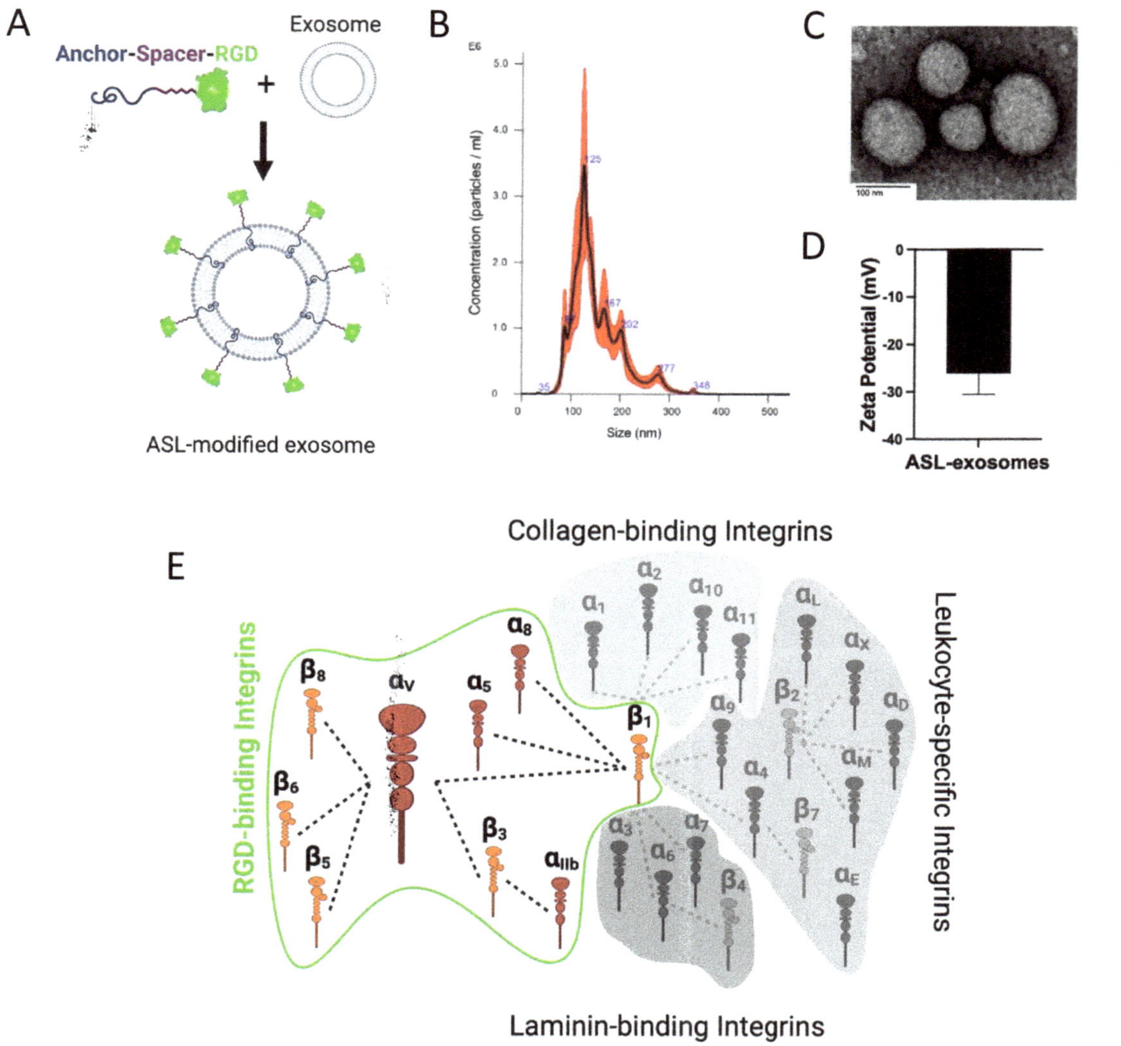

Figure 3. Engineering of Anchor, Spacer, and RGD Ligand (ASL)-modified exosomes, ASL exosomes. Retinal multicell-derived exosomes were engineered with ASL modification composed of a membrane Anchor (BODIPY), Spacer (PEG), and targeting Ligand, and cyclic Arg-Gly-Asp (RGD) peptide (**A**). The size and morphology of ASL exosomes were confirmed by NTA (mean size: 152 nm) (**B**). RGD is one of the major binding ligands for integrin families including $\alpha_5\beta_1$, $\alpha_8\beta_1$, $\alpha_V\beta_1$, $\alpha_V\beta_3$, $\alpha_V\beta_5$, $\alpha_V\beta_6$, $\alpha_V\beta_8$, and $\alpha_{IIb}\beta_3$ (**C**). Zeta potential of exosomes was measured (**D**). Major integrin binding ligands (**E**).

3.4. Increased Integrin α_v Expression in Choroidal Neovascularization

Reliable CNV induction in laser-induced CNV mouse models was confirmed by fundus photography, optical coherence tomography (OCT), fluorescent angiography, and H&E staining of the retinal section seven days after laser photocoagulation (Figure 4A). Expression of integrin α_v in the retina of the laser-induced CNV mouse model was increased by over 100 times compared with expression of integrin α_v in the non-lasered retina (Figure 4B).

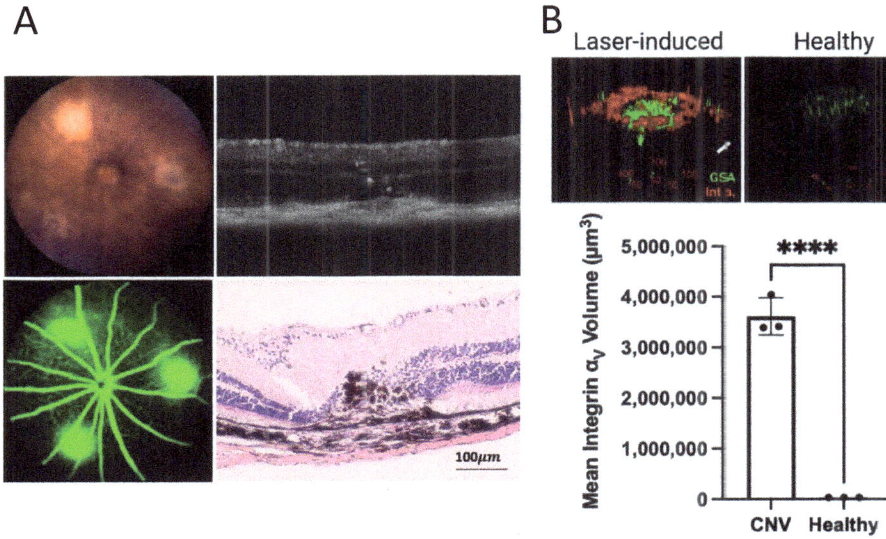

Figure 4. Increased integrin α_v expression in choroidal neovascularization in a laser-induced mouse model. CNV induction in a laser-induced mouse model was confirmed by color fundus photography, fluorescein angiography, OCT, and H&E histology (**A**). Volumetric analysis of integrin α_v expression within CNV lesions ($n = 3$) shows an over 100 times increase in integrin α_v expression within CNV lesions compared with the control retina, scale bar, 100 µm (bar represents mean ± SD, **** $p < 0.0001$) (**B**).

3.5. Active Uptake of ASL Exosomes to CNV Sites

In the laser-induced CNV mouse retina, intravitreally delivered ASL exosomes were predominantly found in CNV lesions and rarely seen in non-CNV lesions. This was confirmed in both choroid/RPE flat-mount and retinal sections. In retinal sections, ASL exosomes were observed at the site of CNV in outer retinal layers including ONL, subretinal space, RPE, and choroid. Immunohistochemistry (IHC) of the retinal sections showed 89.5% of colocalization between fluorescent ASL exosomes and integrin α_v within CNV lesions (Figure 5A–D). HRECs exposed to hypoxic stress induced by $CoCl_2$ in vitro increased the expression of integrin α_v (Figure 6A) and the intracellular uptake of ASL exosomes (Figure 6B).

Some of the local delivery of exosomes to CNV sites in vivo was done intracellularly. IHC with anti-ICAM1 (vascular endothelial cell marker) and anti-F4/80 (macrophage marker) antibodies to determine the identification of these cells showed that ASL exosomes were taken up by both vascular endothelial cells (D) and macrophages (C) (Figure 6C,D).

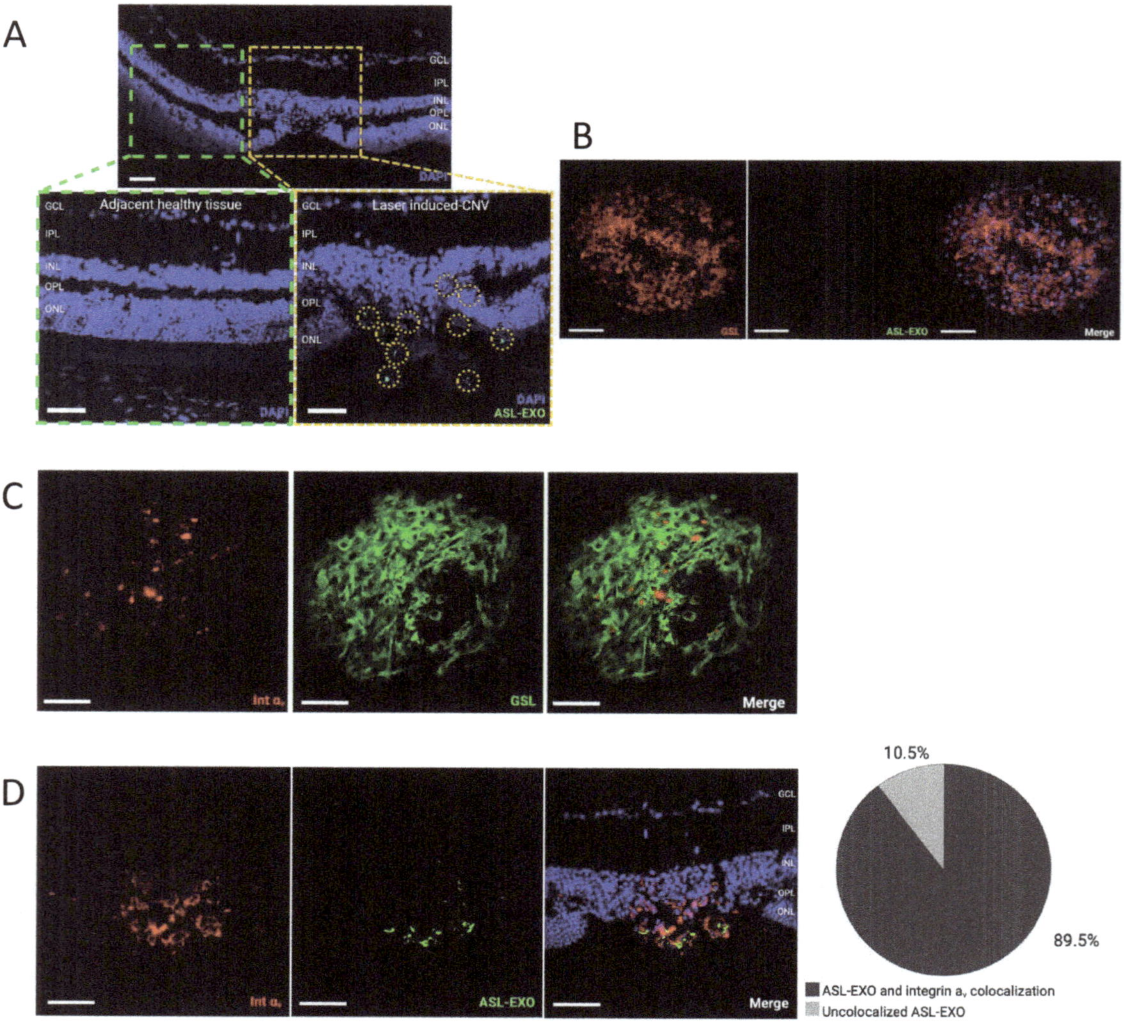

Figure 5. Intravitreally injected ASL exosomes were primarily delivered to the site of CNV. Intravitreal injection of ASL exosomes (1 μL) was given 3 days after laser treatment in a mouse model of CNV and the eyes were harvested 7 days after injection to obtain retinal sections (8 μm) and RPE/choroid flat-mounts. Representative eye sections showing ASL exosomes (green) exclusively delivered to the area of CNV (**A**). Representative RPE/choroid flat-mount of a CNV area showing ASL exosomes (green) delivered to the newly formed vessels (rhodamine-conjugated GSL, red) (**B**) and increased integrin α_v expression (red) within CNV area (rhodamine-conjugated GSL, green) (**C**). The ASL exosomes (green) are colocalized with integrin a_v (red) with 89.5 % of overlapping ($n = 3$) (**D**), scale bar, 100 μm..

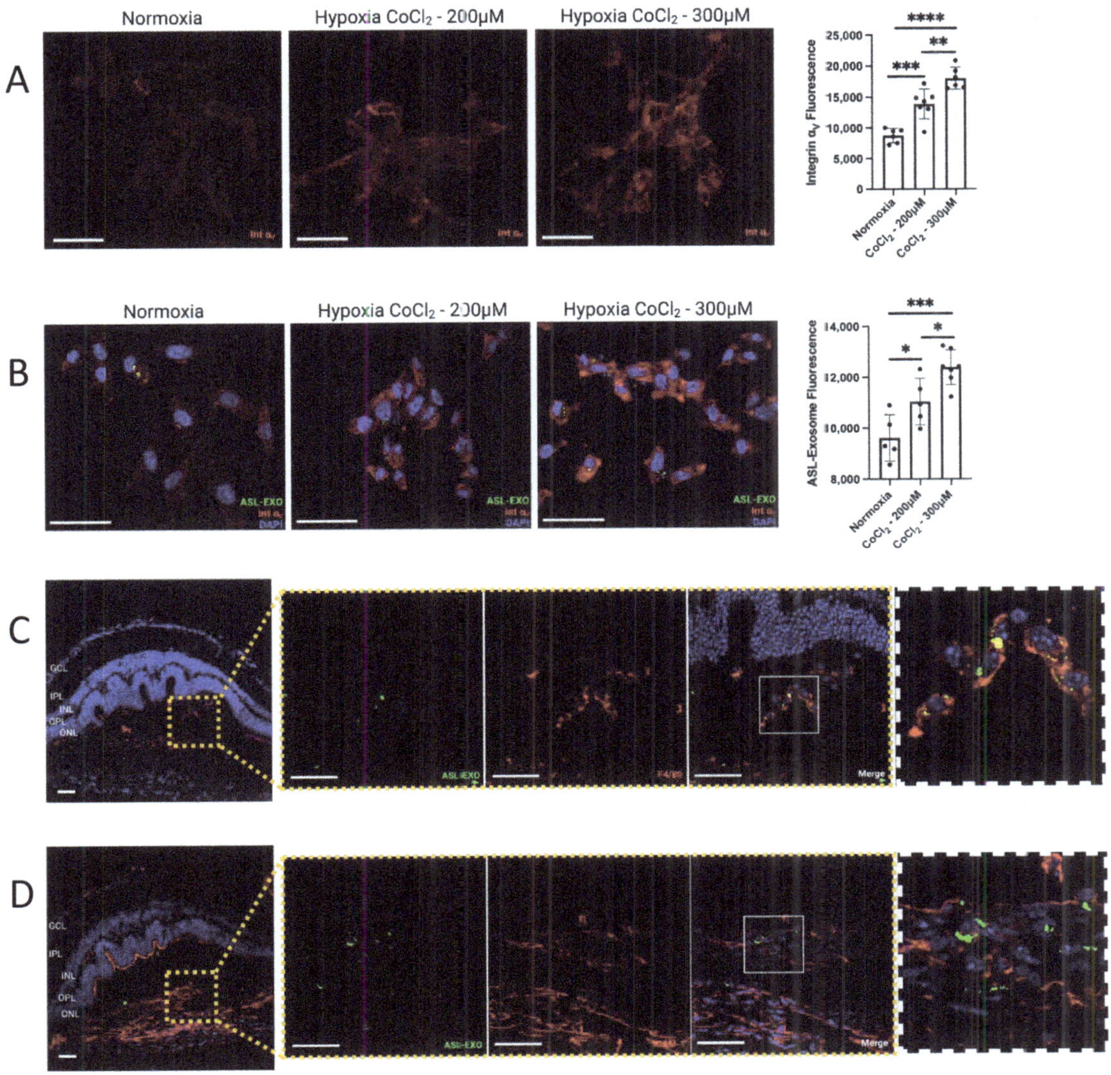

Figure 6. Intracellular uptake of ASL exosomes. Human retinal endothelial cell (HRECs) exposed to hypoxic stress induced by either 200 µM or 300 µM $CoCl_2$ increased expression of integrin α_V (**A**) and increased intracellular uptake of ASL exosomes in vitro (**B**) (bar represents mean ± SD, * $p < 0.05$; ** $p < 0.01$; *** $p < 0.001$; **** $p < 0.0001$). Intravitreal injection of ASL exosomes (1 µL) was given 3 days after laser treatment in a mouse model of CNV and the eyes were harvested 7 days after injection to obtain retinal sections (8 µm) and RPE/choroid flat-mounts. Representative eye sections showing ASL exosomes (green) internalized by macrophages (F4/80, red) (**C**) and endothelial cells (ICAM-1, red) (**D**), respectively, at the area of CNV in vivo. Scale bar, 50 µm.

3.6. No Reactive Retinal Gliosis after Intravitreal Injection of ASL Exosomes

GFAP expression was not increased within the retina after an intravitreal injection of either naïve exosomes (1 day) or ALS-exosomes (7 days), suggesting that intravitreal injection of naïve or ASL-modified exosomes did not induce an immediate reactive retinal gliosis (Figure 7).

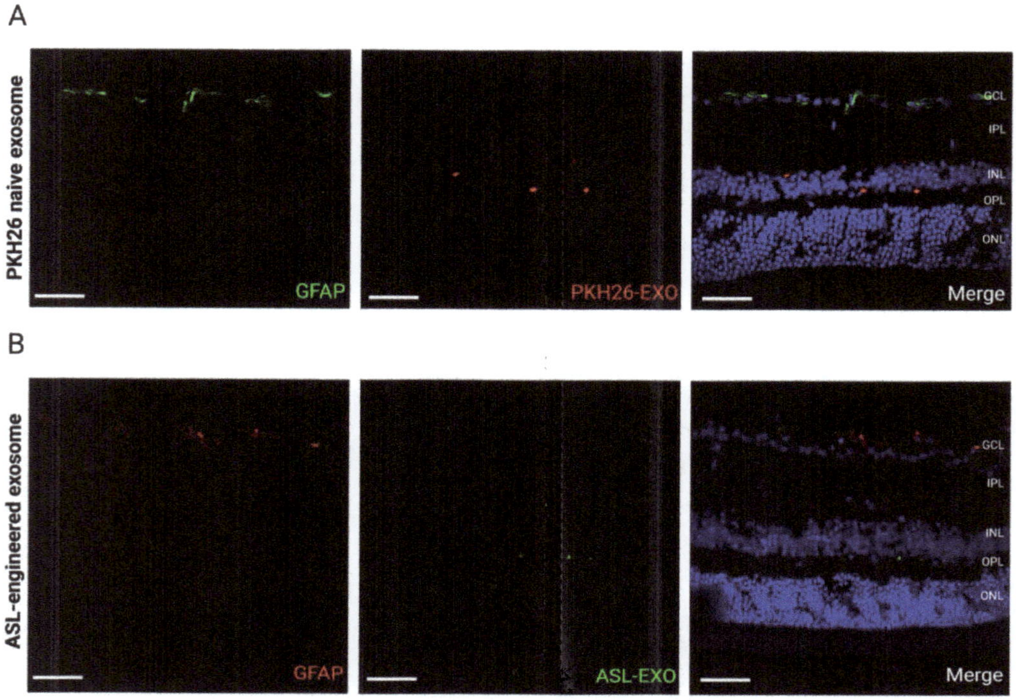

Figure 7. No immediate reactive retinal gliosis by intraocular exosome treatment. Retinal sections show no immediate reactive retinal gliosis stained with GFAP antibody staining 1 day after intravitreal injection of PKH-26-labeled exosomes (**A**) and 7 days after the intravitreal injection of ASL exosomes (**B**) (Scale bar, 50 µm). Note that GFAP is only expressed within ganglion cells in a physiologic state.

4. Discussion

In this study, we describe ASL composed of the Anchor, Spacer, and Arg-Gly-Asp acid (RGD) Ligand modification as an uncomplicated modification of exosomes to promote active targeting of CNV. We developed an exosome-based novel targeted strategy for the treatment of CNV that has many advantages over synthetic nanoparticles, including (1) full-thickness retinal penetration of the retinal multicell-derived exosomes delivered by intravitreal injection as well as bypassing the blood-retinal barrier (BRB), (2) highly efficient active targeting of CNV by RGD-modified exosomes, (3) multifaceted exosome uptake to extracellular and intracellular spaces including retinal vascular endothelial cells and macrophages, and (4) no immediate reactive gliosis from the intravitreal injection of retinal multicell line-derived exosomes or ASL-modified exosomes.

We showed that exosomes administered by intravitreal injection reached the retina, were taken to the inner and outer retina, as well as near the RPE, supporting that intravitreal injection is a good route for retinal penetration in intraocular exosome therapy. Our data support the utilization of exosomes as an intraocular cargo delivery carrier because penetrating the ocular barriers is one of the significant obstacles to the application of thoroughly well-studied synthetic nanocarriers such as polymeric nanoparticles, and liposomes for the treatment of posterior eye diseases. For example, subretinal injection of nanocarriers was often necessary to achieve cargo deliveries to photoreceptors, RPE, or choroids for efficient penetration. However, subretinal injections requiring an intraoperative procedure are more demanding than office-based intravitreal injections. We also confirmed that systemically administered exosomes could cross the blood-retinal barrier (BRB). It has been shown that exosomes can cross natural tissue barriers such as the blood-brain barrier (BBB) and there has been speculation that exosomes will pass the BRB [28,29]. Initially, the size of the nanoparticles was recognized as one of the key factors in tissue penetration, and especially in the bypassing of natural tissue barriers. With an improved understanding of the mechanism of tissue uptake in exosomes, there has been increasing evidence that exosome origin, cargo load, and surface protein may have a more significant influence on tissue uptake of exosomes [30,31]. For example, the molecular composition of exosomes, such as cell surface protein or exosomal cargo, reflects their cell of origin, and cellular or tissue tropism has been shown in exosome uptake [30,31]. While there are only a few reports with a comprehensive study of retinal uptake of intravitreally delivered exosomes; Mathew et al. previously showed that intravitreally delivered bone marrow-derived mesenchymal stem cell-derived exosomes were rapidly cleared from the vitreous and taken to the retina, however exosomes were found mostly no deeper than INL and they were predominantly taken from RGC and microglial cells [32]. Different retinal penetration between previous studies and our study may be due to the different cellular origination of the exosomes. In our study, we tested age matched (4–6-week-old) multiretinal cell-derived exosomes to optimize the potential benefit of favorable retinal penetration of exosomes. To translate intraocular exosome therapy, further studies are needed to determine the source of exosomes.

Although exosomes are naturally occurring cell-secreted nano-sized extracellular vesicles, exosomes are also highly engineerable like synthetic nanocarriers. To establish exosome-based active targeting, we have developed an ASL system composed of an Anchor (BODIPY), Spacer, and Arg-Gly-Asp acid (RGD) Ligand modification. In our previous study, we demonstrated that a fluorescent lipophilic boron-dipyrromethene (BODIPY), used as an exosome anchor, benefited the biodistribution and pharmacokinetic study of ASL exosomes and provided a conjugate domain for a spacer [19]. A PEG spacer conjugated to BODIPY anchors prevented exosome fusions and aggregation at higher concentrations and minimized systemic clearance [19]. RGD is one of the major integrin-binding ligands (Figure 3C) [20–24]. Integrins are essential in VEGF signaling in ocular NV. Although a few previous studies reported that RGD conjugation could actively target CNV, applying synthetic nanoparticles to conjugate with RGD or systemic administration of these particles diminished their clinical utility [33–36]. Here, we demonstrated that multiretinal cell-derived exosomes modified with an ASL system actively targeted CNV in a laser-induced CNV mouse model by promoting their uptake to overexpressed integrin α_v at the site of CNV. In contrast, naïve exosomes penetrated the inner and outer retina in a scattered fashion. Applying an active targeting strategy for CNV can potentially improve the therapeutic efficacy (1) by increasing the local concentration of therapeutic cargo and (2) by enhancing the penetration of anatomical barriers such as retinal fibrosis or gliosis at the site of CNV through the direct intravitreal injection of therapeutic agents, which may overcome the limitation of current intravitreal injection of anti-VEGF agents. Previously increased expression of integrins α_V has been shown in human ocular tissues with CNV and proliferative diabetic retinopathy (PDR) patients, further supporting that an RGD-mediated active targeting strategy for CNV may benefit the treatment of a broad spectrum of retinal neovascularization such as diabetic retinopathy and retinal vein occlusion [25].

In our study, we further confirmed that some of the intravitreally delivered ASL exosomes were taken up intracellularly at the site of CNV lesions. These cells included vascular endothelial cells and macrophages within CNV lesions. Our data support that ASL exosomes have great potential to deliver therapeutic cargo both extracellularly (e.g., cell membrane) and intracellularly. This result suggests that ASL exosomes can be utilized for multimolecular and extracellular or intracellular target treatments. For example, ASL exosomes can pack with multiple cargos such as anti-VEGF and microRNAs (miR), which can potentially increase the therapeutic efficacy of CNV treatment compared to anti-VEGF monotherapy.

Despite the perceived low immunogenicity of exosomes due to their physiological origin, the immunogenicity of any recombinant compound needs to be carefully tested prior to its application to intraocular treatment. In our study, we confirmed that neither naïve exosomes nor ASL exosomes delivered by intravitreal injection induced immediate reactive retinal gliosis. While evaluating the immune response, intraocular exosome treatments from various engineering processes will be needed before translating intraocular exosome treatment in humans; our data is encouraging for the advancement of exosome-based active targeting strategies for the treatment of CNV and other retinal neovascular diseases.

The present study had limitations. ASL exosome-mediated active targeting of CNV was only tested in a fully developed CNV in a laser-induced CNV mouse model. The efficacy of the active targeting of ocular NV needs to be tested in the early stages of CNV as well as other types of ocular NV such as diabetic retinopathy. For the translational application of intraocular exosome therapy coupled with an active targeting strategy, further studies in the cellular tropism of exosome uptake in retinal tissues are needed to determine the source of the exosome for human application. Optimizing intended cargo packing into the exosomes and pharmacokinetics are also needed for future translation and the application of intraocular exosome therapy.

Author Contributions: Conceptualization, D.P., D.K. and S.Y.L.; methodology, D.P., D.K., G.K.G.N., C.K., A.V.N. and S.Y.L.; validation, D.P., D.K., G.K.G.N., C.K., A.V.N. and S.Y.L.; formal analysis, D.P., D.K., G.K.G.N., C.K., A.V.N. and S.Y.L.; resources, D.K. and S.Y.L.; data curation, D.P., D.K., G.K.G.N., C.K. and S.Y.L.; writing—original draft preparation, D.P. and S.Y.L.; writing—review and editing, D.P., D.K., G.K.G.N., C.K. and S.Y.L.; visualization, D.P. and C.K. All authors have read and agreed to the published version of the manuscript.

Funding: This research was supported by the Presbyterian Health Foundation (PHF), Oklahoma City, OK (Clinician Scientist Development Grant #GRF 00003650) (to S.Y.L.), Oklahoma Shared Clinical and Translational Resources (OSCTR), Oklahoma City, OK (Pilot Program #U54 GM104938) (to S.Y.L.); NIH/NEI (USC Roski Eye K12, #K12EY028873) (to S.Y.L.). NIH (#SP20GM10363909) (to D.K.).

Institutional Review Board Statement: Not applicable.

Informed Consent Statement: Not applicable.

Data Availability Statement: This study did not report any data.

Acknowledgments: The authors thank Feng Li and Mark Dittmar (OUHSC P30 Live Animal Imaging Core, Dean A. McGee Eye Institute, Oklahoma City, OK, USA) and the OUHSC P30 Cellular Imaging Core (Dean A. McGee Eye Institute, Oklahoma City, OK, USA). The authors thank Jiyang Cai from the Department of Physiology at OUHSC for sharing anti-F4/80 and anti-ICAM-1 antibodies.

Conflicts of Interest: Lee, Kim, and Kang have filed a patent application entitled "Surface-modified exosomes and methods of use".

References

1. Trams, E.G.; Lauter, C.J.; Salem, N., Jr.; Heine, U. Exfoliation of membrane ecto-enzymes in the form of micro-vesicles. *Biochim. Biophys. Acta* **1981**, *645*, 63–70. [CrossRef]
2. Johnstone, R.M.; Adam, M.; Hammond, J.R.; Orr, L.; Turbide, C. Vesicle formation during reticulocyte maturation. Association of plasma membrane activities with released vesicles (exosomes). *J. Biol. Chem.* **1987**, *262*, 9412–9420. [CrossRef]
3. Jeppesen, D.K.; Fenix, A.M.; Franklin, J.L.; Higginbotham, J.N.; Zhang, Q.; Zimmerman, L.J.; Liebler, D.C.; Ping, J.; Liu, Q.; Evans, R.; et al. Reassessment of Exosome Composition. *Cell* **2019**, *177*, 428–445.e18. [CrossRef] [PubMed]
4. Kalluri, R.; LeBleu, V.S. The biology, function, and biomedical applications of exosomes. *Science* **2020**, *367*, eaau6977. [CrossRef] [PubMed]
5. Kim, M.S.; Haney, M.J.; Zhao, Y.; Yuan, D.; Deygen, I.; Klyachko, N.L.; Kabanov, A.V.; Batrakova, E.V. Engineering macrophage-derived exosomes for targeted paclitaxel delivery to pulmonary metastases: In vitro and in vivo evaluations. *Nanomedicine* **2018**, *14*, 195–204. [CrossRef] [PubMed]
6. Alvarez-Erviti, L.; Seow, Y.; Yin, H.; Betts, C.; Lakhal, S.; Wood, M.J.A. Delivery of siRNA to the mouse brain by systemic injection of targeted exosomes. *Nat. Biotechnol.* **2011**, *29*, 341–345. [CrossRef] [PubMed]
7. Quah, B.J.; O'Neill, H.C. The immunogenicity of dendritic cell-derived exosomes. *Blood Cells Mol. Dis.* **2005**, *35*, 94–110. [CrossRef] [PubMed]

8. Bian, B.; Zhao, C.; He, X.; Gong, Y.; Ren, C.; Ge, L.; Zeng, Y.; Li, Q.; Chen, M.; Weng, C.; et al. Exosomes derived from neural progenitor cells preserve photoreceptors during retinal degeneration by inactivating microglia. *J. Extracell. Vesicles* **2020**, *9*, 1748931. [CrossRef]
9. Li, D.; Zhang, J.; Liu, Z.; Gong, Y.; Zheng, Z. Human umbilical cord mesenchymal stem cell-derived exosomal miR-27b attenuates subretinal fibrosis via suppressing epithelial–mesenchymal transition by targeting HOXC6. *Stem Cell Res. Ther.* **2021**, *12*, 24. [CrossRef]
10. Liang, Y.; Duan, L.; Lu, J.; Xia, J. Engineering exosomes for targeted drug delivery. *Theranostics* **2021**, *11*, 3183–3195. [CrossRef]
11. Xu, M.; Yang, Q.; Sun, X.; Wang, Y. Recent Advancements in the Loading and Modification of Therapeutic Exosomes. *Front. Bioeng. Biotechnol.* **2020**, *8*, 586130. [CrossRef]
12. Rosenfeld, P.J.; Brown, D.M.; Heier, J.S.; Boyer, D.S.; Kaiser, P.K.; Chung, C.Y.; Kim, R.Y.; MARINA Study Group. Ranibizumab for neovascular age-related macular degeneration. *N. Engl. J. Med.* **2006**, *355*, 1419–1431. [CrossRef]
13. Heier, J.S.; Brown, D.M.; Chong, V.; Korobelnik, J.F.; Kaiser, P.K.; Nguyen, Q.D.; Kirchhof, B.; Ho, A.; Ogura, Y.; Yancopoulos, G.D.; et al. Intravitreal Aflibercept (VEGF Trap-Eye) in Wet Age-related Macular Degeneration. *Ophthalmology* **2012**, *119*, 2537–2548. [CrossRef]
14. Mettu, P.S.; Allingham, M.J.; Cousins, S.W. Incomplete response to Anti-VEGF therapy in neovascular AMD: Exploring disease mechanisms and therapeutic opportunities. *Prog. Retin. Eye Res.* **2021**, *82*, 100906. [CrossRef]
15. Khanani, A.M.; Skelly, A.; Bezlyak, V.; Griner, R.; Torres, L.R.; Sagkriotis, A. SIERRA-AMD: A Retrospective, Real-World Evidence Study of Patients with Neovascular Age-Related Macular Degeneration in the United States. *Ophthalmol. Retina* **2020**, *4*, 122–133. [CrossRef]
16. Hutton-Smith, L.A.; Gaffney, E.A.; Byrne, H.M.; Maini, P.K.; Gadkar, K.; Mazer, N.A. Ocular Pharmacokinetics of Therapeutic Antibodies Given by Intravitreal Injection: Estimation of Retinal Permeabilities Using a 3-Compartment Semi-Mechanistic Model. *Mol. Pharm.* **2017**, *14*, 2690–2696. [CrossRef]
17. Teo, K.Y.C.; Joe, A.W.; Nguyen, V.; Invernizzi, A.; Arnold, J.J.; Barthelmes, D.; Gillies, M. Prevalence and Risk Factors for the Development of Physician-Graded Subretinal Fibrosis in Eyes Treated for Neovascular Age-Related Macular Degeneration. *Retina* **2020**, *40*, 2285–2295. [CrossRef]
18. Hajrasouliha, A.R.; Jiang, G.; Lu, Q.; Lu, H.; Kaplan, H.J.; Zhang, H.G.; Shao, H. Exosomes from Retinal Astrocytes Contain Antiangiogenic Components That Inhibit Laser-induced Choroidal Neovascularization. *J. Biol. Chem.* **2013**, *288*, 28058–28067. [CrossRef]
19. Kang, C.; Han, P.; Lee, J.S.; Lee, D.; Kim, D. Anchor, Spacer, and Ligand-Modified Engineered Exosomes for Trackable Targeted Therapy. *Bioconjug. Chem.* **2020**, *31*, 2541–2552. [CrossRef]
20. Pierschbacher, M.D.; Ruoslahti, E. Cell attachment activity of fibronectin can be duplicated by small synthetic fragments of the molecule. *Nature* **1984**, *309*, 30–33. [CrossRef]
21. Chen, H.; Niu, G.; Wu, H.; Chen, X. Clinical Application of Radiolabeled RGD Peptides for PET Imaging of Integrin $\alpha v\beta 3$. *Theranostics* **2016**, *6*, 78–92. [CrossRef]
22. Reardon, D.A.; Nabors, L.B.; Stupp, R.; Mikkelsen, T. Cilengitide: An integrin-targeting arginine–glycine–aspartic acid peptide with promising activity for glioblastoma multiforme. *Expert Opin. Investig. Drugs* **2008**, *17*, 1225–1235. [CrossRef]
23. Shan, D.; Li, J.; Cai, P.; Prasad, P.; Liu, F.; Rauth, A.M.; Wu, X.Y. RGD-conjugated solid lipid nanoparticles inhibit adhesion and invasion of $\alpha v\beta 3$ integrin-overexpressing breast cancer cells. *Drug Deliv. Transl. Res.* **2015**, *5*, 15–26. [CrossRef]
24. Wang, C.; Bao, C.; Liang, S.; Fu, H.; Wang, K.; Deng, M.; Liao, Q.; Cui, D. RGD-conjugated silica-coated gold nanorods on the surface of carbon nanotubes for targeted photoacoustic imaging of gastric cancer. *Nanoscale Res. Lett.* **2014**, *9*, 264. [CrossRef]
25. Friedlander, M.; Theesfeld, C.L.; Sugita, M.; Fruttiger, M.; Thomas, M.A.; Chang, S.; Cheresh, D.A. Involvement of integrins alpha v beta 3 and alpha v beta 5 in ocular neovascular diseases. *Proc. Natl. Acad. Sci. USA* **1996**, *93*, 9764–9769. [CrossRef]
26. Nair, G.K.G.; Pollalis, D.; Wren, J.D.; Georgescu, C.; Sjoelund, V.; Lee, S.Y. Proteomic Insight into the Role of Exosomes in Proliferative Vitreoretinopathy Development. *J. Clin. Med.* **2022**, *11*, 2716. [CrossRef]
27. Saishin, Y.; Saishin, Y.; Takahashi, K.; Silva, R.L.E.; Hylton, D.; Rudge, J.S.; Wiegand, S.J.; Campochiaro, P.A. VEGF-TRAP(R1R2) suppresses choroidal neovascularization and VEGF-induced breakdown of the bloodretinal barrier. *J. Cell. Physiol.* **2003**, *195*, 241–248. [CrossRef]
28. Yang, T.; Martin, P.; Fogarty, B.; Brown, A.; Schurman, K.; Phipps, R.; Yin, V.P.; Lockman, P.; Bai, S. Exosome Delivered Anticancer Drugs Across the Blood-Brain Barrier for Brain Cancer Therapy in Danio Rerio. *Pharm. Res.* **2015**, *32*, 2003–2014. [CrossRef]
29. Elliott, R.O.; He, M. Unlocking the Power of Exosomes for Crossing Biological Barriers in Drug Delivery. *Pharmaceutics* **2021**, *13*, 122. [CrossRef]
30. Emam, S.E.; Lila, A.S.A.; Elsadek, N.E.; Ando, H.; Shimizu, T.; Okuhira, K.; Ishima, Y.; Mahdy, M.A.; Ghazy, F.S.; Ishida, T. Cancer cell-type tropism is one of crucial determinants for the efficient systemic delivery of cancer cell-derived exosomes to tumor tissues. *Eur. J. Pharm. Biopharm.* **2019**, *145*, 27–34. [CrossRef]
31. Zhang, H.; Wu, J.; Wu, J.; Fan, Q.; Zhou, J.; Wu, J.; Liu, S.; Zang, J.; Ye, J.; Xiao, M.; et al. Exosome-mediated targeted delivery of miR-210 for angiogenic therapy after cerebral ischemia in mice. *J. Nanobiotechnol.* **2019**, *17*, 29. [CrossRef] [PubMed]
32. Mathew, B.; Torres, L.; Gamboa Acha, L.; Tran, S.; Liu, A.; Patel, R.; Chennakesavalu, M.; Aneesh, A.; Huang, C.C.; Feinstein, D.; et al. Uptake and Distribution of Administered Bone Marrow Mesenchymal Stem Cell Extracellular Vesicles in Retina. *Cells* **2021**, *10*, 730. [CrossRef] [PubMed]

33. Singh, S.R.; Grossniklaus, H.E.; Kang, S.J.; Edelhauser, H.F.; Ambati, B.K.; Kompella, U.B. Intravenous transferrin, RGD peptide and du-al-targeted nanoparticles enhance anti-VEGF intraceptor gene delivery to laser-induced CNV. *Gene Ther.* **2009**, *16*, 645–659. [CrossRef] [PubMed]
34. Zhang, X.; Bohner, A.; Bhuvanagiri, S.; Uehara, H.; Upadhyay, A.K.; Emerson, L.L.; Bondalapati, S.; Muddana, S.K.; Fang, D.; Li, M.; et al. Targeted Intraceptor Nanoparticle for Neovascular Macular Degeneration: Preclinical Dose Optimization and Toxicology Assessment. *Mol. Ther.* **2017**, *25*, 1606–1615. [CrossRef] [PubMed]
35. Uehara, H.; Muddana, S.K.; Zhang, X.; Das, S.K.; Bhuvanagiri, S.; Liu, J.; Wu, Y.; Choi, S.; Carroll, L.S.; Archer, B.; et al. Targeted Delivery of FLT-Morpholino Using Cyclic RGD Peptide. *Transl. Vis. Sci. Technol.* **2017**, *6*, 9. [CrossRef] [PubMed]
36. Zhang, J.; Zhu, J.; Zhao, L.; Mao, K.; Gu, Q.; Li, D.; Zhao, J.; Wu, X. RGD-modified multifunctional nanoparticles encapsulating salvianolic acid A for targeted treatment of choroidal neovascularization. *J. Nanobiotechnol.* **2021**, *19*, 196. [CrossRef] [PubMed]

cells

MDPI

Article

Polarized RPE Secretome Preserves Photoreceptors in Retinal Dystrophic RCS Rats

Kabir Ahluwalia [1], Juan-Carlos Martinez-Camarillo [2,3], Biju B. Thomas [2,3], Aditya Naik [1], Alejandra Gonzalez-Calle [2,3], Dimitrios Pollalis [2,3], Jane Lebkowski [4], Sun Young Lee [2,3,5], Debbie Mitra [2], Stan G. Louie [1,2,*] and Mark S. Humayun [2,3,*]

1 Mann School of Pharmacy and Pharmaceutical Sciences, University of Southern California, Los Angeles, CA 90089, USA; kahluwal@usc.edu (K.A.); aanaik@usc.edu (A.N.)
2 USC Ginsburg Institute of for Biomedical Therapeutics, University of Southern California, Los Angeles, CA 90033, USA; juan.martinez@med.usc.edu (J.-C.M.-C.); biju.thomas@med.usc.edu (B.B.T.); gonz762@usc.edu (A.G.-C.); pollalis@usc.edu (D.P.); sunyoung.lee@med.usc.edu (S.Y.L.); debbiemitrax@gmail.com (D.M.)
3 USC Roski Eye Institute, Department of Ophthalmology, Keck School of Medicine, University of Southern California, Los Angeles, CA 90033, USA
4 Regenerative Patch Technologies LLC, Menlo Park, CA 94028, USA; jane@regenerativepatch.com
5 Department of Physiology & Neuroscience, Keck School of Medicine, University of Southern California, Los Angeles, CA 90089, USA
* Correspondence: slouie@usc.edu (S.G.L.); humayun@med.usc.edu (M.S.H.); Tel.: +1-323-442-3646 (S.G.L.); +1-323-865-3937 (M.S.H.)

Abstract: Retinal degenerative diseases, including age-related macular degeneration (AMD) and retinitis pigmentosa, lack effective therapies. Conventional monotherapeutic approaches fail to target the multiple affected pathways in retinal degeneration. However, the retinal pigment epithelium (RPE) secretes several neurotrophic factors addressing diverse cellular pathways, potentially preserving photoreceptors. This study explored human embryonic stem cell-derived, polarized RPE soluble factors (PRPE-SF) as a combination treatment for retinal degeneration. PRPE-SF promoted retinal progenitor cell survival, reduced oxidative stress in ARPE-19 cells, and demonstrated critical antioxidant and anti-inflammatory effects for preventing retinal degeneration in the Royal College of Surgeons (RCS) rat model. Importantly, PRPE-SF treatment preserved retinal structure and scotopic b-wave amplitudes, suggesting therapeutic potential for delaying retinal degeneration. PRPE-SF is uniquely produced using biomimetic membranes for RPE polarization and maturation, promoting a protective RPE secretome phenotype. Additionally, PRPE-SF is produced without animal serum to avoid immunogenicity in future clinical development. Lastly, PRPE-SF is a combination of neurotrophic factors, potentially ameliorating multiple dysfunctions in retinal degenerations. In conclusion, PRPE-SF offers a promising therapeutic candidate for retinal degenerative diseases, advancing the development of effective therapeutic strategies for these debilitating conditions.

Keywords: retina; age-related macular degeneration; retinitis pigmentosa; photoreceptor; retinal pigmented epithelium; retinal degeneration; secretome; Royal College of Surgeons rat

Citation: Ahluwalia, K.; Martinez-Camarillo, J.-C.; Thomas, B.B.; Naik, A.; Gonzalez-Calle, A.; Pollalis, D.; Lebkowski, J.; Lee, S.Y.; Mitra, D.; Louie, S.G.; et al. Polarized RPE Secretome Preserves Photoreceptors in Retinal Dystrophic RCS Rats. *Cells* **2023**, *12*, 1689. https://doi.org/10.3390/cells12131689

Academic Editor: Karl Matter

Received: 3 June 2023
Revised: 19 June 2023
Accepted: 20 June 2023
Published: 22 June 2023

1. Introduction

Retinal dystrophies, such as age-related macular degeneration (AMD), are the leading cause of vision loss in industrialized countries. Globally, AMD is projected to reach 288 million cases by 2040 [1]. The molecular pathogenesis of AMD includes the accumulation of cellular debris and oxidative stress triggering chronic inflammation, retinal cell atrophy, and ultimately vision loss [2,3]. Currently, anti-vascular endothelial growth factor (VEGF) treatment is used for neovascular AMD (nAMD), which accounts for 10–15% of AMD cases. Until recently, there was no approved treatment for dry AMD (dAMD) and the only options to slow disease progression were dietary supplements high in antioxidants [4–8]. The first intraocular administration of a complement inhibitor (i.e., C3 inhibitor: pegcetacoplan) was approved by the FDA in February 2023 for the treatment of geographic atrophy (GA), the advanced form of dAMD. Although pegcetacoplan can ameliorate GA progression, it is unable to halt or reverse vision loss [9].

Due to the complex etiology of dAMD, investigational modalities have included antioxidants, visual cycle modulators, mitochondrial modulators, anti-inflammatory agents, complement inhibitors, neuroprotective agents, and stem cell therapies [10]. Limitations associated with monotherapeutic approaches are that they are unable to fully address the multiple pathways promoting AMD development, which may account for limited effectiveness. Stem cell-based therapies directly target GA and other retinal degenerative diseases by replacing atrophic tissue [11–14]. CPCB-RPE1, a subretinal implant containing polarized human embryonic stem cell (hESC)-derived retinal pigment epithelium (RPE) grown on ultrathin parylene membranes was implanted in GA patients in a phase I/IIa clinical trial and showed improvement in visual acuity [12,15,16].

The RPE has long been speculated as a source of various neuroprotective factors. Characterization of the RPE secretome has identified several proteins with neuroprotective properties in retinal degeneration models [17–23]. Unfortunately, the administration of these purified components has not led to an effective therapy in human trials [24,25]. During the development of CPCB-RPE1, photoreceptor preservation was seen beyond the borders of the implant [26]. These findings suggest that the RPE secretes factors capable of promoting neuronal survival [18,27–33], and we propose to study the RPE secretome as a combination therapy for retinal degenerative conditions.

Oxidative stress and inflammation are central to AMD pathogenesis, and it has been shown they significantly alter the RPE secretome, resulting in increased secretion of pro-angiogenic factors and decreased anti-inflammatory factors including complement factor H (CFH), a critical inhibitor of complement activation and inflammation, and the strongest genetic risk for AMD [34–39]. Following the approval of pegcetacoplan, it can be surmised that the RPE secretome, containing a complement inhibitor, is directly linked to disease pathology and treatment. Oxidative damage to the RPE secretome results in a hostile microenvironment, and we hypothesize that the enriched secretome from healthy hESC-RPE can restore the retinal microenvironment and promote photoreceptor preservation in retinal degeneration models. In this report, we demonstrate that polarized RPE soluble factors (PRPE-SF), specifically concentrated PRPE-SF (SF3), promote fetal retinal progenitor cell (fRPC) viability and survival. SF3 also delayed retinal degeneration in the Royal College of Surgeons (RCS) rat model, which was associated with reduced oxidative stress and inflammation. These findings support the conclusion that SF3 can alter the diseased retinal microenvironment, promoting preservation of photoreceptors and visual function, offering an alternative therapeutic strategy for AMD and other complex retinopathies.

2. Materials and Methods

2.1. hESC-RPE Cell Culture and PRPE-SF1 Production

PRPE-SF1 was produced at USC using hESC-RPE cells designated as an intermediate cell bank (ICB), manufactured at the City of Hope under cGMP using the methods established by the CPCB [40–44]. The culture apparatus and PRPE-SF1 production scheme can be found in Figure 1. Briefly, H9 hESCs were expanded in mTeSR1 medium (Stemcell Technologies, Vancouver, BC, Canada), then the medium was replaced with XVIVO10 medium (Lonza, Basel, Switzerland) for 12 weeks to drive spontaneous RPE cell differentiation. Pigmented RPE-like cells were mechanically isolated followed by dissociation with TrypLE (Life Technologies, Grand Island, NY, USA). RPE-like cells were cultured on human vitronectin (AMS Biotechnology, Lake Forest, CA, USA)-coated plates with XVIVO10 medium and frozen at passage two. hESC-derived RPE cells were seeded onto vitronectin (Corning, Glendale, AZ, USA)-coated 10 mm × 20 mm parylene C membrane similar to the CPCB-RPE1 implant. The parylene C membrane has identical characteristics to CPCB-RPE1 as it is specially designed to mimic the Bruch's membrane with a 0.3 μm thickness supported on a

6.0 µm thick mesh frame with diffusion regions for small molecules (manufactured at Leap Biomed Innovators, LLC, Pasadena, CA, USA) [41]. Parylene membranes were placed into 4-chamber wells (Corning Inc., Corning, NY, USA) prior to hESC-RPE seeding at a final seeding density of 1.43×10^5 cells/cm^2 and maintained at 37 °C in a 5% CO_2 incubator. The medium was exchanged every 4 days, PRPE-SF was collected every 4 days starting on day 28 until day 40. The collected PRPE-SF was pooled and filtered with a 0.2 µm syringe filter system followed by 3-fold (3×) concentration, by volume, using Amicon centrifugal filter devices (Millipore Sigma, Burlington, MA, USA) with a 3 kD Ultracel regenerated cellulose membrane. Both unconcentrated (SF1) and concentrated (SF3) PRPE-SF were stored at −80 °C until use. Secreted proteins were characterized using a quantitative 40 human growth factor array Q1 (RayBiotech Inc., Nacross, GA, USA), and PEDF and LIF were determined by ELISA (Boster Biological Technology, Pleasanton, CA, USA).

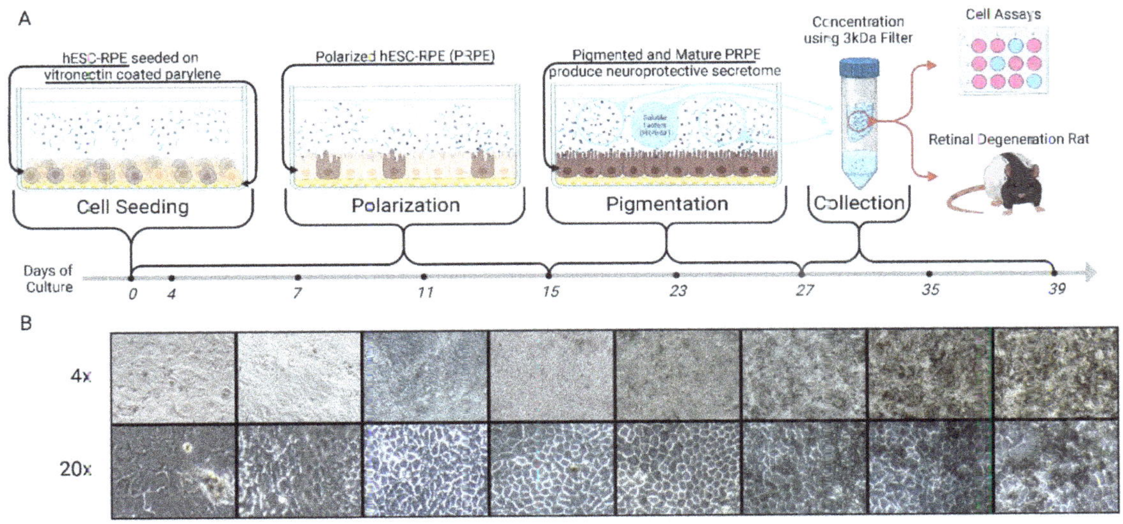

Figure 1. Diagram and timeline of polarized retinal pigmented epithelial soluble factor (PRPE-SF) production (**A**) with corresponding 4× and 20× images of PRPE (**B**). Human embryonic stem cells (hESC) differentiated into RPE were plated onto vitronectin-coated ultrathin parylene membranes and incubated in XVIVO10 medium. Significant morphological changes occurred during the first 15 days of cell culture resulting in polarization and the cobblestone shape expected of mature RPE. In the following 12 days there is was a rapid increase in pigmentation signifying the maturation of the hESC-RPE cells. PRPE-SF (SF1) was collected, concentrated (SF3) using a 3 kDa filter, and stored at −80 °C on days 27, 31, 35, and 39. Subsequently, SF1 and SF3 were characterized in cells and the Royal College of Surgeons rat model of retinal degeneration. Created with BioRender.com (accessed on 15 June 2023).

2.2. Fetal Retinal Progenitor Cell (RPC) and ARPE-19 Culture

The biological activity of SF1 was characterized by using 18–20 weeks gestation of human fRPC. Fetal donor eyes were obtained from Lion Eye Institute under appropriate USC investigational review board approval. Following dissection, the retina was rinsed with PBS and broken up by passing the cells through a 27-gauge needle. The cells were washed with RPC media (DMEM/F12 [1:1], with 10% knockout replacement serum (KRS, Invitrogen), 1 N2 supplement (Invitrogen), 1 B-27 (Invitrogen), 20 ng/mL FGF2 (R&D Systems), and 20 ng/mL EGF (R&D Systems)). Cells were plated into Matrigel-coated plates and cultured for 24 h in RPC media then replaced with experimental media. After 24 h, plates were used in cell viability assays, apoptosis assays, immunocytochemistry staining, and RT-qPCR.

ARPE-19 cells were purchased from American Type Culture Collection (ATCC, Manassas, VA, USA) and were maintained in culture media composed of DMEM/F-12 (Corning Life Sciences, Corning, NY, USA) and 10% fetal bovine serum. Cell cultures were maintained at 37 °C in a 5% CO_2 incubator. ARPE-19 cell assays are described below.

2.3. *ARPE-19 Cell Assays*

The 2′,7′ –dichlorofluorescin diacetate (DCFDA) assay is an intracellular assay for non-selectively detecting ROS. ARPE-19 cells were seeded into 96-well plates at a density of 1.25×10^5 cells/cm^2 in maintenance media. Following 24 h, cells were washed with HBSS (Cytiva, Marlborough, MA, USA) and the medium was replaced with experimental medium diluted 1:1 with fresh XVIVO10. After 24 h, cells were washed with HBSS and media replaced with 25 µM DCFDA (Sigma-Aldrich, St. Louis, MO, USA) in HBSS, and then placed in a CO_2 incubator for 45 min prior to measuring fluorescence on a Biotek Synergy H1 microplate reader (BioTek, Winooski, VT, USA) using 485 nm excitation and 535 nm emission wavelengths. Percent change was calculated from blank-subtracted data using cells in HBSS with no DCFDA. Similarly, ARPE-19 cells were treated in 24-well plates (1.25×10^5 cells/cm^2) for 24 h, RNA was extracted as described below, and ROS-generating and -eliminating genes were probed.

2.4. *Rhodopsin Staining*

Following treatment and media removal, slides were fixed using 10% neutral buffered formalin for 10 min. Heat-induced antigen retrieval was performed with pH 6.0 citrate buffer. Samples were blocked with 5% BSA with 0.1% Triton X-100 for 30 min at room temperature (RT). Rhodopsin antibody was diluted 1:400 with 5% BSA and incubated overnight at 4 °C. Secondary antibody was diluted 1:100 with 5% BSA and incubated for 1 h at RT. Coverslips were mounted with 2 drops of fluorescent enhance mounting medium with DAPI (VECTASHIELD HardSet, Vector Labs, Newark, CA, USA). Slides were imaged on Keyence BZ-X700 Microscope (Keyence, Osaka, Japan).

2.5. *TUNEL Assay*

Following incubation with the various treatments, apoptosis assays were performed using Promega's DeadEnd™ fluorometric TUNEL system (Promega, Madison, WI, USA) according to the manufacturer's protocol. The reaction mix consisted of 45 µL of equilibration buffer, 5 µL nucleotide mix, and 1 µL rTdT enzyme, incubated at 37 °C for 1 h protected from light. The nuclear counterstain utilized 1 µg/mL propidium iodide in PBS for 15 min at RT protected from light. Wells were then washed 3 times with deionized water and replaced with PBS for imaging. The plate was then imaged using a Keyence BZ-X700 microscope (Keyence, Osaka, Japan).

2.6. *RNA-Extraction and RT-qPCR*

Following treatment, total RNA was isolated using a RNeasy Mini Kit (QIAGEN, Hilden, Germany) following the manufacturer's protocol. RNA concentration and purity were assessed using a NanoDrop™ One (Thermo Fisher Scientific, Waltham, MA, USA). Reverse transcription was performed using a reverse transcription system (Promega, Madison, WI, USA). In this process, 500 ng of total RNA was used for transcription and the other components of the reaction mix were as described in the manufacturer's protocol. RT-qPCR was performed using 10 ng cDNA, 300 nM forward and reverse primers, and 1× PowerUp SYBR Green Master Mix (Thermo Fisher Scientific, Waltham, MA, USA) in a 10 µL reaction. Forward and reverse primers were designed by Sigma-Aldrich (Supplementary Table S1) for fRPC studies. ARPE-19 primers were designed using Primer-Blast from NCBI and sequences are listed in Supplementary Table S2. RT-qPCR was performed on an Applied Biosystems Quantstudio 12 K Flex (Thermo Fisher Scientific, Waltham, MA, USA) with SYBR fluorescence with ROX passive reference. The thermal cycling conditions followed the standard cycling for primers with melting temperature greater than 60 °C

UDG activation for 2 min at 50 °C, dual-lock DNA polymerase for 2 min at 95 °C, followed by 40 cycles of denature for 15 s at 95 °C then annealing/extending for 1 min at 60 °C. Melt curve analysis was performed by (1) ramping to 95 °C at 1.6 °C/second and holding for 15 s, (2) ramping to 60 °C at 1.6 °C/second and holding for 1 min, and (3) ramping to 95 °C at 0.15 °C/s and holding for 15 s. Cycle thresholds and baselines were calculated using the automatic settings in the Quantstudio 12 K Flex software. Fold change was calculated using the $2^{-\Delta\Delta CT}$ method [44].

2.7. Animals and Experimental Design

The immunodeficient Royal College of Surgeons (iRCS) rats breeding pair were as previously described [45]. Rats were group housed under specific pathogen-free conditions and had access to water and food ad libitum. At post-natal day 21 (±2 days), iRCS rats received 10 µL intravitreal injections of XVIVO10 (XV1), 3X concentrated XV1 (XV3), 20 µg/mL pigmented epithelial derived factor (PEDF, Bio-Techne, Minneapolis, MN), unconcentrated PRPE-SF (SF1), and concentrated SF1 (SF3) followed by general health and clinical ocular observations. This level of PEDF is equivalent to the concentrations found in SF3. The SF3 treatment group had a higher number of samples due to repeated studies performed over a 2-year period in which SF3 was tested against the various controls in agreement with the approved grant from the California Institute for Regenerative Medicine (TRAN1-11532). The data collected over this period were combined for analysis. Prior to each IVT injection, functional and structural assessment was performed by ERG and OCT, respectively. Animals received the second and third serial injections every 14 days at p35 (+2 days) and p49 (+2 days). Before each IVT injection, anesthesia was administered by abdominal injection of ketamine (37.5 mg/kg) and xylazine (5 mg/kg). Topical anesthesia and pupil dilation were induced by 0.5% proparacaine hydrochloride ophthalmic solution (Akorn, Inc., Lake Forest, IL, USA) and 2.5% phenylephrine hydrochloride and 0.5% tropicamide (Akorn, Inc.), respectively.

2.8. Electroretinogram (ERG) Evaluation

For ERG assessment, animals were placed in dark adaptation overnight the day before the functional testing. Under dark conditions and using a dim red light, animals were anesthetized as described above, along with application of pupil dilation and topical anesthesia eye drops While the animal was under anesthesia, a heating table and monitor were used to monitor body temperature. Reference and ground electrodes were inserted into the infraorbital (malar) area and between the ears, respectively. Scotopic testing was conducted with flash stimuli intensities ranging from 0.1–25 candela (cd) and recorded from both eyes. Subsequently, rats were light-adapted and photopic testing was conducted with intensities from 0.01–25 cd (HMsERG Rodent System, OcuScience, NV, USA). ERG raw data were collected and evaluated.

2.9. Ocular Coherence Tomography (OCT) Evaluation

Imaging scanning was performed at the end of the ERG session. Using the Spectralis OCT (Heidelberg Engineering Inc., Franklin, MA, USA), the animal was placed over the animal-modified head support used for patients. Balanced salt solution (BSS) was applied regularly to moisten the cornea during the imaging acquisition. A full set of b-scans was acquired from both sides of the optic nerve. Each set of images included a high-resolution b-scan and a volume scan composed of 30 b-scans from the temporal and nasal sides of the optic nerve, from both eyes. Morphological analysis of the retina included a grading score based on the outer retina changes, as described in Supplementary Figure S1. By using the Heidelberg Eye Explorer software, the retinal thickness and ONL characteristics were evaluated. After the completion of the OCT imaging session, animals were injected with test articles.

2.10. Intravitreal Injection

Under sterile conditions, using a 30-gauge (G) needle, a scleral incision was performed behind the limbus in the temporal superior quadrant of the left eye. Subsequently, intraocular pressure was reduced by a puncture into the anterior chamber through the periphery of the cornea, also with a 30 G needle. Then, 10 µL of the tested article was injected through the scleral incision by using a 30 G blunt steel needle attached to a 50 µL Hamilton syringe. After the completion of the IVT injection, a clinical assessment of the posterior pole was carried out by clinical visualization through the surgical microscope. A self-sealed healing of the sclera and the conjunctiva was observed, and no sutures were needed. Topical application of antibiotic ointment was performed at the end of the procedure.

2.11. Euthanasia and Tissue Collection

At p60, rats were euthanized by intraperitoneal injection of 0.5 mL pentobarbitol sodium 390 mg and phenytoin sodium 50 mg (Euthasol; Virbac AH, Inc., Fort Worth, TX, USA) and their eyes were enucleated and fixed in Davidson's solution. After 24 h of fixation, the Davidson's solution was replaced with 70% ethanol and sent to the USC Ginsburg Institute for Biomedical Therapeutics Core for paraffin embedding, sectioning, and hematoxylin and eosin (H&E) staining. Anterior segment structures, including cornea, iris, and lens were removed and the posterior pole was exposed. The cut of the eye was through the optic nerve on its sagittal plane. After dissection, all eyes were embedded in paraffin and cut in a microtome starting from the center of the optic nerve. Serial sections of 5 µm in thickness were performed throughout the entire eyeball. Approximately 4 consecutive/serial retinal sections were placed on every glass slide.

2.12. Photoreceptor Counting

Surviving PRs were determined in rats treated with various test and control articles. Since injection of both the test and control articles were not localized to one area of the retina, as they were deployed within the vitreous cavity, slides representing the central area of the retina were selected for enumeration. Specifically, the slide for enumeration was chosen based on the presence of the optic nerve as a landmark. The preserved ONL cell numbers were enumerated from scanned images of the sections. Photoreceptor counting was initiated 1 mm superior or inferior to the optic nerve and continued for 1 mm. The "Nuclear V.09" algorithm (a nucleus counting algorithm) provided in the Aperio ScanScope CS microscope software was used to enumerate the cells. Two sections were counted for each animal and the results were averaged.

2.13. Immunofluorescence Staining

Eye tissue sections were first deparaffinized and rehydrated via immersion in a series of xylene, ethanol, and PBS solutions. Heat-induced epitope retrieval was performed using 1× citrate buffer (pH = 6.0) in a pressure cooker. Tissues sections were then permeabilized with 0.3% Triton X-100 in PBS for 10 min then blocked for 30 min with blocking buffer (PBS containing 2.5% normal goat serum (*v*/*v*)). Primary antibodies (Supplementary Table S3) were incubated overnight at 4 °C then incubated for 45 min at RT with secondary antibody diluted 1:500 with blocking buffer. Nuclear staining was performed using 1 µg/mL DAPI in PBS for 10 min at RT. Coverslips were mounted using VECTASHIELD Vibrance Antifade Mounting Medium (Vector Laboratories, Newark, CA, USA). Fluorescent images were taken on the Olympus BX43 microscope using a 40× objective. Immunofluorescence images were analyzed in ImageJ using 5 images per animal. ImageJ analysis was performed by manually selecting the retinal layers and using the DAPI channel to refine the nuclear layer regions of interest with a histogram threshold of 50–255 on an 8-bit scale. For 4HNE and MDA staining, automatic thresholding using the Phansalkar method was applied and then the percentage area per retinal layer was measured. This was similarly applied for PAD4 in the ONL and outer segments. For CitH3, histogram thresholds were set from 30–255 on an 8-bit scale and the percentage area within the DAPI mask was measured. CD68 cell counts were performed on images with threshold set from 30–255 on an 8-bit scale and particle analysis using 5–300 px^2 size range for the entire image field.

2.14. Statistical Analysis

Statistical analysis was performed in GraphPad Prism 9 (Graphpad Software Inc., La Jolla, CA, USA). All graphs are plotted as mean ± standard error of the mean, unless otherwise noted in the figure legend. Appropriate statistical analyses were performed for each dataset, including unpaired *t*-tests for comparing two treatments. For comparing more than two groups, one-way ANOVA with Tukey's correction, Welch's ANOVA with Dunnett's T3 correction, and Kruskal–Wallis with Dunn's correction were performed based on the dataset. For genetic expression data, ddCT was compared using two-way ANOVA with Tukey's correction. The statistical tests used in each individual analysis are indicated in the figure legends.

2.15. Study Approval

Regulatory approval for use of the hESC lines was obtained from the University of Southern California Stem Cell Research Oversight Committee (SCRO). Animal experiments were conducted in full accordance with University of Southern California Institutional Animal Care and Use Committee (IACUC)-approved protocols, National Institutes of Health Guide for the Care and Use of Laboratory Animals, and the ARVO Statement for the Use of Animals in Ophthalmic and Vision Research. The graphical abstract and Figure 1 were created with Biorender.com (accessed on 15 June 2023).

3. Results

3.1. PRPE-SF Increases fRPC Cellular Viability and Rhodopsin Expression

Cellular apoptosis and proliferation of fRPC were compared among XV1, XV3, 20 μg/mL PEDF, SF1, and SF3. PEDF was chosen as a control as it is highly concentrated in the mature RPE secretome and is known to promote retinal cell survival [26]. The concentration of PEDF used reflects the levels found in SF3 as measured using ELISA. TUNEL staining was used to quantify the number of apoptotic cells, and propidium iodide was used as the counterstain to measure total cell area. The apoptotic/total cell ratio was then normalized to XV1 (Figure 2A). The results showed that the SF3 treatment significantly reduced the apoptotic/total cell ratio, while no reduction was observed with other treatments, including 20 μg/mL PEDF. In addition, both SF1 and SF3 improved cellular viability compared with XV1 and PEDF, as determined by the resazurin assay (Figure 2B).

The impact of SF1 and SF3 on photoreceptor survival was evaluated using rhodopsin expression in fRPCs, comparing XV1, XV3, and 20 μg/mL PEDF (Figure 3A–F). Freshly isolated fRPCs displayed intense rhodopsin immunofluorescent staining (Figure 3A), however, 24 h after treatment with XV1, XV3, PEDF, or SF1, the level of rhodopsin was minimally detectable (Figure 3B–E). In contrast, SF3-treated fRPC maintained high levels of rhodopsin staining (Figure 3F). Correspondingly, increased gene expression of rhodopsin and recoverin were observed for both SF1- and SF3-treated fRPC ($p < 0.05$) when compared with XV1 treatment (Figure 3G,H).

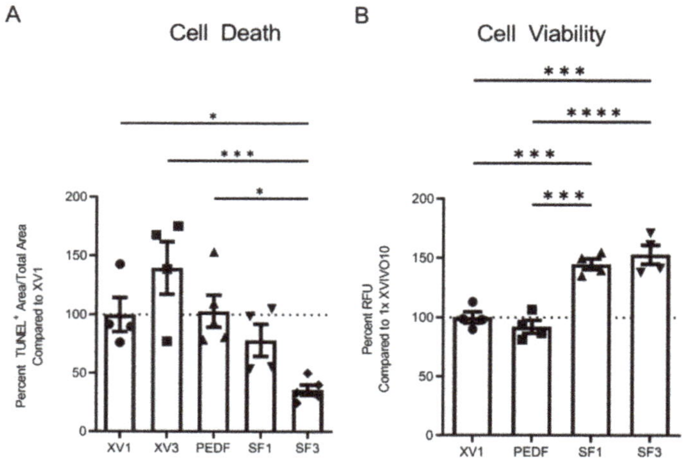

Figure 2. fRPCs treated with SF3 showed significant decrease in cell death compared with all othe treatments as determined via the apoptosis TUNEL assay following 24 h incubation (n = 4 XV1, XV3 and SF1, n = 5 PEDF and SF3) (**A**). Both SF1 and SF3 showed significantly improved cell viability compared with XV1 and PEDF, assessed by resazurin metabolism following 24 h incubation (n = 4 each group) (**B**). Data represented as mean ± SEM. * $p < 0.05$, *** $p < 0.001$, **** $p < 0.0001$, one-way ANOVA with Tukey's correction.

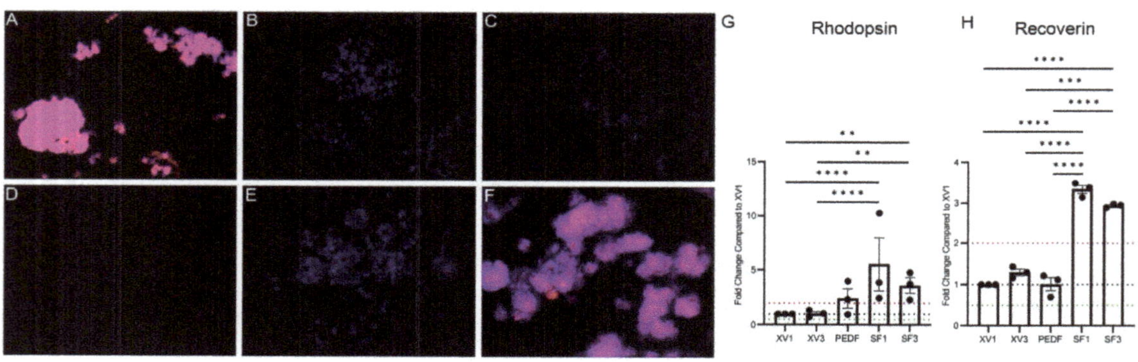

Figure 3. Rhodopsin and DAPI staining of fRPC immediately after isolation (**A**) and after 24-h incubation with XV1 (**B**), XV3 (**C**), PEDF (**D**), SF1 (**E**) and SF3 (**F**). RT-qPCR shows SF1 and SF3 induce rhodopsin (**G**) and recoverin (**H**) gene expression. Data represented as mean ± SEM. ** $p < 0.01$ *** $p < 0.001$, **** $p < 0.0001$, statistically analyzed ddCT values with one-way ANOVA with Tukey's correction using data from n = 3 independent experiments with triplicate samples in each experiment

3.2. PRPE-SF Induces Gene Expression in fRPC

To characterize the biological activities associated with SF1 and SF3, the impact of these treatments on fRPC was evaluated using targeted gene expression (Figure 4A Supplementary Figure S2). Major constituents of SF1 and SF3, determined by microarray and ELISA, are highlighted in Figure 4B showing that NGF, BDNF, PEDF, LIF, TGFβ1 and BMP7 were all present and were enriched by our processing. SF1 and SF3 were able to upregulate progenitor cell proliferation, maturation, and differentiation-associated gene expression.

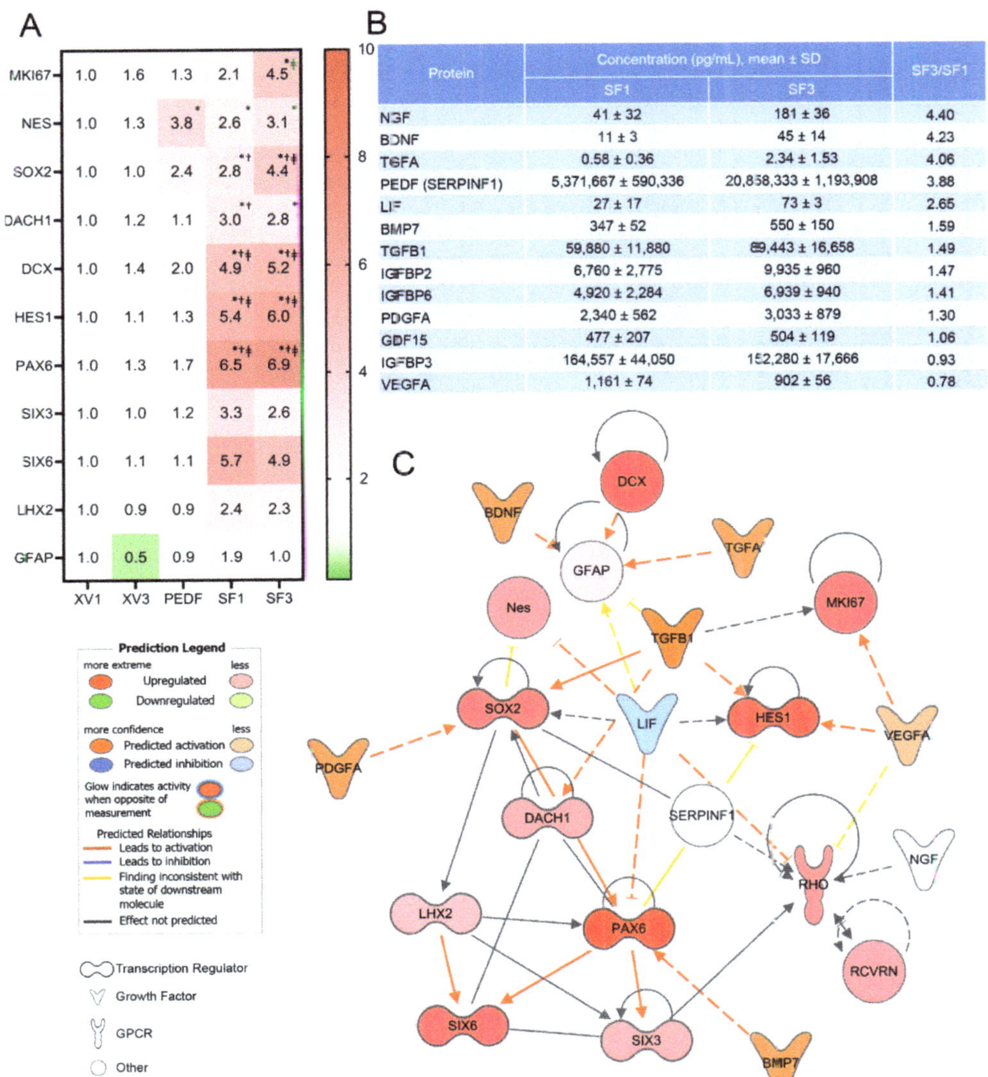

Protein	Concentration (pg/mL), mean ± SD		SF3/SF1
	SF1	SF3	
NGF	41 ± 32	181 ± 36	4.40
BDNF	11 ± 3	45 ± 14	4.23
TGFA	0.58 ± 0.36	2.34 ± 1.53	4.06
PEDF (SERPINF1)	5,371,667 ± 590,336	20,858,333 ± 1,193,908	3.88
LIF	27 ± 17	73 ± 3	2.65
BMP7	347 ± 52	550 ± 150	1.59
TGFB1	59,880 ± 11,880	89,443 ± 16,658	1.49
IGFBP2	6,760 ± 2,775	9,935 ± 960	1.47
IGFBP6	4,920 ± 2,284	6,939 ± 940	1.41
PDGFA	2,340 ± 562	3,033 ± 879	1.30
GDF15	477 ± 207	504 ± 119	1.06
IGFBP3	164,557 ± 44,050	152,280 ± 17,666	0.93
VEGFA	1,161 ± 74	902 ± 56	0.78

Figure 4. Mean fold change of fetal RPC gene expression after 24 h incubation compared with XV1 (**A**). RT-qPCR shows that SF1 and SF3 induce several eye-field transcription factor genes relating to proliferation and retinal development. Mean ± SD concentration of proteins in SF1 and SF3 determined by ELISA or microarray (**B**). Relationships between transcription factors (TFs), mature retinal cell markers, and RPE neurotrophic factors using IPA (**C**). TFs show a complex network of positive and negative regulations on each other, and PRPE-SF components additionally show both positive and negative regulations on TFs. Statistically analyzed ddCT values using two-way ANOVA and Tukey's multiple comparisons of data from $n = 3$ independent experiments with triplicate samples in each experiment, no comparisons were performed for SIX3, SIX6, LHX2, and GFAP which had only $n = 2$ experiments. (* $p < 0.05$ compared to XV1, † $p < 0.05$ compared to XV3, ‡ $p < 0.05$ compared to PEDF.) Individual graphs can be viewed in Supplementary Figure S2. Networks generated with QIAGEN IPA (QIAGEN Inc., https://digitalinsights.qiagen.com/IPA, accessed on 16 March 2023).

To further dissect the molecular interactions, an ingenuity pathway analysis (IPA) was undertaken. Factors such as PEDF, IGFBP3, and TGFβ1 were detected in high concentrations in both SF1 and SF3. TGFβ1, a pleiotropic factor, can upregulate SOX2, HES1, MKI67, and DACH1 (Figure 4A,C). SF1 and SF3 both upregulated MKI67 expression coinciding with the ability to promote fRPC viability. TGFβ1 can increase HES1 expression which regulates cell division, gliogenesis, and maintenance of tissue compartment boundaries [46]. Other properties ascribed to levels of TGFβ1 found in SF1 and SF3 include upregulation of DACH1, which has been shown to promote transition from progenitor cells to neuronal precursor cells achieved through cell cycle synchronization and interaction with cyclin D1 [47]. In addition, SF1 and SF3 were able to upregulate SOX2 expression by 2.8- to 4.4-fold, respectively, which can in turn trigger downstream upregulation of PAX6 and LHX2 expression. IPA also shows the contribution of BMP7 to increase expression of PAX6 where there were 6.5- and 6.9-fold increases compared with fRPC treated with XV1. LHX2, a central factor coordinating optic cup development, is required for BMP signaling and interacts with PAX6 to regulate SIX3 and SIX6 expression [48,49]. SF1 and SF3 upregulated DCX and NES, which are microtubule and cytoskeleton associated enzymes that play a role in the migration of progenitors and the structural orientation of retinal layers [50,51].

While SF1 and SF3 promote progenitor cell proliferation, maturation, and differentiation gene expression, they do not alter GFAP expression, astrocyte markers, or Müller development and reactivity [52]. IPA suggests that upregulation of the transcription factors is consistent with cells fated for the eye, proliferation of progenitor cells, and differentiation pathways [53–55]. While this is a targeted analysis, the relationships identified in IPA are evidence of critical components to monitor in future manufacturing development for a clinical grade product.

3.3. PRPE-SF Preserves Retinal Structure and Function in RCS Rats

The highest SF3/SF1 proteins are known to have antioxidant effects, including NGF, BDNF, PEDF, and LIF [56–59], suggesting that SF3 would have greater antioxidant effects and efficacy in models of retinal degeneration. Experimental media were administered via intravitreal injection (IVT) on p21, p35, and p49 in athymic RCS (iRCS) rats, and on p60, eyes were collected and subjected to molecular dissection. H&E staining revealed photoreceptor preservation in iRCS rats treated with SF3, with a statistically significant increase in photoreceptor nuclei counts when compared with XV1, XV3, and 20 μg/mL PEDF (Figure 5A–E,K). Furthermore, photoreceptor inner segments appeared to be better organized in SF3-treated retinas compared with other treatments. This portion of the photoreceptor is where protein synthesis machinery and mitochondria are localized. Photoreceptor preservation was also evaluated non-invasively throughout the study using OCT and confirmed with histological findings on p60 (Figure 5F–J,L).

To investigate whether photoreceptor preservation translates to preservation of retinal function, scotopic and photopic ERG was performed prior to IVT injection at p21, p35, p49, and p60 in the XV3 and SF3 groups. The scotopic a-wave and b-wave amplitudes, used to measure rod photoreceptor response, showed significantly preserved retinal function in SF3-treated retinas at p49 and p60 (Figure 6). Additionally, the photopic b-wave amplitudes were significantly higher in SF3-treated retinas than XV3-treated retinas at p49 and p60, while the a-wave amplitudes were negligible and did not decline over time (Figure 7). Together, this demonstrates the effectiveness of SF3 in preserving retinal function.

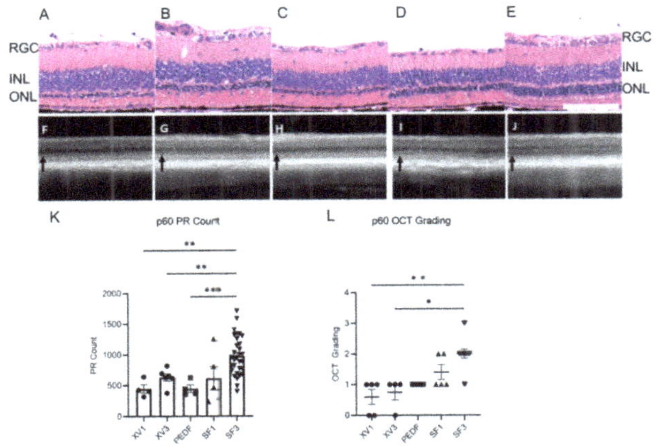

Figure 5. H&E and OCT images of p60 RCS retinas treated with XV1 (**A**,**F**), XV3 (**B**,**G**), PEDF (**C**,**H**), SF1 (**D**,**I**), and SF3 (**E**,**J**). H&E-stained retinas show outer nuclear layer (ONL) thickness correlating to OCT ONL (black arrow). The ONL was minimally detected in all treatment groups except for SF3 (J) of p60 iRCS retinas. Photoreceptor counts from H&E images ((**K**), XV1 $n = 4$, XV3 $n = 6$, PEDF $n = 4$, SF1 $n = 5$, SF3 $n = 31$) and OCT grading scores ((**L**), XV1 $n = 5$, XV3 $n = 4$, PEDF $n = 4$, SF1 $n = 4$, SF3 $n = 14$) show SF3 treatment significantly preserves the ONL compared with XV1 and XV3. Data represented as mean ± SEM, * $p < 0.05$, ** $p < 0.01$, *** $p < 0.001$, Photoreceptor counts: Welch's ANOVA with Dunnett T3 correction; OCT grading: Kruskal–Wallis with Dunn's correction. Scale bar is 100 µm.

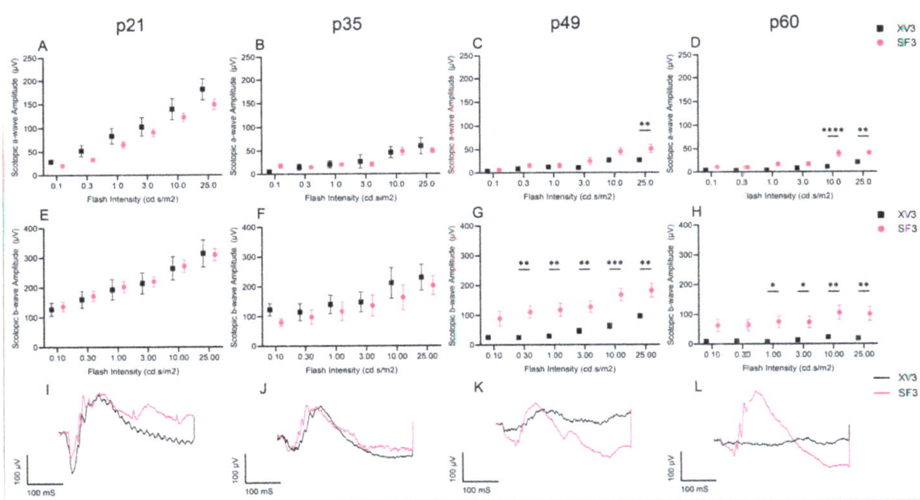

Figure 6. Scotopic ERGs of XV3- and SF3-treated eyes were measured using flash intensities of 0.1–25 cd·s/m^2. Scotopic a-wave (**A**–**D**) and b-wave (**E**–**H**) amplitudes were measured prior to intravitreal injection at p21–p60. Scotopic a-wave and b-wave were both significantly higher in SF3-treated rats at p49 and p60. A representative 25 cd·s/m^2 scotopic ERG is shown for each timepoint (**I**–**L**). Data represented as mean ± SEM, * $p < 0.05$, ** $p < 0.01$, *** $p < 0.001$, **** $p < 0.0001$, two-way ANOVA with Bonferroni's correction (p21: XV3 $n = 8$, SF3 $n = 16$; p35: $n = 4$ both groups; p49: $n = 4$ both groups; p60: XV3 $n = 6$, SF3 $n = 10$).

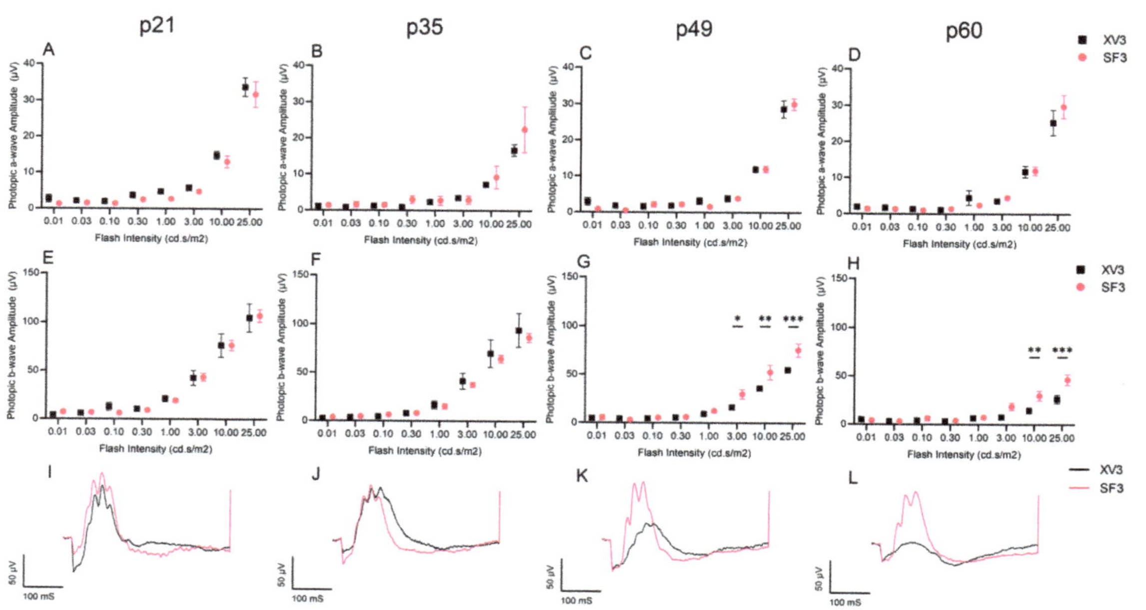

Figure 7. Photopic ERGs of XV3- and SF3-treated eyes were measured using flash intensities of 0.01–25 cd·s/m^2. Photopic a-wave (**A–D**) and b-wave (**E–H**) amplitudes were measured prior to intravitreal injection at p21-p60. Photopic b-wave amplitudes were significantly higher in SF3 treated rats at p49 and p60. A representative 25 cd·s/m^2 photopic ERG is shown for each timepoint (**I–L**). Data represented as mean ± SEM, * $p < 0.05$, ** $p < 0.01$, *** $p < 0.001$, two-way ANOVA with Bonferroni's correction (p21: XV3 $n = 8$, SF3 $n = 16$; p35: $n = 4$ both groups; p49: $n = 4$ both groups; p60: XV3 $n = 6$, SF3 $n = 10$).

3.4. PRPE-SF Reduces Reactive Oxygen Species and Reactive Glial Activation

As shown above, SF3 was able to preserve photoreceptor and retina structure while SF1 did not. Immunostaining of RCS retinas as well as in vitro experiments in ARPE-19 cells further demonstrated that the increased efficacy of SF3 is linked to reduced Müller and microglial cell activation and reduced oxidative stress markers, which are also known to induce inflammation [2]. The in vivo administration of SF3 reduced the activation of Müller and microglial cells, as indicated by decreased levels of GFAP (Figure 8A,B) and CD68 (Figure 8C–G) staining, respectively. XV3-treated iRCS retinas showed evidence of gliosis as indicated by strong GFAP staining as well as subretinal glial membranes below the outer nuclear layer (ONL), which was reduced or not present in SF3-treated retinas. GFAP is commonly upregulated in retinal degeneration models, where the reduction in GFAP may indicate reduced gliosis [60]. XV1- and XV3-treated iRCS retinas showed significant CD68+ staining in the outer segments, while retinas treated with PEDF and SF3 demonstrated a dramatic reduction in CD68+ cells. However, the sample size was low and the reduction in CD68+ cells was not found to be statistically significant.

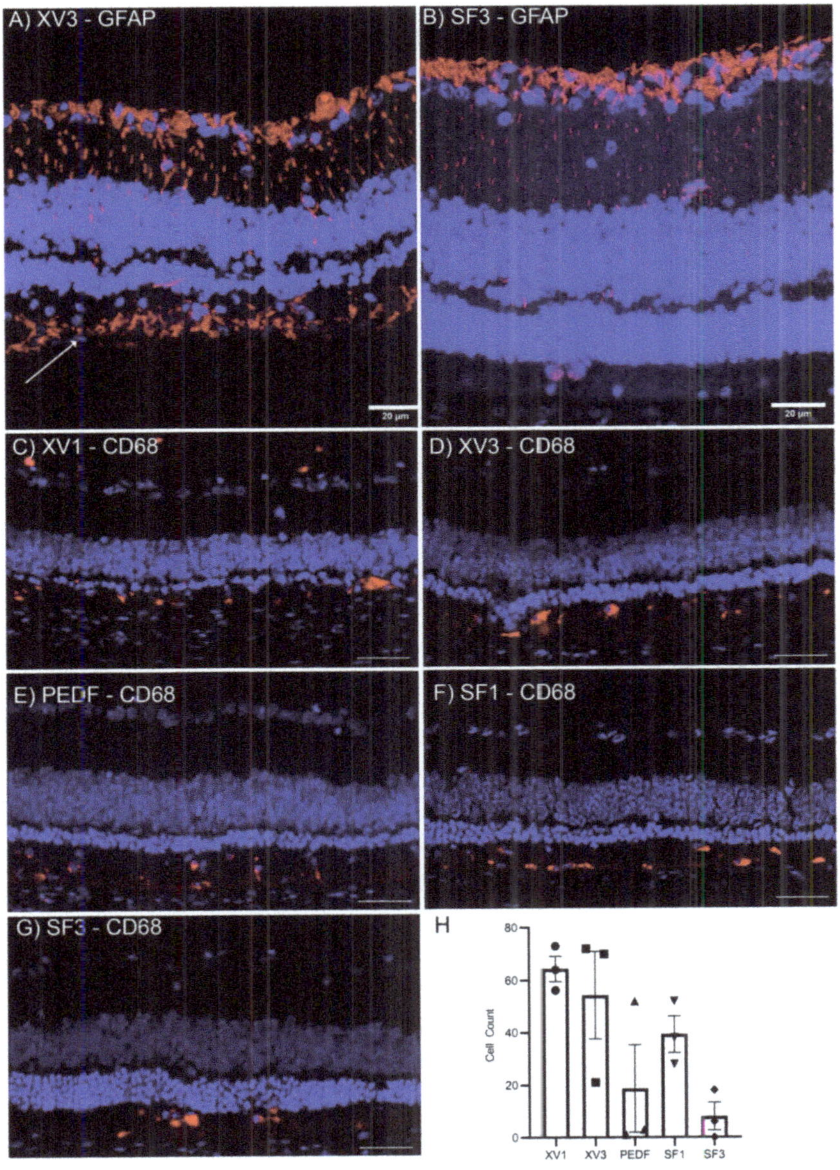

Figure 8. Immunofluorescence images of treated p60 iRCS rat retinas. XV3-treated retinas displayed significant gliosis as marked by GFAP staining throughout the retina and below the ONL ((**A**), white arrow) while SF3 retinas demonstrated reduced GFAP staining (**B**). Anti-CD68 immunofluorescence staining of p60 iRCS retinas (**C–G**). PEDF and SF3 show non-significantly reduced CD68+ cells ((**H**) $n = 3$ each group). Blue: DAPI. Red: GFAP (**A**,**B**) or CD68 (**C–G**). Scale bars are 20 μm (**A**,**B**) and 100 μm (**C–G**).

The effects of SF3 treatment on the levels of ROS were also evaluated (Figure 9). In ARPE-19 cells, DCFDA was used to measure the cytoplasmic levels of ROS following 24 h incubation of XV1, XV3, PEDF, SF1, and SF3. SF1 had little effect on ROS levels, while concentrated SF3 was found to reduce ROS levels by approximately 20–30% compared with XV3 (Figure 9A). ARPE-19 treated with decreased dilutions of SF3 showed alteration in genes regulating ROS production and elimination (Figure 9B). Few to no changes in NADPH-oxidase (NOX) expression of multiple isoforms were found. However, increased SOD2 expression was seen. While SOD3 was also increased in a concentration-dependent manner, the expression changes did not reach significance. In addition, the glutathione pathway was also probed, and changes were not found.

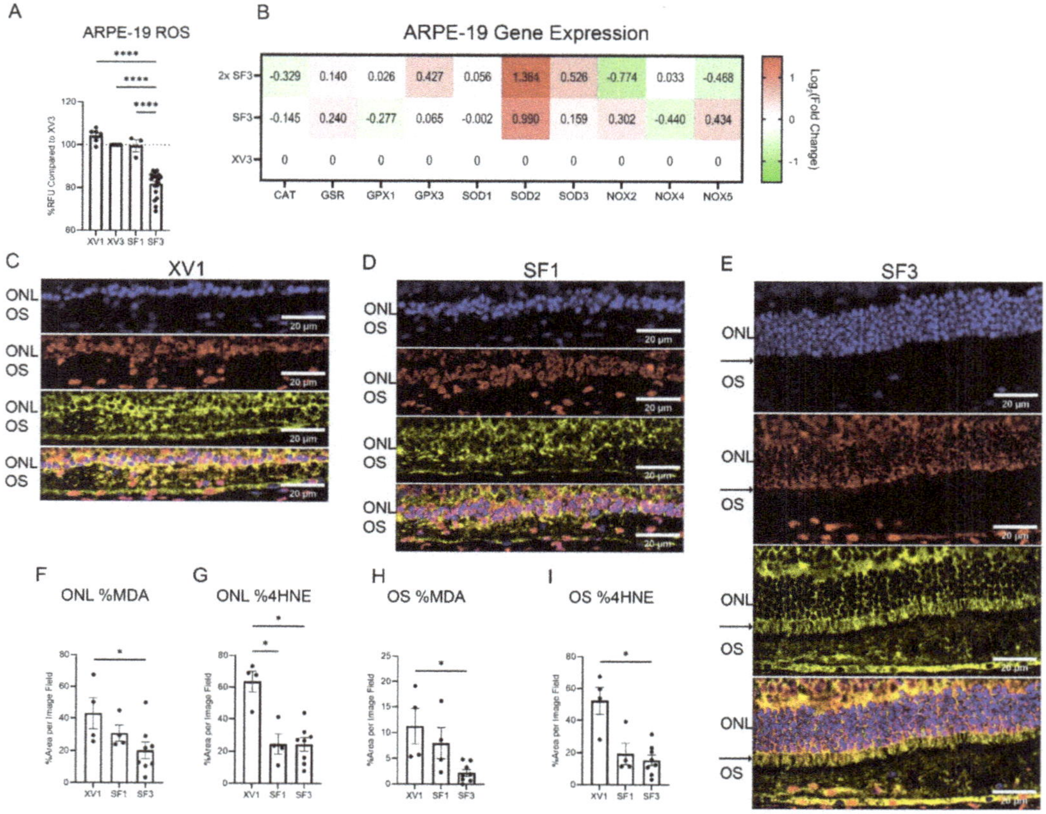

Figure 9. Effects of PRPE-SF on retinal reactive oxygen species. In vitro, SF3 significantly reduces DCFDA oxidation, a non-specific marker for reactive oxygen species, in ARPE−19 cells following 24 h incubation compared with all other treatments (**A**) (one-way ANOVA with Tukey's correction using data from 17 independent experiments with triplicate samples in each experiment). It was observed that 2x SF3 increases significantly increased SOD2 expression following 24 h of treatment (mean Log_2(Fold Change) > 1, duplicates) (**B**). Immunofluorescence images of p60 iRCS rats treated with XV1 ((**C**) n = 4), SF1 ((**D**) n = 4), and SF3 ((**E**) n = 8). Malondialdehyde (MDA; red) shows strong nuclear staining while 4-hydroxynonenal (4HNE; yellow) is primarily non-nuclear. Due to preservation, SF3 images appear larger, but each image shows the outer nuclear layer (ONL) and outer segment (OS) at the same magnification. SF3 treatment shows preservation of inner segments (black arrows in (**E**)) with strong staining for 4HNE. SF3 shows reduced photoreceptor nuclei staining of MDA and reduced ONL and OS staining of 4HNE. ImageJ analysis shows a significant decrease in both MDA and 4HNE percent area in the ONL and OS (Kruskal–Wallis with Dunn's multiple comparisons) (**F–I**). * $p < 0.05$, **** $p < 0.0001$. Blue: DAPI. Red: MDA. Yellow: 4HNE. Scale bar is 20 µm.

To determine whether these in vitro and molecular findings correspond to in vivo studies, the effects of SF3 on oxidative products, such as malondialdehyde (MDA) and 4-hydroxy-2-nonenal (4HNE), were also studied in treated iRCS retinas. Results showed that iRCS rats treated with XV1 or SF1 had high levels of MDA and 4HNE in all retinal layers, while iRCS rats treated with SF3 showed reduced retinal levels of MDA and 4HNE, particularly in the ONL and outer segments (OS) (Figure 9C–I). The areas of MDA and 4HNE were measured as a percentage of the respective layer to account for increased retinal thickness in SF3 retinas. Staining showed that MDA had a stronger nuclear staining, while 4HNE appeared to be cytoplasmic and extracellular, suggesting separate roles for each lipid peroxide in retinal pathologies. Furthermore, SF3-treated retinas showed 4HNE localized within preserved inner segments, which links the ROS produced by photoreceptor mitochondria to the 4HNE staining of the inner segments. A longitudinal study was performed to verify that both MDA and 4HNE increase with the age of the iRCS rat (Supplementary Figure S3). The results showed that MDA showed significant increased expression in the ONL at p49 and p60, while 4HNE showed pan-retinal increases with the strongest increases in the outer plexiform layer, ONL, OS, and RPE layers.

3.5. PRPE-SF Reduces NETosis Markers PAD4 and CitH3

In recent years, sterile neutrophil extracellular trap (NET) formation has emerged as a crucial factor in retinal disease and is known to result from increased oxidative stress and play a role in retinal autoimmunity [61]. Since increased ROS production can lead to PAD4 activation [62], the expression of PAD4 was evaluated in retinas of iRCS rats treated with XV1, PEDF, SF1, and SF3 at p60 (Figure 10A–E). XV1- and PEDF-treated retinas revealed high PAD4 expression present in all retinal layers, including the inner nuclear layer (INL), ONL, and OS. However, retinas treated with SF3 exhibited reduced PAD4 expression, particularly in the ONL and OS. These findings indicate the protective effect of SF3 in reducing PAD4 expression. In addition, the presence of PAD4 in the choroid of retinas treated with XV1 and PEDF highlights the presence of both intra- and extra-retinal sources of PAD4.

Retinas were also analyzed for citrullinated histones (CitH3) which are produced through PAD4-mediated deimination. Results showed that retinas treated with XV1 had high levels of CitH3 in various retinal layers including the ganglion cell layer (GCL), INL, ONL, RPE, and choroid. However, retinas treated with SF3 demonstrated significantly reduced levels of CitH3 (Figure 10F–H). These findings provide evidence that reduction in PAD4 expression and citrullination correspond with delayed retinal degeneration, highlighting the potential therapeutic benefits of SF3 in retinal disease pertaining to autoimmunity.

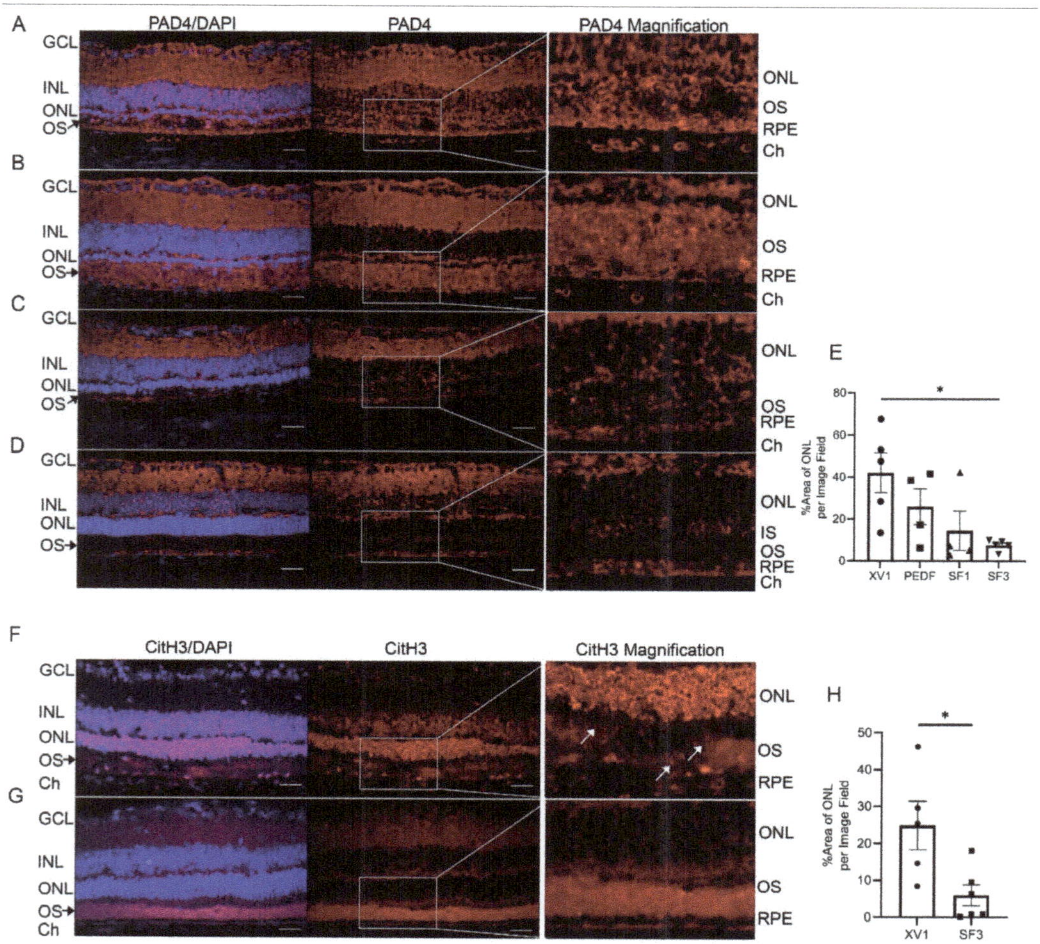

Figure 10. PAD4 immunofluorescence images of p60 iRCS rats treated with XV1 ((**A**), $n = 5$), PEDF ((**B**), $n = 4$), SF1 ((**C**), $n = 4$), and SF3 ((**D**), $n = 5$). PAD4 shows pan-retinal staining with strong staining in the ganglion cell layer (GCL), inner nuclear layer (INL), outer nuclear layer (ONL), outer segments (OS), retinal pigment epithelium (RPE), and choroid (Ch) in XV1- and PEDF-treated retinas. SF1- and SF3-treated retinas show reduced staining intensity; ImageJ analysis shows SF3-treated retinas have significantly reduced PAD4 percent area of ONL compared with XV1 (**E**). (* $p < 0.05$ one-way ANOVA with Tukey's multiple comparisons). Similar to 4HNE, light PAD4 staining is seen in the preserved inner segment region (IS) of SF3-treated retinas (D). Citrullinated histone H3 (CitH3) immunofluorescence images of iRCS rats at p60 treated with XV1 ((**F**), $n = 5$) and SF3 ((**G**), $n = 6$) XV1-treated retinas show GCL, Ch, and extensive ONL staining with punctate debris in the OS (white arrows). ImageJ analysis of ONL CitH3 shows SF3-treated retinas have reduced ONL CitH3 expression compared with XV1 (**H**). (* $p < 0.05$, unpaired t-test). Blue: DAPI. Red: PAD4 or CitH3 Scale bars are 20 μm.

4. Discussion

Therapeutic options for retinal degenerative diseases are extremely limited, whether for AMD or genetic conditions such as retinitis pigmentosa. However, stem cell implantation has emerged as a promising therapeutic approach for directly replacing atrophic cells [11]. Previously, the implantation of hESC-RPE showed that these cells were able to re-epithelialize the dystrophic areas and extend preservation beyond the borders of the implant [16,44]. This effect may be associated with the secretome elaborated by the implanted RPE that is capable of maintaining ocular function, coinciding with reduced ocular oxidative stress and inflammation. The key aspects of PRPE-SF are: (1) PRPE-SF utilizes the biomimetic membrane to induce human ESC-RPE to polarize and mature, thus modulating the secretome towards a protective phenotype [63,64]; (2) PRPE-SF does not use animal serum, which will aid in clinical development to avoid immunogenicity; and (3) PRPE-SF is a combination of neuroprotective factors which can address multiple pathological mechanisms of retinal degeneration. This study examined the therapeutic potential of SF3 for treatment of retinal degenerative conditions by promoting neuronal survival while mitigating oxidative stress and inflammation.

The RPE is a source of neuroprotective factors, however, purified factors when given alone have failed in clinical evaluation. Previous research demonstrated that conditioned medium derived from rat RPE was able to promote rat RPC survival and differentiation [30,31,65]. Similarly, human-derived RPE-conditioned medium was able to reduced cellular death of porcine retinal explants [18]. Previously, we expanded on these findings showing that hESC-derived RPE-conditioned medium can promote fRPC proliferation while inhibiting cell death [63]. Here, we further showed that PRPE-SF treatment of fRPCs was able to upregulate genes associated with progenitor cell proliferation and multipotency through upregulation of MKI67, SOX2, and PAX6. Furthermore, PRPE-SF promoted eye fate determinants (DACH1 and LHX2) and promoted neuronal migration genes (NES and DCX). The enriched secretome was able to upregulate photoreceptor developmental genes, rhodopsin and recoverin, but did not affect glial genes such as GFAP. These findings are consistent with studies where RPE-conditioned media preferentially differentiated rat retinal explants and immortalized RPCs into photoreceptors [33,66].

Upregulation of these genes suggests that PRPE-SF may induce proliferation and differentiation of latent progenitor cells towards neuronal fates, which is being explored in current studies. According to upstream analysis and predictive modeling in IPA, there is a direct relationship between PRPE neurotrophic factors and the effected transcription factor genes in our fRPC dataset. However, SF3 showed a stronger effect on reduced cell death and preserved rhodopsin staining, potentially through protective mechanisms rather than proliferation. Several growth factors in SF1 were proportionally higher in SF3. Proteins with high SF3/SF1 ratios, including NGF, BDNF, PEDF, and LIF, are known to have protective effects for neurodegenerative disorders and probably contribute to blockage of pro-apoptotic pathways [58,59,67]. NGF and PEDF are particularly interesting as these trophic factors have been individually investigated in clinical trials for retinal conditions, including NGF for retinitis pigmentosa, cystoid macular edema, and glaucoma, and PEDF for macular degeneration [68–70].

Previous research has demonstrated that RPE cells secrete a mixture of anti-angiogenic and pro-angiogenic factors, and the balance is shifted towards pro-angiogenic factors in oxidatively stressed RPE cells [36]. While SF3 contains a mixture of these factors, we hypothesized that the unstressed, healthy RPE cells produce these factors at a concentration which can restore this imbalance and promote retinal preservation. A link between these neurotropic factors and oxidative stress led us to interrogate the impact of SF3 in iRCS. IVT administration of SF3 starting on p21 and given bi-weekly till p60 was able to significantly preserve photoreceptors. More importantly, photoreceptor preservation corresponded with the preservation of scotopic a-wave and scotopic and photopic b-wave amplitudes, suggesting that SF3 can also ameliorate and delay decline in retinal function. This study also suggests that antioxidant mechanisms of SF3 contribute to retinal preservation. ARPE-19 treated with SF3 was consistently able to reduce cytoplasmic ROS by 20%. Amongst the ROS-generating and -eliminating genes investigated, SF3 significantly increased SOD2 and marginally increased SOD3 gene expression in ARPE-19. Increased SOD2 and SOD3 expression would increase the rate of superoxide elimination resulting in the observed ROS reduction. These findings coincide with reduction of ROS in vivo, where reduced levels of MDA and 4NHE were seen in SF3-treated iRCS retinas compared with XV1. The MDA reduction was particularly dramatic in the ONL and OS, suggesting that SF3 reduced oxidative stress in areas of photoreceptor degeneration and subsequent microglia activation and translocation. The level of 4HNE, a cytotoxic byproduct of arachidonic acid metabolism, was also lower in the ONL and OS layers of iRCS treated with SF3. Both peroxides form protein and DNA adducts that can induce additional inflammatory responses and perpetuate chronic inflammation as seen in retinal degeneration [2]. Additionally, MDA accumulation has been linked to RPE dysfunction and VEGF secretion in AMD [71]. Microglia and macrophage cells infiltrate in response to these damage-associated molecular patterns (DAMPs) released by damaged photoreceptors in an effort to clear cellular debris and detoxify the local microenvironment [72]. The reduction in ROS and oxidative byproducts may partially explain why SF3 treatment was associated with reduced infiltration of CD68+ cells into the retina.

In this study, ROS elevation was present in iRCS retinas corresponding to PAD4 activation, which is consistent with studies showing elevated intraretinal citrullination in human AMD donor eyes [73]. SF3 was able to reduce ONL and OS PAD4 staining coinciding with reduced CitH3. Additionally, PAD4 and CitH3 choroidal staining was observed in XV1-treated retinas and reduced with SF3. Choroidal neutrophils have been implicated in AMD and the degeneration of the RPE barrier [74], which suggests retinal remodeling associated with sterile inflammation initiated by aberrant oxidative stress and photoreceptor cell debris. Binet et al. suggested that vascular endothelial cells in diseased blood vessels engage in molecular pathways similar to those in aging. Cellular damage ultimately lead to cellular senescence-associated cytokine secretion, recruiting neutrophils that can subsequently trigger NETs [75]. While Binet et al. proposed that NETs can facilitate elimination of diseased senescent vasculature as a protective mechanism, chronic NETosis overactivation probably results in further destruction of the retinal vasculature. This is corroborated in studies demonstrating that NETosis contributes to the chronic inflammatory microenvironment of colorectal tumors [76]. In addition to infiltrating neutrophils in retinal vasculature, Müller cells may be responsible for PAD4 and CitH3 expression in the ONL and OS due to their known expression of PAD4 during reactive gliosis [77]. While this mechanism requires further exploration, SF3 treatment can ameliorate PAD4 and CitH3 staining in iRCS, which is paralleled by the reduction of 4HNE and MDA in treated eyes and retinal preservation.

This dataset shows clear concentration-dependent relationships with photoreceptor preservation using both in vitro and in vivo models. Because SF1 and SF3 had similar effects on viability and proliferation, the antioxidant and anti-inflammatory effects of SF3 are likely to be more important than the mobilization of progenitor cells for prevention of retinal degeneration. Additionally, the preservation of ERG at only p49 and p60 may be an additional indicator of the primary mechanism of PRPE-SF. The RCS rat is a Mertk mutant model of retinal degeneration, in which the primary cause of degeneration is the failure to phagocytose photoreceptor outer segments which require daily renewal [78]. Photoreceptors begin to die in response to accumulating debris and the oxidative microenvironment. As shown with MDA and 4HNE staining, significant staining is not observed until p49, and this accumulating oxidative stress accelerates the degeneration of the retina. Together, this suggests that the p49 ERG preservation is due to the protective effect of PRPE-SF against cytotoxic oxidation.

Our research is primarily limited by the availability of adequate models of retinal degeneration for AMD, which is the most prevalent cause of human retinal disease. However, we chose to utilize the RCS model as it is well characterized and has been utilized in the development of the CPCB-RPE1 implant currently under clinical investigation [12]. The RCS rat model is an aggressive model of retinal degeneration, making long-term efficacy studies untenable. In general, the predictivity of animal studies for human translation is low as many developed therapeutics fail in clinical settings [79]. As such, long-term efficacy studies are best evaluated in clinical studies and a stringent safety profile is developed in pre-clinical stages. We have conducted our work using SF3 stored for roughly a year, indicating the stability of the product when stored properly. In addition, we are performing scalability and long-term safety studies to support advancement to clinical stages. The delivery route and frequency of SF3 is also agreeable, with future clinical applications in which monthly IVT injections are common in the treatment of human retinal pathologies, although we utilized biweekly injections due to the aggressive nature of the RCS retinal degeneration. Further investigation into the underlying molecular mechanisms of retinal degeneration is still ongoing and will better support our analysis. However, we have demonstrated that SF3 affects key mechanisms related to retinal degeneration, such as oxidative stress and inflammation. We have also identified the potential pathological change in citrullination of the photoreceptor nuclei, which has not been observed in prior research. While this requires further elucidation, we are excited by the potential of a new therapeutic target for retinal dystrophies. Lastly, this study was conducted in a specific model of retinal degeneration, but due to the general antioxidant and anti-inflammatory effects, we are hopeful that it can be generalized to other age-related and neurodegenerative conditions.

5. Conclusions

In conclusion, our study underscores the therapeutic potential of PRPE-SF, particularly SF3, in treating retinal degeneration. The data reveals that SF3, derived from PRPE cells, can (1) hinder retinal cell apoptosis, (2) diminish cellular ROS, and (3) alleviate ocular oxidative and inflammatory stress in the retinopathy of the RCS model. Our IPA predictive modeling results reveal a strong relationship between PRPE-SF and the affected genes in this study. Nonetheless, comprehensive characterization of PRPE-SF is warranted to satisfy regulatory requirements and to pinpoint the pivotal active component(s) responsible for mitigating the hostile microenvironment in degenerative models will enhance the reproducibility and effectiveness of PRPE-SF products.

Crucially, PRPE-SF exhibited the preservation of retinal structure and function, as assessed by clinically relevant methodologies. The development of products such as PRPE-SF represents an innovative frontier, previously inaccessible due to technological constraints. However, with the recent advancements in stem cell technology, PRPE-SF emerges as a promising candidate for addressing retinal pathologies, such as dAMD and retinitis pigmentosa. This development could represent a significant leap towards effective therapeutics for these debilitating conditions and may set a precedent for treating other age-related and neurodegenerative diseases.

Supplementary Materials: The following supporting information can be downloaded at: https://www.mdpi.com/article/10.3390/cells12131689/s1.

Author Contributions: K.A. designed, performed, and analyzed experiments, and wrote the manuscript. J.-C.M.-C. and B.B.T. designed, performed, and analyzed experiments, and revised the manuscript. A.G.-C., A.N. and D.P. performed experiments and revised the manuscript. J.L., S.Y.L., D.M., S.G.L. and M.S.H. gave conceptual advice, designed experiments, and revised the manuscript. All authors have read and agreed to the published version of the manuscript.

Funding: This research was supported/partially supported by a grant from the California Institute for Regenerative Medicine (CIRM) (Grant Number TRAN1-11532); Unrestricted Grant to the Department of Ophthalmology from Research to Prevent Blindness, New York, NY; National Eye Institute of the National Institutes of Health under Award Number P30EY029220; Dr. Allen and Charlotte Ginsburg Institute for Biomedical Therapeutics (GF1000134); Dennis and Michele Slivinski (GF1000135); The USC Roski Eye Institute; The Marcus Foundation, Inc. (2977); Dr. Ramani Nathan (GF10000116); and The Retina Research Foundation's Gertrude D. Pyron Award (GF1024397). Kabir Ahluwalia was a Ginsburg Institute for Biomedical Therapeutics Fellow and an American Foundation for Pharmaceutical Education Fellow. The contents of this publication are solely the responsibility of the authors and do not necessarily represent the official views of the National Institutes of Health, CIRM or any other agency of the state of California.

Institutional Review Board Statement: Animal experiments were conducted in full accordance with University of Southern California Institutional Animal Care and Use Committee (IACUC)-approved protocols (protocol code 21045, approved 3 March 2017).

Informed Consent Statement: Not Applicable.

Data Availability Statement: The data presented in this study are available on request from the corresponding author.

Acknowledgments: The authors would like to acknowledge the late David Hinton who was a senior investigator in this project. Hinton dedicated his life to developing treatments for retinal disorders and his friendship and guidance will be greatly missed.

Conflicts of Interest: J.L. and M.S.H. are founders of RPT. The funders had no role in the design of the study; in the collection, analyses, or interpretation of data; in the writing of the manuscript; or in the decision to publish the results.

References

1. Wong, W.L.; Su, X.; Li, X.; Cheung, C.M.G.; Klein, R.; Cheng, C.Y.; Wong, T.Y. Global prevalence of age-related macular degeneration and disease burden projection for 2020 and 2040: A systematic review and meta-analysis. *Lancet Glob. Health* **2014**, *2*, e106–e116. [CrossRef]
2. van Lookeren Campagne, M.; LeCouter, J.; Yaspan, B.L.; Ye, W. Mechanisms of age-related macular degeneration and therapeutic opportunities. *J. Pathol.* **2014**, *232*, 151–164. [CrossRef] [PubMed]
3. Rozing, M.P.; Durhuus, J.A.; Krogh Nielsen, M.; Subhi, Y.; Kirkwood, T.B.L.; Westendorp, R.G.J.; Sørensen, T.L. Age-related macular degeneration: A two-level model hypothesis. *Prog. Retin. Eye Res.* **2020**, *76*, 100825. [CrossRef] [PubMed]
4. Rein, D.B.; Wittenborn, J.S.; Burke-Conte, Z.; Gulia, R.; Robalik, T.; Ehrlich, J.R.; Lundeen, E.A.; Flaxman, A.D. Prevalence of Age-Related Macular Degeneration in the US in 2019. *JAMA Ophthalmol.* **2022**, *140*, 1202–1208. [CrossRef]
5. Boyer, D.S.; Schmidt-Erfurth, U.; van Lookeren Campagne, M.; Henry, E.C.; Brittain, C. The Pathophysiology of Geographic Atrophy Secondary to Age-Related Macular Degeneration and the Complement Pathway as a Therapeutic Target. *Retina* **2017**, *37*, 819–835. [CrossRef]
6. Chandramohan, A.; Stinnett, S.S.; Petrowski, J.T.; Schuman, S.G.; Toth, C.A.; Cousins, S.W.; Lad, E.M. Visual Function Measures in Early and Intermediate Age-Related Macular Degeneration. *Retina* **2016**, *36*, 1021–1031. [CrossRef] [PubMed]
7. Chew, E.Y.; Lindblad, A.S.; Clemons, T. Summary results and recommendations from the age-related eye disease study. *Arch. Ophthalmol.* **2009**, *127*, 1678–1679. [CrossRef] [PubMed]
8. Aronow, M.; Chew, E. AREDS2: Perspectives, Recommendations, and Unanswered Questions. *Curr. Opin. Ophthalmol.* **2015**, *25*, 186–190. [CrossRef]
9. Liao, D.S.; Grossi, F.V.; El Mehdi, D.; Gerber, M.R.; Brown, D.M.; Heier, J.S.; Wykoff, C.C.; Singerman, L.J.; Abraham, P.; Grassmann, F.; et al. Complement C3 Inhibitor Pegcetacoplan for Geographic Atrophy Secondary to Age-Related Macular Degeneration: A Randomized Phase 2 Trial. *Ophthalmology* **2020**, *127*, 186–195. [CrossRef]
10. Meleth, A.D.; Wong, W.T.; Chew, E.Y. Treatment for atrophic macular degeneration. *Curr. Opin. Ophthalmol.* **2011**, *22*, 190–193. [CrossRef]
11. Nazari, H.; Zhang, L.; Zhu, D.; Chader, G.J.; Falabella, P.; Stefanini, F.; Rowland, T.; Clegg, D.O.; Kashani, A.H.; Hinton, D.R.; et al. Stem cell based therapies for age-related macular degeneration: The promises and the challenges. *Prog. Retin. Eye Res.* **2015**, *48*, 1–39. [CrossRef] [PubMed]
12. Kashani, A.H.; Lebkowski, J.S.; Hinton, D.R.; Zhu, D.; Faynus, M.A.; Chen, S.; Rahhal, F.M.; Avery, R.L.; Salehi-Had, H.; Chan, C.; et al. Survival of an HLA-mismatched, bioengineered RPE implant in dry age-related macular degeneration. *Stem Cell Rep.* **2022**, *17*, 448–458. [CrossRef] [PubMed]
13. Mandai, M.; Kurimoto, Y.; Takahashi, M. Autologous Induced Stem-Cell-Derived Retinal Cells for Macular Degeneration. *N. Engl. J. Med.* **2017**, *377*, 792–793. [CrossRef] [PubMed]
14. da Cruz, L.; Fynes, K.; Georgiadis, O.; Kerby, J.; Luo, Y.H.; Ahmado, A.; Vernon, A.; Daniels, J.T.; Nommiste, B.; Hasan, S.M.; et al. Phase 1 clinical study of an embryonic stem cell-derived retinal pigment epithelium patch in age-related macular degeneration. *Nat. Biotechnol.* **2018**, *36*, 328–337. [CrossRef]
15. Kashani, A.H.; Lebkowski, J.S.; Rahhal, F.M.; Avery, R.L.; Salehi-Had, H.; Chen, S.; Chan, C.; Palejwala, N.; Ingram, A.; Dang, W.; et al. One-Year Follow-Up in a Phase 1/2a Clinical Trial of an Allogeneic RPE Cell Bioengineered Implant for Advanced Dry Age-Related Macular Degeneration. *Transl. Vis. Sci. Technol.* **2021**, *10*, 13. [CrossRef] [PubMed]
16. Kashani, A.H.; Lebkowski, J.S.; Rahhal, F.M.; Avery, R.L.; Salehi-Had, H.; Dang, W.; Lin, C.-M.M.; Mitra, D.; Zhu, D.; Thomas, B.B.; et al. A bioengineered retinal pigment epithelial monolayer for advanced, dry age-related macular degeneration. *Sci. Transl. Med.* **2018**, *10*, eaao4097. [CrossRef]
17. Inanc Tekin, M.; Sekeroglu, M.A.; Demirtas, C.; Tekin, K.; Doguizi, S.; Bayraktar, S.; Yilmazbas, P. Brain-Derived Neurotrophic Factor in Patients With Age-Related Macular Degeneration and Its Correlation With Retinal Layer Thicknesses. *Investig. Ophthalmol. Vis. Sci.* **2018**, *59*, 2833–2840. [CrossRef]

18. Kolomeyer, A.M.; Sugino, I.K.; Zarbin, M.A. Characterization of conditioned media collected from cultured adult versus fetal retinal pigment epithelial cells. *Investig. Ophthalmol. Vis. Sci.* **2011**, *52*, 5973–5986. [CrossRef]
19. Mysona, B.A.; Shanab, A.Y.; Elshaer, S.L.; El-Remessy, A.B. Nerve growth factor in diabetic retinopathy: Beyond neurons. *Expert Rev. Ophthalmol.* **2014**, *9*, 99–107. [CrossRef]
20. Meyer, J.G.; Garcia, T.Y.; Schilling, B.; Gibson, B.W.; Lamba, D.A. Proteome and Secretome Dynamics of Human Retinal Pigment Epithelium in Response to Reactive Oxygen Species. *Sci. Rep.* **2019**, *9*, 15440. [CrossRef]
21. Yabe, T.; Kanemitsu, K.; Sanagi, T.; Schwartz, J.P.; Yamada, H. Pigment epithelium-derived factor induces pro-survival genes through cyclic AMP-responsive element binding protein and nuclear factor kappa B activation in rat cultured cerebellar granule cells: Implication for its neuroprotective effect. *Neuroscience* **2005**, *133*, 691–700. [CrossRef] [PubMed]
22. LaVail, M.M.; Unoki, K.; Yasumura, D.; Matthes, M.T.; Yancopoulos, G.D.; Steinberg, R.H. Multiple growth factors, cytokines, and neurotrophins rescue photoreceptors from the damaging effects of constant light. *Proc. Natl. Acad. Sci. USA* **1992**, *89*, 11249–11253. [CrossRef] [PubMed]
23. Yang, P.B.; Seiler, M.J.; Aramant, R.B.; Yan, F.; Mahoney, M.J.; Kitzes, L.M.; Keirstead, H.S. Trophic factors GDNF and BDNF improve function of retinal sheet transplants. *Exp. Eye Res.* **2010**, *91*, 727–738. [CrossRef] [PubMed]
24. Kolomeyer, A.M.; Zarbin, M.A. Trophic factors in the pathogenesis and therapy for retinal degenerative diseases. *Surv. Ophthalmol.* **2014**, *59*, 134–165. [CrossRef]
25. Kanu, L.N.; Ciolino, J.B. Nerve Growth Factor as an Ocular Therapy: Applications, Challenges, and Future Directions. *Semin. Ophthalmol.* **2021**, *36*, 224–231. [CrossRef]
26. Diniz, B.; Thomas, P.; Thomas, B.; Ribeiro, R.; Hu, Y.; Brant, R.; Ahuja, A.; Zhu, D.; Liu, L.; Koss, M.; et al. Subretinal Implantation of Retinal Pigment Epithelial Cells Derived From Human Embryonic Stem Cells: Improved Survival When Implanted as a Monolayer. *Investig. Ophthalmol. Vis. Sci.* **2013**, *54*, 5087–5096. [CrossRef]
27. Sheedlo, H.J.; Li, L.; Turner, J.E. Effects of RPE-cell factors secreted from permselective fibers on retinal cells in vitro. *Brain Res.* **1992**, *587*, 327–337. [CrossRef]
28. Sheedlo, H.J.; Li, L.; Turner, J.E. Effects of RPE age and culture conditions on support of photoreceptor cell survival in transplanted RCS dystrophic rats. *Exp. Eye Res.* **1993**, *57*, 753–761. [CrossRef]
29. Sheedlo, H.J.; Fan, W.; Li, L.; Turner, J.E. Retinal pigment epithelial cell support of photoreceptor survival in vitro. *Vitr. Cell. Dev. Biol.-Anim. J. Soc. Vitr. Biol.* **1995**, *31*, 330–333. [CrossRef]
30. Sheedlo, H.J.; Turner, J.E. Influence of a retinal pigment epithelial cell factor(s) on rat retinal progenitor cells. *Dev. Brain Res.* **1996**, *93*, 88–99. [CrossRef]
31. Sheedlo, H.J.; Turner, J.E. Effects of retinal pigment epithelial cell-secreted factors on neonatal rat retinal explant progenitor cells. *J. Neurosci. Res.* **1996**, *44*, 519–531. [CrossRef]
32. Sheedlo, H.J.; Wordinger, R.J.; Fan, W.; Turner, J.E. A transformed neonatal rat retinal pigment epithelial cell line: Secreted protein analysis and fibroblast growth factor and receptor expression. *Curr. Eye Res.* **1997**, *16*, 116–126. [CrossRef] [PubMed]
33. Sheedlo, H.J.; Nelson, T.H.; Lin, N.; Rogers, T.A.; Roque, R.S.; Turner, J.E. RPE secreted proteins and antibody influence photoreceptor cell survival and maturation. *Dev. Brain Res.* **1998**, *107*, 57–69. [CrossRef] [PubMed]
34. Landowski, M.; Kelly, U.; Klingeborn, M.; Groelle, M.; Ding, J.D.; Grigsby, D.; Bowes Rickman, C. Human complement factor H Y402H polymorphism causes an age-related macular degeneration phenotype and lipoprotein dysregulation in mice. *Proc. Natl Acad. Sci. USA* **2019**, *116*, 3703–3711. [CrossRef]
35. Tong, Y.; Wang, S. Not All Stressors Are Equal: Mechanism of Stressors on RPE Cell Degeneration. *Front. Cell Dev. Biol.* **2020**, *8*, 591067. [CrossRef]
36. Chen, L.; Perera, N.D.; Karoukis, A.J.; Feathers, K.L.; Ali, R.R.; Thompson, D.A.; Fahim, A.T. Oxidative stress differentially impacts apical and basolateral secretion of angiogenic factors from human iPSC-derived retinal pigment epithelium cells. *Sci Rep.* **2022**, *12*, 12694. [CrossRef]
37. Atienzar-Aroca, S.; Flores-Bellver, M.; Serrano-Heras, G.; Martinez-Gil, N.; Barcia, J.M.; Aparicio, S.; Perez-Cremades, D.; Garcia-Verdugo, J.M.; Diaz-Llopis, M.; Romero, F.J.; et al. Oxidative stress in retinal pigment epithelium cells increases exosome secretion and promotes angiogenesis in endothelial cells. *J. Cell Mol. Med.* **2016**, *20*, 1457–1466. [CrossRef]
38. Sreekumar, P.G.; Kannan, R.; Kitamura, M.; Spee, C.; Barron, E.; Ryan, S.J.; Hinton, D.R. αB Crystallin Is Apically Secreted within Exosomes by Polarized Human Retinal Pigment Epithelium and Provides Neuroprotection to Adjacent Cells. *PLoS ONE* **2010**, *5*, e12578. [CrossRef]
39. Rabin, D.M.; Rabin, R.L.; Blenkinsop, T.A.; Temple, S.; Stern, J.H. Chronic oxidative stress upregulates Drusen-related protein expression in adult human RPE stem cell-derived RPE cells: A novel culture model for dry AMD. *Aging* **2012**, *5*, 51–66. [CrossRef]
40. Pennington, B.O.; Clegg, D.O.; Melkoumian, Z.K.; Hikita, S.T. Defined culture of human embryonic stem cells and xeno-free derivation of retinal pigmented epithelial cells on a novel, synthetic substrate. *Stem Cells Transl. Med.* **2015**, *4*, 165–177. [CrossRef]
41. Hu, Y.; Liu, L.; Lu, B.; Zhu, D.; Ribeiro, R.; Diniz, B.; Thomas, P.B.; Ahuja, A.K.; Hinton, D.R.; Tai, Y.C.; et al. A novel approach for subretinal implantation of ultrathin substrates containing stem cell-derived retinal pigment epithelium monolayer. *Ophthalmic. Res.* **2012**, *48*, 186–191. [CrossRef] [PubMed]
42. Rowland, T.J.; Blaschke, A.J.; Buchholz, D.E.; Hikita, S.T.; Johnson, L.V.; Clegg, D.O. Differentiation of human pluripotent stem cells to retinal pigmented epithelium in defined conditions using purified extracellular matrix proteins. *J. Tissue Eng. Regen. Med.* **2013**, *7*, 642–653. [CrossRef]

43. Lu, B.; Zhu, D.; Hinton, D.; Humayun, M.S.; Tai, Y.C. Mesh-supported submicron parylene-C membranes for culturing retinal pigment epithelial cells. *Biomed Microdevices* **2012**, *14*, 659–667. [CrossRef] [PubMed]
44. Livak, K.J.; Schmittgen, T.D. Analysis of relative gene expression data using real-time quantitative PCR and the 2(-Delta Delta C(T)) Method. *Methods* **2001**, *25*, 402–408. [CrossRef] [PubMed]
45. Thomas, B.B.; Zhu, D.; Lin, T.C.; Kim, Y.C.; Seiler, M.J.; Martinez-Camarillo, J.C.; Lin, B.; Shad, Y.; Hinton, D.R.; Humayun, M.S. A new immunodeficient retinal dystrophic rat model for transplantation studies using human-derived cells. *Graefe's Arch. Clin. Exp. Ophthalmol.* **2018**, *256*, 2113–2125. [CrossRef]
46. Gao, F.; Zhang, Y.; Wang, S.; Liu, Y.; Zheng, L.; Yang, J.; Huang, W.; Ye, Y.; Luo, W.; Xiao, D. Hes1 is involved in the self-renewal and tumourigenicity of stem-like cancer cells in colon cancer. *Sci. Rep.* **2014**, *4*, 3963. [CrossRef]
47. Bras-Pereira, C.; Casares, F.; Janody, F. The retinal determination gene Dachshund restricts cell proliferation by limiting the activity of the Homothorax-Yorkie complex. *Development* **2015**, *142*, 1470–1479. [CrossRef] [PubMed]
48. Tetreault, N.; Champagne, M.P.; Bernier, G. The LIM homeobox transcription factor Lhx2 is required to specify the retina field and synergistically cooperates with Pax6 for Six6 trans-activation. *Dev. Biol.* **2009**, *327*, 541–550. [CrossRef] [PubMed]
49. Yun, S.; Saijoh, Y.; Hirokawa, K.E.; Kopinke, D.; Murtaugh, L.C.; Monuki, E.S.; Levine, E.M. Lhx2 links the intrinsic and extrinsic factors that control optic cup formation. *Development* **2009**, *136*, 3895–3906. [CrossRef]
50. Lee, J.H.; Park, H.S.; Shin, J.M.; Chun, M.H.; Oh, S.J. Nestin expressing progenitor cells during establishment of the neural retina and its vasculature. *Anat. Cell Biol.* **2012**, *45*, 38–46. [CrossRef]
51. Fernández-López, B.; Romaus-Sanjurjo, D.; Senra-Martínez, P.; Anadón, R.; Barreiro-Iglesias, A.; Celina Rodicio, M. Spatiotemporal pattern of doublecortin expression in the retina of the sea lamprey. *Front. Neuroanat.* **2016**, *10*, 5. [CrossRef] [PubMed]
52. Bernardos, R.L.; Barthel, L.K.; Meyers, J.R.; Raymond, P.A. Late-stage neuronal progenitors in the retina are radial Muller glia that function as retinal stem cells. *J. Neurosci.* **2007**, *27*, 7028–7040. [CrossRef] [PubMed]
53. Lamba, D.A.; McUsic, A.; Hirata, R.K.; Wang, P.R.; Russell, D.; Reh, T.A. Generation, purification and transplantation of photoreceptors derived from human induced pluripotent stem cells. *PLoS ONE* **2010**, *5*, e8763. [CrossRef] [PubMed]
54. Klassen, H.J.; Ng, T.F.; Kurimoto, Y.; Kirov, I.; Shatos, M.; Coffey, P.; Young, M.J. Multipotent retinal progenitors express developmental markers, differentiate into retinal neurons, and preserve light-mediated behavior. *Investig. Ophthalmol. Vis. Sci.* **2004**, *45*, 4167–4173. [CrossRef]
55. Schmitt, S.; Aftab, U.; Jiang, C.; Redenti, S.; Klassen, H.; Miljan, E.; Sinden, J.; Young, M. Molecular characterization of human retinal progenitor cells. *Investig. Ophthalmol. Vis. Sci.* **2009**, *50*, 5901–5908. [CrossRef]
56. Salinas, M.; Diaz, R.; Abraham, N.G.; Ruiz de Galarreta, C.M.; Cuadrado, A. Nerve Growth Factor Protects against 6-Hydroxydopamine-induced Oxidative Stress by Increasing Expression of Heme Oxygenase-1 in a Phosphatidylinositol 3-Kinase-dependent Manner*. *J. Biol. Chem.* **2003**, *278*, 13898–13904. [CrossRef]
57. Hacioglu, G.; Senturk, A.; Ince, I.; Alver, A. Assessment of oxidative stress parameters of brain-derived neurotrophic factor heterozygous mice in acute stress model. *Iran. J. Basic Med. Sci.* **2016**, *19*, 388–393.
58. Dong, S.; Zhen, F.; Xu, H.; Li, Q.; Wang, J. Leukemia inhibitory factor protects photoreceptor cone cells against oxidative damage through activating JAK/STAT3 signaling. *Ann. Transl. Med.* **2021**, *9*, 152. [CrossRef]
59. He, X.; Cheng, R.; Benyajati, S.; Ma, J.X. PEDF and its roles in physiological and pathological conditions: Implication in diabetic and hypoxia-induced angiogenic diseases. *Clin. Sci.* **2015**, *128*, 805–823. [CrossRef]
60. Sardar Pasha, S.P.B.; Münch, R.; Schäfer, P.; Oertel, P.; Sykes, A.M.; Zhu, Y.; Karl, M.O. Retinal cell death dependent reactive proliferative gliosis in the mouse retina. *Sci. Rep.* **2017**, *7*, 9517. [CrossRef]
61. Hollingsworth, T.J.; Gross, A.K. Innate and Autoimmunity in the Pathogenesis of Inherited Retinal Dystrophy. *Cells* **2020**, *9*, 630. [CrossRef] [PubMed]
62. Rohrbach, A.S.; Slade, D.J.; Thompson, P.R.; Mowen, K.A. Activation of PAD4 in NET formation. *Front. Immunol.* **2012**, *3*, 360. [CrossRef] [PubMed]
63. Zhu, D.; Deng, X.; Spee, C.; Sonoda, S.; Hsieh, C.-L.L.; Barron, E.; Pera, M.; Hinton, D.R. Polarized Secretion of PEDF from Human Embryonic Stem Cell–Derived RPE Promotes Retinal Progenitor Cell Survival. *Investig. Opthalmology Vis. Sci.* **2011**, *52*, 1573–1575. [CrossRef] [PubMed]
64. Danhong Zhu, J.H. Retinal Pigment Epithelial Cell Conditioned Medium Enhances the Yield of RPE Cells Differentiated from Human Embryonic Stem Cells. *J. Clin. Exp. Pathol.* **2015**, *5*, 216. [CrossRef]
65. Gaur, V.P.; Liu, Y.; Turner, J.E. RPE conditioned medium stimulates photoreceptor cell survival, neurite outgrowth and differentiation in vitro. *Exp. Eye Res.* **1992**, *54*, 645–659. [CrossRef] [PubMed]
66. Dutt, K.; Douglas, P.; Cao, Y. RPE-secreted factors: Influence differentiation in human retinal cell line in dose- and density-dependent manner. *J. Ocul. Biol. Dis. Infor.* **2010**, *3*, 144–160. [CrossRef]
67. Kimura, A.; Namekata, K.; Guo, X.; Harada, C.; Harada, T. Neuroprotection, Growth Factors and BDNF-TrkB Signalling in Retinal Degeneration. *Int. J. Mol. Sci.* **2016**, *17*, 1584. [CrossRef]
68. Raffaele, O.S.; Dompé Farmaceutici, S.p.A. Nerve Growth Factor Eye Drops Treatment in Patients With Retinitis Pigmentosa and Cystoid Macular Edema. 2015. Available online: https://clinicaltrials.gov/ct2/show/NCT02609165 (accessed on 6 March 2023).
69. GenVec. Study of AdGVPEDF.11D in Neovascular Age-related Macular Degeneration (AMD). Available online: https://clinicaltrials.gov/ct2/show/NCT00109499 (accessed on 6 March 2023).

70. Dompé Farmaceutici, S.p.A.; Cromsource. Study to Evaluate Safety and Efficacy of rhNGF Eye Drops Solution Versus Vehicle in Patients With Glaucoma. 2016. Available online: https://clinicaltrials.gov/ct2/show/NCT02855450 (accessed on 6 March 2023).
71. Ye, F.; Kaneko, H.; Hayashi, Y.; Takayama, K.; Hwang, S.J.; Nishizawa, Y.; Kimoto, R.; Nagasaka, Y.; Tsunekawa, T.; Matsuura, T.; et al. Malondialdehyde induces autophagy dysfunction and VEGF secretion in the retinal pigment epithelium in age-related macular degeneration. *Free. Radic. Biol. Med.* **2016**, *94*, 121–134. [CrossRef]
72. Papac-Milicevic, N.; Busch, C.J.; Binder, C.J. Malondialdehyde Epitopes as Targets of Immunity and the Implications for Atherosclerosis. *Adv. Immunol.* **2016**, *131*, 1–59. [CrossRef]
73. Bonilha, V.L.; Shadrach, K.G.; Rayborn, M.E.; Li, Y.; Pauer, G.J.; Hagstrom, S.A.; Bhattacharya, S.K.; Hollyfield, J.G. Retinal deimination and PAD2 levels in retinas from donors with age-related macular degeneration (AMD). *Exp. Eye Res.* **2013**, *111*, 71–78. [CrossRef]
74. Hinton, D.R.; Zhou, J.; He, S.; Zhang, N.; Spee, C.; Zhou, P.; Ryan, S.J.; Kannan, R. Neutrophils compromise retinal pigment epithelial barrier integrity. *J. Biomed. Biotechnol.* **2010**, *2010*, 289360. [CrossRef]
75. Binet, F.; Cagnone, G.; Crespo-Garcia, S.; Hata, M.; Neault, M.; Dejda, A.; Wilson, A.M.; Buscarlet, M.; Mawambo, G.T.; Howard, J.P.; et al. Neutrophil extracellular traps target senescent vasculature for tissue remodeling in retinopathy. *Science* **2020**, *369*, eaay5356. [CrossRef] [PubMed]
76. Dong, T.; Dave, P.; Yoo, E.; Ebright, B.; Ahluwalia, K.; Zhou, E.; Asante, I.; Salimova, M.; Pei, H.; Lin, T.; et al. NAP1051, a Lipoxin A4 Biomimetic Analogue, Demonstrates Antitumor Activity Against the Tumor Microenvironment. *Mol. Cancer Ther.* **2021**, *20*, 2384–2397. [CrossRef]
77. Palko, S.I.; Saba, N.J.; Mullane, E.; Nicholas, B.D.; Nagasaka, Y.; Ambati, J.; Gelfand, B.D.; Ishigami, A.; Bargagna-Mohan, P.; Mohan, R. Compartmentalized citrullination in Muller glial endfeet during retinal degeneration. *Proc. Natl. Acad. Sci. USA* **2022**, *119*, e2121875119. [CrossRef]
78. D'Cruz, P.M. Mutation of the receptor tyrosine kinase gene Mertk in the retinal dystrophic RCS rat. *Hum. Mol. Genet.* **2000**, *9*, 645–651. [CrossRef] [PubMed]
79. Leenaars, C.H.C.; Kouwenaar, C.; Stafleu, F.R.; Bleich, A.; Ritskes-Hoitinga, M.; De Vries, R.B.M.; Meijboom, F.L.B. Animal to human translation: A systematic scoping review of reported concordance rates. *J. Transl. Med.* **2019**, *17*, 223. [CrossRef]

cells

MDPI

Article

Retinal Vascular Physiology Biomarkers in a 5XFAD Mouse Model of Alzheimer's Disease

Nathanael Matei [1], Sophie Leahy [1], Norman P. Blair [2], James Burford [1], Mansour Rahimi [1] and Mahnaz Shahidi [1,*]

1 Department of Ophthalmology, University of Southern California, Los Angeles, CA 90033, USA
2 Department of Ophthalmology and Visual Sciences, University of Illinois at Chicago, Chicago, IL 60612, USA
* Correspondence: mshahidi@usc.edu

Abstract: Background: Alzheimer's disease (AD) is a neurodegenerative disorder that affects the brain and retina and lacks reliable biomarkers for early diagnosis. As amyloid beta (Aβ) manifestations emerge prior to clinical symptoms and plaques of amyloid may cause vascular damage, identification of retinal vascular biomarkers may improve knowledge of AD pathophysiology and potentially serve as therapeutic targets. The purpose of the current study was to test the hypothesis that retinal hemodynamic and oxygen metrics are altered in 5XFAD mice. Methods: Thirty-two male mice were evaluated at 3 months of age: sixteen 5XFAD transgenic and sixteen wild-type mice. Spectral-domain optical coherence tomography, vascular oxygen tension, and blood flow imaging were performed in one eye of each mouse. After imaging, the imaged and fellow retinal tissues were submitted for histological sectioning and amyloid protein analysis, respectively. Protein analysis was also performed on the brain tissues. Results: Retinal physiological changes in venous diameter and blood velocity, arterial and venous oxygen contents, coupled with anatomical alterations in the thickness of retinal cell layers were detected in 5XFAD mice. Moreover, an increase in Aβ42 levels in both the retina and brain tissues was observed in 5XFAD mice. Significant changes in retinal oxygen delivery, metabolism, or extraction fraction were not detected. Based on compiled data from both groups, arterial oxygen content was inversely related to venous blood velocity and nerve fiber/ganglion cell layer thickness. Conclusions: Concurrent alterations in retinal hemodynamic and oxygen metrics, thickness, and tissue Aβ42 protein levels in 5XFAD mice at 3 months of age corresponded to previously reported findings in human AD. Overall, these results suggest that this mouse model can be utilized for studying pathophysiology of AD and evaluating potential therapies

Keywords: Alzheimer's disease; retinal vascular physiology biomarkers; 5XFAD; amyloid β

Citation: Matei, N.; Leahy, S.; Blair, N.P.; Burford, J.; Rahimi, M.; Shahidi, M. Retinal Vascular Physiology Biomarkers in a 5XFAD Mouse Model of Alzheimer's Disease. *Cells* **2022**, *11*, 2413. https://doi.org/10.3390/cells11152413

Academic Editor: Paola Bagnoli

Received: 14 July 2022
Accepted: 30 July 2022
Published: 4 August 2022

1. Background

As life expectancy continues to increase, the prevalence of dementia has grown to ~47 million people worldwide in 2015 [1]. Alzheimer's disease (AD), the most common type of dementia, is a neurodegenerative disorder that affects the brain and retina, and it lacks effective treatments as well as reliable biomarkers for early diagnosis [2]. As an extension of the diencephalon, the retina displays many pathological characteristics of the Alzheimer brain, e.g., accumulation of amyloid beta (Aβ) oligomers, increased p-Tau, and loss of function [3,4]. To date, diagnosis is based on clinical and neuropsychological examinations in combination with neuroimaging studies, but the abnormalities revealed by these methods occur late in the disease and limit interventions. Therefore, retinal biomarkers may provide therapeutic targets to intervene early and attenuate or even prevent the pathophysiologic process that progresses to symptomatic dementia.

The five-familial AD (5XFAD) transgenic mouse model closely mirrors human AD pathophysiology as early as 2 months of age with clear neuronal degeneration in the brain [5]. Aβ42 levels are specifically increased in 5XFAD mice. In the retina of 5XFAD mice, studies have reported a marked increase in Aβ42/Aβ40 [6,7] and pattern electroretinography impairments [8] at 3 months of age, followed by cognitive impairments as early as

4 months of age [5]. Furthermore, histological evaluation of the retina in a 5XFAD mouse model at 12 months has reported thickening of the Bruch's membrane and thinning of the retinal pigment epithelium [9]. Using optical coherence tomography (OCT), Lim et al. reported retinal nerve fiber layer (RNFL) thinning at 6 months of age in the 5XFAD mouse model [10]. However, cellular apoptosis was detected in the hippocampus of 5XFAD mice as early as 4 months [11].

As future treatments for AD become available, early-stage diagnosis and timely initiation of the treatments may attenuate or even prevent unfavorable outcomes. Given that Aβ and tau manifestations emerge ~20 years before clinical symptoms [12] and that plaques of amyloid may cause vascular damage, retinal vascular physiological biomarkers may improve AD diagnosis prior to significant neurodegeneration. Increased oxygen saturation in retinal arterioles and venules was shown in AD patients with mild cognitive impairment, suggesting vascular abnormalities to be present with degeneration of retinal cell layers [13,14]; this suggests the potential of oxygen saturation as an ocular biomarker of AD. Moreover, reduced retinal venous blood velocity [15–17] and blood flow [15,18] due to the narrowing of the lumen as a result of Aβ deposition in the vessel walls have been reported in patients with early AD. Studies have also documented retinal anatomical changes, namely retinal ganglion cell loss in foveal and parafoveal retinas from postmortem AD patients [19,20].

Optimally, a comprehensive assessment of retinal physiological biomarkers in AD would include hemodynamic metrics (vessel diameter, blood velocity, and blood flow) and oxygen metrics [vascular oxygen contents, oxygen metabolism (MO_2), delivery (DO_2), and extraction fraction (OEF)]. As abnormalities in blood flow and oxygen metrics have been shown in human AD, retinal physiological biomarkers in animal models of AD have the potential to improve knowledge of AD pathophysiology. Such knowledge may enable predicting irreversible neurodegeneration and identify a window of opportunity for intervention. The purpose of the current study was to test the hypothesis that retinal hemodynamic and oxygen metrics are altered in 5XFAD mice.

2. Materials and Methods

2.1. Animals

All experiments were approved by the University of Southern California Institutional Animal Care and Use Committee (IACUC: 20785). The experiments complied with the guidelines of the statement from Use of Animals in Ophthalmic and Vision research by the Association for Research in Vision and Ophthalmology, and they were reported according to the Animal Research: Reporting of In Vivo Experiments guidelines. Thirty-two mice were evaluated. Sixteen male wild-type mice (weight = 26.4 ± 1.7 g; age = 3 months) were obtained from Jackson Laboratory (C57BL/6J; N = 10) or Charles River Laboratories (C57BL/6; N = 6). Sixteen male 5XFAD transgenic mice (the MMRRC stock 24848, Stock Number: 034848-JAX, and Citation ID: RRID:MMRRC_034848-JAX) (weight = 26.5 ± 2.6 g; age = 3 months) were obtained from the Jackson Laboratory (Bar Harbor, ME, USA). According to the MMRRC description, the homozygote phenotype is likely to be similar to the hemizygote phenotype; mice are viable and fertile, and generate almost entirely Aβ42. This strain does not carry the retinal degeneration allele Pde6b^{rd1}. The 5XFAD strain is hemizygous for the amyloid precursor protein and presenilin 1 transgenes on a congenic C57BL/6 background. Given that studies on 5XFAD mice have reported a marked increase in Aβ42 [6,7] at 3 months of age followed by some cognitive impairments as early as 4 months of age [5], we chose to evaluate hemodynamic and oxygen metrics in 5XFAD mice at 3 months. We also chose this model to evaluate the impact of Aβ42 on the retina in order to reduce the number of confounding variables seen in other genetic models of AD, such as tauopathy.

One eye per mouse was selected at random for imaging. Prior to imaging, anesthesia was induced via intraperitoneal injection of ketamine (100 mg/kg) and xylazine (5 mg/kg), and the pupil was dilated. Prior to imaging, 3 mice in each group died. Spectral-domain

OCT, vascular oxygen tension (PO_2), and blood flow imaging were performed in 13 mice in each group, as described below. After euthanasia, eyes from all mice were enucleated. The imaged and fellow eyes were used for histological sectioning and protein analysis, respectively. Additionally, the brain was removed for protein analysis. The experimental outline of this study is presented in Figure 1.

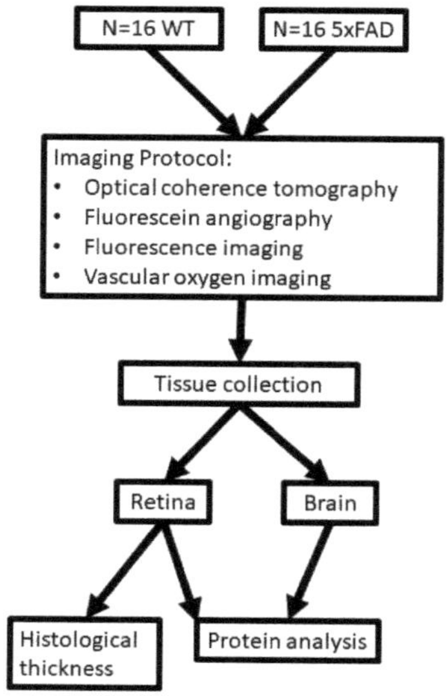

Figure 1. Outline for study design. Diagram showing imaging protocol, tissue collection, and protein analysis for wild-type (WT) and five-familial Alzheimer's disease (5XFAD) groups.

2.2. Spectral-Domain Optical Coherence Tomography

As previously described [21,22], spectral-domain OCT imaging (Spectralis, Heidelberg Engineering, Heidelberg, Germany) was performed in two regions adjacent to the optic nerve head: nasal and temporal. Images were processed using the Heidelberg Eye Explorer software (Heidelberg Eye Explorer 1.9.10.0; Heidelberg Engineering) to measure total retinal thickness (TRT) as an average of measurements obtained in nasal and temporal regions. Due to poor image quality, TRT data were available for 12 WT and 11 5XFAD mice eyes.

2.3. Vascular PO_2 and Blood Flow Imaging

For retinal vascular PO_2 imaging, we administered fluorescent microspheres and Pd-porphine (an oxygen-sensitive molecular probe) through a femoral catheter. As previously described [21,22], a phosphorescence lifetime imaging system was used to determine PO_2. Briefly, a frequency-domain approach was used in retinal vessels to evaluate the phosphorescence lifetime of Pd-porphine in the retinal vessels. Using the Stern-Volmer equation, PO_2 data were derived from phosphorescence lifetime measurements. To determine O_2 content, we summed the oxygen bound to hemoglobin and dissolved in blood. The mouse hemoglobin dissociation curve was used to calculate oxygen saturation from PO_2 values. Arterial (O_{2A}) and venous (O_{2V}) oxygen contents were derived by averaging values from all vessels, and the arteriovenous oxygen content difference (O_{2AV}) was calculated as O_{2A}-O_{2V}. As previously described [21,22], a series of images rapidly acquired at 104 Hz was used to measure blood velocity (V) in each vein based on the displacement of fluorescent microspheres over time. Blood vessel diameter (D) was determined using fluorescein angiography via injection of 10% fluorescein sodium (5 mg/kg, AK-FLUOR; Akorn, Decatur, IL, USA). In each eye, measurements of D in arteries and veins were averaged to provide mean arterial diameter (D_A), venous diameter (D_V), and venous velocity

(V_V), respectively. Blood flow in each vein was calculated at the product of blood V, D^2, and $\pi/4$, and measurements were summed to calculate the total retinal blood flow (TRBF). The following equations were used to calculate DO_2, MO_2, and OEF: $DO_2 = TRBF \times O_{2A}$, $MO_2 = TRBF \times O_{2AV}$, $OEF = MO_2/DO_2$. Data were obtained from 13 WT and 13 5XFAD mice eyes.

2.4. Cell Layer Thickness

The eyes were sectioned to include the pupil, optic nerve, and retina, both nasally and temporally. This ensured that the sections were oriented perpendicular to the retinal surface so that measurements were not obliquely made through the retina. Using established methods, retinal sections were prepared and stained with hematoxylin and eosin (H&E). Retinal layer thickness measurements were made using a standard procedure [23]. Two digital images, nasal and temporal to the optic nerve head, were acquired and analyzed using ImageJ software (ImageJ 1.53; https://imagej.nih.gov/ij/, accessed on 13 July 2022). Layer boundaries were identified and used to calculate the thickness of the inner retina (IRL), inner plexiform layer (IPL), inner nuclear layer (INL), outer plexiform layer (OPL), outer nuclear layer (ONL), photoreceptor layer (PRL), and total retinal thickness (TRT). The thickness of the combined nerve fiber layer (NFL) and retinal ganglion cell layer (RGCL) was calculated as: NFL/RGCL = IRL − (IPL + INL). The NFL/RGCL cannot be separated in mice as in humans. Thickness data for each layer were averaged over the two regions. Due to tissue processing issues, retinal layer thickness data were available for 10 WT and 8 5XFAD mice eyes.

2.5. Protein Analysis

Enzyme-linked immunosorbent assays (ELISA) were performed on brain and retinal tissue extracts to quantitate the concentrations of Aβ42 and Aβ40 proteins. Briefly, mice were perfused with cold PBS, with a pH of 7.4, by intracardiac injection, followed by dissection of the retina and whole brain, which were snap-frozen in liquid nitrogen and stored at −80 °C. The retina was suspended in 70 µL of RIPA lysis buffer (Santa Cruz Biotechnology, Dallas, TX, USA), sonicated, and centrifuged at 14,000× *g* at 4 °C for 30 min. The supernatant was used for ELISA testing. Similarly, the whole brains from these mice were homogenized in RIPA lysis buffer using a tissue homogenizer and then centrifuged, and the supernatant was used as whole-cell protein extract. The protein concentration was determined using a BCA protein assay kit (23225; ThermoFisher Scientific, Waltham, MA, USA). Aβ levels in the extracts were quantified using commercial ELISA kits: Mouse Aβ40 Elisa kit (KMB3481; Invitrogen, Waltham, MA, USA) and Human Aβ42 ELISA kit (KHB3441; Invitrogen). Standard curves for Aβ40 and Aβ42 were determined in all experiments using the provided standards. Duplicate samples were used for Aβ40 and Aβ42 experiments. Due to the small volume of retinal tissue in a single eye, protein level data were available for 4 WT and 9 5XFAD mice brains and retinas.

2.6. Immunoreactivity for Amyloid Angiopathy

Immunofluorescence staining was performed on transverse retinal sections, as previously described [23]. Briefly, the slides were deparaffinized, boiled for 20 min in a citrate antigen retrieval buffer (10 mM sodium citrate, 0.05% Tween 20, pH 6.0), and blocked with 10% normal donkey serum. According to previous studies investigating retinal vascular amyloidosis, CD31 was used as a marker for vascular endothelium [24,25]. The retinal tissue was incubated with the primary antibodies: rat anti-CD31 (1:100; Abcam ab56299, RRID: AB_940884) and rabbit anti-Aβ42 (1:100; Cell Signaling Technology Cat# 14974, RRID: AB_2798671), and incubated with a corresponding fluorescence-conjugated secondary antibody and 4′,6-diamidino-2-phenylindole dihydrochloride (DAPI) (nuclear marker, color blue) (Jackson ImmunoResearch, West Grove, PA, USA). No staining was observed in the imaging of the negative control: staining performed without primary antibody. The sections were visualized with a confocal microscope (LSM880, ZEISS Microscopy Jena, Germany) using 63× magnification.

2.7. Statistical Analysis

No outliers were identified, and normality of the data distributions was confirmed. Unpaired, two-tailed Student's *t*-tests were used for statistical analysis of two-group comparisons. Pearson's correlation analysis was used to relate O_{2A} to V_V, and O_{2A} (and O_{2V}) to NFL/RGCL thickness. Statistical significance was accepted at $p < 0.05$. We used SPSS Statistics, version 24 (IBM Armonk, New York, NY, USA) for all data analyses. Statistical power analysis was performed only on measured metrics. With a sample size of 13, differences in O_{2A}, O_{2V}, D_V, and V_V could be detected with 72%, 55%, 53%, and 98% power at the alpha level of 0.05, respectively. The study had 89% and 68% power to detect a correlation coefficient of 0.57 and 0.53 at the alpha level of 0.05 with a sample size of 26 (V_V and O_{2A}) and 19 (O_{2A} and NFL/RGCL thickness), respectively. Power calculations were performed using G*power 3.1.9.4.

3. Results

3.1. Hemodynamic Metrics

The mean and standard deviation of D_A, D_V, V_V, and TRBF for each group are shown in Figure 2. In the WT group, D_A, D_V, V_V, and TRBF were 27 ± 2 μm, 30 ± 3 μm 10.5 ± 3.1 mm/s, and 1.73 ± 0.52 /μLmin, respectively. D_V (33 ± 3 μm) was significantly increased, whereas V_V (6.6 ± 1.4 mm/s) was reduced in the 5XFAD group ($p \leq 0.04$). No significant differences were detected in D_A (27 ± 3 μm) and TRBF (1.63 ± 0.50 mm/s) in the 5XFAD group compared with the WT group ($p \geq 0.7$).

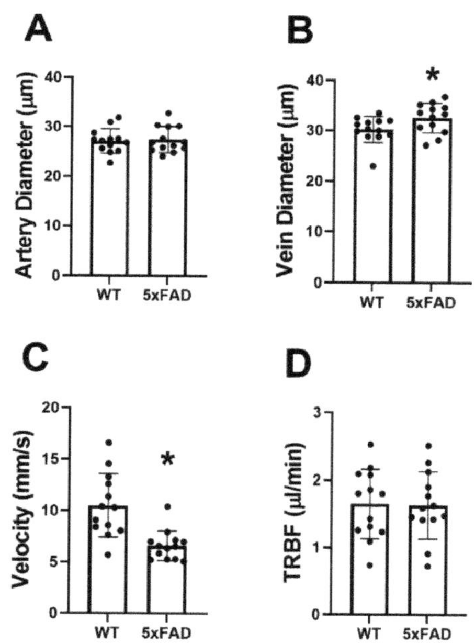

Figure 2. Retinal vessel diameter, venous blood velocity, and total retinal blood flow (TRBF) in wild-type (WT) and five-familial Alzheimer's disease (5XFAD) mice. (**A**) Retinal arterial diameter, (**B**) venous diameter, (**C**) venous velocity, and (**D**) TRBF in WT and 5XFAD groups. The data are presented as mean ± SD. * $p < 0.05$.

3.2. Oxygen Metrics

The mean and standard deviation of O_{2A}, O_{2V}, and O_{2AV} for each group are presented in Figure 3. O_{2A}, O_{2V}, and O_{2AV} of the WT group were 6.5 ± 1.4, 2.8 ± 1.9, and 3.7 ± 1.3 mLO$_2$/dL, respectively. O_{2A} (7.8 ± 1.2 mLO$_2$/dL) and O_{2V} (4.3 ± 1.6 mLO$_2$/dL) were significantly increased in the 5XFAD group ($p \leq 0.04$). No significant difference was detected in O_{2AV} between the two groups ($p = 0.83$).

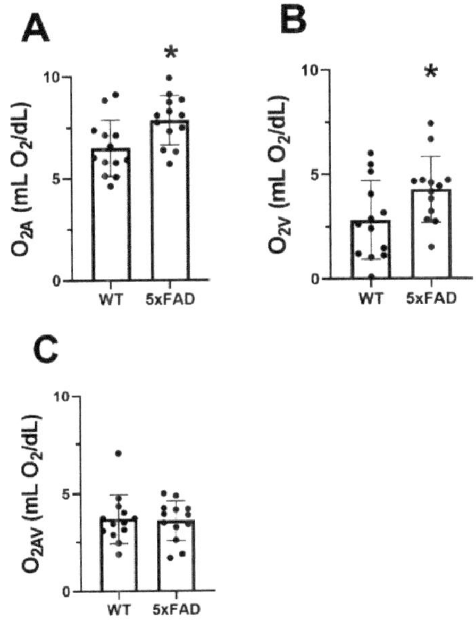

Figure 3. Vascular oxygen contents in wild-type (WT) and five-familial Alzheimer's disease (5XFAD) mice. (**A**) Retinal arterial oxygen content (O_{2A}), (**B**) venous oxygen content (O_{2V}), and (**C**) arteriovenous oxygen content difference (O_{2AV}) in WT and 5XFAD groups. The data are presented as mean ± SD. * $p < 0.05$.

In the WT group, MO_2, DO_2, and OEF were 57 ± 17 nLO$_2$/min, 107 ± 39 nLO$_2$/min, and 0.59 ± 0.21, respectively. Compared with the WT group, no differences were detected in MO_2 (60 ± 20 nLO$_2$/min), DO_2 (133 ± 31 nLO$_2$/min), and OEF (0.46 ± 0.15) in the 5XFAD group ($p \geq 0.1$).

3.3. Retinal Layer Thickness

Region of imaging and a representative OCT image acquired in a WT mouse are displayed in Figure 4A. The mean and standard deviation of TRT measurements, determined by OCT imaging for each group, are displayed in Figure 4B. In the WT group, TRT was 242 ± 11 μm ($N = 12$). Compared with the WT group, a marginally significant difference was detected in TRT (235 ± 6 μm) ($N = 11$) in the 5XFAD group ($p = 0.07$).

Representative H&E sections used to compare layer thickness are displayed in Figure 4C. The mean and standard deviation of retinal layer thickness measurements by histology evaluation, stratified by group, is shown in Figure 4D. In the WT group ($N = 10$), the thickness of each retinal layer was as follows: NFL/RGCL: 16.2 ± 4.2, IPL: 36.7 ± 4.0, INL: 31.0 ± 4.4, OPL: 14.6 ± 2.0, ONL: 44.3 ± 4.6, and PRL: 29.1 ± 3.9 μm. Compared with the WT group, the thickness of OPL (17.2 ± 1.3) and ONL (47.6 ± 2.7 μm) were increased, whereas the NFL/RGCL thickness (15.1 ± 2.8 μm) was decreased in the 5XFAD group ($N = 9$) ($p \leq 0.01$). Compared with the WT group, no significant difference was detected in thickness of INL, IPL, or PR ($p > 0.06$).

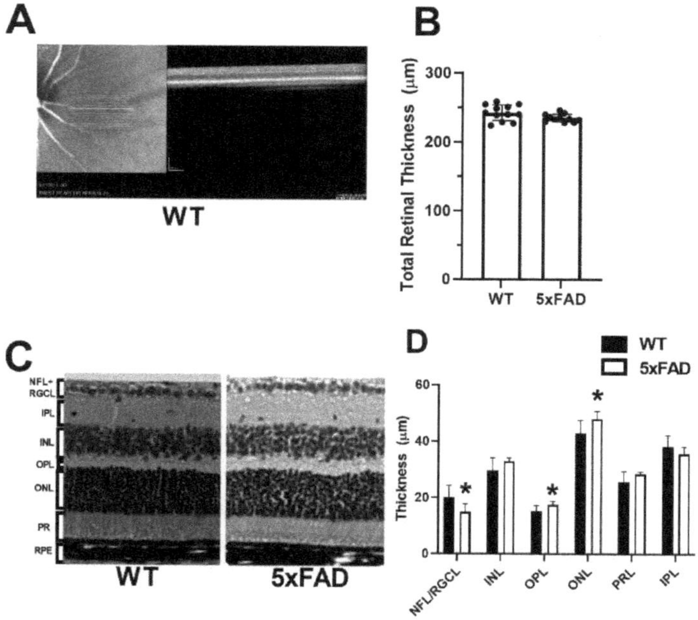

Figure 4. Measurements of total and retinal layer thickness in wild-type (WT) and five-familial Alzheimer's disease (5XFAD) mice. (**A**) Region of imaging and representative optical coherence tomography image of the retina. (**B**) Total retinal thickness measured by spectral domain optical coherence tomography. (**C**) Representative retinal layer thicknesses measured from histological sections. (**D**) Thickness measurements of nerve fiber layer/retinal ganglion cell (NFL/RGCL), inner plexiform layer (IPL), inner nuclear layer (INL), outer plexiform layer (OPL), outer nuclear layer (ONL), and photoreceptor layer (PRL). The data are presented as mean ± SD. * $p < 0.05$.

3.4. *Aβ42/40 Protein Evaluation in the Retina and Brain*

The mean and standard deviation of Aβ42 and AB40 protein levels in the retina are shown in Figure 5A,C, respectively. In the WT group ($N = 4$), Aβ42 and Aβ40 were 2.4 ± 0.8 and 2.5 ± 0.2 pg/mg, respectively. Compared with the WT group, Aβ42 (5.4 ± 2.6 pg/mg) and Aβ40 (4.9 ± 1.5 pg/mg) were increased in the 5XFAD group ($N = 9$) ($p \leq 0.04$).

The mean and standard deviation of Aβ42 and Aβ40 protein levels in the brain are shown in Figure 5B,D, respectively. In the WT group ($N = 4$), Aβ42 and Aβ40 were 1.4 ± 0.5 and 9.3 ± 3.0 pg/mg, respectively. Compared with the WT group, Aβ42 (6.1 ± 1.9 pg/mg) was increased ($p = 0.001$) in the 5XFAD group ($N = 9$), whereas no significant difference was observed in levels of Aβ40 (8.2 ± 1.2 pg/mg) ($p = 0.4$).

3.5. *Retinal Amyloid Angiopathy*

Immunofluorescence staining of retinal transverse sections from 5XFAD and WT mice was used to qualitatively evaluate amyloid angiopathy. Representative examples of retinal vessels (depicted in green), colocalized with Aβ protein (depicted in red) are shown in Figure 6. Aβ42 protein expression was present in the 5XFAD retina, whereas no staining of the Aβ42 protein was observed in the WT retina. The displayed size difference in vessels between 5XFAD and WT does not represent group differences.

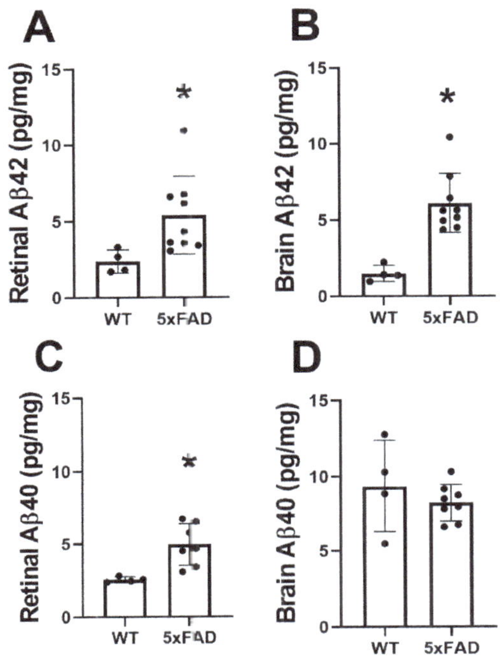

Figure 5. Amyloid-beta (Aβ) protein levels in retinas and brains of wild-type (WT) and five-familial Alzheimer's disease (5XFAD) mice. Aβ42 levels in the (**A**) retina and (**B**) brain. Aβ40 levels in the (**C**) retina and (**D**) brain. The data are presented as mean ± SD. * $p < 0.05$.

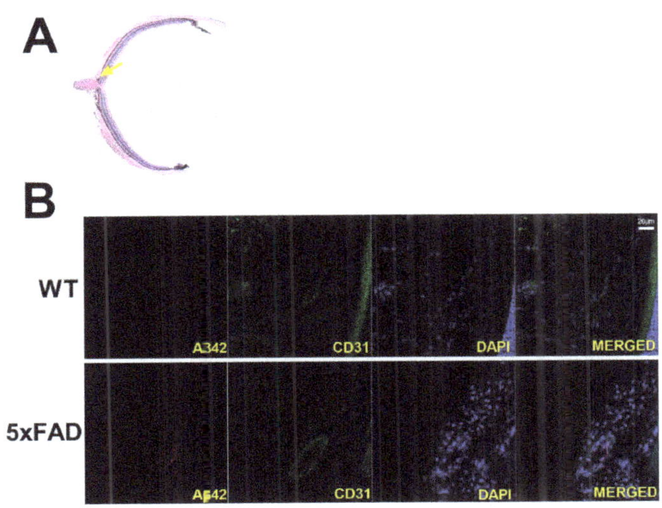

Figure 6. (**A**) Representative H&E-stained transverse retinal section and yellow arrow highlighting area of interest. Amyloid angiopathy is presented in transverse retinal sections of wild-type (WT) and five-familial Alzheimer's disease (5XFAD) mice (**B**). Amyloid-beta 42 (Aβ42) is presented in red, endothelial cells in green, and nuclei in blue. Scale bar, 20 µm.

3.6. Associations

The relation between O_{2A} and V_V is displayed in Figure 7A. Based on the compiled data from both groups, an inverse linear association between O_{2A} and V_V was observed ($r = -0.57$; $N = 26$; $p = 0.002$). The relationships of NFL/RGCL thickness to O_{2A} and O_{2V} are shown in Figure 7B,C, respectively. NFL/RGCL thickness was inversely related to both O_{2A} and O_{2V} ($r = -0.53$; $N = 19$; $p = 0.02$).

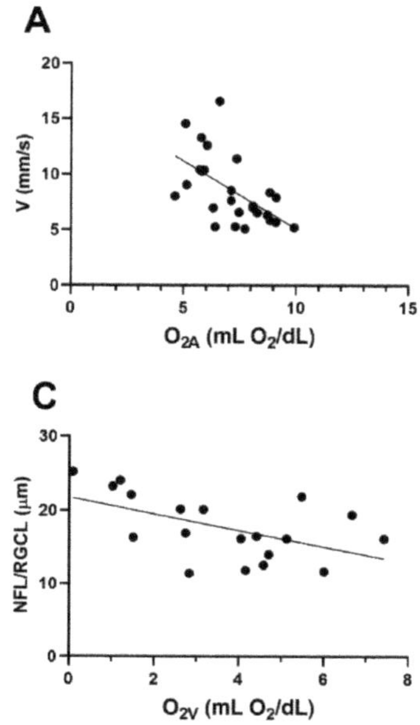

Figure 7. Based on compiled data from wild-type (WT) and five-familial Alzheimer's disease (5XFAD) mice: (**A**) relationship between venous velocity (V_V) and arterial oxygen content (O_{2A}). (**B**) and (**C**) The relationships of nerve fiber layer/retinal ganglion cell (NFL/RGCL) thickness to O_{2A} and venous oxygen content (O_{2V}), respectively.

4. Discussion

Retinal vascular physiological changes in D_V, V_V, O_{2A}, and O_{2V} coupled with tissue anatomical alterations in OPL, ONL, PRL, and NFL/RGCL thickness were detected in 5XFAD mice at 3 months of age, concurrent with an increase in Aβ42 levels in both the retina and brain. With several novel findings, we confirmed our hypothesis that alterations in retinal vessel caliber, velocity, and oxygen content were present in 5XFAD mice; however, we did not detect changes in TRBF, MO_2, DO_2, or OEF.

The first major finding was that D_V increased, whereas V_V decreased, in the 5XFAD mice at 3 months of age. In AD patients, cerebral amyloid angiopathy is characterized by the deposition of Aβ in the walls of cerebral and leptomeningeal vessels and results in the abnormal distribution of blood flow due to the thickening of vessel walls and consequent narrowing of the lumen [26]. Studies have shown that collagen and Aβ accumulate in capillaries [27,28] and veins [29,30] in postmortem AD brains [31]. In an evaluation of postmortem retinas, correlations were found between retinal vascular (major retinal vessels and capillaries) abnormalities (Aβ plaques and pericyte loss) and cerebral Aβ plaques, cerebral amyloid angiopathy, and clinical status [25]. Consistent with our findings, reduced retinal venous blood velocity was observed in AD patients with cerebral and retinal amyloid angiopathy [15–17]. Moreover, reported pericyte loss in capillaries of postmortem AD retinas [25] and capillary degeneration in a mouse model of AD [24] likely affect blood flow regulation. Therefore, we suggest that increased Aβ42 deposition and possibly collagen-thickened vein vessel walls resulted in the subsequent narrowing of the lumina and decreased pliability of vessels, reducing their capacity to normally dilate or contract. These pathophysiological changes may have contributed to an autoregulatory increase in D_V and decrease in V_V to maintain TRBF.

The second major finding was that O_{2A} and O_{2V} were increased in the 5XFAD mice at 3 months of age. Similarly, others have reported an increase in retinal oxygen saturation in arterioles and venules in AD patients [13,14]. Given that AD results in amyloid angiopathy, the thickening of vessel walls may lead to a decrease in diffusion of glucose and oxygen from the circulatory system, including the retina. For example, oxygen saturation in the retina was increased in patients with diabetic retinopathy, attributed in part to the thickening of capillary vessel walls, which increases the distance oxygen must travel to reach its destination [32]. The observed thinning of NFL/RGCL in the current study also supports reduced oxygen utilization in this layer. These factors would tend to increase O_{2V} but not O_{2A}, and O_{2AV} and MO_2 were not found to differ in the two groups of mice. Additional investigation is needed to fully understand these vascular oxygen content findings.

The third major finding was that no statistically significant changes in MO_2, DO_2, and OEF were observed in 5XFAD mice at 3 months of age. Given the evidence for retinal amyloid angiopathy and degeneration of RGCL in AD, it is expected that these will become abnormal at some point. For instance, using fluorodeoxyglucose positron emission tomography (FDG-PET), a method that measures glucose metabolism, previous studies have found consistent patterns of brain metabolic dysfunction that occur prior to cell death in AD patients [33,34]. In the 5XFAD mouse model, FDG-PET has detected reduced cerebral metabolic rates by 13 months [35]. Given that the NFL/RGCL make up a small part (approximately 19%) of the inner retina, MO_2 was likely reduced to some extent but did not reach significance. Additionally, because we found no difference in TRBF and an increase in O_{2A}, DO_2 should have mathematically been increased in the 5XFAD group, but due to the variability in our data, we did not detect a difference in DO_2. In sum, 3 months may have been too early to detect impairments in MO_2, DO_2, and OEF.

The fourth major finding was reduced NFL/RGCL thickness and increased OPL, ONL, and PRL thickness, coupled with no significant change in TRT in the 5XFAD mice. Past research has reported thinning of RGCL in AD patients, whereas thickening was observed in the ONL and PRL, using ultrahigh-resolution OCT [36]. Others have reported RGCL loss in patients [37,38], as well as in other AD animal models, such as APP-PS1ΔE9 [39] and 3xTg-AD mouse models [40]. Although our understanding of the molecular mechanisms for inner retinal thinning remains incomplete, studies have shown that the activation of the JNK pathway, a stress modulator that is activated by Aβ, plays a part in the degeneration of neurons in the brain and retina [41]. Using an inhibitor of JNK, D-JNKI1—in a phase III clinical trial for treating intraocular inflammation resulting from surgery or trauma [42]—Buccarello et al. reported that monthly intraperitoneal injections of D-JNKI1 preserved the RGCL in a mouse model of AD [41]. Given that patients also tolerate this drug, it may have a significant translational impact in treating AD. If oxygen diffusion from retinal circulation is decreased, outer and inner retinal tissue may be hypoxic to different degrees, so the ONL/PRL thickening and NFL/RGCL thinning could in part be due to cytotoxic edema and cell death, respectively. We propose three possible contributors to this: (1) disproportionate deposition of amyloid in the innermost retina and in and along vessels, (2) suboptimal regulation of TRBF to fully compensate for impeded oxygen diffusion that appears to result in a slight reduction in MO_2 (this is based on thinning in the NFL/RGCL), and (3) altered distribution of blood flow such that there are greater reductions in flow in the superficial than the deep capillary plexus [43]. These factors affect the NFL/RGCL more severely, which accounts for cell loss; however, there is still swelling found in the OPL and ONL. Further research is needed to understand the pathophysiology of NFL/RGCL thinning and OPL/ONL/PRL thickening. In the current study, TRT was not significantly different between AD and WT mice. However, TRT measured by histology was lower than measurements obtained in vivo by OCT imaging. These differences in TRT measurements may be attributed to histological processing causing shrinkage in tissue or software limitations that are not adjusting for differences in optical properties between mouse and human models.

The fifth major finding was that the expression of Aβ42 was increased in both the brain and retina at 3 months of age in the 5XFAD mice, consistent with previous studies of both the retina [6,7] and brain [6,44]. The dominant pathogenesis theory of AD suggests that the production and clearance of Aβ42 and related Aβ peptides is most likely the initiator of AD disease and its downstream physiological effects [45]. We are the first to report increased Aβ42 levels, coupled with changes in V_V, O_{2A}, O_{2V}, and retinal cell layers in the 5XFAD mice. To date, there is a dearth of non-invasive biomarkers for AD. Ocular imaging that detects physiological and anatomical abnormalities associated with AD onset may become an essential modality for early diagnosis. Typically, Aβ42 is deposited in senile plaques, whereas Aβ40 normally deposits in the vascular wall of cerebral amyloid angiopathy (CAA) [46]. Previous literature has reported that CAA, in which plaques are increased in cerebral blood vessels and tissue, was observed in more than 80% of AD patients [47]. Moreover, Aβ deposits in the brain and retina of AD patients were associated with accumulated deposits within retinal vasculature [25,48]. However, further research is needed to understand the link between Aβ42 and its impact on V_V, O_{2A}, and retinal thickness.

A sixth major finding of the current study was the association of retinal vascular oxygen content with blood velocity and retinal layer thickness. There was a significant inverse relationship between O_{2A} and V_V in the retina. One possible explanation for this finding is the presence of amyloid angiopathy and thickening of the vessel walls due to the deposition of Aβ. Under physiological conditions, elevated O_{2A} triggers a regulatory response that leads to vasoconstriction [49] and reduced TRBF [50]. However, under pathological conditions as in 5XFAD, the observed increase in O_{2A} may be attributed to reduced oxygen diffusion across thickened vessel walls upstream of the measured site, in vessels traveling to the eye or even farther along toward the heart. If inner retinal vessels are thickened, we would anticipate a decrease in oxygen diffusion that could result in hypoxic conditions, supported by the increased D_V observed in vascular regulation. In addition to oxygen regulation, retinal vessel caliber can be affected by VEGF levels. Blair et al. found a significant increase in D_V in response to exogenous VEGF administration [51]. Although VEGF levels were not measured in the current study, the literature has shown a marked increase in plasma levels of VEGF in AD patients [52]. This vasodilatory response is counteracted by a reduction in V_V in order to maintain blood flow. Nevertheless, if hypoxia is prolonged, cells may ultimately undergo death as evidenced by our finding of decreased NFL/RGCL thickness and elevated O_{2V}, although O_{2AV} showed no significant change.

There were several limitations in the current study. First, these results may not be generalizable to other mice models of AD or human AD, as there may be variations in disease progression and phenotype according to genetic models, species, sex, and age. Future studies are warranted to determine the effect of sex and age on retinal hemodynamics and oxygen metrics in WT and 5XFAD mice. Second, factors such as the constants used in the Stern-Volmer equation may be different within the retinal tissue environment; moreover, hemoglobin concentrations, the oxygen–hemoglobin dissociation curve, and blood pH were not measured. However, given that the same values were used in both groups, these inaccuracies likely did not impact the relative changes observed in the 5XFAD mouse model compared with WT. Third, WT mice were not littermates to the 5XFAD mice, and six WT mice purchased from Charles River Laboratories were not screened for retinal degeneration mutations. Therefore, genetic variances may have contributed to differences in the WT group as well as between groups. Fourth, given the small volume of retinal tissue from a single eye, protein analysis was limited such that other proteins such as amyloid precursor protein could not be evaluated. Finally, small sample size may have limited our ability to detect some differences (type-II error) between groups and correlations between parameters.

5. Conclusions

Concurrent alterations in retinal vascular physiology, anatomy, and tissue Aβ42 protein levels in 5XFAD mice at 3 months of age corresponded to previously reported findings in human AD. Overall, these findings suggest that this mouse model can be utilized for studying the pathophysiology of AD and evaluating potential therapies.

Author Contributions: The idea and experimental design of the present study were made by M.S. Imaging and image analysis were performed by S.L., N.M., J.B. and M.R. The manuscript was drafted by N.M., and critical revisions of the manuscript were made by N.P.B. and M.S. All authors have read and agreed to the published version of the manuscript.

Funding: This work was supported by NEI grants EY017918 and EY029220, and the Research to Prevent Blindness Foundation.

Institutional Review Board Statement: All animal procedures and experiments were approved by the University of Southern California Institutional Animal Care and Use Committee.

Informed Consent Statement: All authors have approved the manuscript and agreed with its submission.

Data Availability Statement: All data generated during this study have been included in the manuscript. Further data supporting the findings of this study are available from the corresponding authors on request.

Acknowledgments: Not applicable.

Conflicts of Interest: MS holds a patent for the oxygen imaging technology. The other authors have no conflict of interest.

Abbreviations

Aβ	beta-amyloid
AD	Alzheimer's disease
CAA	cerebral amyloid angiopathy
D_A	arterial diameter
D_V	venous diameter
DO_2	oxygen delivery
ELISA	Enzyme-linked immunosorbent assays
5XFAD	five-familial Alzheimer's disease
FDG-PET	fluorodeoxyglucose positron emission tomography
INL	inner nuclear layer
IPL	inner plexiform layer
NFL/RGCL	nerve fiber layer/retinal ganglion cell layer
MO_2	oxygen metabolism
O_{2A}	arterial oxygen content
O_{2AV}	arteriovenous oxygen content difference
O_{2V}	venous oxygen content
OCT	optical coherence tomography
OEF	oxygen extraction fraction
ONL	outer nuclear layer
OPL	outer plexiform layer
PO_2	vascular oxygen tension
PRL	photoreceptor layer
RNFL	retinal nerve fiber layer
TRBF	total retinal blood flow
TRT	total retinal thickness
V	venous blood velocity
WT	wild type

References

1. Martin, J.P. *World Alzheimer Report 2015: The Global Impact of Dementia: An Analysis of Prevalence, Incidence, Cost and Trends*; Alzheimer's Disease International: London, UK, 2015.
2. Liao, H.; Zhu, Z.; Peng, Y. Potential Utility of Retinal Imaging for Alzheimer's Disease: A Review. *Front. Aging Neurosci.* **2018**, *10*, 188. [CrossRef]
3. Guo, L.; Duggan, J.; Cordeiro, M.F. Alzheimer's disease and retinal neurodegeneration. *Curr. Alzheimer Res.* **2010**, *7*, 3–14. [CrossRef]
4. MacCormick, I.J.; Czanner, G.; Faragher, B. Developing retinal biomarkers of neurological disease: An analytical perspective. *Biomark. Med.* **2015**, *9*, 691–701. [CrossRef]
5. Oakley, H.; Cole, S.L.; Logan, S.; Maus, E.; Shao, P.; Craft, J.; Guillozet-Bongaarts, A.; Ohno, M.; Disterhoft, J.; Van Eldik, L.; et al. Intraneuronal β-Amyloid Aggregates, Neurodegeneration, and Neuron Loss in Transgenic Mice with Five Familial Alzheimer's Disease Mutations: Potential Factors in Amyloid Plaque Formation. *J. Neurosci.* **2006**, *26*, 10129–10140. [CrossRef]
6. Pogue, A.; Dua, P.; Hill, J.; Lukiw, W. Progressive inflammatory pathology in the retina of aluminum-fed 5xFAD transgenic mice. *J. Inorg. Biochem.* **2015**, *152*, 206–209. [CrossRef]
7. Parthasarathy, R.; Chow, K.M.; Derafshi, Z.; Fautsch, M.P.; Hetling, J.R.; Rodgers, D.W.; Hersh, L.B.; Pepperberg, D.R. Reduction of amyloid-beta levels in mouse eye tissues by intra-vitreally delivered neprilysin. *Exp. Eye Res.* **2015**, *138*, 134–144. [CrossRef]
8. Criscuolo, C.; Cerri, E.; Fabiani, C.; Capsoni, S.; Cattaneo, A.; Domenici, L. The retina as a window to early dysfunctions of Alzheimer's disease following studies with a 5xFAD mouse model. *Neurobiol. Aging* **2018**, *67*, 181–188. [CrossRef]
9. Park, S.W.; Kim, J.H.; Mook-Jung, I.; Kim, K.W.; Park, W.J.; Park, K.H. Intracellular amyloid beta alters the tight junction of retinal pigment epithelium in 5XFAD mice. *Neurobiol. Aging* **2014**, *35*, 2013–2020. [CrossRef]
10. Lim, J.K.H.; Li, Q.X.; He, Z.; Vingrys, A.J.; Chinnery, H.R.; Mullen, J.; Bui, B.V.; Nguyen, C.T.O. Retinal Functional and Structural Changes in the 5xFAD Mouse Model of Alzheimer's Disease. *Front. Neurosci.* **2020**, *14*, 862. [CrossRef]
11. Ziegler-Waldkirch, S.; d'Errico, P.; Sauer, J.F.; Erny, D.; Savanthrapadian, S.; Loreth, D.; Katzmarski, N.; Blank, T.; Bartos, M.; Prinz, M.; et al. Seed-induced Abeta deposition is modulated by microglia under environmental enrichment in a mouse model of Alzheimer's disease. *EMBO J.* **2018**, *37*, 167–182. [CrossRef]
12. Holtzman, D.M.; Morris, J.C.; Goate, A.M. Alzheimer's disease: The challenge of the second century. *Sci. Transl. Med.* **2011**, *3*, 77sr71. [CrossRef] [PubMed]
13. Einarsdottir, A.B.; Hardarson, S.H.; Kristjansdottir, J.V.; Bragason, D.T.; Snaedal, J.; Stefansson, E. Retinal oximetry imaging in Alzheimer's disease. *J. Alzheimers Dis.* **2016**, *49*, 79–83. [CrossRef] [PubMed]
14. Olafsdottir, O.B.; Saevarsdottir, H.S.; Hardarson, S.H.; Hannesdottir, K.H.; Traustadottir, V.D.; Karlsson, R.A.; Einarsdottir, A.B.; Jonsdottir, K.D.; Stefansson, E.; Snaedal, J. Retinal oxygen metabolism in patients with mild cognitive impairment. *Alzheimers Dement.* **2018**, *10*, 340–345. [CrossRef] [PubMed]
15. Berisha, F.; Feke, G.T.; Trempe, C.L.; McMeel, J.W.; Schepens, C.L. Retinal abnormalities in early Alzheimer's disease. *Investig. Ophthalmol. Vis. Sci.* **2007**, *48*, 2285–2289. [CrossRef]
16. Jiang, H.; Liu, Y.; Wei, Y.; Shi, Y.; Wright, C.B.; Sun, X.; Rundek, T.; Baumel, B.S.; Landman, J.; Wang, J. Impaired retinal microcirculation in patients with Alzheimer's disease. *PLoS ONE* **2018**, *13*, e0192154. [CrossRef]
17. Gameiro, G.R.; Jiang, H.; Liu, Y.; Deng, Y.; Sun, X.; Nascentes, B.; Baumel, B.; Rundek, T.; Wang, J. Retinal tissue hypoperfusion in patients with clinical Alzheimer's disease. *Eye Vis.* **2018**, *5*, 21. [CrossRef]
18. Feke, G.T.; Hyman, B.T.; Stern, R.A.; Pasquale, L.R. Retinal blood flow in mild cognitive impairment and Alzheimer's disease. *Alzheimers Dement* **2015**, *1*, 144–151. [CrossRef]
19. Marziani, E.; Pomati, S.; Ramolfo, P.; Cigada, M.; Giani, A.; Mariani, C.; Staurenghi, G. Evaluation of retinal nerve fiber layer and ganglion cell layer thickness in Alzheimer's disease using spectral-domain optical coherence tomography. *Investig. Ophthalmol. Vis. Sci.* **2013**, *54*, 5953–5958. [CrossRef]
20. Blanks, J.C.; Hinton, D.R.; Sadun, A.A.; Miller, C.A. Retinal ganglion cell degeneration in Alzheimer's disease. *Brain Res.* **1989**, *501*, 364–372. [CrossRef]
21. Blair, N.P.; Wanek, J.; Felder, A.E.; Brewer, K.C.; Joslin, C.E.; Shahidi, M. Inner Retinal Oxygen Delivery, Metabolism, and Extraction Fraction in Ins2Akita Diabetic Mice. *Investig. Ophthalmol. Vis. Sci.* **2016**, *57*, 5903–5909. [CrossRef]
22. Rahimi, M.; Leahy, S.; Matei, N.; Blair, N.P.; Jeong, S.; Craft, C.M.; Shahidi, M. Assessment of inner retinal oxygen metrics and thickness in a mouse model of inherited retinal degeneration. *Exp. Eye Res.* **2021**, *205*, 108480. [CrossRef]
23. Matei, N.; Leahy, S.; Auvazian, S.; Thomas, B.; Blair, N.P.; Shahidi, M. Relation of Retinal Oxygen Measures to Electrophysiology and Survival Indicators after Permanent, Incomplete Ischemia in Rats. *Transl. Stroke Res.* **2020**, *11*, 1273–1286. [CrossRef]
24. Shi, H.; Koronyo, Y.; Fuchs, D.T.; Sheyn, J.; Wawrowsky, K.; Lahiri, S.; Black, K.L.; Koronyo-Hamaoui, M. Retinal capillary degeneration and blood-retinal barrier disruption in murine models of Alzheimer's disease. *Acta Neuropathol. Commun.* **2020**, *8*, 202. [CrossRef]
25. Shi, H.; Koronyo, Y.; Rentsendorj, A.; Regis, G.C.; Sheyn, J.; Fuchs, D.T.; Kramerov, A.A.; Ljubimov, A.V.; Dumitrascu, O.M.; Rodriguez, A.R.; et al. Identification of early pericyte loss and vascular amyloidosis in Alzheimer's disease retina. *Acta Neuropathol.* **2020**, *139*, 813–836. [CrossRef]
26. Thal, D.R.; Griffin, W.S.; de Vos, R.A.; Ghebremedhin, E. Cerebral amyloid angiopathy and its relationship to Alzheimer's disease. *Acta Neuropathol.* **2008**, *115*, 599–609. [CrossRef]

27. Kalaria, R.N.; Pax, A.B. Increased collagen content of cerebral microvessels in Alzheimer's disease. *Brain Res.* **1995**, *705*, 349–352. [CrossRef]
28. De Jong, G.I.; De Vos, R.A.; Steur, E.N.; Luiten, P.G. Cerebrovascular hypoperfusion: A risk factor for Alzheimer's disease? Animal model and postmortem human studies. *Ann. N. Y. Acad. Sci.* **1997**, *826*, 56–74. [CrossRef]
29. Brown, W.R.; Moody, D.M.; Thore, C.R.; Challa, V.R. Cerebrovascular pathology in Alzheimer's disease and leukoaraiosis. *Ann. N. Y. Acad. Sci.* **2000**, *903*, 39–45. [CrossRef]
30. Brown, W.R.; Moody, D.M.; Challa, V.R.; Thore, C.R.; Anstrom, J.A. Venous collagenosis and arteriolar tortuosity in leukoaraiosis. *J. Neurol. Sci.* **2002**, *203-204*, 159–163. [CrossRef]
31. van Horssen, J.; Wilhelmus, M.M.; Heljasvaara, R.; Pihlajaniemi, T.; Wesseling, P.; de Waal, R.M.; Verbeek, M.M. Collagen XVIII: A novel heparan sulfate proteoglycan associated with vascular amyloid depositions and senile plaques in Alzheimer's disease brains. *Brain Pathol.* **2002**, *12*, 456–462. [CrossRef]
32. Guduru, A.; Martz, T.G.; Waters, A.; Kshirsagar, A.V.; Garg, S. Oxygen Saturation of Retinal Vessels in All Stages of Diabetic Retinopathy and Correlation to Ultra-Wide Field Fluorescein Angiography. *Investig. Ophthalmol. Vis. Sci.* **2016**, *57*, 5278–5284. [CrossRef]
33. Herholz, K.; Salmon, E.; Perani, D.; Baron, J.C.; Holthoff, V.; Frolich, L.; Schonknecht, P.; Ito, K.; Mielke, R.; Kalbe, E.; et al. Discrimination between Alzheimer dementia and controls by automated analysis of multicenter FDG PET. *Neuroimage* **2002**, *17*, 302–316. [CrossRef]
34. Anchisi, D.; Borroni, B.; Franceschi, M.; Kerrouche, N.; Kalbe, E.; Beuthien-Beumann, B.; Cappa, S.; Lenz, O.; Ludecke, S.; Marcone, A.; et al. Heterogeneity of brain glucose metabolism in mild cognitive impairment and clinical progression to Alzheimer disease. *Arch. Neurol.* **2005**, *62*, 1728–1733. [CrossRef]
35. Macdonald, I.R.; DeBay, D.R.; Reid, G.A.; O'Leary, T.P.; Jollymore, C.T.; Mawko, G.; Burrell, S.; Martin, E.; Bowen, C.V.; Brown, R.E.; et al. Early detection of cerebral glucose uptake changes in the 5XFAD mouse. *Curr. Alzheimer Res.* **2014**, *11*, 450–460. [CrossRef]
36. Shao, Y.; Jiang, H.; Wei, Y.; Shi, Y.; Shi, C.; Wright, C.B.; Sun, X.; Vanner, E.A.; Rodriguez, A.D.; Lam, B.L.; et al. Visualization of Focal Thinning of the Ganglion Cell-Inner Plexiform Layer in Patients with Mild Cognitive Impairment and Alzheimer's Disease. *J. Alzheimers Dis.* **2018**, *64*, 1261–1273. [CrossRef]
37. Hinton, D.R.; Sadun, A.A.; Blanks, J.C.; Miller, C.A. Optic-nerve degeneration in Alzheimer's disease. *N. Engl. J. Med.* **1986**, *315*, 485–487. [CrossRef]
38. Blanks, J.C.; Schmidt, S.Y.; Torigoe, Y.; Porrello, K.V.; Hinton, D.R.; Blanks, R.H. Retinal pathology in Alzheimer's disease. II. Regional neuron loss and glial changes in GCL. *Neurobiol. Aging* **1996**, *17*, 385–395. [CrossRef]
39. Gupta, V.K.; Chitranshi, N.; Gupta, V.B.; Golzan, M.; Dheer, Y.; Wall, R.V.; Georgevsky, D.; King, A.E.; Vickers, J.C.; Chung, R.; et al. Amyloid beta accumulation and inner retinal degenerative changes in Alzheimer's disease transgenic mouse. *Neurosci. Lett.* **2016**, *623*, 52–56. [CrossRef]
40. Chiquita, S.; Campos, E.J.; Castelhano, J.; Ribeiro, M.; Sereno, J.; Moreira, P.I.; Castelo-Branco, M.; Ambrosio, A.F. Retinal thinning of inner sub-layers is associated with cortical atrophy in a mouse model of Alzheimer's disease: A longitudinal multimodal in vivo study. *Alzheimers Res. Ther.* **2019**, *11*, 90. [CrossRef]
41. Buccarello, L.; Sclip, A.; Sacchi, M.; Castaldo, A.M.; Bertani, I.; ReCecconi, A.; Maestroni, S.; Zerbini, G.; Nucci, P.; Borsello, T. The c-jun N-terminal kinase plays a key role in ocular degenerative changes in a mouse model of Alzheimer disease suggesting a correlation between ocular and brain pathologies. *Oncotarget* **2017**, *8*, 83038–83051. [CrossRef]
42. Beydoun, T.; Deloche, C.; Perino, J.; Kirwan, B.A.; Combette, J.M.; Behar-Cohen, F. Subconjunctival injection of XG-102, a JNK inhibitor peptide, in patients with intraocular inflammation: A safety and tolerability study. *J. Ocul. Pharmacol. Ther.* **2015**, *31*, 93–99. [CrossRef]
43. Kornfield, T.E.; Newman, E.A. Regulation of blood flow in the retinal trilaminar vascular network. *J. Neurosci.* **2014**, *34*, 11504–11513. [CrossRef]
44. Maarouf, C.L.; Kokjohn, T.A.; Whiteside, C.M.; Macias, M.P.; Kalback, W.M.; Sabbagh, M.N.; Beach, T.G.; Vassar, R.; Roher, A.E. Molecular Differences and Similarities Between Alzheimer's Disease and the 5XFAD Transgenic Mouse Model of Amyloidosis. *Biochem. Insights* **2013**, *6*, 1–10. [CrossRef]
45. Selkoe, D.J.; Hardy, J. The amyloid hypothesis of Alzheimer's disease at 25 years. *EMBO Mol. Med.* **2016**, *8*, 595–608. [CrossRef]
46. Qi, X.M.; Ma, J.F. The role of amyloid beta clearance in cerebral amyloid angiopathy: More potential therapeutic targets. *Transl. Neurodegener.* **2017**, *6*, 22. [CrossRef]
47. Jellinger, K.A. Alzheimer disease and cerebrovascular pathology: An update. *J. Neural. Transm.* **2002**, *109*, 813–836. [CrossRef]
48. Koronyo, Y.; Biggs, D.; Barron, E.; Boyer, D.S.; Pearlman, J.A.; Au, W.J.; Kile, S.J.; Blanco, A.; Fuchs, D.T.; Ashfaq, A.; et al. Retinal amyloid pathology and proof-of-concept imaging trial in Alzheimer's disease. *JCI Insight* **2017**, *2*, e93621. [CrossRef]
49. Hommer, N.; Kallab, M.; Sim, Y.C.; Lee, A.X.; Chua, J.; Tan, B.; Schlatter, A.; Werkmeister, R.M.; Schmidl, D.; Schmetterer, L.; et al. Effect of hyperoxia and hypoxia on retinal vascular parameters assessed with optical coherence tomography angiography. *Acta Ophthalmol.* 2021, *early view*. [CrossRef]
50. Riva, C.E.; Grunwald, J.E.; Sinclair, S.H. Laser Doppler Velocimetry study of the effect of pure oxygen breathing on retinal blood flow. *Investig. Ophthalmol. Vis. Sci.* **1983**, *24*, 47–51.

51. Blair, N.P.; Wanek, J.; Teng, P.Y.; Shahidi, M. The effect of intravitreal vascular endothelial growth factor on inner retinal oxygen delivery and metabolism in rats. *Exp. Eye Res.* **2016**, *143*, 141–147. [CrossRef]
52. Cho, S.J.; Park, M.H.; Han, C.; Yoon, K.; Koh, Y.H. VEGFR2 alteration in Alzheimer's disease. *Sci. Rep.* **2017**, *7*, 17713. [CrossRef] [PubMed]

cells MDPI

Article

Insights into Metabolic Activity and Structure of the Retina through Multiphoton Fluorescence Lifetime Imaging Microscopy in Mice

Niranjana Kesavamoorthy [1], Jason A. Junge [2], Scott E. Fraser [2] and Hossein Ameri [1,*]

1 Department of Ophthalmology, USC Roski Eye Institute, Keck School of Medicine, University of Southern California, Los Angeles, CA 90033, USA; niranjana.kesavamoorthy@med.usc.edu
2 Department of Biological Sciences, David Dornsife College of Letters Arts and Sciences, University of Southern California Dana, Los Angeles, CA 90089, USA; junge@usc.edu (J.A.J.); sfraser@provost.usc.edu (S.E.F.)
* Correspondence: ameri@med.usc.edu

Abstract: Fluorescence lifetime imaging microscopy (FLIM) evaluates the metabolic state of tissue based on reduced nicotinamide adenine dinucleotide (NAD(P)H) and flavin adenine dinucleotide (FAD). Fluorescence lifetime imaging ophthalmoscopy (FLIO) can image the fundus of the eyes, but cannot detect NAD(P)H. We used multiphoton FLIM to study the metabolic state of the retina in fixed eyes of wild-type mice C57BL6/J. We sectioned the eye using a polyacrylamide gel-embedding technique and estimated the percentage of bound NAD(P)H. We found that oxidative phosphorylation was the predominant metabolic state, particularly in the inner retina, when a fixed retina was used. We also demonstrated the feasibility of FAD imaging of the retina. In addition, we demonstrated that autofluorescence and various FLIM channels, such as hemoglobin, melanin and collagen, can be used to evaluate the structure of the retina and other parts of the eye without any special staining.

Keywords: retina; multiphoton fluorescence lifetime imaging microscopy; FLIM; NAD(P)H; metabolic imaging; glycolysis; oxidative phosphorylation

Citation: Kesavamoorthy, N.; Junge, J.A.; Fraser, S.E.; Ameri, H. Insights into Metabolic Activity and Structure of the Retina through Multiphoton Fluorescence Lifetime Imaging Microscopy in Mice. *Cells* **2022**, *11*, 2265. https://doi.org/10.3390/cells11152265

Academic Editor: Yvan Arsenijevic

Received: 18 June 2022
Accepted: 20 July 2022
Published: 22 July 2022

1. Introduction

Fluorescence lifetime imaging microscopy (FLIM) displays the fluorescence lifetimes of various fluorophores in a sample as a two-dimensional image. Unlike intensity-based images, fluorescence lifetimes depend not on the fluorophore concentration but on the local environment. The same fluorophore could have different lifetimes based on its interacting molecule [1]. Using FLIM, one can estimate the metabolic state of the tissue based on the fluorescence lifetimes of the free and bound reduced nicotinamide adenine dinucleotide (NAD(P)H) [1,2]. The lifetime of free NAD(P)H is smaller than the protein-bound NAD(P)H [1]. The presence of free NAD(P)H in tissue suggests glycolysis, whereas the presence of bound NAD(P)H signals oxidative phosphorylation [3]. Previous studies have used FLIM to determine a predominance of glycolysis or oxidative phosphorylation by measuring the amounts of free and bound NAD(P)H in various tissues, including human organoid [2,4,5]. FLIM can be performed using a multiphoton or confocal approach [6]. The advantages of the multiphoton approach over the confocal method are that the light penetrates deeper and photodamage is limited because only the required sample volume is excited, preventing light scatter and absorption before reaching the region of interest [7–9].

FLIM has been used to study eyes. Some studies have used fluorescence lifetime imaging ophthalmoscopy (FLIO), a combination of FLIM and confocal laser scanning ophthalmoscopy, to image eyes in vivo; however, NAD(P)H cannot be imaged by FLIO [10–13]. Previous investigators have imaged ex vivo, unfixed porcine eyes, using two-photon FLIM to investigate the fundus and to assess the morphology and metabolism of the cornea and lens [14,15]. In this article, we demonstrate that FLIM can be used to study the metabolic status of the retina as well as the structure of the retina and other parts of the eye.

2. Materials and Methods

2.1. Animals

All procedures complied with Institutional Animal Care and Use Committee (IACUC guidelines at the University of Southern California. Wild-type mice (C57BL6/J) were purchased from Jackson Laboratories (Bar Harbor, ME, USA). Mice were housed in a 12-h light/dark cycle and food and water were provided ad libitum.

2.2. Enucleation and Fixation of Eyes

Animals were euthanized, and the eyeballs were enucleated. Eleven eyes were included in the study. Before enucleation, the superior portion of the cornea was marked with cautery to preserve the orientation of the eyeballs. The eyeballs were punctured and placed in 4% paraformaldehyde for one hour. The corneas were then removed and the eyecups were placed in 4% paraformaldehyde for two hours.

2.3. Embedding and Sectioning

The eyecups were embedded in polyacrylamide gel, based on Hayaran et al.'s method [16] The gel was prepared by mixing 3 mL 40% Bisacrylamide, 6 mL ultrapure Milli-Q water, 1 mL 10× Tris Buffered Saline, and 47 μL 10% Ammonium Persulfate. The eyecups were placed in this gel mixture, their lens removed within the gel mixture, and 30 μL Tetramethylethylenediamine was then added (Supplementary Figure S1). Sectioning of the embedded eyecups was performed using Leica Microtome VT1200 (Leica Microsystems, Deerfield, MA, USA). Sections between 100 μm and 200 μm thick were made. In half of the samples, the eyes were oriented to obtain superior–inferior sections; in the other half, nasal-temporal sections were obtained. In each eye, the largest central sections that included the optic nerve and peripheral retina were selected for imaging.

2.4. Multiphoton FLIM Imaging

The Leica SP8 DIVE FALCON (Leica Microsystems, Deerfield, MA, USA) was used for imaging; the samples were excited at 740 nm for metabolic and intrinsic autofluorescence signal imaging and 860 nm for second harmonic generation (SHG) at ~500 μW laser power for each wavelength. Autofluorescence was collected on three channels of hybrid detectors: 425–475 nm for NAD(P)H, 502–577 nm for flavin adenine dinucleotide (FAD), and 600–650 nm for hemoglobin and melanin. The NAD(P)H signal was excited through non-linear, multiphoton excitation by a Spectra Physics InSight X3 tunable IR laser (MKS Spectra-Physics, Milpitas, CA, USA) and emission was collected from 425–475 nm through adjustment of the Leica 4Tune NDD filtering system (Leica Microsystems, Deerfield, MA, USA). Light of wavelengths 475+ nm was excluded from detection, because other autofluorescent molecules, such as FAD, are multiphoton excitable at 740 nm and we wanted to exclude those signals in our analysis of NAD(P)H. The SHG of collagen was collected from 411 to 463 nm. The choice of excitation and emission spectra were based on our previous work and the work of others [4,17–19]. We used a Leica 40×/1.2 NA CSII water immersion objective (Leica Microsystems, Deerfield, MA, USA), and a single focal plane image was chosen ten microns deep in the retina sections for standardization. For three-dimensional retina images to detect vascular networks, we collected z-stacks from the entire thickness of the section with optical sectioning at Nyquist, and displayed in Imaris 64 bit 9.7.0 software (Bitplane, Belfast, UK).

2.5. FLIM Analysis

The vertical sections were divided into a superior and an inferior region, and the horizontal sections were split into temporal and nasal regions. Each region was classified into five zones: far periphery (near the ora serrata), mid periphery (halfway between the ora serrata and the optic disc), peripapillary (200 µm from the optic disc), central (400 µm from the optic disc), and paracentral (600 µm from the optic disc). Each zone was divided into outer (photoreceptor inner (IS), outer segments (OS), and outer nuclear layer (ONL)), inner (outer plexiform layer (OPL)), inner nuclear layer (INL), inner plexiform layer (IPL), ganglion cell layer (GCL), and retinal nerve fiber layer (RNFL)). The phasor approach to FLIM analysis [20,21] to determine percentages of bound NAD(P)H, was accomplished in Leica's LASX FLIM FCS software. In this software package, a metabolic trajectory was drawn from one edge of the unit circle, from 0.4 ns lifetime (lifetime constant, τ, of 100% free NAD(P)H), passing through the center of mass, and extrapolated to an intersection with the edge of the unit circle (representing 100% bound NAD(P)H), using the ratiometric analysis tool. Next, regions of interest (ROI) were created in the retinal images to determine phasor positions for the pixels within the ROIs; we standardized ROIs to have a width of 150 µm. We then moved the ratiometic slider along the metabolic trajectory we drew in the phasor, over the center of mass for the phasor distribution for each ROI. (Supplementary Figure S2). The bound NAD(P)H %, which is the ratio of bound NAD(P)H / total NAD(P)H multiplied by 100, is calculated from the position of each ROI's center of mass along the metabolic trajectory. The phasor position for melanin was determined based on a previous study [22]. SHG Leica's DIVE FALCON is designed as a spectral + lifetime imaging instrument and can therefore separate collagen autofluorescence from second harmonic signal. The second harmonic signal of collagen has an extremely short lifetime, approaching instantaneous, and therefore, a distinct phasor position which is far from collagen autofluorescence [18]. Thus, we can infer that the signal we obtained from collagen is due to second harmonic and not due to autofluorescence.

To determine the phasor position for hemoglobin, we selected an ROI in images corresponding to a retinal blood vessel, and by using the phasor ROI selection tool, we then highlighted all pixels within a retinal image containing that specific phasor lifetime position for hemoglobin.

2.6. *Statistical Analysis*

The mean, standard deviation (SD), and SEM for the percentage of bound NAD(P)H were obtained. Two-tailed paired *t*-tests were used to compare the percentage of bound NAD(P)H.

3. Results

3.1. *Metabolic Imaging*

3.1.1. NAD(P)H Channel

In the NAD(P)H channel (425–475 nm emission spectra), we compared the percentage of bound NAD(P)H between the outer retina (photoreceptor inner and outer segments and outer nuclear layer) and the inner retina (outer plexiform layer, inner nuclear layer, inner plexiform layer, ganglion cell layer, and retinal nerve fiber layer) (Figure 1 and Table S1). The zones with artifacts, damages and folds were excluded. The mean ± SDs of the bound NAD(P)H percentage in the outer and inner retina were 58.9 ± 3.7 and 60.9 ± 2.6, respectively. The total number of images used was 86. The total number of eyes imaged in the NAD(P)H channel was 11 (Figure 2A). Statistical analysis revealed a highly statistically significant difference in the percentage of bound NAD(P)H between the outer and inner retina ($p < 0.0001$).

Next, we compared the percentage of bound NAD(P)H in the far periphery with the central zones to determine whether the density of photoreceptors and ganglion cells has any effects on metabolic activity (Table S2 and Supplementary Figure S3). The comparison was made in two different ways: the percentage of bound NAD(P)H in the overall retina of the two zones and the percentage of bound NAD(P)H in the outer retina of the two zones. In the overall retina, the mean ± SD bound NAD(P)H percentage was 59.8 ± 2.5 in the far periphery and 58.6 ± 4.3 in the central zones ($p = 0.36$). In the outer retina, the mean ± SD bound NAD(P)H percentage was 58.8 ± 2.6 in the far periphery and 58.2 ± 4.1

in the central zones ($p = 0.64$) (n = 15) (Figure 2B). Neither comparisons demonstrated a statistically significant difference between the far periphery and central retina.

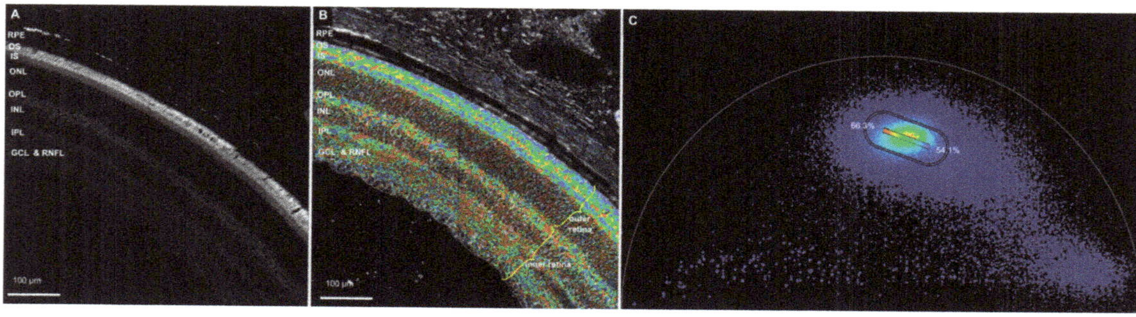

Figure 1. Retina in the NAD(P)H channel. (**A**) Autofluorescence image of the retina. (**B**) The NAD(P)H channel. The blue spots indicate areas with 54.1% bound NAD(P)H, and the red spots show areas with 66.3% bound NAD(P)H. As is evident, the outer retina is more towards the blue spectrum, and the inner retina is more towards the red spectrum, indicating less bound NAD(P)H in the outer retina. (**C**) Phasor plot for the above image showing rainbow spectrum from red to blue. (RPE—retinal pigment epithelium, OS—photoreceptor outer segment, IS—photoreceptor inner segment, ONL—outer nuclear layer, OPL—outer plexiform layer, INL—inner nuclear layer, IPL—inner plexiform layer, GCL—ganglion cell layer, RNFL—retinal nerve fiber layer, NAD(P)H—reduced nicotinamide adenine dinucleotide).

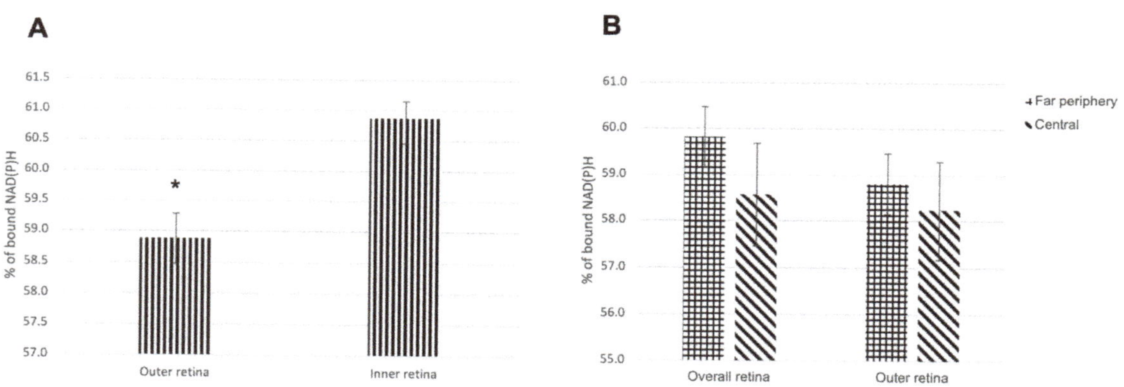

Figure 2. (**A**) Percentage of bound NAD(P)H in the outer retina vs. the inner retina. (**B**) Percentage of bound NAD(P)H in the far periphery compared to the central zones in the overall and the outer retina. Data are presented as mean ± SEM. (NAD(P)H-reduced nicotinamide adenine dinucleotide). * $p < 0.0001$.

3.1.2. FAD Channel

We were also able to image FAD, which could also be used to study metabolic state. For the FAD imaging, we used 502–577-nm emission spectra (Figure 3).

3.2. Structural Imaging

Autofluorescence images could determine the retinal structure by delineating the retinal layers. All the retina layers could be visualized, except the internal limiting membrane and the external limiting membrane. We also visualized a distinct band of relative hypo-autofluorescence in the inner plexiform layer, which is not visible on bright-field microscopy or common structural retinal staining, such as Haematoxylin and Eosin, and

Periodic Acid Schiff (Figure 4). Using multiphoton FLIM, we were able to evaluate other fluorophores, such as melanin, hemoglobin and collagen.

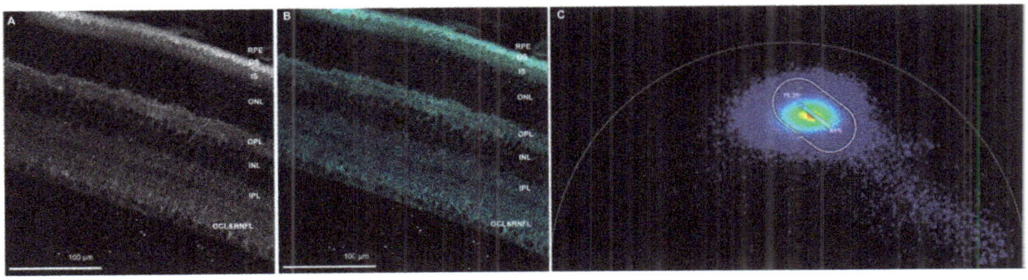

Figure 3. Retina in the FAD channel. (**A**) Autofluorescence image of the retina. (**B**) The FAD channel The dark blue indicates less bound FAD than cyan. The dark blue spots indicate areas with 75.3% free FAD, and the cyan spots show areas with 65% free FAD. (**C**) Phasor plot showing spectrum from dark blue to cyan. A metabolic trajectory was drawn from one edge of the unit circle from 2.6 ns lifetime (lifetime constant, τ, of 100% free FAD), passing through the center of mass, and extrapolated to an intersection with the edge of the unit circle (representing 100% bound FAD), using the ratiometric analysis tool. The color bar was marked on top of the trajectory to indicate the free and bound FAD. (RPE-retinal pigment epithelium, OS—photoreceptor outer segment, IS—photoreceptor inner segment, ONL—outer nuclear layer, OPL—outer plexiform layer, INL—inner nuclear layer, IPL—inner plexiform layer, GCL—ganglion cell layer, RNFL—retinal nerve fiber layer, FAD—flavin adenine dinucleotide). Scale bar: 100 µm.

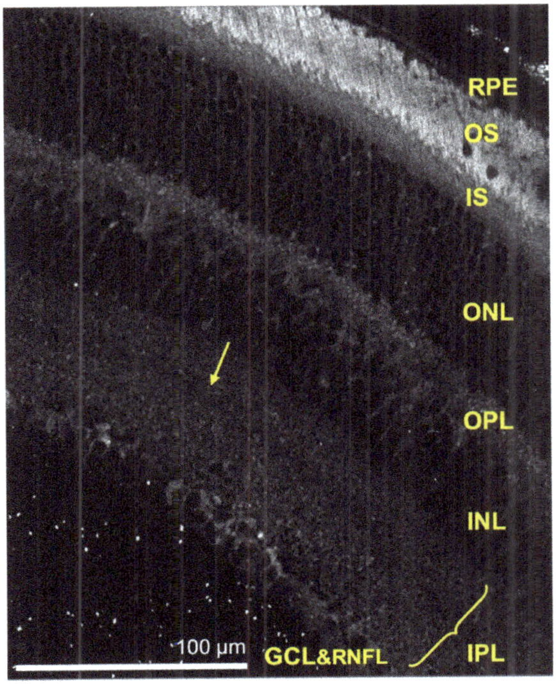

Figure 4. Autofluorescence image of the retina showing the retinal layers. The yellow arrow indicates the hypo-autofluorescent band in the inner plexiform layer. (RPE—retinal pigment epithelium, OS—photoreceptor outer segment, IS—photoreceptor inner segment, ONL—outer nuclear layer, OPL—outer plexiform layer, INL—inner nuclear layer, IPL—inner plexiform layer, GCL—ganglion cell layer and RNFL—retinal nerve fiber layer). Scale bar: 100 µm.

3.2.1. Hemoglobin and Melanin Channel

We could detect hemoglobin signals within the retinal and choroidal vessels when imaged with 600–650 nm emission spectra. A 3D image of the retina with several Z-sections demonstrated a deep capillary plexus in the outer plexiform layer, a superficial capillary plexus in the junction of the inner nuclear layer and inner plexiform layer, and a superficial layer composed of retinal vessels in the ganglion cell layer (Figure 5). In addition, we found two distinct bands of hemoglobin signal in the basal and apical portions of the retinal pigment epithelium (RPE) (Figure 6C).

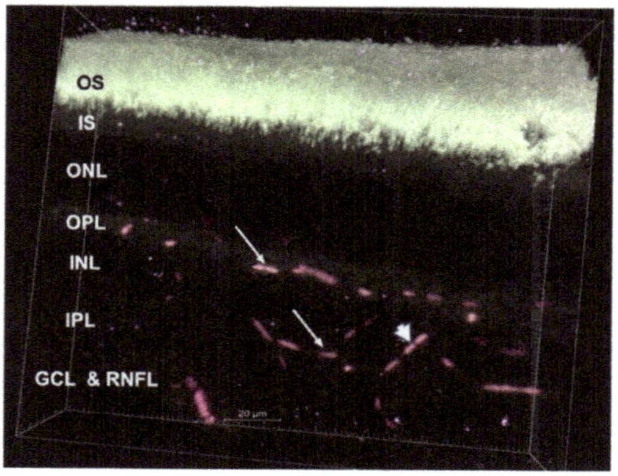

Figure 5. 3D image of the retina obtained by z-stacking (thickness—200 μm). The arrows indicate the deep and superficial capillary plexus, and the arrowhead shows a penetrating vessel connecting the two plexuses. The image reveals all the retinal layers except RPE (RPE was clipped out). (RPE—retinal pigment epithelium, OS—photoreceptor outer segment, IS—photoreceptor inner segment, ONL—outer nuclear layer, OPL—outer plexiform layer, INL—inner nuclear layer, IPL—inner plexiform layer, GCL-ganglion cell layer, RNFL-retinal nerve fiber layer). Imaris software was used to generate the 3D image.

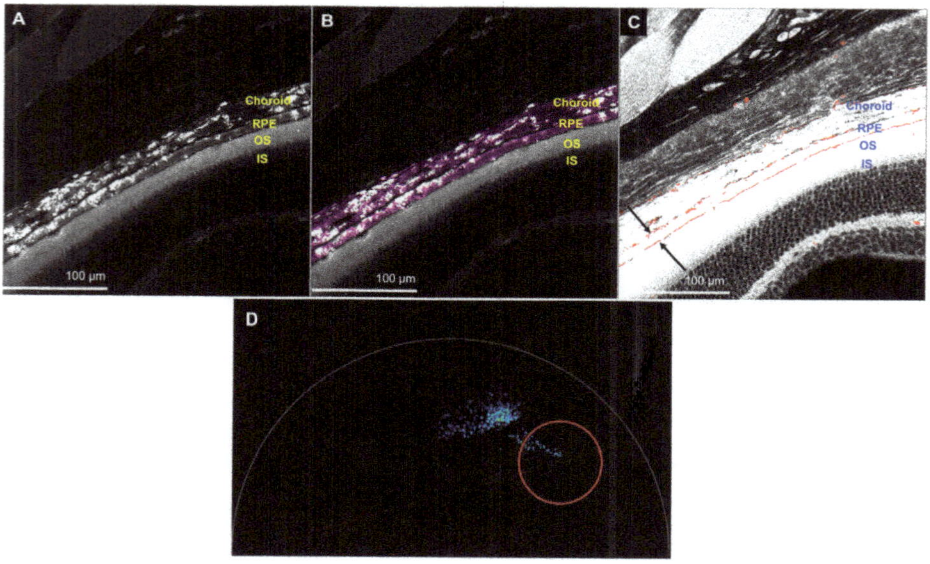

Figure 6. Representative images of the retina and choroid showing melanin and hemoglobin signals. (**A**) Autofluorescence image to show outline of the RPE and photoreceptors and determine the exact

location of the melanin and hemoglobin signals. (**B**) Image displaying the melanin signal in the choroid and the RPE (shown in pink). (**C**) Image showing the hemoglobin signal in basal portion of the RPE (top arrow), apical portion of the RPE (bottom arrow) and the choroid (shown in red). To better visualize the hemoglobin signal, the brightness of the image was increased. The above three images were obtained from the most superficial z-section, since the choroid and RPE appear darker in the deeper z-sections because of blockage. (**D**) Phasor position for hemoglobin signal (red circle). The blue stain corresponds to the region of interest selected. (RPE—retinal pigment epithelium, OS—photoreceptor outer segment, IS—photoreceptor inner segment).

FLIM could also be used to study melanin. The RPE and the choroid displayed the melanin signal with the same emission spectra (600–650 nm) (Figures 6B and 7A).

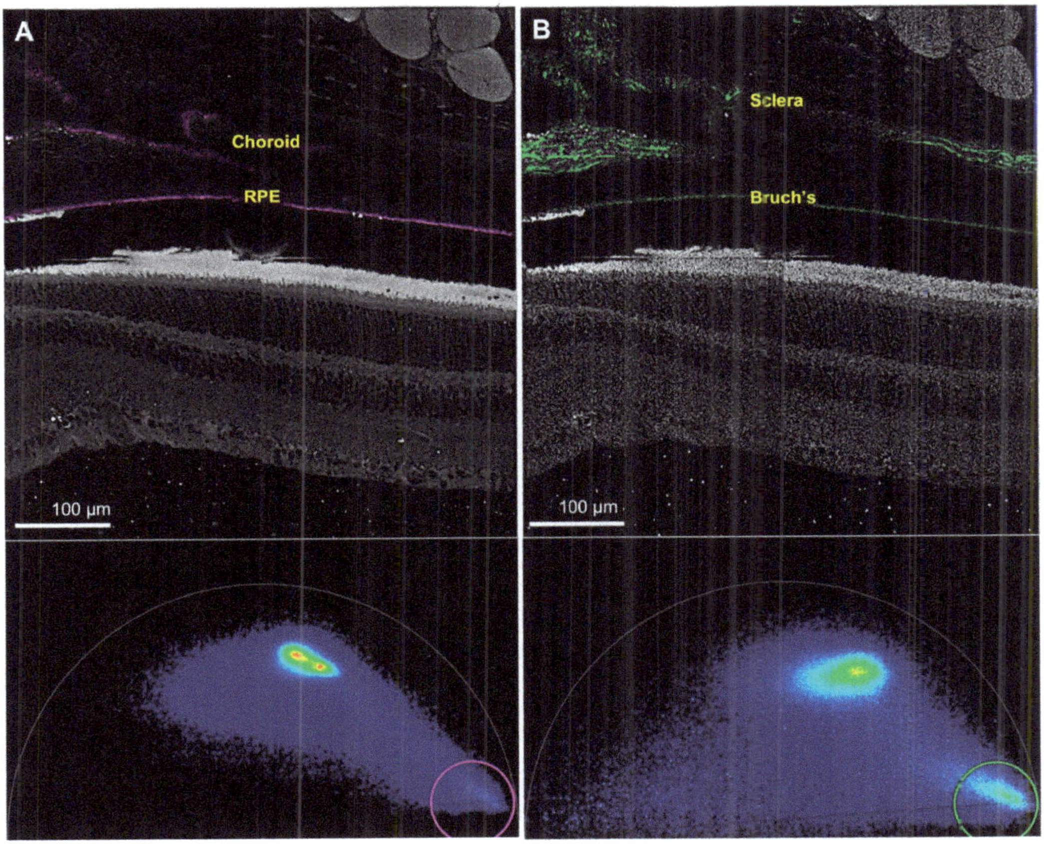

Figure 7. Representative images of the retina, choroid and sclera in the melanin and collagen channels. The same region was imaged in two different channels. (**A**) Top. Image showing the signal for melanin in the RPE and the choroid (in pink). Bottom. Phasor plot showing the phasor position of melanin (pink circle). (**B**) Top. The sclera and the Bruch's membrane generate a second harmonic corresponding to collagen (in green). Bottom. Phasor representation of collagen (green circle). (RPE—retinal pigment epithelium). The gap between the RPE and photoreceptors is an artifact separation.

3.2.2. Collagen Channel

The collagen channel could display the sclera, cornea, conjunctiva, episcleral tissue, optic nerve sheath, Bruch's membrane, and iris stroma, all of which generate a second harmonic (Figures 7B and 8) due to the fibrillar collagen [23]. The scleral fibers had an uneven thickness and irregular fiber arrangement, with the greatest thickness in the perilimbal region, which gradually diminished towards the posterior pole (Figure 8A,C). In contrast, the cornea had a uniform thickness, and the fibers had a regular structure (Figure 8B).

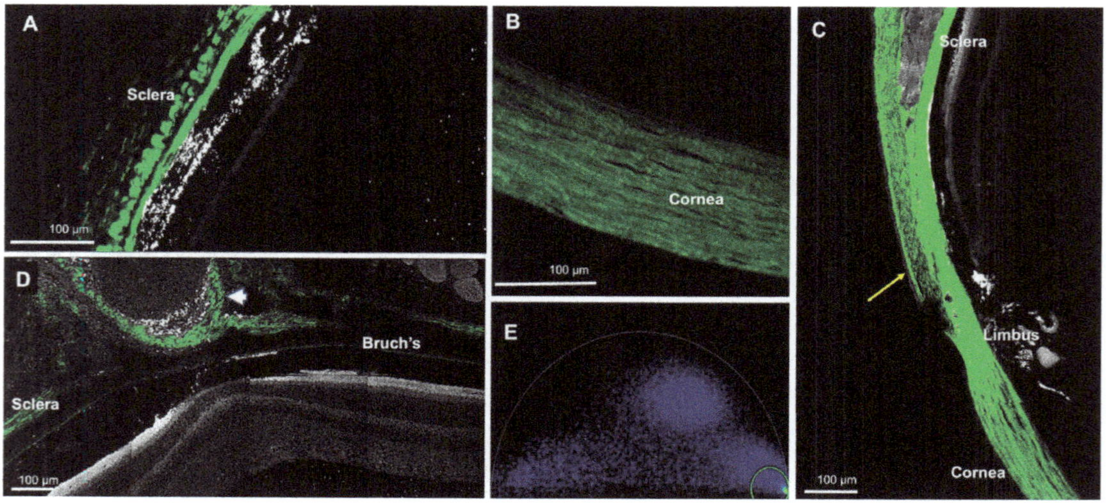

Figure 8. Representative collagen images from different regions generated by the second harmonic. (**A**) The sclera. Note the irregular arrangement and wavy fibers. (**B**) The cornea. The fibers are uniformly arranged. (**C**) The conjunctiva, sclera and cornea combined. The conjunctiva is indicated with a yellow arrow. The thickness of the sclera is greatest at the perilimbal region and decreases posteriorly. (**D**) The optic nerve sheath (indicated by a white arrowhead). (**E**) Representative phasor image for collagen (signal demonstrated in green).

4. Discussion

In this article, we demonstrated that fixed mice eyes could be used for metabolic studies of the retina by measuring the percentages of free and bound NAD(P)H. We observed higher percentages of bound NAD(P)H in both the outer and inner retina, indicating a predominance of oxidative phosphorylation throughout the retinal thickness. Comparing the inner and outer retina, we noted a greater number of free NAD(P)H in the outer retina, indicating a higher level of glycolysis in the outer retina relative to the inner retina. The bound NAD(P)H percentages obtained in our study are specific to fixed retina. Since a previous study demonstrated that fixation increases fluorescence lifetimes, further studies are required to evaluate any differences between fresh and fixed retina [24]. Similar to our study, the article by Rueda et al. showed more glycolysis in the outer retina in mice (except in the photoreceptor outer segment). They also found an overall high oxidative phosphorylation (except in the outer photoreceptor segments and the inner nuclear soma), but the percentages of glycolysis and oxidative phosphorylation were not estimated [25]. Although fixation increases the fluorescence lifetimes, our findings are partially in agreement with a study by Browne et. al on human retinal organoids, which indicated a predominance of oxidative phosphorylation in the inner regions, but a predominance of glycolysis in the outer regions of the organoids [4]. Future studies on the human retina will determine whether its metabolic state is more similar to human retinal organoids, or to that found in mice. A study by Volland et al. found that in mice retina, the photoreceptor density was higher in the central regions compared to the periphery, with the highest photoreceptor density located 400 µm temporal to the optic disc; this region was defined as the area centralis [26]. Another study revealed that ganglion cell density was higher in the central retina than in the periphery [27]. To assess the potential correlation between the metabolic states of the retina and the cell density of photoreceptors and ganglion cells, we compared the metabolic states of the far periphery (near ora serrata), which has the lowest cell density, and the central zone (400 μm from the optic disc). These zones demonstrated no significant difference between them.

We showed that a metabolic image of the retina could also be obtained with the FAD channel. However, in our study, we did not determine the fraction of bound and free FAD for our samples, since our focus was to determine the metabolic state of the retina using NAD(P)H. Previous studies have used FAD to study the metabolic state of cancer cells. Imaging with the FAD channel and the NAD(P)H channel can potentially be used to determine the ratio of bound NAD(P)H to bound FAD, which estimates the redox state of the tissue [2,28,29].

Using multiphoton FLIM, in addition to metabolic imaging, we performed structural imaging of the retina. We also showed that FLIM can be used for structural imaging of the sclera, cornea, choroid, optic nerve sheath, and the tissues surrounding the eyes, without any staining. FLIO can also be used to study the structure of the retina; however, unlike FLIM it cannot detect NAD(P)H in the fundus. This is due to the short emission spectra and excitation wavelength of NAD(P)H, and these short wavelengths are obstructed by the cornea and lens [10,11]. The autofluorescence images of the retina helped us to delineate the retinal layers. Interestingly, we noted a relative band of hypo-autofluorescence within the inner plexiform layer. Comparing our images to published immunochemical studies, which demonstrated specific immunostaining in the same region as our hypo-autofluorescent band, we speculate that the band corresponds to dendrites of type 2 catecholaminergic amacrine cells, neuronal junctions of some amacrine cells, ganglion cells, and bipolar cells [30–32]. Further histological studies are required to confirm this, which is beyond the scope of this study.

In FLIM, the hemoglobin channel can display retinal blood vessels well, particularly when a 3D video is generated from a thick retinal section. However, the vessels appear to be interrupted; this is due to the segmentation of the blood in the retinal vessels and capillaries, which occurs in tissues postmortem [33]. Similar to previous studies, we observed three vascular networks in the retina, a deep capillary plexus in the outer plexiform layer, a superficial capillary plexus in the junction of the inner nuclear layer and inner plexiform layer, and a superficial layer composed of retinal vessels in the ganglion cell layer [34,35]. Zimmerman et al. also demonstrated the same, but classified them into two layers [36]. We were also able to detect penetrating vessels between the layers. In addition to the retinal blood vessels, we found the presence of hemoglobin signal in the choriocapillaris. Although not performed in this study, intracardiac dye before euthanasia may allow for a better demonstration of vessels. Hemoglobin could also be present outside the blood vessels and, using FLIM, we detected its presence in the RPE. Tezel et al. have previously reported the presence of hemoglobin in the basal portion of the RPE and speculated that it could act as an oxygen reservoir during periods of increased oxygen demand [37]. Interestingly, in our study, in addition to a band of hemoglobin in the basal portion of the RPE, we noted an additional band in the apical portion.

FLIM also allows for the study of melanin distribution. In our study, a band of melanin signal corresponded to the RPE. Melanin signal was also detected in the choroid. Imaging melanin can be used to study diseases that affect the RPE, such as age-related macular degeneration and inherited retinal diseases.

This study demonstrated the additional possibility of imaging collagen by generating a second harmonic. We most likely detected type I collagen, since fibrillar collagen generates a second harmonic, and type I produces a prominent second harmonic [23,38]. In addition, type I collagen is the primary collagen in the cornea and sclera [39,40]. Thus, FLIM can illustrate the collagen architecture of the cornea and the sclera without any special stains. Collagen was detected in the choroid, including Bruch's membrane, optic nerve sheath, episcleral tissue, conjunctiva, cornea, and iris stroma, but it was most prominent in the sclera. In line with these findings, previous studies have also imaged collagen in the sclera by generating a second harmonic in human and rabbit eyes [41,42].

A limitation of this study is that the imaging was performed on fixed tissues. We used fixed eyes for two reasons: (1) the soft consistency of the unfixed retina made it cumbersome to section, and (2) imaging the fresh retina resulted in artifacts; the fresh retinal sections disintegrated and the retinal architecture became distorted. We fixed the eyes by immersing in 4% paraformaldehyde for a total of 3 h.

In summary, our study determined the metabolic states and structure of fixed eyes from wild-type mice. We found that oxidative phosphorylation is the predominant metabolic state in the retina, but the outer retina shows significantly less oxidative phosphorylation than the inner retina. In addition, we were able to better localize hemoglobin that is present in the RPE. Multiphoton FLIM allowed for the study of the whole eye using NAD(P)H, FAD, hemoglobin, melanin, and collagen channels. In the future, with additional strategies, metabolic states of various disease models in mice can be determined using this study as a baseline.

Supplementary Materials: The following supplementary materials can be downloaded at https://www.mdpi.com/article/10.3390/cells11152265/s1. Table S1: Percentage of bound NAD(P)H in the outer retina and inner retina; Table S2: Percentage of bound NAD(P)H in the overall retina and the outer retina (far periphery and central zones). Figure S1: Image showing the central section of the eye ball through the dissecting microscope; Figure S2: Illustration of phasor analysis; Figure S3: Intensity images of the far periphery and central zones; Figure S4: A2 distribution images of far periphery and central zones.

Author Contributions: Conceptualization, J.A.J., S.E.F. and H.A.; writing—original draft, N.K. writing—review and editing J.A.J., S.E.F. and H.A.; methodology J.A.J.; project administration, N.K and H.A.; investigation, N.K. and J.A.J.; formal analysis N.K., J.A.J. and H.A.; supervision H.A. validation H.A. All authors have read and agreed to the published version of the manuscript.

Funding: This research was funded by an Unrestricted Grant to the Department of Ophthalmology from Research to Prevent Blindness, New York, NY; NIH Grant P30EY029220; NIH Grant K12 EY028873.

Institutional Review Board Statement: The animal study protocol was approved by the Institutional Animal Care and Use Committee of the University of Southern California (protocol code 20758 and date of approval: 3 May 2021).

Informed Consent Statement: Study does not involve human subjects.

Data Availability Statement: The data presented in this study are available in the supplementary materials and within the article. The supplementary data are also available in zenodo: https://doi.org/10.5281/zenodo.6820505.

Conflicts of Interest: The authors declare no conflict of interest.

References

1. Lakowicz, J.R.; Szmacinski, H.; Nowaczyk, K.; Johnson, M.L. Fluorescence lifetime imaging of free and protein-bound NADH. *Proc. Natl. Acad. Sci. USA* **1992**, *89*, 1271–1275. [CrossRef] [PubMed]
2. Skala, M.C.; Riching, K.M.; Gendron-Fitzpatrick, A.; Eickhoff, J.; Eliceiri, K.W.; White, J.G.; Ramanujam, N. In vivo multiphoton microscopy of NADH and FAD redox states, fluorescence lifetimes, and cellular morphology in precancerous epithelia. *Proc. Natl. Acad. Sci. USA* **2007**, *104*, 19494–19499. [CrossRef] [PubMed]
3. Chance, B.; Baltscheffsky, H. Respiratory Enzymes in Oxidative Phosphorylation. *J. Biol. Chem.* **1958**, *233*, 736–739. [CrossRef]
4. Browne, A.W.; Arnesano, C.; Harutyunyan, N.; Khuu, T.; Martinez, J.C.; Pollack, H.A.; Koos, D.S.; Lee, T.C.; Fraser, S.E.; Moats, R.A.; et al. Structural and Functional Characterization of Human Stem-Cell-Derived Retinal Organoids by Live Imaging. *Investig. Opthalmology Vis. Sci.* **2017**, *58*, 3311–3318. [CrossRef]
5. Sameni, S.; Syed, A.; Marsh, J.L.; Digman, M.A. The phasor-FLIM fingerprints reveal shifts from OXPHOS to enhanced glycolysis in Huntington Disease. *Sci. Rep.* **2016**, *6*, 34755. [CrossRef]
6. Coletta, V.C.; Rezende, C.A.; da Conceição, F.R.; Polikarpov, I.; Guimarães, F.E.G. Mapping the lignin distribution in pretreated sugarcane bagasse by confocal and fluorescence lifetime imaging microscopy. *Biotechnol. Biofuels* **2013**, *6*, 43. [CrossRef]
7. Denk, W.; Strickler, J.H.; Webb, W.W. Two-photon laser scanning fluorescence microscopy. *Science* **1990**, *248*, 73–76. [CrossRef]

8. Stelzer, E.H.; Hell, S.; Lindek, S.; Stricker, R.; Pick, R.; Storz, C.; Ritter, G.; Salmon, N. Nonlinear absorption extends confocal fluorescence microscopy into the ultra-violet regime and confines the illumination volume. *Opt. Commun.* **1994**, *104*, 223–228. [CrossRef]
9. Centonze, V.E.; White, J.G. Multiphoton Excitation Provides Optical Sections from Deeper within Scattering Specimens than Confocal Imaging. *Biophys. J.* **1998**, *75*, 2015–2024. [CrossRef]
10. Schweitzer, D.; Schenke, S.; Hammer, M.; Schweitzer, F.; Jentsch, S.; Birckner, E.; Becker, W.; Bergmann, A. Towards metabolic mapping of the human retina. *Microsc. Res. Tech.* **2007**, *70*, 410–419. [CrossRef]
11. Schweitzer, D.; Hammer, M.; Schweitzer, F.; Anders, R.; Doebbecke, T.; Schenke, S.; Gaillard, E.R.; Gaillard, E.R. In vivo measurement of time-resolved autofluorescence at the human fundus. *J. Biomed. Opt.* **2004**, *9*, 1214–1222. [CrossRef] [PubMed]
12. Dysli, C.; Schuerch, K.; Escher, P.; Wolf, S.; Zinkernagel, M.S. Fundus Autofluorescence Lifetime Patterns in Retinitis Pigmentosa. *Investig. Opthalmol. Vis. Sci.* **2018**, *59*, 1769–1778. [CrossRef] [PubMed]
13. Dysli, C.-S.; Dysli, M.; Enzmann, V.; Wolf, S.; Zinkernagel, M.S. Fluorescence Lifetime Imaging of the Ocular Fundus in Mice. *Investig. Opthalmol. Vis. Sci.* **2014**, *55*, 7206–7215. [CrossRef] [PubMed]
14. Peters, S.; Hammer, M.; Schweitzer, D. Two-photon excited fluorescence microscopy application for ex vivo investigation of ocular fundus samples. In Proceedings of the European Conferences on Biomedical Optics, Munich, Germany, 22–26 May 2011; Volume 8086, p. 808605. [CrossRef]
15. Batista, A.; Breunig, H.G.; König, A.; Morgado, A.M.; König, K. Assessment of the metabolism and morphology of the porcine cornea, lens and retina by 2-photon imaging. *J. Biophotonics* **2018**, *11*, e201700324. [CrossRef] [PubMed]
16. Hayaran, A.; Bijlani, V. Polyacrylamide as an infiltrating and embedding medium for vibratome sectioning of human fetal cerebellum containing Dil-filled axons. *J. Neurosci. Methods* **1992**, *42*, 65–68. [CrossRef]
17. Zipfel, W.R.; Williams, R.M.; Christie, R.; Nikitin, A.Y.; Hyman, B.T.; Webb, W.W. Live tissue intrinsic emission microscopy using multiphoton-excited native fluorescence and second harmonic generation. *Proc. Natl. Acad. Sci. USA* **2003**, *100*, 7075–7080. [CrossRef]
18. Ranjit, S.; Dvornikov, A.; Stakic, M.; Hong, S.-H.; Levi, M.; Evans, R.M.; Gratton, E. Imaging Fibrosis and Separating Collagens using Second Harmonic Generation and Phasor Approach to Fluorescence Lifetime Imaging. *Sci. Rep.* **2015**, *5*, srep13378. [CrossRef]
19. Gao, S.; Lan, X.; Liu, Y.; Shen, Z.; Lu, J.; Ni, X. Characteristics of blood fluorescence spectra using low-level, 457.9-nm excitation from Ar+ laser. *Chin. Opt. Lett.* **2004**, *2*, 160–161.
20. Digman, M.A.; Caiolfa, V.R.; Zamai, M.; Gratton, E. The Phasor Approach to Fluorescence Lifetime Imaging Analysis. *Biophys. J.* **2008**, *94*, L14–L16. [CrossRef]
21. Stringari, C.; Cinquin, A.; Cinquin, O.; Digman, M.A.; Donovan, P.J.; Gratton, E. Phasor approach to fluorescence lifetime microscopy distinguishes different metabolic states of germ cells in a live tissue. *Proc. Natl. Acad. Sci. USA* **2011**, *108*, 13582–13587. [CrossRef]
22. Sitiwin, E.; Madigan, M.C.; Gratton, E.; Cherepanoff, S.; Conway, R.M.; Whan, R.; Macmillan, A. Shedding light on melanins within in situ human eye melanocytes using 2-photon microscopy profiling techniques. *Sci. Rep.* **2019**, *9*, 18585. [CrossRef] [PubMed]
23. Roth, S.; Freund, I. Second harmonic generation in collagen. *J. Chem. Phys.* **1979**, *70*, 1637–1643. [CrossRef]
24. Chacko, J.V.; Eliceiri, K.W. NAD(P)H fluorescence lifetime measurements in fixed biological tissues. *Methods Appl. Fluoresc.* **2019**, *7*, 044005. [CrossRef] [PubMed]
25. Rueda, E.M.; Johnson, J.E., Jr.; Giddabasappa, A.; Swaroop, A.; Brooks, M.J.; Sigel, I.; Chaney, S.Y.; Fox, D.A. The cellular and compartmental profile of mouse retinal glycolysis, tricarboxylic acid cycle, oxidative phosphorylation, and ~P transferring kinases. *Mol. Vis.* **2016**, *22*, 847–885.
26. Volland, S.; Esteve-Rudd, J.; Hoo, J.; Yee, C.; Williams, D.S. A Comparison of Some Organizational Characteristics of the Mouse Central Retina and the Human Macula. *PLoS ONE* **2015**, *10*, e0125631. [CrossRef]
27. Dräger, U.C.; Olsen, J.F. Origins of crossed and uncrossed retinal projections in pigmented and albino mice. Mouse Retinal Ganglion Cells. *J. Comp. Neurol.* **1980**, *191*, 383–412. [CrossRef]
28. Alam, S.R.; Wallrabe, H.; Svindrych, Z.; Chaudhary, A.K.; Christopher, K.G.; Chandra, D.; Periasamy, A. Investigation of Mitochondrial Metabolic Response to Doxorubicin in Prostate Cancer Cells: An NADH, FAD and Tryptophan FLIM Assay. *Sci. Rep.* **2017**, *7*, 2977. [CrossRef]
29. Cao, R.; Wallrabe, H.K.; Periasamy, A. Multiphoton FLIM imaging of NAD(P)H and FAD with one excitation wavelength. *J. Biomed. Opt.* **2020**, *25*, 014510. [CrossRef]
30. Yamagata, M.; Sanes, J.R. Expression and Roles of the Immunoglobulin Superfamily Recognition Molecule Sidekick1 in Mouse Retina. *Front. Mol. Neurosci.* **2019**, *11*, 485. [CrossRef]
31. Contini, M.; Lin, B.; Kobayashi, K.; Okano, H.; Masland, R.H.; Raviola, E. Synaptic input of ON-bipolar cells onto the dopaminergic neurons of the mouse retina. *J. Comp. Neurol.* **2010**, *518*, 2035–2050. [CrossRef]
32. Knop, G.C.; Feigenspan, A.; Weiler, R.; Dedek, K. Inputs Underlying the ON–OFF Light Responses of Type 2 Wide-Field Amacrine Cells in TH::GFP Mice. *J. Neurosci.* **2011**, *31*, 4780–4791. [CrossRef]
33. Clark, M.A.; Worrell, M.B.; Pless, J.E. Postmortem Changes in Soft Tissues. In *Forensic Taphonomy: The Postmortem Fate of Human Remains*; Haglund, W.D., Sorg, M.H., Eds.; CRC Press: Boca Raton, FL, USA, 1997; pp. 156–164.

34. Prahst, C.; Ashrafzadeh, P.; Mead, T.; Figueiredo, A.; Chang, K.; Richardson, D.; Venkaraman, L.; Richards, M.; Russo, A.M.; Harrington, K.; et al. Mouse retinal cell behaviour in space and time using light sheet fluorescence microscopy. *eLife* **2020**, *9*, e49779. [CrossRef] [PubMed]
35. Wang, M.-H.; Hsiao, G.; Al-Shabrawey, M. Eicosanoids and Oxidative Stress in Diabetic Retinopathy. *Antioxidants* **2020**, *9*, 520. [CrossRef] [PubMed]
36. Giannakaki-Zimmermann, H.; Kokona, D.; Wolf, S.; Ebneter, A.; Zinkernagel, M.S. Optical Coherence Tomography Angiography in Mice: Comparison with Confocal Scanning Laser Microscopy and Fluorescein Angiography. *Transl. Vis. Sci. Technol.* **2016**, *5*, 11. [CrossRef] [PubMed]
37. Tezel, T.H.; Geng, L.; Lato, E.B.; Schaal, S.; Liu, Y.; Dean, D.; Klein, J.B.; Kaplan, H.J.; Ellis, D.Z.; Dismuke, W.M.; et al. Synthesis and Secretion of Hemoglobin by Retinal Pigment Epithelium. *Investig. Opthalmol. Vis. Sci.* **2009**, *50*, 1911–1919. [CrossRef]
38. Cox, G.; Kable, E.; Jones, A.; Fraser, I.; Manconi, F.; Gorrell, M.D. 3-Dimensional imaging of collagen using second harmonic generation. *J. Struct. Biol.* **2003**, *141*, 53–62. [CrossRef]
39. Harnisch, J.-P.; Buchen, R.; Sinha, P.K.; Barrach, H.J. Ultrastructural identification of type I and II collagen in the cornea of the mouse by means of enzyme labeled antibodies. *Albrecht Von Graefes Arch. Für Klin. Exp. Ophthalmol.* **1978**, *208*, 9–13. [CrossRef]
40. Keeley, F.W.; Morin, J.D.; Vesely, S. Characterization of collagen from normal human sclera. *Exp. Eye Res.* **1984**, *39*, 533–542. [CrossRef]
41. Park, C.Y.; Marando, C.; Liao, J.A.; Lee, J.K.; Kwon, J.; Chuck, R.S. Details of the Collagen and Elastin Architecture in the Human Limbal Conjunctiva, Tenon's Capsule and Sclera Revealed by Two-Photon Excited Fluorescence Microscopy. *Investig. Opthalmology Vis. Sci.* **2016**, *57*, 5602–5610. [CrossRef]
42. Zyablitskaya, M.; Takaoka, A.; Munteanu, E.L.; Nagasaki, T.; Trokel, S.L.; Paik, D.C. Evaluation of Therapeutic Tissue Crosslinking (TXL) for Myopia Using Second Harmonic Generation Signal Microscopy in Rabbit Sclera. *Investig. Opthalmol. Vis. Sci.* **2017**, *58*, 21–29. [CrossRef]

cells MDPI

Article

Non-Invasive Evaluation of Retinal Vascular Alterations in a Mouse Model of Optic Neuritis Using Laser Speckle Flowgraphy and Optical Coherence Tomography Angiography

Seth E. Buscho [1,†], Fan Xia [1,†], Shuizhen Shi [1], Jonathan L. Lin [1,2], Bartosz Szczesny [1,3], Wenbo Zhang [1,4], Massoud Motamedi [1] and Hua Liu [1,*]

1 Department of Ophthalmology and Visual Sciences, University of Texas Medical Branch, Galveston, TX 77555, USA; sebuscho@utmb.edu (S.E.B.); faxia@utmb.edu (F.X.); shushi@utmb.edu (S.S.); jolin@utmb.edu (J.L.L.); baszczes@utmb.edu (B.S.); we2zhang@utmb.edu (W.Z.); mmotamed@utmb.edu (M.M.)
2 Department of Pharmacology and Toxicology, University of Texas Medical Branch, Galveston, TX 77555, USA
3 Department of Anesthesiology, University of Texas Medical Branch, Galveston, TX 77555, USA
4 Department of Neurobiology, University of Texas Medical Branch, Galveston, TX 77555, USA
* Correspondence: hualiu@utmb.edu; Tel.: +1-(409)-742-3793; Fax: +1-(409)-747-5402
† These authors contributed equally to this research.

Abstract: Optic neuritis, a characteristic feature of multiple sclerosis (MS), involves the inflammation of the optic nerve and the degeneration of retinal ganglion cells (RGCs). Although previous studies suggest that retinal blood flow alterations occur during optic neuritis, the precise location, the degree of impairment, and the underlying mechanisms remain unclear. In this study, we utilized two emerging non-invasive imaging techniques, laser speckle flowgraphy (LSFG) and optical coherence tomography angiography (OCTA), to investigate retinal vascular changes in a mouse model of MS, known as experimental autoimmune encephalomyelitis (EAE). We associated these changes with leukostasis, RGC injury, and the overall progression of EAE. LSFG imaging revealed a progressive reduction in retinal blood flow velocity and increased vascular resistance near the optic nerve head in the EAE model, indicating impaired ocular blood flow. OCTA imaging demonstrated significant decreases in vessel density, number of junctions, and total vessel length in the intermediate and deep capillary plexus of the EAE mice. Furthermore, our analysis of leukostasis revealed a significant increase in adherent leukocytes in the retinal vasculature of the EAE mice, suggesting the occurrence of vascular inflammation in the early development of EAE pathology. The abovechanges preceded or were accompanied by the characteristic hallmarks of optic neuritis, such as RGC loss and reduced visual acuity. Overall, our study sheds light on the intricate relationship between retinal vascular alterations and the progression of optic neuritis as well as MS clinical score. It also highlights the potential for the development of image-based biomarkers for the diagnosis and monitoring of optic neuritis as well as MS, particularly in response to emerging treatments.

Keywords: multiple sclerosis (MS); optic neuritis; laser speckle flowgraphy (LSFG); optical coherence tomography angiography (OCTA); retinal vasculature; retinal blood flow; retinal vascular inflammation; biomarker

Citation: Buscho, S.E.; Xia, F.; Shi, S.; Lin, J.L.; Szczesny, B.; Zhang, W.; Motamedi, M.; Liu, H. Non-Invasive Evaluation of Retinal Vascular Alterations in a Mouse Model of Optic Neuritis Using Laser Speckle Flowgraphy and Optical Coherence Tomography Angiography. *Cells* **2023**, *12*, 2685. https://doi.org/10.3390/cells12232685

Academic Editor: Paola Bagnoli

Received: 14 August 2023
Revised: 4 November 2023
Accepted: 20 November 2023
Published: 22 November 2023

1. Introduction

Multiple sclerosis (MS) is a progressive autoimmune disorder marked by the demyelination of the central nervous system (CNS). It primarily presents in young adulthood and is estimated to affect 2.8 million people worldwide [1]. MS severely affects the quality of life of patients by inhibiting their ability to work, socialize, pursue recreational activities, and maintain self-care [2]. Symptoms of MS observed in a clinical setting encompass a range of issues, such as impaired vision, movement disorders, dizziness, cognitive dysfunction, slurred speech, and difficulties related to sexual, bowel, and bladder function [1]. Among them, optic neuritis (ON) is a major clinical feature and can lead to visual impairment or even blindness. Around 30–70% of MS patients develop optic neuritis during their disease course, and up to 94–99% of patients with post-mortem examinations show demyelinating plaques in the optic nerves [2].

ON is characterized by inflammation of the optic nerve [3]. It was previously thought that immune cells, particularly activated T cells, were responsible for attacking the myelin sheath surrounding the optic nerve, leading to the demyelination, dysfunction, and degeneration of the optic nerve [3]. The thinning of the retinal nerve fiber layer (RNFL) and the dysfunction and degeneration of retinal ganglion cells (RGCs) occur as a result of the degeneration of the optic nerve, ultimately leading to an irreversible loss of vision [4,5]. Moreover, demyelination-independent neurodegeneration also contributes to the pathology of ON during MS [6]. Beyond RGC degeneration, clinical studies using optical coherence tomography angiography (OCTA) suggest that retinal vessel density and blood perfusion are altered in ON. However, the data are often contradictory. Some studies indicated that eyes from MS patients had significantly decreased retinal vessel density in the superficial vascular complex (SVC) and deep vascular complex (DVC) compared with healthy controls and that vessel density reduction was even greater in eyes with ON [7–11], associated with reduced oxygen extraction and retinal blood flow [12]; however, other studies did not find any vessel density reduction in MS eyes compared with healthy eyes or a difference between MS eyes with vs. without ON [8–10]. There are also many inconsistencies regarding the location of and mechanisms underlying these vascular changes [8–10]. Considering the close interaction between retinal vessels and neurons, and that vascular dysfunction may contribute to neurodegeneration, it is important to further clarify whether vascular alterations occur in ON or not.

In addition to the use of OCTA for the non-invasive monitoring and quantification of MS-induced alterations in retinal vascular features, laser speckle flowgraphy (LSFG) has emerged as a valuable non-invasive tool to monitor changes in retinal blood flow under various pathological conditions [13–18]. By utilizing the laser speckle phenomenon, LSFG enables the quantitative estimation of blood flow in ocular tissues, including the retina. This technology provides a convenient and highly sensitive method for assessing alterations in retinal blood flow with promising capabilities to capture dynamic changes in blood flow, yielding improvement in the current understanding of the pathophysiology of retinal vascular disorders, thus offering valuable insights for diagnosis, monitoring, and potential treatment strategies for conditions such as MS.

The current understanding of retinal vascular changes in animal models accompanying ON is limited. To address this, we conducted a longitudinal evaluation of retinal vascular alterations in the mouse model of experimental autoimmune encephalomyelitis (EAE), a well-established model for studying ON [19]. Utilizing two clinically relevant non-invasive imaging techniques, OCTA and LSFG, we characterized retinal vascular changes in the EAE model during disease progression. Furthermore, we performed a retinal leukostasis study to characterize the pathophysiology of ON involving inflammation and vascular dysfunction, which allowed us to relate these changes to the findings from non-invasive assessments of retinal vascular alterations in the EAE model using OCTA and LSFG imaging techniques. This study contributes to a deeper understanding of the retinal vascular dynamics associated with ON and has the potential to impact the development of more effective diagnosis and treatment strategies for ON as well as MS.

2. Materials and Methods

2.1. Animals

In our current pre-clinical study, which received approval from the Institutional Animal Care and Use Committee at the University of Texas Medical Branch, we utilized C57BL/6J wild type (WT) mice (Stock No: 000664) and C57BL/6-EGFP (Stock No: 006567) from the Jackson Laboratory (Bar Harbor, ME, USA). The mice were bred in the animal care facility at the University of Texas Medical Branch. All of the experimental procedures and animal usage strictly adhered to the guidelines provided by the Association for Research in Vision and Ophthalmology Statement for the Use of Animals in Ophthalmic and Vision Research.

2.2. EAE Model

We chose to exclusively use female mice for this study, because C57BL/6J males do not display severe clinical symptoms and reach the maximum clinical score [20]. At 13 weeks of age, we induced the EAE model using the MOG 35-55/CFA EMULSION PTX kit (#EK2110, Hooke Laboratories, Lawrence, MA, USA), as previously described [4]. Briefly, an emulsion containing myelin oligodendrocyte glycoprotein (MOG) 35-55 peptide and complete Freund's adjuvant was subcutaneously injected into each hind leg (0.1 mL/leg). At 2 and 24 h after injection, pertussis toxin (PTX) (100 ng/mouse) was administered intraperitoneally. The control mice were injected with an emulsion containing complete Freund's adjuvant without MOG 35-55 peptide and PTX. After the induction of immunization, optic neuritis typically starts at 2 weeks and peaks at 3 weeks [5]. The clinical score for EAE (n = 7) and control (n = 5) mice was obtained at 0, 7, 10, 14, 21, and 28 days post-immunization (dpi), in accordance with the manual of the MOG 35-55/CFA EMULSION PTX kit and based on changes in the motor functions of the tail, legs, and neck. Rapid increases in the clinical score usually occur between 12 and 18 dpi [21].

2.3. Bone Marrow (BM) Transplantation and Scanning Laser Ophthalmoscope (SLO)

BM transplantation was performed as previously described [22,23]. Briefly, 8-week-old WT mice were irradiated at a dose of 850 rads (8.5 Gy) with a Gammacell 40 irradiator (MDS Nordion, Ottawa, ON, Canada). A lead shield was placed on the head and eyes of each mouse to protect the head and eyes from potential radiation injury. We subsequently obtained BM cells from the femurs and tibias of donor GFP mice, followed by washing, counting, and resuspension in phosphate-buffered saline (PBS). Within 24 h of irradiation, we injected 200 μL of the cell suspension (1.0×10^7 cells) into WT recipient mice via tail vein. After 4 weeks, the EAE model was induced in WT recipient mice. Right before or 7 days after EAE induction, SLO imaging was performed. Briefly, the mice were injected intraperitoneally with 100 mg/kg ketamine and 10 mg/kg xylazine for anesthesia, and their eyes were dilated with topical tropicamide and phenylephrine. The mice were then placed on a heated platform, and the Heidelberg Spectralis HRA system (Heidelberg Engineering, Franklin, MA, USA) was used to record in vivo movement of GFP-tagged leukocytes in the retinas. The frame rate was set to 5 frames/second with a 55° field of view, and each image consisted of 100 frames. Leukocytes were counted for each retina (n = 6 pre-EAE, n = 6 EAE).

2.4. Leukostasis

Retinal leukostasis was performed as previously described [22,24,25]. In short, 14 days after EAE induction, the anesthesia of the mice was achieved with an intraperitoneal injection, consisting of 100 mg/kg ketamine and 10 mg/kg xylazine. The chest cavity was opened, and the right atrium was incised to allow for outflow. A perfusion catheter was inserted into the left ventricle. Following the removal of erythrocytes and nonadherent leukocytes using PBS, adherent leukocytes and retinal vasculature were labeled by perfusion with rhodamine-coupled concanavalin A (Con A) lectin (40 g/mL in PBS, pH 7.4; Vector Laboratories, Burlingame, CA, USA). PBS was subsequently re-perfused to eliminate any remaining unbound Con A. The eyeballs of the mice were collected and fixed overnight with 4% paraformaldehyde (PFA) at 4 °C, and then their retinas (n = 6 Control, n = 6 EAE) were dissected for the quantification of the total number of adherent leukocytes.

2.5. LSFG Imaging

The retinas of the EAE and control mice were imaged using an LSFG-Micro System (Softcare Co., Ltd., Fukuoka, Japan) at 7, 10, 14, and 28 dpi (n = 9–10 Control, n = 14–16 EAE). Prior to LSFG imaging, the anesthesia of the mice was achieved with an intraperitoneal injection, consisting of 100 mg/kg ketamine and 10 mg/kg xylazine. The eyes were dilated with one drop of phenylephrine and one drop of tropicamide. GenTeal® Tears Eye Gel (Alcon Laboratories, Inc., Fort Worth, TX, USA) was lightly applied to the mouse corneas, a circular glass slide was placed on top of the gel, and the mice were then positioned on the LSFG imaging platform. Images were acquired with the optic nerve head (ONH) in the center over a 4 s (118 frames) period at a rate of 30 frames per second to produce a 3.2×2.5 mm^2 composite area. At each timepoint, each eye was measured three consecutive times without changing positions between each scan. The average of these values was used for subsequent statistical analyses.

2.6. LSFG Analysis

LSFG images were analyzed using the LSFG Analyzer software (version 3.00) (Softcare Co., Ltd., Fukuoka, Japan) (Figure 1). Recorded parameters are described in Table 1. First, the mean blur rate (MBR) of the vessel area (MV), tissue area (MT), and overall area (MA) was quantified within a 200-pixel diameter circle centered around the ONH. Second, waveform analysis was performed, which yielded the beat strength (BS), a marker of the fluctuation in blood flow at the onset of a heartbeat; the BS over MBR (BOM), a marker of peripheral vascular resistance; the maximum MBR over a cardiac cycle (Max MBR); the minimum MBR over a cardiac cycle (Min MBR); and the Max–Min MBR over a cardiac cycle. Third, the volume of blood flow through retinal arteries and veins was quantified as the total retinal flow index divided by the vessel width (TRFI/w), using the total retinal artery and vein analyzer (TRAVA). All of the examiners were blinded to the mouse treatment during LSFG image acquisition and analysis.

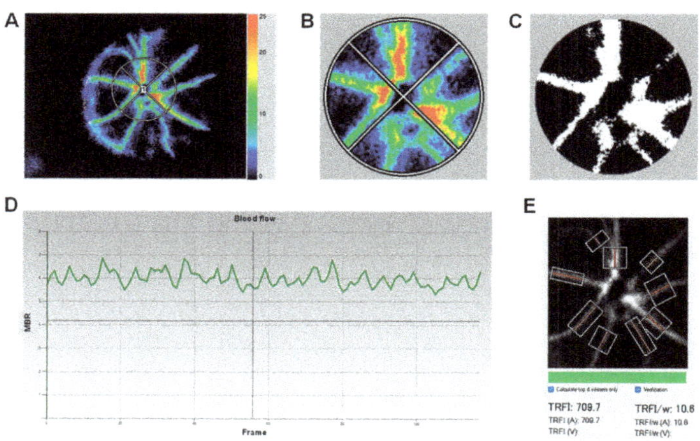

Figure 1. Illustration of the analysis of MBR and waveform values from LSFG. (**A**) Representative composite color maps using the MBR as measured with LSFG with a 200-pixel diameter encircling the optic nerve head. (**B**) Enlargement of the optic nerve head area. The red color indicates high MBR, and the blue color indicates low MBR. (**C**) LSFG Analyzer Software automated binarization of the image into vessels (white area) and tissue (black area). (**D**) Fluctuations of MBR throughout a 118-frame scan with a calculation of waveform parameters: beat strength (BS), beat strength over MBR (BOM), Max MBR, Min MBR, and Max–Min MBR. (**E**) The total retinal artery and vein analyzer (TRAVA) automated the calculation of the total retinal flow index divided by the vessel width (TRFI/w) for the top four highest velocity arteries only (top four in bright red).

Table 1. LSFG parameters with their associated descriptions.

Parameter	Description
MV	MV indicates intravascular flow velocity for large retinal vessels (i.e., arteries, veins, arterioles, venules) determined by an automated MBR threshold.
MT	MT indicates blood flow velocity within tissue (i.e., capillaries) as predefined by the automated MBR threshold.
MA	MA indicates blood flow velocity in both vascular and tissue areas.
BS	BS indicates fluctuation of blood flow at the onset of a cardiac cycle and is proportional to the peripheral vascular resistance.
BOM	The BOM is defined as the BS divided by the average MBR value in a rubber band. This parameter expresses peripheral vascular resistance.
Max MBR	The maximum MBR obtained during the 4 s (118 frames) scan.
Min MBR	The minimum MBR obtained during the 4 s (118 frames) scan.
Max–Min MBR	The minimum MBR obtained is subtracted from the maximum MBR to obtain the Max–Min MBR. An elevated Max–Min MBR suggests large variation in blood flow.
TRFI/w	The volume of blood flow occurring in four vessels of the retina with the highest blood flow.

2.7. OCTA Imaging

OCTA scans were performed for each retina (n = 14 Control, n = 16 EAE) at 14 dpi. Prior to imaging, the anesthesia of the mice was achieved with 3.8% isoflurane using a Digital Low-Flow Anesthesia System (Kent Scientific, Torrington, CT, USA), and phenylephrine and tropicamide were administered to their eyes to induce mydriasis. Next, the mice were wrapped in cotton gauze to inhibit their movement and maintain their body temperature. They were subsequently placed on a heating pad on the imaging platform with a continuous supply of isoflurane. GenTeal® Tears Eye Gel was applied to the eye that was not actively being imaged to prevent corneal clouding. OCTA images were acquired using the Heidelberg HRA + OCT Spectralis (Heidelberg Engineering, Heidelberg, Germany), as previously described [26]. Briefly, a 20° by 20° volume was captured of each quadrant of the retina using the 30° lens and mouse adaptor lens.

2.8. OCTA Analysis

OCTA scans with significant image artifacts or a quality-control index rated at 25 dB or less, as determined by the Heidelberg Eye Explorer version 6.12.4.0, were omitted from data analysis. The vessel density, total vessel length, and total number of junctions were recorded for each OCTA image using the AngioTool software (National Cancer Institute, Bethesda, MD, USA) [26,27]. The mean value was subsequently calculated across all imaged quadrants for one retina to yield each data point for statistical analysis. The examiners were blinded to the mouse treatments during OCTA image acquisition and analysis.

2.9. Visual Acuity

The visual acuity of the mice (n = 9 Control, n = 14, EAE) was measured based on an innate visual–motor reflex using a plexiglass testing chamber and the software OptoMotry [28]. In brief, the mice were placed on a platform, which was surrounded by four identical screens displaying changeable moving grids. The mice were able to move freely with a limited range on the platform, and an experimenter tracked the mouse's head movement with a crosshair. The animal was determined to be able to visualize the stimulus if its head tracked the stimulus resulting from the cylinder's rotation relative to the stationary arms of the crosshair.

2.10. Retinal Flatmounts

At 28 dpi, the mouse eyeballs were fixed in 4% PFA at 4 °C overnight. Following dissection, the retinas were washed with PBS, blocked, and then permeabilized, using PBS that contained 5% donkey serum and 0.3% Triton-X-100 (ThermoFisher Scientific Waltham, MA, USA), for a period of 3 h. Next, they were incubated with an antibody against RBPMS (1:200; MilliporeSigma, Burlington, MA, USA), a marker for RGCs, at 4 °C overnight. The next day, the retinas were incubated with an Alexa Fluor 594-conjugated secondary antibody (1:400; ThermoFisher Scientific, Waltham, MA, USA) at 4 °C for 4 h. In the end, the retinas (n = 9 Control, n = 14 EAE) were mounted, and images were captured using confocal microscopy (LSM 800, Carl Zeiss Inc., Thornwood, NY, USA).

2.11. Statistical Analysis

The mean and the standard error of the mean (SEM) for both groups were reported. To compare the EAE and control mice, the student's t-test was undertaken using the GraphPad Prism program (GraphPad Software Inc., La Jolla, CA, USA). The alpha value was designated at 0.05, such that a $p < 0.05$ was considered statistically different.

3. Results

3.1. Establishment of the EAE Model

First, we sought to establish the EAE model in the mice. After disease induction, the mice were monitored for changes in body weight and EAE clinical scores, which were evaluated based on the appearance of clinical signs of EAE at 0, 7, 10, 14, 21, and 28 dpi (Figure 2A,B). While there were no significant differences in body weight and EAE clinical scores for the first 10 days, the body weight and EAE clinical scores of the EAE mice were significantly different from the control mice at 14 dpi and became progressively more severe thereafter. At 28 dpi, the EAE mice exhibited marked weight loss and severely restricted movement, as demonstrated by their high EAE score. Moreover, the staining of retinal flatmounts for RBPMS, which is an RGC marker, demonstrated a significant loss of RGCs in the EAE mice compared to the control mice (Figure 2C). These data indicate the successful establishment of the EAE model.

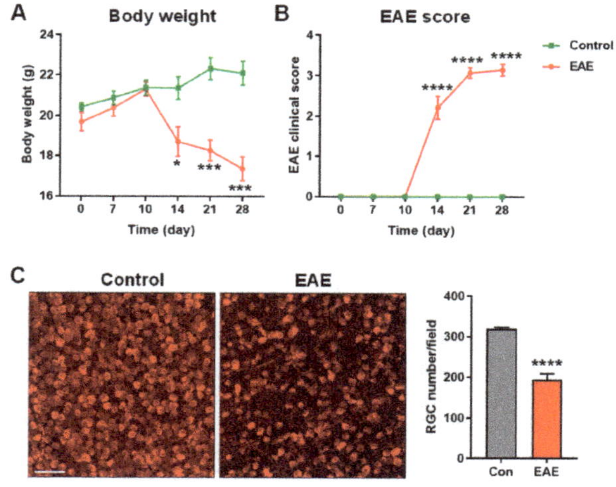

Figure 2. Evaluation of the EAE mice model. The EAE model was induced in WT mice. Body weight (**A**) and EAE clinical score (**B**) were measured at 0, 7, 10, 14, 21, and 28 dpi. n = 5–7. (**C**) Representative images of RBPMS-positive RGCs from the control and EAE mice at 28 dpi. The bar graph represents the quantification of the RGC number. Scale bar = 50 μm. n = 9–14; * $p < 0.05$, *** $p < 0.001$, **** $p < 0.0001$.

3.2. Characterization of Retinal Vasculature in EAE Using OCTA

OCTA is a dye-free, non-invasive, and cost-effective imaging modality that generates high resolution images of the retinal vasculature. Currently, it is widely used in clinical settings to examine vascular alterations in retinopathies. OCTA distinguishes vasculature from background tissue based on the movement of red blood cells, thus only detecting vasculature that is both structurally present and functionally intact [29,30]. At 14 dpi, we obtained images of the retinal vasculature of the control vs. EAE mice with OCTA and then analyzed the vessel density, counted the number of junctions, and measured the total vessel length within the superficial vascular complex (SVC), intermediate capillary plexus (ICP), and deep capillary plexus (DCP), respectively (Figure 3). Although there was no statistically significant difference in the vascular parameters in the SVC, the EAE mice displayed significantly decreased vessel density, number of junctions, and total vessel length in the ICP and DCP. These data indicate that retinal vascular alterations can be detected non-invasively using OCTA in the EAE model.

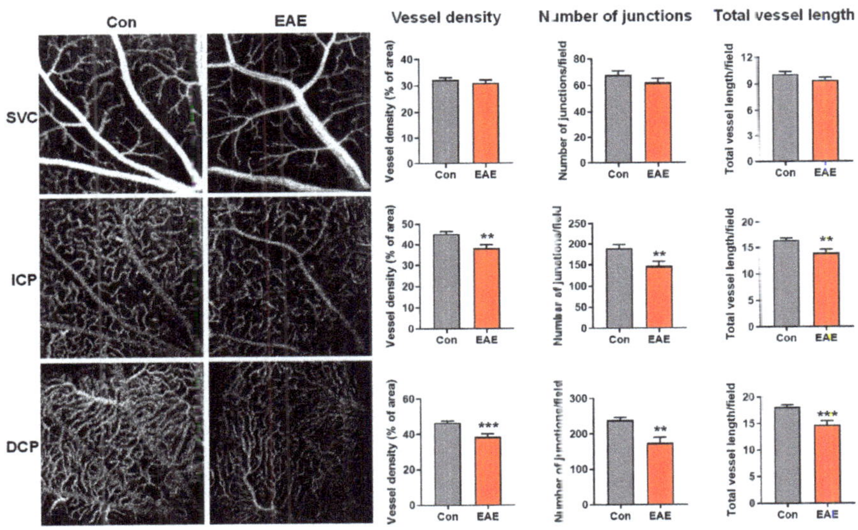

Figure 3. EAE impairs retinal vascular function. The EAE model was induced in WT mice. Retinal vasculature was examined via OCTA at 14 dpi, and representative OCTA images were shown. Vessel density, vessel branch points (number of junctions), and total vessel length were quantified with AngioTool. $n = 14–16$; ** $p < 0.01$, *** $p < 0.001$. SVC: superficial vascular complex; ICP: intermediate capillary plexus; DCP: deep capillary plexus.

3.3. Retinal Blood Flow Is Impaired in the EAE Mice

To further characterize retinal vascular alterations in EAE, we utilized the LSFG-Micro System, another non-invasive and dye-free imaging technology, to quantitatively assess ocular blood flow near the optic nerve head (Figure 4). The main output parameter of LSFG is the mean blur rate (MBR), which is calculated based on the rate of speckle pattern blurring resulting from moving blood cells, providing a quantitative index for relative blood flow velocity. In our study, the MBR was calculated over a 4 s (118 frames) period for the vascular area (MV), tissue area (MT), and all of the areas (MA) (Figure 5A). The MV was not greatly affected at 7 and 10 dpi; however, the MT and the MA were significantly reduced in the EAE mice as early as 7 dpi or 10 dpi, respectively. With the progression of EAE, the MV, MT, and MA continued to decline at 14 and 28 dpi.

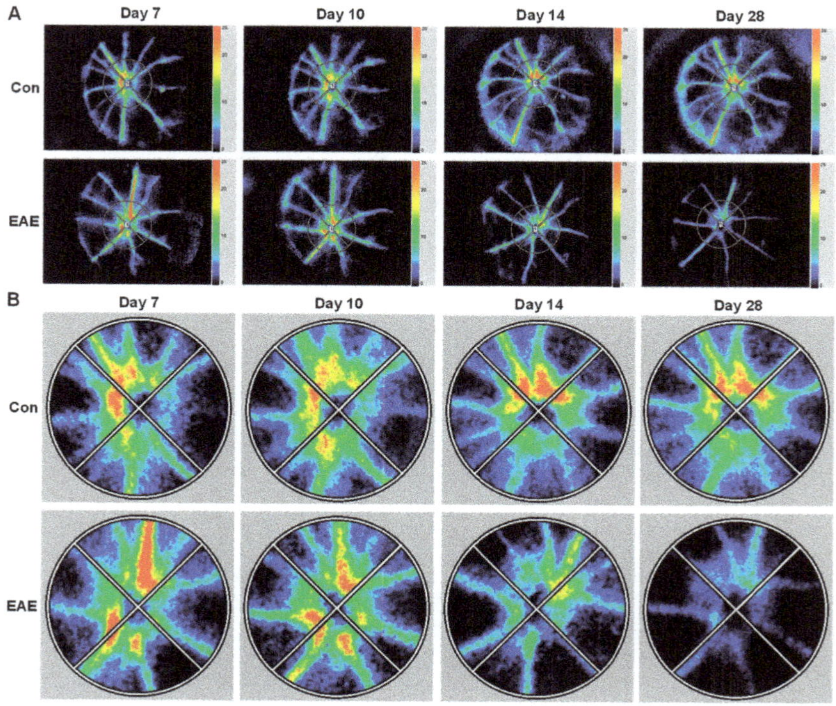

Figure 4. EAE impairs retinal blood flow. The EAE model was induced in WT mice. Retinal blood flow was examined with LSFG at 7, 10, 14, and 28 dpi. (**A**) Representative composite color maps using the MBR via LSFG with a 200-pixel diameter encircling the optic nerve head. (**B**) Enlargement of the optic nerve head area. The red color indicates high MBR, and the blue color indicates low MBR.

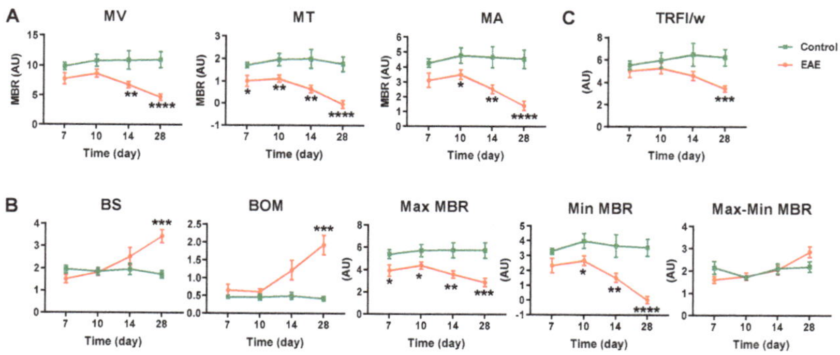

Figure 5. EAE decreases retinal blood flow velocity and increases vascular resistance. LSFG images were analyzed using the LSFG Analyzer software (version 3.00) for the MBR in MV, MT, and MA (**A**), waveform parameters BS, BOM, Max MBR, Min MBR, and Max–Min MBR in the overall area of the ONH (**B**); and TRFI/w (**C**). n = 9–16; * $p < 0.05$, ** $p < 0.01$, *** $p < 0.001$, **** $p < 0.0001$. MV: MBR in the vessel area; MT: MBR in the tissue area; MA: MBR in all areas.

In addition to the MBR, waveform analysis was also conducted to assess the BS, BOM, Max MBR, Min MBR, and the Max–Min MBR, as described in Table 1. The BS and BOM were found to not be altered early during EAE progression; however, they were significantly increased in the EAE mice at 28 dpi, indicating increased retinal vascular resistance (Figure 5B). The Max MBR and Min MBR were altered beginning at 7 dpi or

10 dpi, but the Max–Min MBR demonstrated no difference at any time point between the control and EAE mice (Figure 5B).

Finally, the total retinal flow index divided by vessel width (TRFI/w) was calculated as the retinal blood flow volume in the top four highest velocity arteries (Figures 1E and 5C). A trend of decrease in TRFI/w was observed in the EAE mice at 14 dpi, and a significant reduction was found at 28 dpi, indicating that peak blood flow volume in large retinal arteries dropped during EAE progression. In sum, our data demonstrated that retinal blood flow velocity is significantly and progressively reduced, while vascular resistance is increased during EAE progression.

3.4. Visual Acuity Is Decreased in the EAE Mice

Visual acuity assessment is one method used to monitor disease progression and assess functional impairment in MS patients [31–33]. To understand retinal neuronal changes relative to vascular dysfunction, we assessed visual acuity at different time points in the EAE model. We found that visual acuity was significantly decreased in the EAE mice starting at 10 dpi with progressive decline at 14 and 28 dpi (Figure 6). These results support the notion that visual function is impaired in EAE prior to the onset of systemic changes such as weight loss and EAE clinical score, which were observed around at 14 dpi. However, visual acuity was not impaired before the earliest detectable vascular changes, which could be identified with LSFG at 7 dpi.

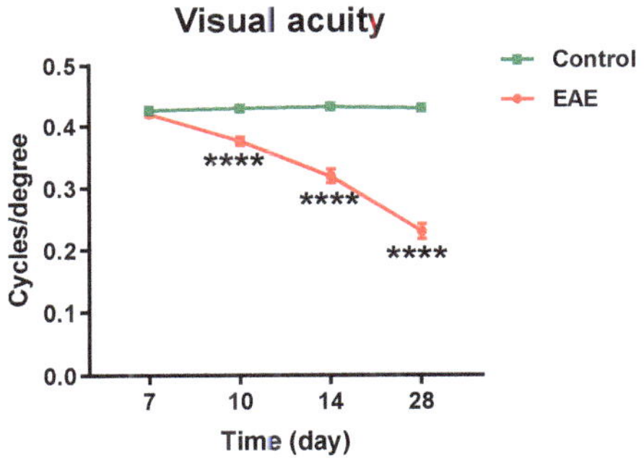

Figure 6. Visual acuity in the EAE mice. The EAE model was induced in WT mice. Visual acuity was evaluated at 7, 10, 14, and 28 dpi. n = 9–14; **** $p < 0.0001$.

3.5. Retinal Vascular Inflammation Is Increased in EAE

To understand the potential mechanisms underlying reduced blood flow in EAE, we focused on vascular inflammation, as extensive leukocyte recruitment with subsequent adhesion to the vascular endothelium could block blood flow [34,35]. Additionally, cytokines produced by leukocytes or local cells could induce vascular dysfunction, resulting in impaired vessel dilation and increased vessel constriction. Thus, we generated BM chimeric mice, which had leukocytes labeled with GFP, in order to examine leukocyte actions in the retinal vasculature via high resolution SLO imaging. Given that each image was composed of 100 frames, only fixed elements and features within the retina, such as retinal vessels and firmly attached leukocytes, were documented while capturing images. We observed a minimal number of leukocytes adherent to retinal vasculature prior to the onset of EAE, indicating there was no overt retinal vascular inflammation in pre-EAE retina. However, at 7 dpi, the number of attached leukocytes was significantly increased, particularly in retinal veins and capillaries (Figure 7A,B). The marked increase in leukostasis is consistent with a previous report that, in the optic nerve, an upregulation of inflammatory cytokines occurs as early as 7 dpi [36]. Moreover, we assessed leukocyte attachment to retinal vessels using a leukostasis assay at 14 dpi. We perfusion-labeled retinal vasculature and leukocytes with rhodamine-coupled Con A and visualized leukocyte adherence via confocal microscopy.

We found that, at 14 dpi, the number of adherent leukocytes in the central and peripheral retinal vasculature was also significantly greater in the EAE mice (Figure 7C–E). Indeed the EAE mice had an almost 6-fold higher number of adherent leukocytes compared to the control mice. In sum, using two different approaches, we demonstrated that vascular inflammation occurred in the retina early during EAE pathology development.

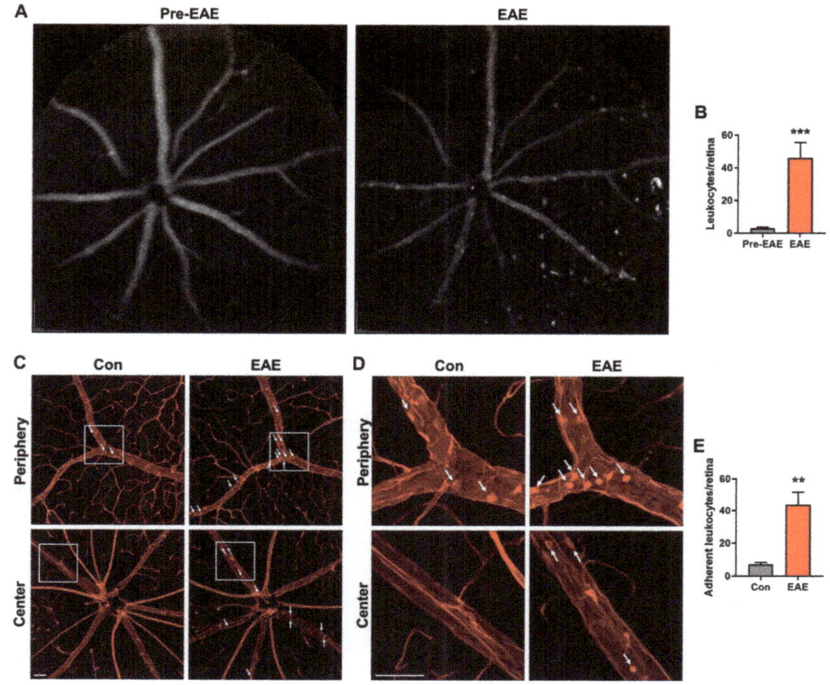

Figure 7. Leukocyte attachment is increased in the retina of the EAE mice. (**A**,**B**) SLO imaging was performed right before and 7 days after EAE induction, and stationary leukocytes in the retinas were quantified. (**C**–**E**) Leukostasis was performed 14 days after EAE induction. Representative images of leukostasis in peripheral and central retinas from the control and EAE mice were shown. Images are enlarged from the square of C. White arrows indicate adherent leukocytes. The bar graph represents the number of adherent leukocytes per retina. Scale bar = 50 μm. $n = 6$; ** $p < 0.01$, *** $p < 0.001$.

4. Discussion

Using non-invasive imaging modality, LSFG, we provide evidence that retinal blood flow velocity is significantly reduced in a mouse EAE model, which is frequently used to study the mechanisms of optic neuritis. We found that profoundly impaired vessel perfusion occurs at very early stages in EAE as demonstrated by decreased MBR in the tissue area. Initially, blood perfusion within small retinal vasculature progressively decreased, followed by alterations in large retinal vessels accompanying changes in the clinical score and body weight as indicators of disease progression. While there are no other studies that have used LSFG to characterize retinal blood flow velocity in EAE, our results are in line with animal studies that demonstrate hypoxia and hypoperfusion of the inflamed spinal cord occur during EAE [37–39]. Consistent with the reduction in retinal blood flow, an analysis of retinal vasculature with OCTA, which constructs high resolution images of retinal vascular structure based on the signal difference generated by erythrocyte motion [29,30], revealed that there was a significant loss of vessel density, total vessel length, and the total number of junctions in the ICP and DCP, two sublayers of the DVC, in the EAE mice. Overall, our results are congruent with several clinical studies that found that decreased retinal blood flow velocity using retinal function imager [40] and decreased retinal vascular density in OCTA scans are present in patients with optic neuritis [40–45]. Our findings in conjunction with the study by Bostan et al. also suggest that the DVC may be more sensitive to MS-induced vascular alterations compared to the SVC [45].

The mechanism underlying this pathognomonic decrease in blood flow velocity has yet to be understood, but it is widely known that inflammation could cause reduced blood flow, vascular blockage, and vasoconstriction [34]. We observed a dramatic increase in the number of adherent leukocytes to retinal vasculature in the EAE mice. Hence, it is likely that the decreased MBR we observed with LSFG could be due to an inhibition or blockage of blood flow by leukocytes. Yet, this mechanism does not explain the decreased MV at later stages in EAE, as leukocyte adhesion and extravasation are largely confined to the level of the post-capillary venule, which is not accounted for by MV measurement [46]. Additionally, our results indicate that the BS, a value proportional to the peripheral vascular resistance, increases at a later stage of EAE, which further supports the role of an additional modulator of retinal blood flow velocity. Therefore, it is plausible that decreased blood flow velocity could be the result of vasoconstricting factors produced during vascular inflammation. For example, immune cells such as macrophages, dendritic cells, and T cells could synthesize endothelin-1 (ET-1), which is a potent vasoconstricting factor [47]. Previous studies suggest that ET-1 may play a role in promoting hypoperfusion in the retina, extraocular arteries, and brain of patients with MS [48–50]. Additionally, the administration of ET-1 receptor antagonists has shown promise in mitigating the progression of EAE and MS and reversing cerebral hypoperfusion [49,51]. Proinflammatory cytokines such as TNF-α and IL-1β can also impair vasodilation and enhance response to vasoconstrictors [52]. Future studies are needed to elucidate the specific role of leukocytes and leukocyte-derived cytokines in the regulation of retinal blood flow in optic neuritis.

Our results also suggest that vascular impairment occurs prior to RGC dysfunction, as evidenced by a significant decrease in MT beginning at 7 dpi vs. visual acuity decline beginning at 10 dpi. It is still unclear whether vascular alterations could precipitate or contribute to retinal neurodegeneration in optic neuritis in MS. Nonetheless, significant research points towards vascular hypoperfusion playing a role in neuronal compromise through a hypoxia-induced mechanism. A decreased availability of oxygen leads to the decreased mitochondrial consumption of oxygen, with a subsequent depletion of ATP and loss of ATP-dependent metabolic processes [53,54]. One of the most significant ATP-consuming processes in the CNS, the Na^+/K^+-ATPase, promotes the extrusion of Na^+ into the extracellular space, following electrical stimulation under physiologic circumstances. In cases of ATP depletion, the excessive accumulation of intracellular sodium from a loss of the Na^+/K^+-ATPase promotes sodium–calcium exchanger activity, resulting in calcium excitotoxicity with subsequent cell death [55]. This has been further explored by studies that used the anti-epileptic sodium channel blockers phenytoin [56], lamotrigine [57], and carbamazepine [58] to treat EAE. One large clinical trial indeed showed that patients treated with phenytoin suffered from 30% less thinning of the RNFL compared to patients treated with a placebo and that phenytoin was neuroprotective against optic neuritis [59]. Another study noted that hypoxic pre-conditioning in EAE resulted in attenuated leukocyte infiltration and a less severe clinical course [60]. These studies, as well as our own, warrant a further exploration of the notion that vascular hypoperfusion is part of the pathogenesis of optic neuritis rather than an epiphenomenon or independent event.

As the number one cause of non-traumatic disabling disease in young adults [61], it is imperative that biomarkers of MS, which can reliably predict the onset and progression of the disease, are identified. At present, the gold standard methods of diagnosing MS, including magnetic resonance imaging (MRI) and cerebrospinal fluid (CSF) analyses, are not suitable for a large-scale screening of populations at risk. MRI is expensive, while CSF analysis is invasive by nature, thus increasing a patient's risk of complications [62,63]. Most of the previously identified biomarkers are either invasive or are inadequate in their reliability, limiting their applicability to clinical settings [64]. As an extension of the brain, the retina offers an ideal platform to diagnose and monitor pathologic changes within the CNS, including those caused by MS, through non-invasive imaging. A quantitative monitoring of the thinning of the RNFL or the composition of the ganglion cell layer and inner plexiform layer using OCT has emerged as a valuable tool for assessing MS-induced structural changes in the retina [65], offering potential for diagnosing and monitoring MS progression. While conventional OCT imaging provides high resolution images of the retinal structure, it is not suitable for assessing MS-induced vascular changes in the retina. However, recent advancements in OCTA technology have allowed for the evaluation of MS-induced retinal vascular alterations, providing a functional imaging modality for monitoring and quantifying these changes [10]. This functional imaging approach, combined with laser speckle analysis enabling the visualization and quantification of vascular dysfunction as demonstrated in this current study, offers a comprehensive assessment of both morphological and functional aspects of the retina, enhancing our ability to detect early signs of MS. Certainly, there is a limitation in using LSFG and OCTA, since optic neuropathies are caused by various disorders—not only MS but also trauma, sinusitis, and so on. Integrity of imaging approach with other biomarkers and patient's history is needed to rule out other possibilities and increase its specificity to diagnose optic neuritis associated with MS.

In summary, we provide original evidence that the retina undergoes significant hypoperfusion during the course of optic neuritis in a mouse EAE model. These pathologic alterations in the retina of the EAE mice clears contradictions in clinical studies and supports clinical studies showing decreased retinal vessel density in optic neuritis, similar to vascular changes observed in the brain and spinal cord of MS patients [45,66–68]. Moreover, we were able to precisely demonstrate that retinal vascular alterations predate the onset of impairment in both the visual function and systemic signs of EAE. As such, the identification of vascular changes with noninvasive imaging offers a unique opportunity for the early initiation of therapeutic intervention and potentially superior patient outcomes. This study is the first to study retinal vascular changes in an animal ON model. As animal models have a much higher degree of reproducibility than clinical studies, it would be advantageous to further explore retinal vascular alterations in an EAE model to unravel its underlying mechanisms and determine the cause-and-effect relationship between vessel dysfunction and neurodegeneration in optic neuritis. Our data using quantitative OCTA and LSFG, two cost-effective imaging modalities suitable for rapid patient screening based on the visualization and quantification of retinal vasculature alterations, have shown promise in assessing MS-induced vascular changes in the retina of a suitable small animal model for MS, potentially yielding image-based biomarkers that can serve as valuable tools for diagnosing and monitoring MS progression. Continued research in this direction holds promise for improving the accuracy and efficiency of MS diagnosis, ultimately leading to better outcomes for patients.

Author Contributions: S.E.B., F.X., W.Z. and H.L. designed the experiments. S.E.B. and F.X. performed the experiments. S.E.B., F.X. and H.L analyzed the data. S.E.B., F.X., S.S., J.L.L., B.S., W.Z., M.M. and H.L. wrote the manuscript. All authors have read and agreed to the published version of the manuscript.

Funding: Supported in part by the National Institutes of Health grants EY031054 (to H.L), EY034376 and EY034616 (to W.Z. and F.X.), EY022694 and EY026629 (to W.Z.), and Retina Research Foundation and UT System Faculty STARs Award (to W.Z.).

Institutional Review Board Statement: The animal study protocol was approved by the Institutional Animal Care and Use Committee at the University of Texas Medical Branch.

Informed Consent Statement: Not applicable.

Data Availability Statement: All data are presented in the main manuscript and the manuscript figures.

Conflicts of Interest: The authors declare no conflict of interest.

References

1. Walton, C.; King, R.; Rechtman, L.; Kaye, W.; Leray, E.; Marrie, R.A.; Robertson, N.; La Rocca, N.; Uitdehaag, B.; Van Der Mei, I.; et al. Rising prevalence of multiple sclerosis worldwide: Insights from the Atlas of MS, third edition. *Mult. Scler. J.* **2020**, *26*, 1816–1821. [CrossRef] [PubMed]
2. De Heredia-Torres, M.P.; Huertas-Hoyas, E.; Sánchez-Camarero, C.; Máximo-Bocanegra, N.; Alegre-Ayala, J.; Sánchez-Herrera-Baeza, P.; Martínez-Piédrola, R.M.; García-Bravo, C.; Mayoral-Martín, A.; Serrada-Tejeda, S. Occupational performance in multiple sclerosis and its relationship with quality of life and fatigue. *Eur. J. Phys. Rehabil. Med.* **2020**, *56*, 148–154. [CrossRef] [PubMed]
3. Toosy, A.T.; Mason, D.F.; Miller, D.H. Optic neuritis. *Lancet Neurol.* **2014**, *13*, 83–99. [CrossRef] [PubMed]
4. Palani, C.D.; Fouda, A.Y.; Liu, F.; Xu, Z.; Mohamed, E.; Giri, S.; Smith, S.B.; Caldwell, R.B.; Narayanan, S.P. Deletion of Arginase 2 Ameliorates Retinal Neurodegeneration in a Mouse Model of Multiple Sclerosis. *Mol. Neurobiol.* **2019**, *56*, 8589–8602. [CrossRef]
5. Huang, H.; Miao, L.; Liang, F.; Liu, X.; Xu, L.; Teng, X.; Wang, Q.; Ridder, W.H.; Shindler, K.S.; Sun, Y.; et al. Neuroprotection by eIF2α-CHOP inhibition and XBP-1 activation in EAE/optic neuritiss. *Cell Death Dis.* **2017**, *8*, e2936. [CrossRef]
6. Manogaran, P.; Samardzija, M.; Schad, A.N.; Wicki, C.A.; Walker-Egger, C.; Rudin, M.; Grimm, C.; Schippling, S. Retinal pathology in experimental optic neuritis is characterized by retrograde degeneration and gliosis. *Acta Neuropathol. Commun.* **2019**, *7*, 116. [CrossRef]
7. Nawar, A.; Khader, S.; Ghali, A.; Ghoneim, A. Evaluation of optical coherence tomography angiography findings in patients with multiple sclerosis. *Indian J. Ophthalmol.* **2021**, *69*, 1457–1463. [CrossRef]
8. Liu, J.; Song, S.; Gu, X.; Li, H.; Yu, X. Microvascular impairments detected by optical coherence tomography angiography in multiple sclerosis patients: A systematic review and meta-analysis. *Front. Neurosci.* **2023**, *16*, 1121899. [CrossRef]
9. Wang, X.; Wang, X.; Chou, Y.; Ma, J.; Zhong, Y. Significant retinal microvascular impairments in multiple sclerosis assessed through optical coherence tomography angiography. *Mult. Scler. Relat. Disord.* **2023**, *70*, 104505. [CrossRef]
10. Mohammadi, S.; Gouravani, M.; Salehi, M.A.; Arevalo, J.F.; Galetta, S.L.; Harandi, H.; Frohman, E.M.; Frohman, T.C.; Saidha, S.; Sattarnezhad, N.; et al. Optical coherence tomography angiography measurements in multiple sclerosis: A systematic review and meta-analysis. *J. Neuroinflamm.* **2023**, *20*, 85. [CrossRef]
11. Aly, L.; Noll, C.; Wicklein, R.; Wolf, E.; Romahn, E.F.; Wauschkuhn, J.; Hosari, S.; Mardin, C.; Berthele, A.; Hemmer, B.; et al. Dynamics of Retinal Vessel Loss After Acute Optic Neuritis in Patients with Relapsing Multiple Sclerosis. *Neurol.-Neuroimmunol. Neuroinflamm.* **2022**, *9*, e1159. [CrossRef]
12. Kallab, M.; Hommer, N.; Schlatter, A.; Bsteh, G.; Altmann, P.; Popa-Cherecheanu, A.; Pfister, M.; Werkmeister, R.M.; Schmidl, D.; Schmetterer, L.; et al. Retinal Oxygen Metabolism and Haemodynamics in Patients with Multiple Sclerosis and History of Optic Neuritis. *Front. Neurosci.* **2021**, *15*, 761654. [CrossRef]
13. Tomita, R.; Iwase, T.; Fukami, M.; Goto, K.; Ra, E.; Terasaki, H. Elevated retinal artery vascular resistance determined by novel visualized technique of laser speckle flowgraphy in branch retinal vein occlusion. *Sci. Rep.* **2021**, *11*, 20034. [CrossRef]
14. Tamplin, M.R.; Broadhurst, K.A.; Vitale, A.H.; Hashimoto, R.; Kardon, R.H.; Grumbach, I.M. Longitudinal Testing of Retinal Blood Flow in a Mouse Model of Hypertension by Laser Speckle Flowgraphy. *Transl. Vis. Sci. Technol.* **2021**, *10*, 16. [CrossRef]
15. Kunikata, H.; Nitta, F.; Aizawa, N.; Omodaka, K.; Shiga, Y.; Yasuda, M.; Nakazawa, T. The effect of intravitreal bevacizumab on ocular blood flow in diabetic retinopathy and branch retinal vein occlusion as measured by laser speckle flowgraphy. *Clin. Ophthalmol.* **2014**, *8*, 1119–1127. [CrossRef]
16. Hanazaki, H.; Yokota, H.; Aso, H.; Yamagami, S.; Nagaoka, T. Evaluation of ocular blood flow over time in a treated retinal arterial macroaneurysm using laser speckle flowgraphy. *Am. J. Ophthalmol. Case Rep.* **2021**, *21*, 101022. [CrossRef]
17. Hanaguri, J.; Yokota, H.; Watanabe, M.; Kuo, L.; Yamagami, S.; Nagaoka, T. Longitudinal stability of retinal blood flow regulation in response to flicker stimulation and systemic hyperoxia in mice assessed with laser speckle flowgraphy. *Sci. Rep.* **2020**, *10*, 19796. [CrossRef]
18. Kunikata, H.; Nakazawa, T. Recent Clinical Applications of Laser Speckle Flowgraphy in Eyes with Retinal Disease. *Asia-Pac. J. Ophthalmol.* **2016**, *5*, 151–158. [CrossRef]
19. Manogaran, P.; Walker-Egger, C.; Samardzija, M.; Waschkies, C.; Grimm, C.; Rudin, M.; Schippling, S. Exploring experimental autoimmune optic neuritis using multimodal imaging. *NeuroImage* **2018**, *175*, 327–339. [CrossRef]
20. McCombe, P.A.; Greer, J.M. Effects of biological sex and pregnancy in experimental autoimmune encephalomyelitis: It's complicated. *Front. Immunol.* **2022**, *13*, 1059833. [CrossRef]
21. Hasselmann, J.P.; Karim, H.; Khalaj, A.J.; Ghosh, S.; Tiwari-Woodruff, S.K. Consistent induction of chronic experimental autoimmune encephalomyelitis in C57BL/6 mice for the longitudinal study of pathology and repair. *J. Neurosci. Methods* **2017**, *284*, 71–84. [CrossRef] [PubMed]
22. Liu, W.; Ha, Y.; Xia, F.; Zhu, S.; Li, Y.; Shi, S.; Mei, F.C.; Merkley, K.; Vizzeri, G.; Motamedi, M.; et al. Neuronal Epac1 mediates retinal neurodegeneration in mouse models of ocular hypertension. *J. Exp. Med.* **2020**, *217*, 4. [CrossRef] [PubMed]
23. Ha, Y.; Liu, H.; Zhu, S.; Yi, P.; Liu, W.; Nathanson, J.; Kayed, R.; Loucas, B.; Sun, J.; Frishman, L.J.; et al. Critical Role of the CXCL10/C-X-C Chemokine Receptor 3 Axis in Promoting Leukocyte Recruitment and Neuronal Injury during Traumatic Optic Neuropathy Induced by Optic Nerve Crush. *Am. J. Pathol.* **2016**, *187*, 352–365. [CrossRef] [PubMed]
24. Xia, F.; Ha, Y.; Shi, S.; Li, Y.; Li, S.; Luisi, J.; Kayed, R.; Motamedi, M.; Liu, H.; Zhang, W. Early alterations of neurovascular unit in the retina in mouse models of tauopathy. *Acta Neuropathol. Commun.* **2021**, *9*, 51. [CrossRef] [PubMed]

25. Liu, W.; Xia, F.; Ha, Y.; Zhu, S.; Li, Y.; Folorunso, O.; Pashaei-Marandi, A.; Lin, P.-Y.; Tilton, R.G.; Pierce, A.P.; et al. Neuroprotective Effects of HSF1 in Retinal Ischemia-Reperfusion Injury. *Investig. Opthalmol. Vis. Sci.* **2019**, *60*, 965–977. [CrossRef]
26. Buscho, S.; Palacios, E.; Xia, F.; Shi, S.; Li, S.; Luisi, J.; Kayed, R.; Motamedi, M.; Zhang, W.; Liu, H. Longitudinal characterization of retinal vasculature alterations with optical coherence tomography angiography in a mouse model of tauopathy. *Exp. Eye Res.* **2022**, *224*, 109240. [CrossRef]
27. Zudaire, E.; Gambardella, L.; Kurcz, C.; Vermeren, S. A Computational Tool for Quantitative Analysis of Vascular Networks. *PLoS ONE* **2011**, *6*, e27385. [CrossRef]
28. Prusky, G.T.; Alam, N.M.; Beekman, S.; Douglas, R.M. Rapid Quantification of Adult and Developing Mouse Spatial Vision Using a Virtual Optomotor System. *Investig. Opthalmol. Vis. Sci.* **2004**, *45*, 4611–4616. [CrossRef]
29. E de Carlo, T.; Romano, A.; Waheed, N.K.; Duker, J.S. A review of optical coherence tomography angiography (OCTA). *Int. J. Retin. Vitr.* **2015**, *1*, 5. [CrossRef]
30. Spaide, R.F.; Fujimoto, J.G.; Waheed, N.K. Optical Coherence Tomography Angiography. *Retina* **2015**, *35*, 2161–2162. [CrossRef]
31. Garcia-Martin, E.; de Gopegui, E.R.; Satue, M.; Gil-Arribas, L.; Jarauta, L.; Ara, J.R.; Martin, J.; Fernandez, F.J.; Vilades, E.; Rodrigo, M.J. Progressive Functional and Neuroretinal Affectation in Patients with Multiple Sclerosis Treated with Fingolimod. *J. Neuro-Ophthalmol.* **2020**, *41*, e415–e423. [CrossRef] [PubMed]
32. Balcer, L.J.; Raynowska, J.; Nolan, R.; Galetta, S.L.; Kapoor, R.; Benedict, R.; Phillips, G.; LaRocca, N.; Hudson, L.; Rudick, R.; et al. Validity of low-contrast letter acuity as a visual performance outcome measure for multiple sclerosis. *Mult. Scler. J.* **2017**, *23*, 734–747. [CrossRef] [PubMed]
33. Balcer, L.J.; Baier, M.L.; Cohen, J.A.; Kooijmans, M.F.; Sandrock, A.W.; Nano–Schiavi, M.L.; Pfohl, D.C.; Mills, M.; Bowen, J.; Ford, C.; et al. Contrast letter acuity as a visual component for the Multiple Sclerosis Functional Composite. *Neurology* **2003**, *61*, 1367–1373. [CrossRef] [PubMed]
34. El Mathari, B.; Sene, A.; Charles-Messance, H.; Vacca, O.; Guillonneau, X.; Grepin, C.; Sennlaub, F.; Sahel, J.-A.; Rendon, A.; Tadayoni, R. Dystrophin Dp71 gene deletion induces retinal vascular inflammation and capillary degeneration. *Hum. Mol. Genet.* **2015**, *24*, 3939–3947. [CrossRef] [PubMed]
35. Hernández, J.C.C.; Bracko, O.; Kersbergen, C.J.; Muse, V.; Haft-Javaherian, M.; Berg, M.; Park, L.; Vinarcsik, L.K.; Ivasyk, I.; Rivera, D.A.; et al. Neutrophil adhesion in brain capillaries reduces cortical blood flow and impairs memory function in Alzheimer's disease mouse models. *Nat. Neurosci.* **2019**, *22*, 413–420. [CrossRef] [PubMed]
36. Liu, Y.; You, C.; Zhang, Z.; Zhang, J.; Yan, H. Roles of Treg/Th17 Cell Imbalance and Neuronal Damage in the Visual Dysfunction Observed in Experimental Autoimmune Optic Neuritis Chronologically. *Neuromol. Med.* **2015**, *17*, 391–403. [CrossRef]
37. Davies, A.L.; Desai, R.A.; Bloomfield, P.S.; McIntosh, P.R.; Chapple, K.J.; Linington, C.; Fairless, R.; Diem, R.; Kasti, M.; Murphy, M.P.; et al. Neurological deficits caused by tissue hypoxia in neuroinflammatory disease. *Ann. Neurol.* **2013**, *74*, 815–825. [CrossRef]
38. Nikić, I.; Merkler, D.; Sorbara, C.; Brinkoetter, M.; Kreutzfeldt, M.; Bareyre, F.M.; Brück, W.; Bishop, D.; Misgeld, T.; Kerschensteiner, M. A reversible form of axon damage in experimental autoimmune encephalomyelitis and multiple sclerosis. *Nat. Med.* **2011**, *17*, 495–499. [CrossRef]
39. Nathoo, N.; Rogers, J.A.; Yong, V.W.; Dunn, J.F. Detecting Deoxyhemoglobin in Spinal Cord Vasculature of the Experimental Autoimmune Encephalomyelitis Mouse Model of Multiple Sclerosis Using Susceptibility MRI and Hyperoxygenation. *PLoS ONE* **2015**, *10*, e0127033. [CrossRef]
40. Wang, L.; Kwakyi, O.; Nguyen, J.; Ogbuokiri, E.; Murphy, O.; Caldito, N.G.; Balcer, L.; Frohman, E.; Frohman, T.; Calabresi, P.A.; et al. Microvascular blood flow velocities measured with a retinal function imager: Inter-eye correlations in healthy controls and an exploration in multiple sclerosis. *Eye Vis.* **2018**, *5*, 29. [CrossRef]
41. Yilmaz, H.; Ersoy, A.; Icel, E. Assessments of vessel density and foveal avascular zone metrics in multiple sclerosis: An optical coherence tomography angiography study. *Eye* **2020**, *34*, 771–778. [CrossRef]
42. Kleerekooper, I.; Houston, S.; Dubis, A.M.; Trip, S.A.; Petzold, A. Optical Coherence Tomography Angiography (OCTA) in Multiple Sclerosis and Neuromyelitis Optica Spectrum Disorder. *Front. Neurol.* **2020**, *11*, 604049. [CrossRef] [PubMed]
43. Murphy, O.C.; Kwakyi, O.; Iftikhar, M.; Zafar, S.; Lambe, J.; Pellegrini, N.; Sotirchos, E.S.; Gonzalez-Caldito, N.; Ogbuokiri, E.; Filippatou, A.; et al. Alterations in the retinal vasculature occur in multiple sclerosis and exhibit novel correlations with disability and visual function measures. *Mult. Scler. J.* **2020**, *26*, 815–828. [CrossRef] [PubMed]
44. Ulusoy, M.O.; Horasanlı, B.; Işık-Ulusoy, S. Optical coherence tomography angiography findings of multiple sclerosis with or without optic neuritis. *Neurol. Res.* **2020**, *42*, 319–326. [CrossRef]
45. Bostan, M.; Chua, J.; Sim, Y.C.; Tan, B.; Bujor, I.; Wong, D.; Garhöfer, G.; Tiu, C.; Schmetterer, L.; Popa-Cherecheanu, A. Microvascular changes in the macular and parafoveal areas of multiple sclerosis patients without optic neuritis. *Sci. Rep.* **2022**, *12*, 13366. [CrossRef] [PubMed]
46. Granger, D.N.; Senchenkova, E. Inflammation and the Microcirculation. In *Integrated Systems Physiology-From Cell to Function*; Morgan & Claypool Life Sciences: Williston, VT, USA, 2010.
47. Shinagawa, S.; Okazaki, T.; Ikeda, M.; Yudoh, K.; Kisanuki, Y.Y.; Yanagisawa, M.; Kawahata, K.; Ozaki, S. T cells upon activation promote endothelin 1 production in monocytes via IFN-γ and TNF-α. *Sci. Rep.* **2017**, *7*, 14500. [CrossRef]

48. Castellazzi, M.; Lamberti, G.; Resi, M.V.; Baldi, E.; Caniatti, L.M.; Galante, G.; Perri, P.; Pugliatti, M. Increased Levels of Endothelin-1 in Cerebrospinal Fluid Are a Marker of Poor Visual Recovery after Optic Neuritis in Multiple Sclerosis Patients. *Dis. Markers* **2019**, *2019*, 9320791. [CrossRef]
49. D'haeseleer, M.; Beelen, R.; Fierens, Y.; Cambron, M.; Vanbinst, A.-M.; Verborgh, C.; Demey, J.; De Keyser, J. Cerebral hypoperfusion in multiple sclerosis is reversible and mediated by endothelin-1. *Proc. Natl. Acad. Sci. USA* **2013**, *110*, 5654–5658. [CrossRef]
50. Pache, M.; Kaiser, H.J.; Akhalbedashvili, N.; Lienert, C.; Dubler, B.; Kappos, L.; Flammer, J. Extraocular Blood Flow and Endothelin-1 Plasma Levels in Patients with Multiple Sclerosis. *Eur. Neurol.* **2003**, *49*, 164–168. [CrossRef]
51. Shin, T.; Kang, B.; Tanuma, N.; Matsumoto, Y.; Wie, M.; Ahn, M.; Kang, J. Intrathecal administration of endothelin-1 receptor antagonist ameliorates autoimmune encephalomyelitis in Lewis rats. *NeuroReport* **2001**, *12*, 1465–1468. [CrossRef]
52. Vila, E.; Salaices, M. Cytokines and vascular reactivity in resistance arteries. *Am. J. Physiol. Circ. Physiol.* **2005**, *288*, H1016–H1021. [CrossRef] [PubMed]
53. Papandreou, I.; Cairns, R.A.; Fontana, L.; Lim, A.L.; Denko, N.C. HIF-1 mediates adaptation to hypoxia by actively downregulating mitochondrial oxygen consumption. *Cell Metab.* **2006**, *3*, 187–197. [CrossRef] [PubMed]
54. Desai, R.A.; Smith, K.J. Experimental autoimmune encephalomyelitis from a tissue energy perspective. *F1000Research* **2017**, *6*, 1973. [CrossRef] [PubMed]
55. Bechtold, D.A.; Smith, K.J. Sodium-mediated axonal degeneration in inflammatory demyelinating disease. *J. Neurol. Sci.* **2005**, *233*, 27–35. [CrossRef] [PubMed]
56. Lo, A.C.; Saab, C.Y.; Black, J.A.; Waxman, S.G. Phenytoin Protects Spinal Cord Axons and Preserves Axonal Conduction and Neurological Function in a Model of Neuroinflammation In Vivo. *J. Neurophysiol.* **2003**, *90*, 3566–3571. [CrossRef] [PubMed]
57. Bechtold, D.A.; Miller, S.J.; Dawson, A.C.; Sun, Y.; Kapoor, R.; Berry, D.; Smith, K.J. Axonal protection achieved in a model of multiple sclerosis using lamotrigine. *J. Neurol.* **2006**, *253*, 1542–1551. [CrossRef] [PubMed]
58. Lidster, K.; Jackson, S.J.; Ahmed, Z.; Munro, P.; Coffey, P.; Giovannoni, G.; Baker, M.D.; Baker, D. Neuroprotection in a Novel Mouse Model of Multiple Sclerosis. *PLoS ONE* **2013**, *8*, e79188. [CrossRef]
59. Raftopoulos, R.; Hickman, S.J.; Toosy, A.; Sharrack, B.; Mallik, S.; Paling, D.; Altmann, D.R.; Yiannakas, M.C.; Malladi, P.; Sheridan, R.; et al. Phenytoin for neuroprotection in patients with acute optic neuritis: A randomised, placebo-controlled, phase 2 trial. *Lancet Neurol.* **2016**, *15*, 259–269. [CrossRef]
60. Dore-Duffy, P.; Wencel, M.; Katyshev, V.; Cleary, K. Chronic mild hypoxia ameliorates chronic inflammatory activity in myelin oligodendrocyte glycoprotein (MOG) peptide induced experimental autoimmune encephalomyelitis (EAE). *Adv. Exp. Med. Biol.* **2011**, *701*, 165–173. [CrossRef]
61. Dobson, R.; Giovannoni, G. Multiple sclerosis—A review. *Eur. J. Neurol.* **2018**, *26*, 27–40. [CrossRef]
62. Sistrom, C.L.; McKay, N.L. Costs, Charges, and Revenues for Hospital Diagnostic Imaging Procedures: Differences by Modality and Hospital Characteristics. *J. Am. Coll. Radiol.* **2005**, *2*, 511–519. [CrossRef] [PubMed]
63. Engelborghs, S.; Niemantsverdriet, E.; Struyfs, H.; Blennow, K.; Brouns, R.; Comabella, M.; Dujmovic, I.; Van Der Flier, W.; Frölich, L.; Galimberti, D.; et al. Consensus guidelines for lumbar puncture in patients with neurological diseases. *Alzheimer's Dement. Diagn. Assess. Dis. Monit.* **2017**, *8*, 111–126. [CrossRef] [PubMed]
64. Kivisäkk, P.; Healy, B.C.; Francois, K.; Gandhi, R.; Gholipour, T.; Egorova, S.; Sevdalinova, V.; Quintana, F.; Chitnis, T.; Weiner, H.L.; et al. Evaluation of circulating osteopontin levels in an unselected cohort of patients with multiple sclerosis: Relevance for biomarker development. *Mult. Scler. J.* **2013**, *20*, 438–444. [CrossRef] [PubMed]
65. Petzold, A.; Balcer, L.J.; Calabresi, P.A.; Costello, F.; Frohman, T.C.; Frohman, E.M.; Martinez-Lapiscina, E.H.; Green, A.J.; Kardon, R.; Outteryck, O.; et al. Retinal layer segmentation in multiple sclerosis: A systematic review and meta-analysis. *Lancet Neurol.* **2017**, *16*, 797–812. [CrossRef] [PubMed]
66. Xia, F.; Lin, J.L.; Zhang, D.L.; Shi, S.; E Buscho, S.; Motamedi, M. Quantification of Leukocyte Trafficking in a Mouse Model of Multiple Sclerosis through In Vivo Imaging. *EC Ophthalmol.* **2022**, *13*, 2–10.
67. Uzuner, N.; Özkan, S.; Gücüyener, D.; Özdemir, G. Cerebral blood flow velocity changes to visual stimuli in patients with multiple sclerosis. *Mult. Scler. J.* **2002**, *8*, 217–221. [CrossRef]
68. West, K.L.; Sivakolundu, D.K.; Zuppichini, M.D.; Turner, M.P.; Spence, J.S.; Lu, H.; Okuda, D.T.; Rypma, B. Altered task-induced cerebral blood flow and oxygen metabolism underlies motor impairment in multiple sclerosis. *J. Cereb. Blood Flow Metab.* **2021**, *41*, 182–193. [CrossRef]

cells

MDPI

Article

Assessment of Choroidal Vasculature and Innate Immune Cells in the Eyes of Albino and Pigmented Mice

Ismail S. Zaitoun [1,2,*], Yong-Seok Song [1,2], Hammam B. Zaitoun [3], Christine M. Sorenson [2,4] and Nader Sheibani [1,2,5,6]

1 Department of Ophthalmology and Visual Sciences, School of Medicine and Public Health, University of Wisconsin, Madison, WI 53705, USA
2 McPherson Eye Research Institute, School of Medicine and Public Health, University of Wisconsin, Madison, WI 53705, USA
3 Faculty of Medicine, Yarmouk University in Irbid, Irbid 21163, Jordan
4 Department of Pediatrics, School of Medicine and Public Health, University of Wisconsin, Madison, WI 53705, USA
5 Department of Cell and Regenerative Biology, School of Medicine and Public Health, University of Wisconsin, Madison, WI 53705, USA
6 Department of Biomedical Engineering, University of Wisconsin, Madison, WI 53706, USA
* Correspondence: iszaitoun@wisc.edu

Abstract: The visualization of choroidal vasculature and innate immune cells in the eyes of pigmented mice has been challenging due to the presence of a retinal pigment epithelium (RPE) layer separating the choroid and retina. Here, we established methods for visualizing the choroidal macrophages, mast cells, and vasculature in eyes of albino and pigmented mice using cell type-specific staining. We were able to visualize the choroidal arterial and venous systems. An arterial circle around the optic nerve was found in mice similar to the Zinn–Haller arterial circle that exists in humans and primates. Three different structural patterns of choriocapillaris were observed throughout the whole choroid: honeycomb-like, maze-like, and finger-like patterns. Choroidal mast cells were relatively few but dense around the optic nerve. Mast cell distribution in the middle and periphery was different among strains. Macrophages were found in all layers of the choroid. Thus, utilizing the simple and reliable methods described herein will allow the evaluation of transgenic and preclinical mouse models of ocular diseases that affect the choroid, including age-related macular degeneration (AMD), diabetic choroidopathy, and retinopathy of prematurity. These studies will advance our understanding of the pathophysiology, and molecular and cellular mechanisms that can be targeted therapeutically, in these diseases.

Keywords: choriocapillaris; mast cells; macrophages; smooth muscle cells; uveal diseases; age-related macular degeneration

Citation: Zaitoun, I.S.; Song, Y.-S.; Zaitoun, H.B.; Sorenson, C.M.; Sheibani, N. Assessment of Choroidal Vasculature and Innate Immune Cells in the Eyes of Albino and Pigmented Mice. *Cells* **2022**, *11*, 3329. https://doi.org/10.3390/cells11203329

Academic Editors: Hossein Ameri and Volker Enzmann

Received: 29 August 2022
Accepted: 17 October 2022
Published: 21 October 2022

1. Introduction

The primary circulation system that surrounds and nourishes the outer retina and retinal pigment epithelium (RPE) is the choroidal vasculature. The choroid is a densely vascularized connective tissue consisting of three vascular layers: the anterior choriocapillaris layer that sits beneath the RPE cell basement membrane and Bruch's membrane, the middle Sattler's layer which is formed from medium size blood vessels, and the outermost Haller's layer which is formed from large blood vessels [1,2]. The choriocapillaris vasculature is a distinctively anastomosed and fenestrated circulation system. The choroidal circulation is responsible for nourishing and maintaining the temperature of the photoreceptors and RPE cells [3–6]. The structural and functional integrity of the choroid is crucial for the health of photoreceptor and RPE cells, as choroidal dysfunction contributes to the pathogenesis of age-related macular degeneration (AMD) [7,8], and is associated with other diseases such as myopia [9] and diabetic retinopathy [10].

In addition to the vasculature, the choroidal stroma or connective tissue is home to different types of innate immune cells, such as mast cells and macrophages [11,12]. Mast cells are the major effector cells that play crucial roles in innate and adapted immunity along with autoimmunity. The activation of mast cells induces their degranulation which results in the release of different chemokines, cytokines, histamine, and proteases [13]. In turn, released proteases, such as tryptase, chymase, and cathepsin G, among others, cause connective tissue and basement membrane degradation. Mast cells also produce proangiogenic factors, including the vascular endothelial growth factor (VEGF) [14] that initiates or modulates angiogenesis. Recent evidence suggested important roles played by choroidal mast cells in the development of AMD [15]. Macrophages normally exist in the whole choroidal connective tissue. During their resting state, macrophages sense their surroundings and interact with choroidal vasculature and RPE cells [16]. Macrophages affect vascular development and homeostasis. Clinical and preclinical studies demonstrated that changes in the normal state of choroidal macrophages are associated with vascular degeneration during aging and in patients with AMD [16–18].

Our understanding of the causes of AMD and other uveal diseases requires further delineation. One main stumbling block is the lack of simple and reliable methods to visualize blood vessels and immune cells in wholemount choroid of both pigmented and albino mice using immunostaining. Here we report a reproducible method to visualize the choroidal blood vessels, mast cells, and macrophages, independently or simultaneously, in albino mice. When combined with melanin bleaching, these methods can be used on pigmented mice. Using these methods, here we visualized and assessed the blood vessels, mast cells, and macrophages in the choroid of three albino (FVBN, CD-1, BALB/cJ) and one pigmented (C57BL/6J) mouse strains.

2. Materials and Methods

2.1. Animals

Six-week-old animals of both sexes from FVBN, CD-1, BALB/cJ, and C57BL/6J mice were used for these studies. CD-1 mice were from Charles River Laboratories (Wilmington, MA, USA). FVBN, BALB/cJ, and C57BL/6J were obtained from Jackson Laboratory (Bar Harbor, MA, USA) and were bred in our mouse colony at UW-Madison. Eyes from a comparable number of males and females were used in these studies. Mice were housed and allowed *ad libitum* access to standard rodent chow and water. Animals were euthanatized via exposure to carbon dioxide.

2.2. Tissue Preparation for Staining

Eyes were enucleated, washed in phosphate-buffered saline (PBS), and fixed in 4% paraformaldehyde (PFA: Cat# 15710; Electron Microscopy Sciences, Hatfield, PA, USA) for 1 h on a rocker at room temperature. Eyes were then washed three times in PBS. The eyeball was first cleaned from all muscle, fat, and conjunctiva tissues attached to the sclera. To attain only the choroidoscleral structure, the cornea, lens, and retina were dissected away. Care was taken to keep the limbus attached to the choroidoscleral structure. This tissue was then kept for 1 h in the blocking solution (1% BSA, 0.3% Triton X-100, 0.05% Azide, all in 1X PBS) at room temperature on a rocker.

2.3. Fluorescence Immunostaining of Choroidal Blood Vessels and Macrophages

The antibodies were prepared in the blocking solution (1% BSA, 0.3% Triton-X100, 0.05% sodium azide, all in PBS) and kept on ice. The goat anti-podocalyxin (R&D Systems, Minneapolis, MN, USA; Cat# AF1556) was used to stain the choroidal vasculature by diluting 1 µL of antibody in 500 µL of blocking solution. The rabbit anti-Iba1 (Wako Pure Chemical Industries, Osaka, Japan; Cat#: 019-19741) was used to stain choroidal macrophages by diluting 1 µL of antibody in 500 µL of blocking solution. Once antibodies were prepared, the blocking solution was removed away from the tissue and incubated with antibody solutions on a rocker in the cold room for 2–3 days. The tissues were then washed with PBS (3 times, 5 min each) at room temperature on a rocker. The appropriate secondary antibodies were prepared by diluting 1 μL of each secondary antibody in 500 μL of blocking solution and were added to the tissues. Secondary antibodies were from Jackson ImmunoResearch Laboratories (West Grove, PA, USA) including donkey anti-goat-Cy5 (Cat#: 705-175-147) and donkey anti-rabbit-Cy2 (Cat#: 711-225-152). Samples were incubated overnight in the secondary antibody solution on a rocker at 4 °C. The stained choroidoscleral structures were flattened by several radial cuts and mounted flat with an RPE layer facing upward on a slide with DAPI Fluoromount-G (Southern Biotech, Birmingham, AL, USA; Cat#: 0100-20). Signals were visualized by fluorescence microscopy and images were taken in digital format using Nikon confocal microscope system A1+. Captured images were then processed using NIS elements software (Nikon, Japan).

2.4. Fluorescence Immunostaining of Choroidal Arteries

The eyeballs from 3-week-old FVBN mice were enucleated, washed in PBS, and fixed in 4% PFA for 1 h at room temperature on a rocker, and then washed three times in PBS. The choroidoscleral structure was dissected and kept for 1 h in 1 mL of the blocking solution (1% BSA, 0.3% Triton X100, 0.05% Azide, all in PBS) at room temperature on a rocker. The tissue was incubated with anti-α-SMA-FITC (Sigma, #F3777, 1:500 in the blocking solution) overnight in the cold room on a rocker. The choroidoscleral structures were then washed in PBS, flat-mounted, and imaged as described above.

2.5. Fluorescence Staining of Choroidal Mast Cells

Heparin is a mixture of sulfated glycosaminoglycan that exists in mast cells. Avidin is a tetrameric glycoprotein that binds to heparin. Fluorescence-conjugated avidin is successfully used to visualize mast cells in both humans and mice [19–23]. We used rhodamine-avidin (Vector Laboratories, Burlingame, CA, USA; Cat#: A-2012) to stain the mast cells. Rhodamine-avidin was diluted (1:500) in the blocking buffer and samples were kept in the cold room on a rocker overnight. For albino mice, tissues were washed three times in PBS and mounted on a slide. When co-staining of mast cells and blood vessels or macrophages was desired, rhodamine-avidin was added to the secondary antibodies mix.

In the case of pigmented eyes, bleaching in hydrogen peroxide was necessary. H_2O_2 has been used by others to bleach melanin from pigmented eyes with the purpose of imaging the intact retina or the subretina [24,25]. To make sure hydrogen peroxide bleaches melanin without affecting heparin integrity in mast cells from C57BL/6J mice, choroidoscleral structures were crosslinked using Sulfo-SMCC (sulfosuccinimidyl 4-(N-maleimidomethyl) cyclohexane-1-carboxylate) (ThermoFisher, A39268). Sulfo-SMCC is an amine-to-sulfhydryl crosslinker and it was used here to stabilize heparin. To ensure successful crosslinking, each choroidoscleral tissue was incubated in 100 μL of Sulfo-SMCC in a 1.5 mL Eppendorf tube for 12–24 h in the cold room on a rocker. Tissues were washed in PBS (3 times, 5 min each) at room temperature on a rocker before fixing the tissue in 4% PFA for 1 h at room temperature on a rocker.

2.6. Bleaching and Redeveloping the Fluorescence Signal

For labeling the choroidoscleral structure from pigmented mice, the same staining steps described above were performed first. Extra steps were then performed, including post-stain fixation and bleaching. When Avidin staining was done, the stained tissue was kept in Sulfo-SMCC for 12–24 h in the cold room on a rocker as detailed above. Tissues were washed with PBS (3 times, 5 min each) at room temperature on a rocker, and then fixed in 4% PFA for 1 h at room temperature on a rocker. Fixed tissues were washed with PBS (3 times, 5 min each) on a rocker at room temperature before bleaching. For bleaching, we used 1% H_2O_2 (in 1xPBS) to maintain the natural epitope and tissue architecture. Tissues were submerged in 1% H_2O_2 and kept in a 55 °C water bath for 4–6 h. The bleaching solution was changed two to three times with warmed fresh 1% H_2O_2 in 1xPBS. Bleaching was completed once the tissue became relatively transparent. The 1% H_2O_2 solution was removed, the tissue was thoroughly washed with PBS (3 times), and then incubated with the 1.5 mL of blocking solution in a 2 mL tube. For re-staining of the bleached tissues, secondary antibodies (and if needed the rhodamine-avidin), prepared as described above, were added to the bleached tissues, and kept overnight in the cold room on a rocker. Tissues were then washed three times in PBS, mounted, visualized, and imaged as described above. To make sure bleaching does not compromise the sample's structural features, choroidoscleral tissues from albino FVBN mice were stained with an anti-podocalyxin antibody and imaged before and after bleaching. With bleaching, no noticeable changes to the choroidal vasculature were noted, thus demonstrating the feasibility of bleaching choroidoscleral tissues with 1% H_2O_2 to study the choroidal vasculature and other cells in the pigmented mice.

2.7. Statistical Analysis

Mast cells were manually counted. To evaluate statistical differences in multiple groups, a one-way ANOVA followed by Tukey's multiple comparison test was performed. Results are presented as mean ± SD. $p < 0.05$ was considered significant. GraphPad Prism 8.0 (GraphPad Software, San Diego, CA, USA) was used to perform statistical analysis and create graphs.

3. Results

The ocular choroidal tissue contains blood vessels and stroma (extravascular tissue). Choroidal blood vessels are composed of arteries, arterioles, veins, venules, and a capillary bed, namely choriocapillaris. Blood vessels in the choroid become smaller as they branch while descending from the sclera border towards the Bruch's membrane. This branching pattern allows the distinction of three vascular layers in humans: the Haller's layer of the large vessels close to the sclera, the Sattler's layer of medium size blood vessels, and the choriocapillaris layer of highly anastomosed capillary bed next to Bruch's membrane. Although on a much smaller scale because of eye size, we think our data show that the three layers are discernible in mice as well (Supplementary Video S1). The stroma is a connective tissue that encompasses the blood vessels and harbors immune cells, including mast cells and macrophages, neural tissues, and other types of cells. Choriocapillaris and Bruch's membrane are fused together via a basement membrane and the intercapillary pillars (also called septa), which are a part of the choroidal connective tissue.

3.1. Visualization of the Blood Vessels in the Choroid

Eyes from the albino FVBN mice were used to screen for blood vessel markers normally used to label the vasculature in the retina and in other tissues. We searched for markers that, according to our criteria, would robustly and exclusively label all blood vessels, but nothing else in the choroid. Anti-podocalyxin antibody was the only antibody that met these criteria. Anti-α-SMA antibody robustly labeled smooth muscle cells in the choroid. The following markers failed to meet the criteria for vascular staining: *Griffonia simplicifolia isolectin* B4 (IB4) and antibodies for intercellular adhesion molecule 2 (ICAM2), type I collagen, and type IV collagen. These markers either did not label the choroidal vasculature in our hands at all, or they labeled only some large blood vessels. The use of anti-α-SMA and anti-podocalyxin together allowed the clear visualization of the arterial system, venous system, and choriocapillaris in the mouse choroid.

3.2. Arterial System

The anti-α-SMA staining labels the major arterial system in the choroid. When combined with podocalyxin staining, the double staining allowed the clear visualization of the major arterial system with respect to the rest of the choroidal vasculature (Figure 1A–I). The posterior ciliary artery (PCA), which branches from the ophthalmic artery, is the major artery that feeds the retina and choroid. Once in the center of the choroid, close to the optic disc, the PCA branches into two long posterior ciliary arteries (LPCAs) that extend nasally and temporally. These LPCAs feed the entire choroidal vasculature and the blood vessel systems in the iris and ciliary body. Right after they enter the iris, the LPCAs give rise to two symmetrical branches forming the circular iris artery. Both nasal and temporal LPCAs send branches around the optic disc, as it appears from the α-SMA staining, forming a vascular circle (Figure 1A–F), similar to the Zinn–Haller arterial circle in the choroids of humans and primates. At least two major LPCA-derived arteries feed the superior and inferior parts of the choroid. LPCAs and their major LPCA-derived arteries run perpendicular to each other, dividing the choroid into four quadrants. From the peripapillary area to the equator, the LPCAs give rise to two to four branches on each side, which feed the associated choriocapillaris. The closer the branch is to the peripapillary area, the larger it is in circumference and length. These branches run parallel to their parent LPCA while branching themselves away from the LPCAs, but they stay in the same plane before these arterioles change direction directly and connect with the choriocapillaris. The major arteries that feed the superior and inferior parts of the choroid branch forward and in all directions. Except for LPCAs, only a few if any major arteries or arterioles exist in the periphery of the choroid. We were unable to identify short posterior ciliary arteries (SPCAs) in the mouse choroids we examined in this study. SPCAs are known to exist in the human choroid.

3.3. Venous System

Podocalyxin staining labeled all the choroidal vasculature well with the exception of the LPCAs along with their major branches in the peripapillary area, which were faintly labeled. Interestingly, these faintly podocalyxin-labeled blood vessels are the blood vessels that are labeled with α-SMA. Double staining with α-SMA and podocalyxin allowed clear visualization of major drainage systems with respect to the rest of the choroidal vasculature (Figure 1A–I). Two major drainage systems could be identified: vortex veins, which drain blood from the iris, the periphery of the choroid, and the portion of the posterior choroid close to the equator. There are four vortex veins close to the equator of the choroid on the dorsal, ventral, nasal, and temporal sides (Figure 1A–C,G–I). For each vortex vein, several large veins converge together to form larger sacs of blood vessels before these sacs come together to form the vortex vein. The upper part of the vortex veins, which is closer to the sclera, was sometimes positive for α-SMA staining (Figure 1A,D). In addition to the vortex veins, we identified veins in the center of the choroid that appear to be the peripapillary vein connecting to the posterior ciliary vein. This venous system drains blood vessels from the center of the choroid (Figure 1E,F).

3.4. Choriocapillaris

The choriocapillaris shows different structural patterns in different locations of the choroid (Figure 2A–K). Generally, three different patterns can be found: a dense honeycomb-like pattern (Figure 2A), a maze-like pattern (Figure 2B), and a finger-like projection pattern (Figure 2C). The finger-like projection pattern with large intercapillary spaces (septa) exists mainly in the far periphery of the choroid close to the anterior segment (Figure 2G). These finger-like vessels form arcades right before connecting with blood vessels of the ciliary body at the ora serrata. The other two patterns can be found in different parts of the choroid. We found the honeycomb pattern, with small septa, in a small strip in the equator of the choroid (Figure 2F), in the vicinity of the tributaries of the vortex vein (Figure 2H) and close to the optic disc (Figure 2K). The maze-like pattern is dominant in the mid periphery, in the equator, and in the peripapillary areas, where their septa area is large, and varies a lot (Figure 2F–K). The caliber of the capillaries varies as well. We found the choriocapillaris of 6-week-old FVBN mice to be light in nature where the septa area is considerably large and the dense choriocapillaris is uncommon. Due to this finding, we next examined whether other mice strains of similar age (6-week-old) have similar choriocapillaris density. We investigated the choroid from two more strains of albino mice, including BALB/cJ and CD-1, and from the pigmented C57BL/6J mice.

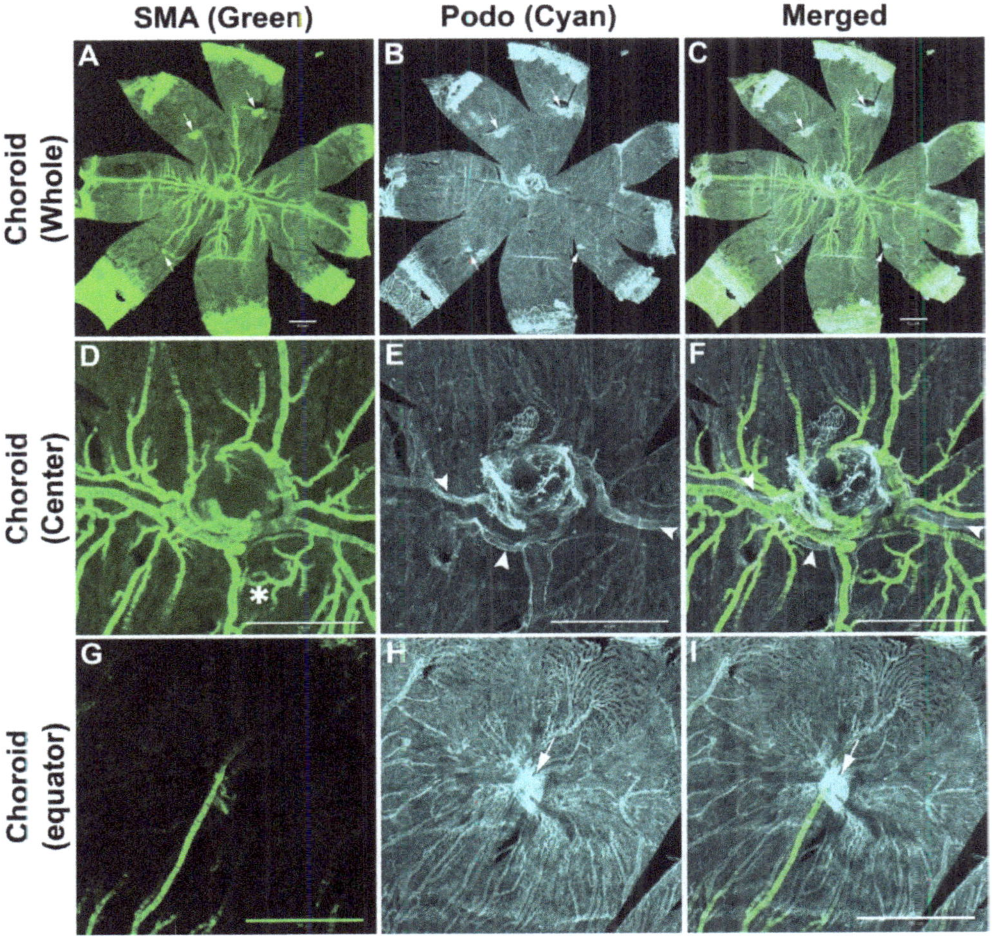

Figure 1. Visualization of the arterial and venous systems of the mouse choroid. (**A**,**D**,**G**) The anti-α-SMA staining labels the major arterial system in the choroid. The two long posterior ciliary arteries (LPCAs) extend nasally and temporally. It is obvious that LPCAs feed the entire choroidal vasculature. LPCAs send branches around the optic disc as it appears from the α-SMA staining, forming a vascular circle, such as the Zinn–Haller arterial circle in the choroids of humans and primates. Secondary arteries sometimes form loops (**D**, asterisk). At least two major LPCA-derived arteries feed the superior and inferior parts of the choroid. LPCAs and their major LPCA-derived arteries run perpendicular to each other, dividing the choroid into four quadrants. (**B**,**E**,**H**) Anti-podocalyxin staining strongly labels most of the choroidal vasculature except the arterial system, especially in the center of the choroid. Four vortex veins are observed at the equator of the choroid (arrows). A second venous system can be seen in the center of the choroid (arrowheads). These central veins appear to drain blood from the center of the choroid. (**C**,**F**,**I**) Merged images from staining with anti-α-SMA and anti-podocalyxin antibodies show the detailed structure of all choroidal components, including arteries, veins, and choriocapillaris. Thus, both venous systems in the choroid are more easily identified with respect to the rest of the vasculature, especially arteries. Number of choroids examined = 6. All representative images were from a 3-week-old FVBN mouse and captured using a confocal microscope. Scale bars: 500 μm.

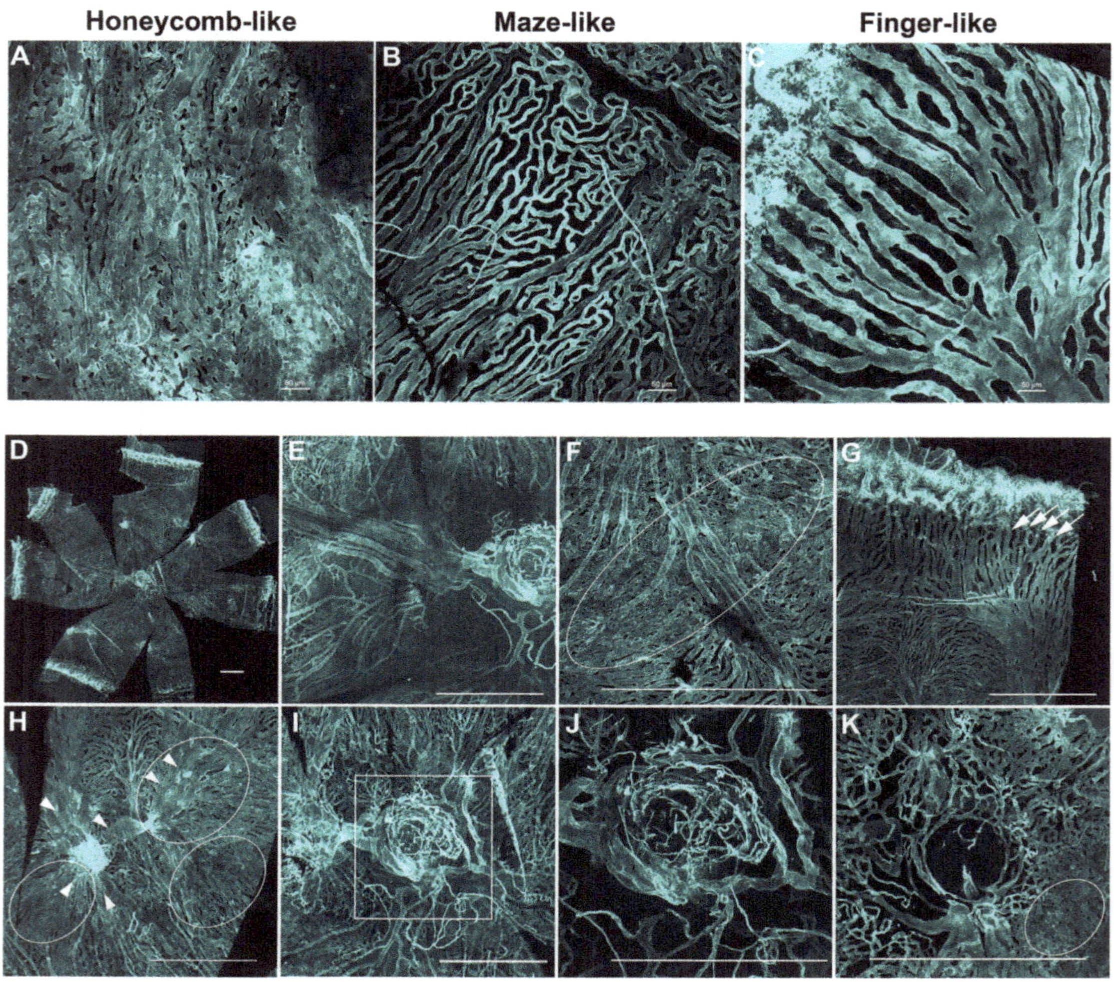

Figure 2. Visualization of the choriocapillaris and veins in the mouse choroid. The choriocapillaris display three different patterns, including the honeycomb-like pattern (**A**), the maze-like pattern (**B**), and the finger-like projections pattern (**C**,**D**). Low magnification image of the whole choroidal vasculature of 6-week-old FVBN mouse. (**E**) High magnification of choroid vasculature that includes the optic disc, the posterior pole, and part of the equator. Other high-magnification images of the choroid show the choriocapillaris in the equator away from the vortex veins (**F**), periphery (**G**) and equator with the vortex vein (**H**), and optic disc area (**I**–**K**). Inset in **I** was focally dissected to separate the large blood vessels (**J**) from the choriocapillaris (**K**). The honeycomb-like pattern observed in the middle of the equator, in the vicinity of the tributaries of the vortex vein, and on one side of the optic disc (**F**,**H**,**K**; oval shapes). Please note the tributaries of the vortex vein (**H**; arrowheads). The finger-like projections pattern was observed in the periphery (**G**; arrows). The maze-like pattern was observed in the equator (**F**), mid-periphery (**G**), and in the peripapillary areas (**K**). Tissues were stained using anti-podocalyxin antibody. Number of choroids examined = 14. All representative images were from 6-week-old FVBN mice and captured using confocal microscopy. Scale bars: 50 (**A**–**C**) and 500 μm (**D**–**K**).

3.5. Choriocapillaris from Albino and Pigmented Mice

To examine the choroid and its cellular components from the pigmented C57BL/6J mice, we first needed to eliminate the melanin pigment that exists in the RPE and choroidal melanocytes. The goal was to bleach the melanin without damaging the tissue. For this purpose, we successfully bleached melanin in the eye in 1% hydrogen peroxide (1% H_2O_2). H_2O_2 has been used by others to bleach melanin from pigmented eyes with the purpose of imaging the intact retina or the subretina [24,25]. Figure 3 shows choroidoscleral flat-mounts from albino FVBN, bleached C57BL/6J, and from unbleached C57BL/6J mice. After successfully developing these novel methods for staining the choroidal vasculature and clearing the pigment from the eye, we were able to compare the choroidal vascular structure from both albino and pigmented mice for the first time.

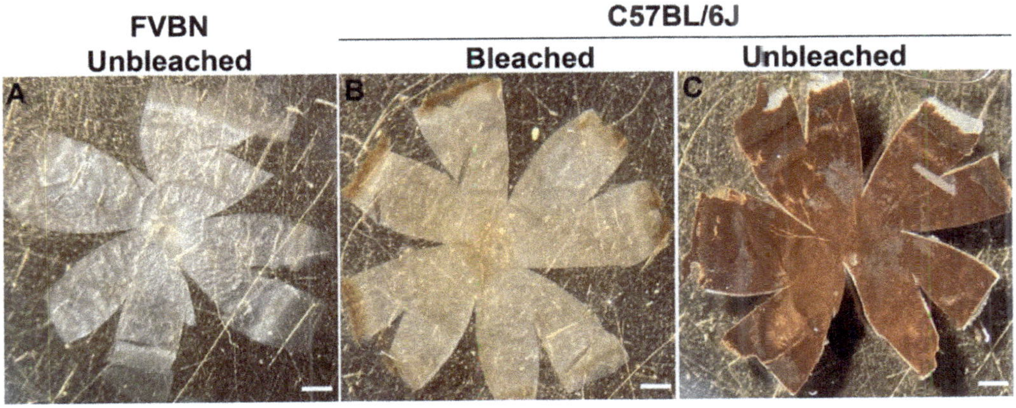

Figure 3. Whole choroidoscleral tissue from albino (FVBN, unbleached) (**A**), bleached C57BL/6J (**B**), and unbleached C57BL/6J (**C**) mice. Exposing the choroidoscleral tissue from pigmented C57BL/6J mice to 1% H_2O_2 for 4–5 h at 55 °C in a water bath was sufficient to bleach almost all melanin in the tissue, as it becomes almost transparent. Representative images were from 6-week-old FVBN and C57BL/6J mice and captured in digital format. Scale bars: 500 μm.

Figure 4 shows representative images of the choroidal vasculature from 6-week-old FVBN, CD-1, BALB/cJ, and C57BL/6J mice. The whole choroid–sclera and the vortex vein are shown. Choriocapillaris from the peripapillary, equator, and periphery regions are shown as well. The choroidal structure from the FVBN is shown in Figures 2 and 4A–E and was described above. In the CD-1 mice, the maze-like pattern is dominant in the peripapillary, equator, and in periphery around the LPCAs (Figure 4F–J). The finger-like and arcade structures exist in the far periphery of the choriocapillaris (Figure 4I). The honeycomb pattern was rarely found in the peripapillary, but it was seen close to the vortex vein (Figure 4J). In the BALB/cJ mice, the honeycomb pattern exists in the peripapillary, equator, and vortex vein areas of the choriocapillaris (Figure 4L,M). The maze-like pattern exists in the equator and mid-periphery areas (Figure 4M,N). The finger-like and arcade structures exist in the far periphery of the choriocapillaris (Figure 4N). The dense nature of the choriocapillaris in BALB/cJ mice is obvious. In the C57BL/6J mice, the honeycomb pattern exists in the peripapillary, the equator areas, and around the vortex vein of the choriocapillaris (Figure 4Q,R,T). The maze-like pattern exists in the equator and mid-periphery areas (Figure 4R,S). The finger-like and arcade structures exist in the far periphery of the choriocapillaris (Figure 4S). The dense nature of the choriocapillaris in this strain is clear. Taken together, the choriocapillaris from CD-1 mice seemed similar to that from FVBN mice. These two albino strains showed choriocapillaris atrophy, at least at this age (6-week-old), where capillary calibers are small and their intercapillary spaces are commonly large. This is in contrast with choriocapillaris from the albino BALB/cJ and pigmented C57BL/6J mice. In the C57BL/6J mice, the choriocapillaris is similar to that in the BALB/cJ mice, but slightly denser.

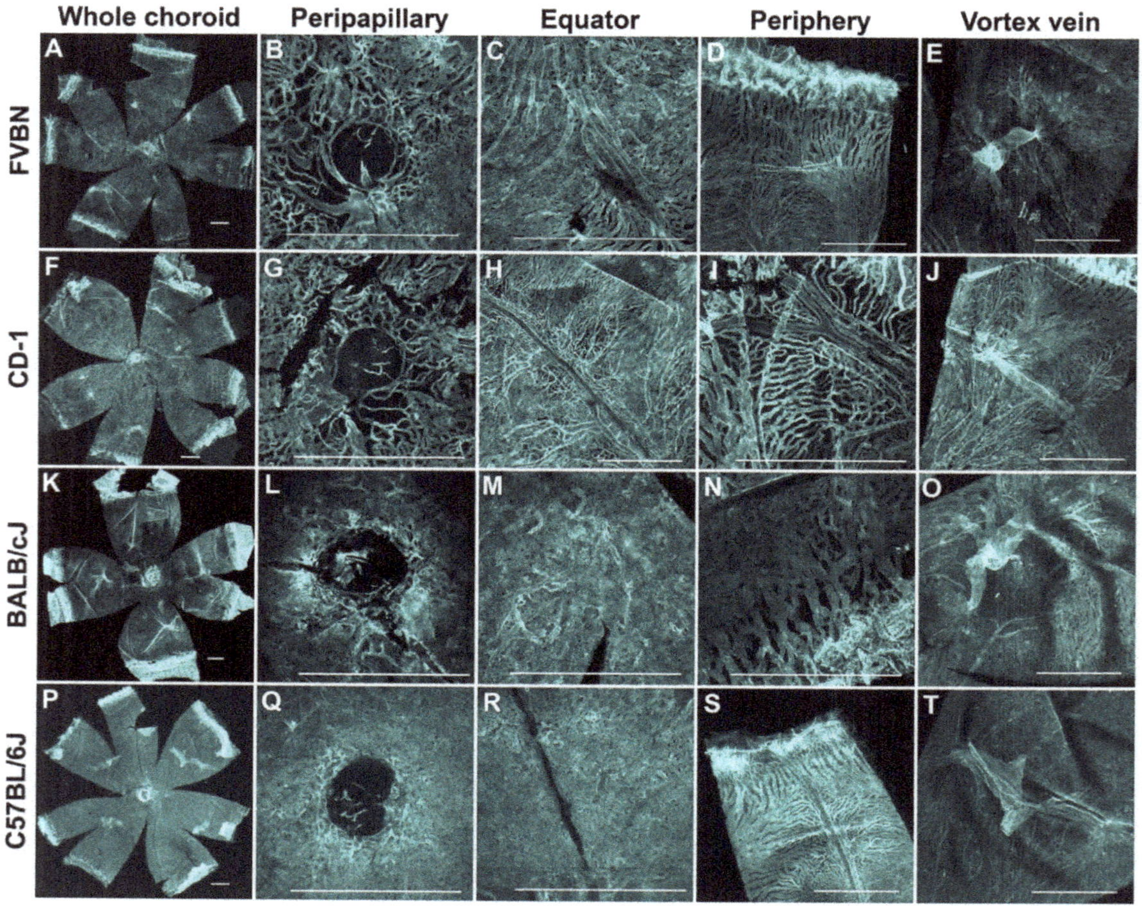

Figure 4. Visualization of the choroidal vasculature from albino and pigmented mice. Representative images from FVBN (**A–E**), CD-1 (**F–J**), BALB/cJ (**K–O**), and bleached C57BL/6J (**P–T**) mice. Shown are representative images from the low magnification of the whole choroidal vasculature, along with high magnification from the peripapillary, equator, periphery, and vortex vein. The choriocapillaris displays three different patterns, including finger-like projections, honeycomb-like, and maze-like patterns. Images show that the choroids from FVBN and CD-1 albino mice are sparse and less dense as compared with choroids from BALB/cJ albino mice. Choriocapillaris of the C57BL/6J mice is similar to that in the BALB/cJ mice, but slightly denser. Tissues were stained using an anti-podocalyxin antibody. Number of choroids examined = 14 (FVBN), 7 (CD-1), 9 (BALB/cJ), and 9 (C57BL/6J). All representative images were from 6-week-old animals and captured using a confocal microscope. Scale bars: 500 μm.

3.6. Visualization of Mast Cells in the Choroid of Albino and Pigmented Mice

To visualize mast cells in the choroid from albino and pigmented mice, we used rhodamine-avidin. Avidin recognizes and binds heparin in mast cells, and fluorescence conjugated avidin is successfully used to visualize mast cells in both human and mice [19–23]. Heparin is a mixture of sulfated glycosaminoglycan. To make sure hydrogen peroxide bleached melanin without affecting heparin integrity in mast cells from C57BL/6J mice, choroidoscleral structures were crosslinked with Sulfo-SMCC before bleaching. Sulfo-SMCC is an amine-to-sulfhydryl crosslinker (please refer to Methods section for details). Figure 5 shows mast cell distribution in choroidoscleral flat-mounts from FVBN (Figure 5A), CD-1(Figure 5B), BALB/cJ (Figure 5C), and C57BL/6J (Figure 5D) mice. Mast cells in these four mouse stains were counted, analyzed, and graphed (Figure 5E–H). Mast cells reside in the vicinity of large blood vessels. Their distribution and density varied among different strains. Mast cells showed a similar distribution pattern in the choroidoscleral tissue from C57BL/6J, CD-1, and FVBN mice. In these three stains, a small but concentrated number of mast cells exist around the optic disc area, but the majority of mast cells reside in the limbus. A similar number to those mast cells in the limbus is distributed throughout the rest of the choroid (referred to here as the middle choroid). Mast cell distribution in the BALB/cJ is unique, where mast cells are distributed equally throughout the whole choroidoscleral tissue. In addition, mast cells in this line of mice appear to be larger in size. Mast cell counts in all four examined strains showed BALB/cJ and FVBN to have the highest number of mast cells. C57BL/6J mice had the lowest number of mast cells.

3.7. Visualization of Macrophages in the Choroid of Pigmented Mice

To visualize macrophages in the mice choroid, we used the anti-ionized calcium-binding adaptor molecule 1 (Iba1) antibody. This marker is used by others to label macrophages in the choroids of humans and albino mice [16,17]. Figure 6 shows representative images of macrophages in the choroid of C57BL/6J mice. When visualized along with blood vessels, macrophages were found in all layers of the choroid, including the choriocapillaris (Figure 6A), large blood vessels (Figure 6B), and the scleral aspect of the choroid (Figure 6C).

3.8. Co-Visualization of Blood Vessels, Mast Cells, and Macrophages in the Choroid of Pigmented Mice

We successfully stained the choroid from C57BL/6J mice with podocalyxin and Iba1 antibodies to visualize the vasculature and macrophages, respectively, and with rhodamine-avidin to visualize mast cells (Figure 7A–D). Thus, the method described here allows the simultaneous examination of a number of cell types in the choroid and evaluates potential changes in their interactions, localization, and density.

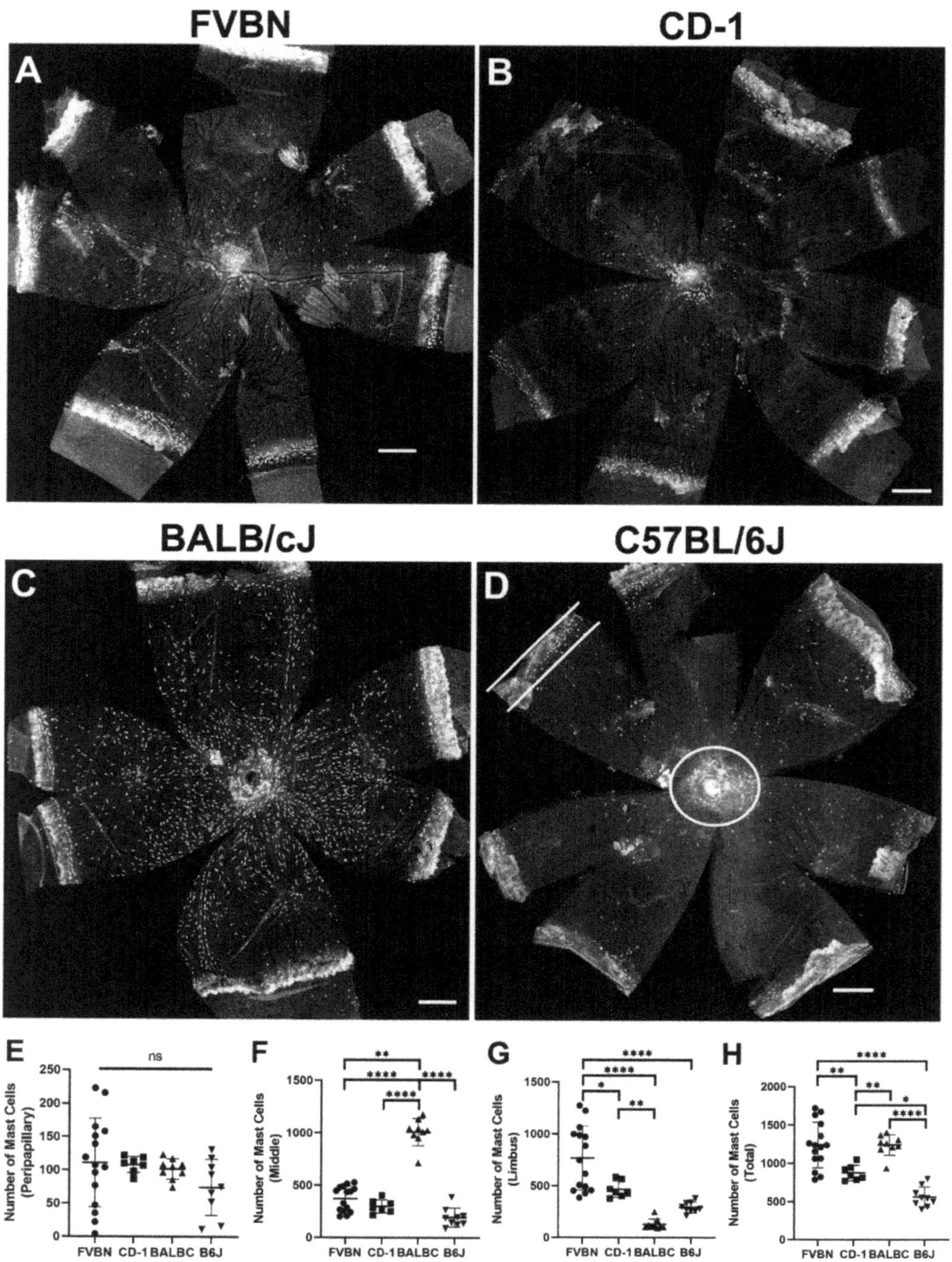

Figure 5. Visualization of mast cells in the choroid of albino and pigmented mice. Representative images showing the distribution of mast cells in choroidoscleral tissues from FVBN (**A**), CD-1 (**B**), BALB/cJ (**C**), and bleached C57BL/6J (**D**) mice. The counts of mast cells in the peripapillary (**E**),

middle (**F**), limbus (**G**), and whole (**H**) choroid of the four examined strains of mice were determined and statistically analyzed. Each data point represents one choroid. A small number of mast cells exist around the optic disc area (**D**; white circle), but the majority of mast cells reside in the limbus (**D**; flanked by two white lines). Mast cells showed similar distribution pattern in the choroidoscleral tissue from C57BL/6J, CD-1, and FVBN mice. While a small number of mast cells exist around the optic disc area in these three strains of mice, the majority of mast cells reside in the limbus. An approximately similar number to those mast cells in the limbus is distributed in the rest of the choroid (referred to here as middle). BALB/cJ mice showed a unique mast cell distribution as compared to the other three examined strains in this study. Mast cells in BALB/cJ appear larger in size and were approximately equally distributed throughout the whole choroidal tissue. Mast cell counts in these four strains showed BALB/cJ and FVBN to have similar highest numbers. C57BL/6J mice had the lowest number. Rhodamine-avidin was used to label mast cells in the choroid. All representative images were from 6-week-old animals and captured using confocal microscopy. Number of choroids examined = 14 (FVBN), 7 (CD-1), 9 (BALB/cJ), and 9 (C57BL/6J). Scale bars: 500 µm. Results are presented as mean ± SD. ns: not significant, * $p < 0.05$, ** $p < 0.01$, and **** $p < 0.0001$.

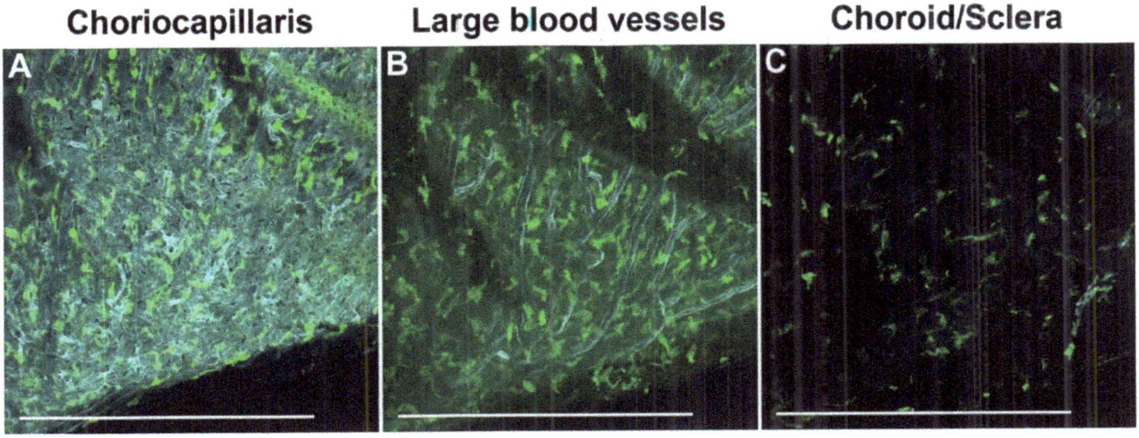

Figure 6. Visualization of macrophages in the choroid of pigmented mice. Macrophages were stained using the anti-ionized calcium binding adaptor molecule 1 (Iba1) antibody. To visualize macrophages with respect to different blood vessel layers, the choroid was co-immunostained with an anti-podocalyxin antibody. Representative images showing macrophages throughout the choroid of C57BL/6J mice, specifically in the choriocapillaris (**A**), large blood vessel layers (**B**), and in the scleral aspect of the choroid (**C**). All representative images were from 6-week-old animals and captured using a confocal microscope. Number of choroids examined = 9. Scale bars: 500 µm.

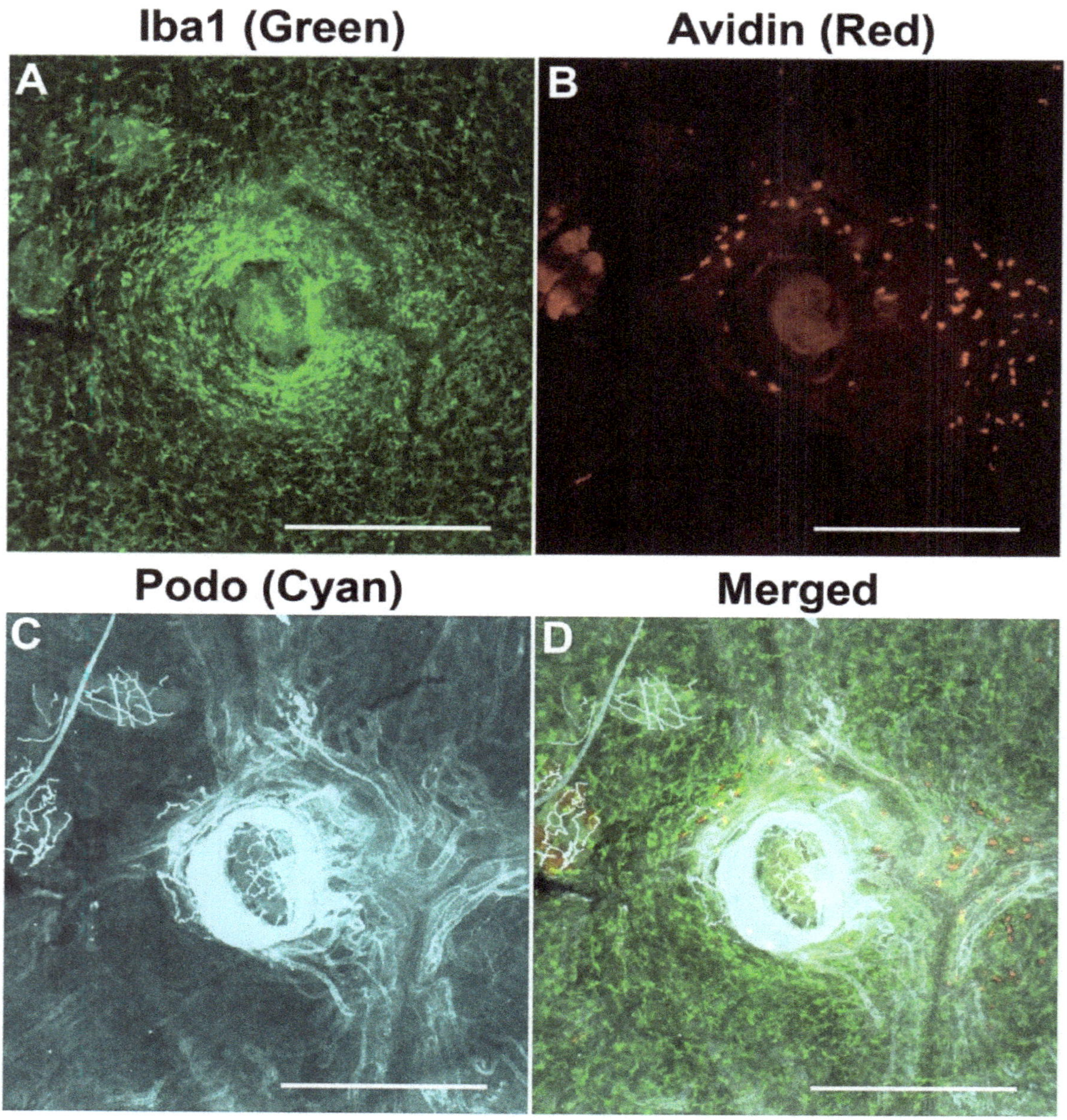

Figure 7. Co-visualization of blood vessels, mast cells, and macrophages in the choroid of pigmented mice. Macrophages, mast cells, and blood vessels in the choroid were labeled with anti-Iba1 antibody, rhodamine-avidin, and anti-podocalyxin, respectively. Representative images from stained and bleached mouse choroid showed robust labeling of macrophages (**A**), mast cells (**B**), and blood vessels (**C**). In addition, merged images (**D**) showed each targeted cell type with respect to other types of cells/blood vessels of the choroid. All representative images were from 6-week-old C57BL/6J animals and captured using confocal microscopy. Number of choroids examined = 9. Scale bars: 500 μm.

4. Discussion

Utilizing the methods described here revealed several novel findings especially in pigmented mice. (i) Staining with α-SMA facilitated the visualization of an arterial circle around the optic nerve in mice that is similar to the Zinn–Haller arterial circle known to exist in humans and primates. (ii) Staining with podocalyxin facilitated the visualization of the choriocapillaris. Three different structural patterns of choriocapillaris were observed throughout the whole choroid: honeycomb-like, maze-like, and finger-like patterns. The distribution of these patterns was highly variable among strains at the age we examined. (iii) To the best of our knowledge, this is the first report of the visualization of mast cells and macrophages in eyes from pigmented mice. Mast cell distribution and density were found to vary in eyes from different strains and in eyes from different animals of the same strain. (iv) Utilizing the staining and bleaching method described in this report allowed the visualization of the vasculature, mast cells, and macrophages, independently or simultaneously, in the choroid of pigmented mice.

The ocular choroidal tissue contains blood vessels and stroma, the extravascular connective tissue. Choroidal blood vessels are composed of arteries, arterioles, veins, venules, and a capillary bed, the choriocapillaris. Blood vessels in the choroid become smaller as they branch while descending from the sclera border towards the Bruch's membrane. This branching pattern allows the distinction of three vascular layers: the Haller's layer of the large vessels close to the sclera; Sattler's layer of medium size vessels; and the choriocapillaris layer of a highly anastomosed capillary bed next to Bruch's membrane. The stroma is a connective tissue that encompasses the blood vessels and harbors immune cells, such as mast cells and macrophages, neural tissues, and other types of cells. Choriocapillaris and Bruch's membrane are fused together via a basement membrane and the intercapillary pillars (also called septa), which is part of the choroidal connective tissue.

The integrity of the choroid is critical for the homeostasis of the RPE and the outer retina. It is well established that the degeneration of the choroid is causally linked to many eye diseases, such as AMD, diabetic choroidopathy, myopia, uveal autoimmune diseases, and retinopathy of prematurity. These diseases cause visual impairments in millions of people each year. However, cellular and molecular pathologic mechanisms responsible for choroidal damage in these diseases are largely unknown. While mouse models have been successfully used to better understand the pathologic mechanisms responsible for human diseases that occur in the retina, their use in diseases related to damage in the choroid have been limited. Investigating the choroid in mice has been challenging because of the lack of simple and reliable methods to visualize blood vessels and inflammatory cells in the choroid. To the best of our knowledge, this is the first detailed report that provides simple methods that enable the visualization of blood vessels, mast cells, and macrophages, independently or simultaneously, in albino and pigmented strains of mice. In this report, we describe in detail the structure of the choroidal vasculature and the distribution of mast cells, and macrophages in three albino (FVBN, CD-1, BALB/cJ) and one pigmented (C57BL/6J) mouse strains.

The choroidal vasculature has been partially visualized in mice using corrosion cast or venous infusion with fluorescent materials [16,26,27]. Lutty's group reported in an ARVO abstract the use of anti-podocalyxin to stain the choroidal vasculature of pigmented mice [28]. The studies presented here provide a complete detailed description of the mouse choroidal vasculature among several strains. Staining with anti-α-SMA antibody showed that the two LPCAs feed the entire choroidal vasculature and the blood vessels in the iris and the ciliary body. We could not identify short posterior ciliary arteries (SPCAs) in the choroid of mice. SPCAs exist in the choroid of humans. Branches from LPCAs were observed to go around the optic disc forming a vascular circle, such as the Zinn–Haller arterial circle in the choroids of humans and primates. Podocalyxin staining made it easy to visualize the venous system and the choriocapillaris in the choroid. Two venous systems could be found, one is located in the center around the optic disc that seems to drain through the posterior ciliary vein. The second venous system is comprised of four vortex veins located close to the equator on the dorsal, ventral, nasal, and temporal sides of the choroid. Vortex veins drain blood from the iris, periphery of the choroid, and the portion of the posterior choroid close to the equator.

Podocalyxin staining also allowed the detailed visualization of choriocapillaris. Three different patterns in different parts of the choroid could be seen: finger-like projections, regular dense honeycomb, and irregular maze-like patterns. These studies demonstrated similarities and differences in choriocapillaris among different mouse strains, including albino and pigmented mice. Our studies showed that the choroidal vasculature in mice is similar to that in rats, which was studied using the corrosion cast and scanning electron microscope [29]. However, this rat study did not find a vascular circle similar to the Zinn-Haller arterial circle that exists in the choroid of humans and primates, which we showed is present in mice. No lobular structures were found in the choroid of mice, which is similar to the choroid of rats [29]. Choroidal lobuli exist in human and primate eyes [5,30].

Mast cells exist in the choroid of the eye in mammalian species including mice. Staining with avidin allowed the visualization of mast cells in the choroid from different albino and pigmented mice. Assessing the distribution and counting of mast cells revealed many facts about this cell type in the choroid of mice. Mast cells are normally located around the major blood vessels. Mast cell density and distribution in the choroid varied at different levels. It varies in different locations in the same choroid, among individuals from the same strain, and among strains. Visualizing the macrophages revealed their distribution in all layers of the choroid in mice. Collectively, the method described here allows the study of the choroid from albino and pigmented mice at cellular levels, which has the potential of advancing our understanding of the molecular and cellular mechanisms contributing to various ocular pathologies including AMD.

Supplementary Materials: The following are available online at https://www.mdpi.com/article/10.3390/cells11203329/s1, Video S1: 3-D Visualization of the Choroidal Vasculature.

Author Contributions: Conceptualization, I.S.Z. and N.S.; methodology, I.S.Z., Y.-S.S., H.B.Z. and C.M.S.; validation, I.S.Z., Y.-S.S. and H.B.Z.; formal analysis, I.S.Z., Y.-S.S. and H.B.Z.; investigation, I.S.Z. and C.M.S.; resources, I.S.Z. and N.S.; data curation, I.S.Z., Y.-S.S. and H.B.Z.; writing—original draft preparation, I.S.Z. and N.S.; writing—review and editing, I.S.Z., Y.-S.S., H.B.Z., C.M.S. and N.S.; visualization, I.S.Z., Y.-S.S. and H.B.Z.; supervision, I.S.Z. and N.S.; project administration, I.S.Z. and N.S.; funding acquisition, I.S.Z., C.M.S. and N.S. All authors have read and agreed to the published version of the manuscript.

Funding: This work and/or the investigator(s) were supported by an unrestricted award from Research to Prevent Blindness to the Department of Ophthalmology and Visual Sciences, Retina Research Foundation, RRF/Daniel M. Albert chair, Carl Marshall Reeves & Mildred Almen Reeves Foundation, and by the National Institutes of Health grants P30 EY016665, EY030076, EY032543, and HL158073. We also thank the University of Wisconsin Translational Research Initiatives in Pathology laboratory (TRIP), supported by the UW Department of Pathology and Laboratory Medicine, UWCCC (P30 CA014520), and the Office of The Director- NIH (S10 OD023526) for use of its facilities and services.

Institutional Review Board Statement: All experiments were conducted in agreement with the National Institutes of Health Guide for the Care and Use of Laboratory Animals and approved by the Institutional Animal Care and Use Committee of the University of Wisconsin School of Medicine and Public Health (IACUC assurance number: D16-00239).

Informed Consent Statement: Not applicable.

Data Availability Statement: All the data presented here are included in the manuscript. Further inquiries should be directed to the corresponding author.

Acknowledgments: We would like to dedicate this publication to the memory of our great colleague and friend Gerard A. Lutty, for his outstanding contributions to the field of ocular vascular biology, and for always being available to share his knowledge and guide us in our studies.

Conflicts of Interest: The authors declare no conflict of interest.

References

1. Nickla, D.L.; Wallman, J. The multifunctional choroid. *Prog. Retin. Eye Res.* **2010**, *29*, 144–168. [CrossRef] [PubMed]
2. Hayreh, S.S. Segmental nature of the choroidal vasculature. *Br. J. Ophthalmol.* **1975**, *59*, 631–648. [CrossRef] [PubMed]
3. Lejoyeux, R.; Benillouche, J.; Ong, J.; Errera, M.-H.; Rossi, E.A.; Singh, S.R.; Dansingani, K.K.; da Silva, S.; Sinha, D.; Sahel, J.-A.; et al. Choriocapillaris: Fundamentals and advancements. *Prog. Retin. Eye Res.* **2022**, *87*, 100997. [CrossRef]
4. Bill, A.; Sperber, G.; Ujiie, K. Physiology of the choroidal vascular bed. *Int. Ophthalmol.* **1983**, *6*, 101–107. [CrossRef] [PubMed]
5. Fryczkowski, A.W. Anatomical and functional choroidal lobuli. *Int. Ophthalmol.* **1994**, *18*, 131–141. [CrossRef]
6. Parver, L.M.; Auker, C.R.; Carpenter, D.O. Choroidal Blood Flow. III. Reflexive control in human eyes. *Arch. Ophthalmol.* **1983**, *101*, 1604–1606. [CrossRef]
7. Sohn, E.H.; Flamme-Wiese, M.J.; Whitmore, S.S.; Workalemahu, G.; Marneros, A.G.; Boese, E.A.; Kwon, Y.H.; Wang, K.; Abramoff, M.D.; Tucker, B.A.; et al. Choriocapillaris Degeneration in Geographic Atrophy. *Am. J. Pathol.* **2019**, *189*, 1473–1480. [CrossRef]
8. Arya, M.; Sabrosa, A.S.; Duker, J.S.; Waheed, N.K. Choriocapillaris changes in dry age-related macular degeneration and geographic atrophy: A review. *Eye Vis.* **2018**, *5*, 22. [CrossRef]
9. Li, J.; Zhou, H.; Feinstein, M.; Wong, J.; Wang, R.K.; Chan, L.; Dai, Y.; Porco, T.; Duncan, J.L.; Schwartz, D.M. Choriocapillaris Changes in Myopic Macular Degeneration. *Transl. Vis. Sci. Technol.* **2022**, *11*, 37. [CrossRef]
10. Regatieri, C.V.; Branchini, L.; Carmody, J.; Fujimoto, J.G.; Duker, J.S. Choroidal thickness in patients with diabetic retinopathy analyzed by spectral-domain optical coherence tomography. *Retina* **2012**, *32*, 563–568. [CrossRef]
11. Reekie, I.R.; Sharma, S.; Foers, A.; Sherlock, J.; Coles, M.C.; Dick, A.D.; Denniston, A.K.; Buckley, C.D. The Cellular Composition of the Uveal Immune Environment. *Front. Med.* **2021**, *8*, 721953. [CrossRef] [PubMed]
12. McMenamin, P.G.; Saban, D.R.; Dando, S.J. Immune cells in the retina and choroid: Two different tissue environments that require different defenses and surveillance. *Prog. Retin. Eye Res.* **2019**, *70*, 85–98. [CrossRef] [PubMed]
13. Mukai, K.; Tsai, M.; Saito, H.; Galli, S.J. Mast cells as sources of cytokines, chemokines, and growth factors. *Immunol. Rev.* **2018**, *282*, 121–150. [CrossRef] [PubMed]
14. Abdel-Majid, R.M.; Marshall, J.S. Prostaglandin E_2Induces Degranulation-Independent Production of Vascular Endothelial Growth Factor by Human Mast Cells. *J. Immunol.* **2004**, *172*, 1227–1236. [CrossRef] [PubMed]
15. Bhutto, I.A.; McLeod, D.S.; Jing, T.; Sunness, J.S.; Seddon, J.M.; Lutty, G.A. Increased choroidal mast cells and their degranulation in age-related macular degeneration. *Br. J. Ophthalmol.* **2016**, *100*, 720–726. [CrossRef] [PubMed]
16. Kumar, A.; Zhao, L.; Fariss, R.N.; McMenamin, P.G.; Wong, W.T. Vascular Associations and Dynamic Process Motility in Perivascular Myeloid Cells of the Mouse Choroid: Implications for Function and Senescent Change. *Investig. Opthalmology Vis. Sci.* **2014**, *55*, 1787–1796. [CrossRef]
17. McLeod, D.S.; Bhutto, I.; Edwards, M.M.; Silver, R.E.; Seddon, J.M.; Lutty, G.A. Distribution and Quantification of Choroidal Macrophages in Human Eyes With Age-Related Macular Degeneration. *Investig. Opthalmology Vis. Sci.* **2016**, *57*, 5843–5855. [CrossRef]
18. Wang, S.; Zaitoun, I.S.; Darjatmoko, S.R.; Sheibani, N.; Sorenson, C.M. Bim Expression Promotes the Clearance of Mononuclear Phagocytes during Choroidal Neovascularization, Mitigating Scar Formation in Mice. *Life* **2022**, *12*, 208. [CrossRef]
19. Liu, J.; Fu, T.; Song, F.; Xue, Y.; Xia, C.; Liu, P.; Wang, H.; Zhong, J.; Li, Q.; Chen, J.; et al. Mast Cells Participate in Corneal Development in Mice. *Sci. Rep.* **2015**, *5*, 17569. [CrossRef]
20. Tharp, M.D.; Seelig, L.L., Jr.; Tigelaar, R.E.; Bergstresser, P.R. Conjugated avidin binds to mast cell granules. *J. Histochem. Cytochem.* **1985**, *33*, 27–32. [CrossRef]
21. McNeil, B.D.; Pundir, P.; Meeker, S.; Han, L.; Undem, B.J.; Kulka, M.; Dong, X. Identification of a mast-cell-specific receptor crucial for pseudo-allergic drug reactions. *Nature* **2015**, *519*, 237–241. [CrossRef] [PubMed]
22. Kunder, C.A.; St John, A.L.; Li, G.; Leong, K.W.; Berwin, B.; Staats, H.F.; Abraham, S.N. Mast cell–derived particles deliver peripheral signals to remote lymph nodes. *J. Exp. Med.* **2009**, *206*, 2455–2467. [CrossRef] [PubMed]
23. Cherwinski, H.M.; Murphy, C.A.; Joyce, B.L.; Bigler, M.E.; Song, Y.S.; Zurawski, S.M.; Moshref, M.M.; Gorman, D.M.; Miller, K.L.; Zhang, S.; et al. The CD200 Receptor Is a Novel and Potent Regulator of Murine and Human Mast Cell Function. *J. Immunol.* **2005**, *174*, 1348–1356. [CrossRef] [PubMed]
24. Ye, Y.; Dinh Duong, T.A.; Saito, K.; Shinmyo, Y.; Ichikawa, Y.; Higashide, T.; Kagami, K.; Fujiwara, H.; Sugiyama, K.; Kawasaki, H. Visualization of the Retina in Intact Eyes of Mice and Ferrets Using a Tissue Clearing Method. *Transl. Vis. Sci. Technol.* **2020**, *9*, 1. [CrossRef] [PubMed]
25. Kim, S.Y.; Assawachananont, J. A New Method to Visualize the Intact Subretina From Retinal Pigment Epithelium to Retinal Tissue in Whole Mount of Pigmented Mouse Eyes. *Transl. Vis. Sci. Technol.* **2016**, *5*, 6. [CrossRef] [PubMed]
26. Jiao, C.; Adler, K.; Liu, X.; Sun, W.; Mullins, R.F.; Sohn, E.H. Visualization of Mouse Choroidal and Retinal Vasculature Using Fluorescent Tomato Lectin Perfusion. *Transl. Vis. Sci. Technol.* **2020**, *9*, 1. [CrossRef] [PubMed]
27. Ninomiya, H.; Inomata, T. Microvasculature of the mouse eye: Scanning electron microscopy of vascular corrosion casts. *J. Exp. Anim. Sci.* **2006**, *43*, 149–159. [CrossRef]

28. Bhutto, I.A.; Tiwari, A.; Thomson, B.R.; Edwards, M.M.; Lutty, G.A. Visualization of choroidal vasculature in pigmented mouse eyes. *Investig. Ophthalmol. Vis. Sci.* **2020**, *61*, 2235.
29. Bhutto, I.A.; Amemiya, T. Microvascular architecture of the rat choroid: Corrosion cast study. *Anat. Rec.* **2001**, *264*, 63–71. [CrossRef]
30. Olver, J.M. Functional anatomy of the choroidal circulation: Methyl methacrylate casting of human choroid. *Eye* **1990**, *4 Pt 2*, 262–272. [CrossRef]

cells MDPI

Communication

Distribution of Copper, Iron, and Zinc in the Retina, Hippocampus, and Cortex of the Transgenic APP/PS1 Mouse Model of Alzheimer's Disease

Seyed Mostafa Hosseinpour Mashkani [1], David P. Bishop [2], Newsha Raoufi-Rad [1], Paul A. Adlard [3], Olga Shimoni [1] and S. Mojtaba Golzan [4,*]

1 Institute for Biomedical Materials and Devices, School of Mathematical and Physical Sciences, Faculty of Science University of Technology Sydney, 15 Broadway, Sydney, NSW 2007, Australia
2 Hyphenated Mass Spectrometry Laboratory (HyMaS), School of Mathematical and Physical Sciences, Faculty of Science, University of Technology Sydney, 15 Broadway, Sydney, NSW 2007, Australia
3 Synaptic Neurobiology Laboratory, The Florey Institute of Neuroscience and Mental Health, The University of Melbourne, Melbourne, VIC 3000, Australia
4 Vision Science Group, Graduate School of Health (GSH), University of Technology Sydney, 15 Broadway, Sydney NSW 2007, Australia
* Correspondence: mojtaba.golzan@uts.edu.au

Abstract: A mis-metabolism of transition metals (i.e., copper, iron, and zinc) in the brain has been recognised as a precursor event for aggregation of Amyloid-β plaques, a pathological hallmark of Alzheimer's disease (AD). However, imaging cerebral transition metals in vivo can be extremely challenging. As the retina is a known accessible extension of the central nervous system, we examined whether changes in the hippocampus and cortex metal load are also mirrored in the retina. Laser ablation inductively coupled plasma-mass spectrometry (LA-ICP-MS) was used to visualise and quantify the anatomical distribution and load of Cu, Fe, and Zn in the hippocampus, cortex, and retina of 9-month-old Amyloid Precursor Protein/Presenilin 1 (APP/PS1, $n = 10$) and Wild Type (WT, $n = 10$) mice. Our results show a similar metal load trend between the retina and the brain, with the WT mice displaying significantly higher concentrations of Cu, Fe, and Zn in the hippocampus ($p < 0.05$, $p < 0.0001$, $p < 0.01$), cortex ($p < 0.05$, $p = 0.18$, $p < 0.0001$) and the retina ($p < 0.001$, $p = 0.01$, $p < 0.01$) compared with the APP/PS1 mice. Our findings demonstrate that dysfunction of the cerebral transition metals in AD is also extended to the retina. This could lay the groundwork for future studies on the assessment of transition metal load in the retina in the context of early AD.

Keywords: Alzheimer's disease; retina; transition metals; laser ablation inductively coupled plasma-mass spectrometry

Citation: Hosseinpour Mashkani, S.M.; Bishop, D.P.; Raoufi-Rad, N.; Adlard, P.A.; Shimoni, O.; Golzan, S.M. Distribution of Copper, Iron, and Zinc in the Retina, Hippocampus, and Cortex of the Transgenic APP/PS1 Mouse Model of Alzheimer's Disease. *Cells* **2023**, *12*, 1144. https://doi.org/10.3390/cells12081144

Academic Editor: Hossein Ameri

Received: 27 February 2023
Revised: 30 March 2023
Accepted: 30 March 2023
Published: 13 April 2023

1. Introduction

Alzheimer's disease (AD) is a common, incurable, and progressive dementia sub-type which is pathologically characterised by the formation of neurofibrillary tangles (NFTs) and senile plaques (SPs) through the hyperphosphorylation of tau protein and deposition of Amyloid-β (Aβ) protein [1]. There is extensive research evidence on the physiological distribution and homeostasis of transition metal ions such as iron, copper, and zinc, and in particular, their role in maintaining normal physiological functions in the brain, including signal transduction, energy production, and neurotransmitter synthesis [2]. Current evidence suggests that dyshomeostasis of transition metals, which leads to the formation of toxic oxidative species, has significant implications for the formation of amyloid plaques [3,4], the tau hyperphosphorylation process [3,5,6], and neuroinflammation associated with the pathological progression of AD [7–9]. The brain is vulnerable to free radical damage and oxidative stress since neuronal cell membranes consist of high levels of polyunsaturated lipids [10].

Transition metals have been identified as potential novel targets for therapeutic intervention [11]. High concentrations of Zn (1 mM), Cu (0.4 mM), and Fe (1 mM) have been reported within amyloid plaques [12] and previous studies have suggested that chelating Fe, Cu, and Zn from amyloid plaques can reduce their toxicity and consequently increase their solubility, which further demonstrates the role of Fe, Cu, and Zn in AD pathophysiology [12–14].

To produce a profound understanding of how transition metals are implicated in neurodegenerative diseases such as Alzheimer's and Parkinson's disease, it is imperative to assess the metal concentration and distribution changes within regions affected by the disease process. A range of brain imaging modalities including computed tomography (CT) [15], magnetic resonance imaging (MRI) [16,17], and positron emission tomography (PET) [17] have been used to establish the presence of AD pathology. Nevertheless, they all suffer from common drawbacks, including high costs, limited accessibility, poor sensitivity and specificity, [18] and the absence of standardization and scalability [19]. Typically, metal concentrations in anatomical regions have been measured through cutting, digestion, and analysis using various analytical techniques [20]. However, these techniques result in losing spatial information which is vital when the disease states being assessed involve small, well-defined regions or specific cell types. In light of this, laser ablation-inductively coupled plasma-mass spectrometry (LA-ICP-MS), as a means of ultra-sensitive chemical analysis with a part per billion detection limit, has received much attention for visualizing metals in biological systems, such as intact samples and tissue sections [21]. LA-ICP-MS has been applied for quantitative imaging of Cu, Fe, Zn, and Mn in the MPTP mouse model of Parkinson's disease [22]; measuring the Mn, Fe, Cu, and Zn concentrations in the brain of a rat model of Parkinson's disease [23]; copper mapping in a zebrafish model of Menkes disease [24]; imaging of Cu, Zn, Pb, and U in human brain tumour resections [25]; measuring Zn and Fe concentrations in the mouse model of traumatic brain injury (TBI) [26]; and imaging of iron in the frontal cortex [27], hippocampus [28], and white and grey matter [29] of healthy controls and Alzheimer's disease patients.

The eye, and specifically the retina, is an extension of the central nervous system [30] A growing body of evidence has linked retinal changes to pathophysiological features of AD, making the eye a strategic roadmap for screening and monitoring AD progression, particularly in its preclinical stages [19,31,32]. The retina is not restricted by some of the aforementioned limitations associated with brain imaging, and thus retinal imaging offers an attractive solution when studying AD-specific biomarkers. Whilst alterations in transition metals in the brain have been characterised before [33], little is known about whether such changes are also mirrored in the retina. Findings from studies that have linked retinal changes to pathological changes in the brain in AD [34] lead us to the hypothesise that a change in cerebral transition metal levels should be paralleled in the retina. Using LA-ICP-MS, we characterised the spatial distribution and quantified expression of Cu, Fe, and Zn in the hippocampus, the cortex, and the retina of 9-month-old Amyloid Precursor Protein/Presenilin 1 (APP/PS1) and Wild Type (WT) mice. The outcomes from this study may provide an insight in understanding transition metal alterations in the retina in AD.

2. Materials and Methods

All experiments were conducted within the Graduate School of Health (GSH) at the University of Technology, Sydney (UTS). APP/PS1 and normal aged mice were obtained from the Florey Institute of Neuroscience and Mental Health. All animal experimental procedures were approved by the Florey Institute of Neuroscience Animal Ethics Committee prior to the commencement of experiments (19-060-FINMH).

2.1. Animals

APP/PS1, a double transgenic mouse expressing a chimeric mouse/human amyloid precursor protein (Mo/HuAPP695swe) and a mutant human presenilin 1 (PS1-dE9) and age-matched C57BL6 WT mice were used in our experiments. We chose to examine animals at 9 months of age, as this reflects Aβ deposition in the hippocampus, cognitive impairment, and also impaired long-term potentiation (LTP) in the CA1 region of the hippocampus [35].

2.2. Tissue Collection

A total of twenty mice with an equal gender distribution were included in this study (10 APP/PS1 and 10 WT; 5 males and 5 females in each group). Following euthanasia with sodium pentobarbitone (80 mg/kg), transcardial perfusion was performed using 0.1 M phosphate buffer saline (PBS). The brain and whole eyes were removed, immediately placed in paraformaldehyde (4% w/v) and stored at 4 °C overnight. They were then cryopreserved in a 30% sucrose solution (PBS) for three days. Tissues were finally placed in an appropriate size mould and filled with optimal cutting temperature compound (OCT) and stored at −80 °C. Tissues were sectioned using the Leica CM1950 (Leica biosystem) at a thickness of 10 µm.

2.3. LA ICP-MS Imaging

LA ICP-MS was employed to measure the concentration of Cu, Fe, and Zn and their spatial distribution in the hippocampus, cortex, and retina of APP/PS1 and WT mice. The study was carried out on an Elemental Scientific Lasers NWR193 laser hyphenated to an Agilent Technologies 7700 ICP-MS, with 3 mL min^{-1} H_2 added in the reaction cell [36] and argon used as the carrier gas. LA-ICP-MS conditions were optimized on NIST 612 Trace Element in Glass CRM. The samples were ablated under 50 µm spot size and a scan speed of 200 µm/s at a frequency of 20 Hz.

2.4. Image Analysis

The data were collated into a single image file using in-house developed software, Pew^2, [37] and imported into ImageJ (National Institute of Health, MD, USA). The Allen Mouse Brain Atlas was used as reference [38] to draw a contour on the boundary of all regions of interest (hippocampus, cortex, and retina). DAPI staining was also used to visualize the gross anatomical morphology and to better guide the process of identifying region-specific anatomical boundaries. Following this process, the mean grey intensity value of each region was measured using the ImageJ built-in function. A minimum of 3 images per region were analysed and the average value taken as representative mean metal load.

2.5. Statistical Analysis

All statistical analysis was performed using Graphpad Prism (Dotmatics, CA, USA). Results are presented as mean ± standard error of the mean (SEM). Normality of data distribution was assessed using the D'Agostino and Pearson test. Unpaired t-test was used to compare differences between the two animal groups for each of the metals and anatomical regions.

3. Results

3.1. Copper (Cu)

The anatomical distribution of Cu in the brains and retinas is shown in Figure 1A,B, respectively. Visual assessment of the calibrated quantitative images demonstrated greater Cu enrichment in the hippocampus and cortex of WT mice compared with APP/PS1 mice (Figure 1A). A similar pattern was also observed in the eye, with the retina of WT mice harbouring higher concentrations of Cu compared to the APP/PS1 mice (Figure 1B). Results from the quantified image analysis were consistent with the visual assessment, with higher concentrations of Cu observed in the hippocampus, cortex, and retina of WT mice compared with APP/PS1 mice (Figure 2). More specifically, retinal Cu concentrations (µg/g) were significantly higher in WT mice compared with APP/PS1 mice (18.2 ± 0.9 vs. 16.2 ± 1.2, $p < 0.001$). In the hippocampus and cortex, Cu levels (µg/g) were also significantly higher in WT mice compared with APP/PS1 mice (20.3 ± 1.4 vs. 19 ± 0.5, $p < 0.05$ for the hippocampus; 14.4 ± 1.2 vs. 13 ± 1.5 $p < 0.05$ for the cortex).

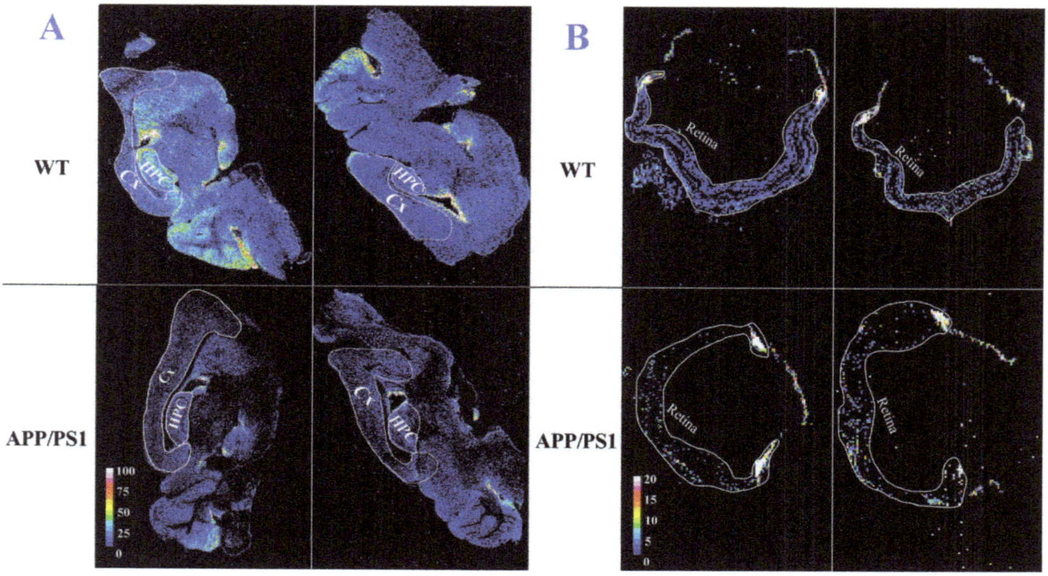

Figure 1. Spatial distribution of Cu in the brain (**A**) and eye (**B**). In each panel: sample map of 63Cu in sagittal brain/eye sections of 9-month-old WT (**upper row**) and APP/PS1 (**lower row**) mice. The scale represents calibrated Cu in ppm. HPC, hippocampus; CX, cortex. Side by side images of the brain and retina belong to two different animals.

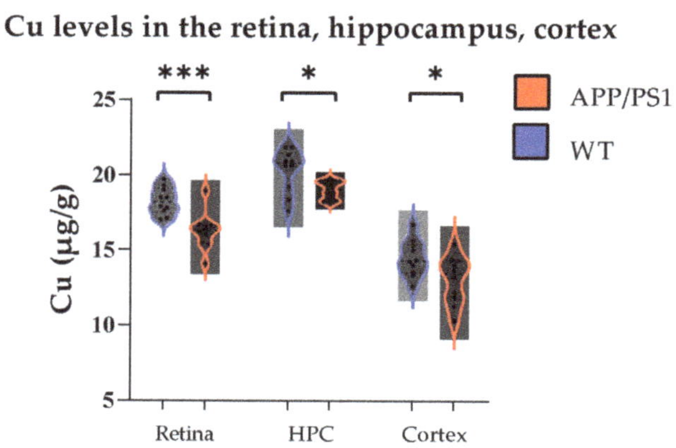

Figure 2. Quantified Cu levels. In all anatomical regions studied, Cu was significantly higher in WT (n = 10) compared with APP/PS1 (n = 10) mice. Error bars represent the Standard Error of the Mean (SEM). (* $p < 0.05$, *** $p < 0.001$).

3.2. *Iron (Fe)*

The spatial distribution of Fe in the brain and retinal sections of a representative WT and APP/PS1 mouse is shown in Figure 3A,B. Visual assessment of the calibrated quantitative images of the brain shows that the hippocampus and cortex of WT mice possess higher concentrations of Fe compared with APP/PS1 mice (Figure 3A). The retina of WT mice displayed a similar pattern (higher Fe content) relative to the age-matched APP/PS1 mice (Figure 3B). The intensity-based image analysis also confirmed these observations:

the hippocampus, the cortex, and the retina of WT mice have higher Fe concentrations than APP/PS1 mice (Figure 4). In the retina and hippocampus, Fe concentrations (μg/g) were significantly higher in WT mice compared with APP/PS1 mice (82.5 ± 6.3 vs. 73.1 ± 5.5, $p < 0.01$ for retina, 53.5 ± 3.6 vs. 37.3 ± 5.7, $p < 0.0001$ for hippocampus). In the cortex, while Fe concentration (μg/g) in the WT mice was higher than that of APP/PS1 mice, the difference was not statistically significant (40.4 ± 5.3 vs. 36.4 ± 6.3, $p = 0.18$).

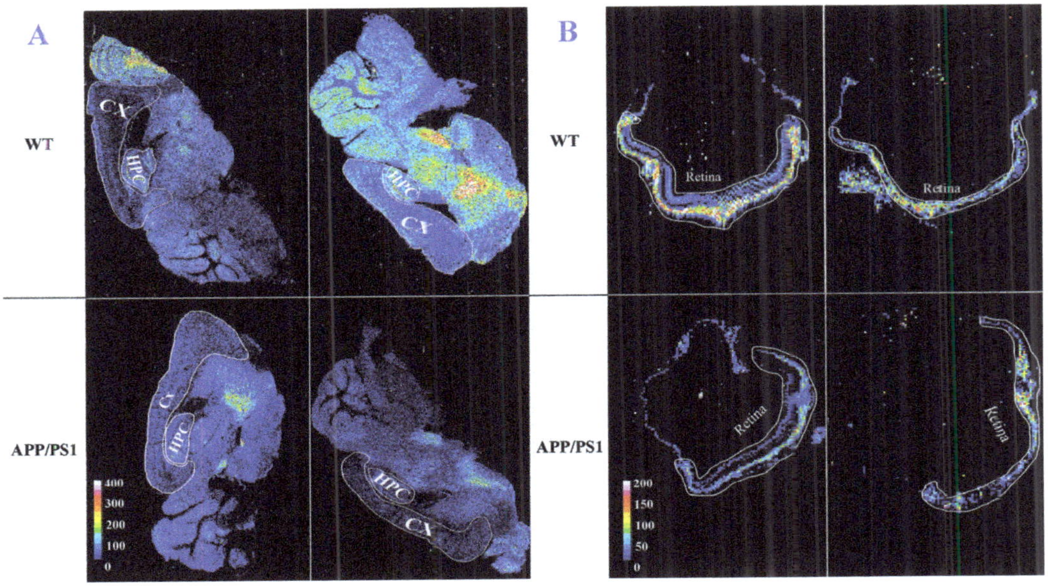

Figure 3. Spatial distribution of Fe in the brain (**A**) and eye (**B**). In each panel: sample map of 56Fe in sagittal brain/eye sections of 9-month-old WT (**upper row**) and APP/PS1 (**lower row**) mice. The scale represents calibrated Fe in ppm. HPC, hippocampus; CX, cortex. Side by side images of the brain and retina belong to two different animals.

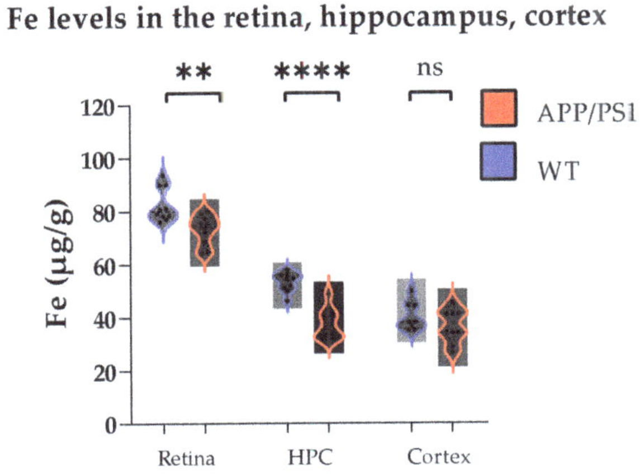

Figure 4. Quantified Fe levels. In the retina and hippocampus, Fe was significantly higher in WT (n = 10) compared with APP/PS1 (n = 10) mice. The Fe load in the cortex was not significantly different between the two groups ($p = 0.18$). Error bars represent the Standard Error of the Mean (SEM). (** $p < 0.01$, **** $p < 0.001$).

3.3. Zinc (Zn)

The quantitative images of Zn distribution in the brains and eyes are shown in Figure 5A,B. Surveying the calibrated quantitative images, they illustrate higher Zn concentration in the hippocampus and cortex of WT mice compared with APP/PS1 mice (Figure 5A). Similarly, the retina of WT mice also displayed the same trend, containing higher zinc levels than the APP/PS1 mice (Figure 5B). The intensity-based image analysis demonstrated significantly higher concentrations of zinc (μg/g) in the hippocampus, cortex and retina of WT mice compared with APP/PS1 mice (59.8 ± 6 vs. 52.3 ± 2.2, $p < 0.01$ for the retina; 52.5 ± 8 vs. 43.3 ± 2.7, $p < 0.01$ for the hippocampus; 20.4 ± 0.9 vs. 17.5 ± 1 $p < 0.0001$ for the cortex) (Figure 6).

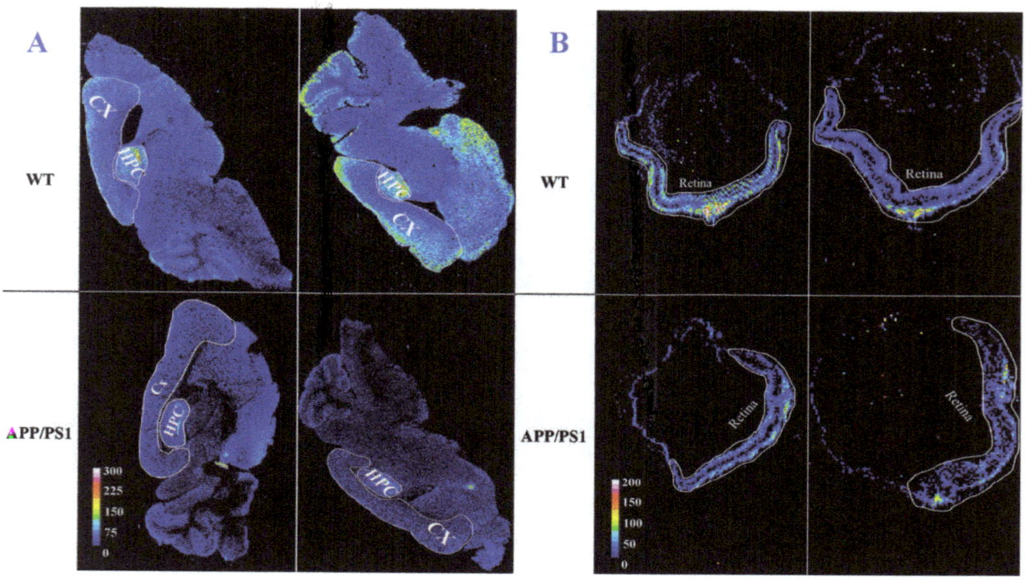

Figure 5. Spatial distribution of Zn in the brain (**A**) and eye (**B**). In each panel: sample map of 66Zn in sagittal brain/eye sections of 9-month-old WT (**upper row**) and APP/PS1 (**lower row**) mice. The scale represents calibrated Zn in ppm. HPC, hippocampus; CX, cortex. Side by side images of the brain and retina belong to two different animals.

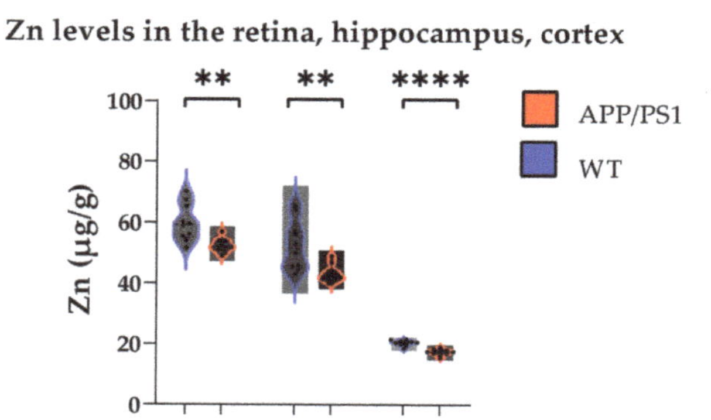

Figure 6. Quantified Zn levels. In all anatomical regions studied, Zn was significantly higher in WT ($n = 10$) compared with APP/PS1 ($n = 10$) mice. Error bars represent the Standard Error of the Mean (SEM). (** $p < 0.01$, **** $p < 0.0001$).

4. Discussion

In this study, we used LA-ICP-MS to study the spatial distribution of Cu, Fe, and Zn in cross-sectional slices obtained from the brains and the eyes of 9-month-old APP/PS1 and WT mice. Calibrated image-based intensity analysis was also employed to quantify transition metal levels in each of these structures. Our findings showed that the retina, hippocampus, and cortex of WT samples possessed higher Cu, Fe, and Zn levels than APP/PS1 mice. Collectively, our findings suggest that metal dyshomeostasis observed in the hippocampus and cortex are also mirrored in the retina.

Various analytical techniques, such as secondary ion mass spectrometry (SIMS) [39], particle-induced X-ray emission (PIXE) [40], X-ray fluorescence (SXRF) microprobe [41], graphite furnace atomic absorption spectrometry (GFAAS) [42], neutron activation analysis (NAA) [43], and instrumental neutron activation analysis (INAA) [44] have been used for bio-elemental dissection of AD pathology. Using PIXE and SXRF, altered levels of iron and zinc has been reported in the hippocampi of human AD brains as well as PSAPP mice [40,45]. LA-ICP-MS, as the biological trace element imaging technique, has advantages over the techniques mentioned above, such as high sample output, high sensitivity, high accuracy and precision of the analytical data, no charging-up effects, and fewer matrix effects, which results in a simple quantification of analytical data at a fraction of the cost of other techniques. These advantages motivated us to employ LA-ICP-MS for visualizing and quantifying Cu, Fe, and Zn in the retina, hippocampus, and cortex of APP/PS1 mice.

4.1. Copper

Copper is a necessary element for maintaining healthy cellular processes in the brain and the eye. In the brain, copper is vital for metalloenzyme functions which are essential in metabolic processes such as energy metabolism, antioxidant-defence mechanisms, and neurotransmitter synthesis [46,47]. In the retina, and similarly in the brain, "free" Cu participates in neurotransmission [48]. While copper overload is toxic, copper deficiency results in morphological changes in retinal structures [49]. The quantified total copper in the human retina has been reported as 6.6 ± 1.4 and 9.0 ± 5.0 μg/g in the RPE/choroid and neuroretina, respectively [50].

An age-associated increase in Cu levels has been reported in the brains and retinas of humans and mice [50,51], and in AD, a mis-metabolism of Cu has been associated with oxidative stress [52]. Copper has an impact on the regulation of other transition metals in AD [53]. Aβ precursor protein (APP) molecules are cleaved into two pathways (non-amyloidogenic and amyloidogenic) in the presence or absence of Cu [54]. The amyloidogenic pathway leads to amyloid generation. Previous studies have reported that Cu^{+2} ions prevent amyloid formation by reacting with a γ-secretase complex [4] or disturbing APP dimerization [55]. Therefore, low Cu concentration in AD raises Aβ production and accumulation by enhancing the amyloidogenic processing of APP [56].

Cu ions [57], through reduction from Cu^{2+} to Cu^{1+}, play a protective role against free radicals, an underlying cause of mitochondrial oxidative damage. Lower Cu content leads to higher levels of free radicals, as observed in amyloid plaques [58]. This could explain our observations of significantly lower Cu levels in the APP/PS1 mouse model of AD compared with WT mice. However, further studies are required to determine how Cu metabolism is disrupted in the brain of APP/PS1 mice and whether a similar process is also responsible for retinal Cu deficiency observed in these animals.

4.2. Iron

Iron is involved in a variety of essential metabolic processes, including oxygen transport, electron transport, DNA synthesis, redox/non-redox reactions, and other cell functions [59]. Iron is a highly redox-active compound, and due to its role in the generation of ROS, it is tightly regulated [60]. Changes in cerebral Fe levels have been proposed as a marker for cognitive impairment in AD [61]. Iron dyshomeostasis can instigate Aβ creation and aggregation because Fe is able to act on the iron regulatory element (IRE) site of APP mRNA, consequently boosting the APP translation and expression [62]. Furthermore, iron can increase the beta-secretase cleavage of APP, which enhances Aβ production by preventing alpha-secretase-induced APP cleavage via furin [63]. Moreover, iron can bind to the tau protein and boost the tau phosphorylation process, resulting in increasing hyperphosphorylated tau aggregation into neurofibrillary tangles.

Iron is also critical for maintaining retinal function [64], with retinal iron dyshomeostasis resulting in ocular diseases, such as glaucoma, cataract, AMD, and conditions causing intraocular haemorrhage [65]. It has been reported that the blood-retinal barrier protects the retina from systemic iron loading [66]. However, higher intraocular iron concentration leads to oxidative damage to the retina, which results in lipid peroxidation of photoreceptors and retinal degeneration [67]. Iron is mainly present in the choroid, the retinal pigment epithelium (RPE), and photoreceptors layer [68,69]. Using Graphite furnace atomic absorption spectrometry (GFAAS) and inductively coupled plasma mass spectrometry (ICP-MS) techniques [70], a mean Fe load of 117.63 ± 14.58 μg/g has been reported in the human neuroretina [71]

Many studies have reported an overload of Fe in AD, which contributes to Aβ deposition [72–75]. Despite this, we observed a uniform Fe deficiency in the retina, hippocampus, and cortex of APP/PS1 mice compared to their counterpart WT animals. This could be explained using a study by Maynard et al. [51] that suggested high levels of the carboxyl-terminal fragment of APP in the AD mouse brain can lead to a reduction in Fe levels [51]. In addition, the lower level of iron regulatory proteins, including hepcidin, iron-homeostatic peptide, and ferroportin, in AD results in lower Fe levels [76]. A study using APP/PS1 mice [77], the same model used in our study, has established that Fe is chemically reduced in the presence of aggregating Aβ plaques under certain physiological conditions. This could also explain the reduction in retinal Fe levels, as Aβ plaques have also been reported in the retina in AD [31]. Nevertheless, further mechanistic studies are required to interrogate this theory.

4.3. Zinc

Zinc, the second most abundant micronutrient in the human body, is a significant player in biological processes such as maintaining protein structure and stability, enzyme activities, adjusting various cellular processes, signal transduction, learning, and memory, and in the development and integrity of the immune system [78,79]. In the brain, Zn^{2+} is synaptically diffused during neuronal activity and plays a vital role in axonal and synaptic transmission and is obligatory for nucleic acid metabolism and brain tubulin growth and phosphorylation [80]. High levels of Zn are also present in the retina, suggesting a critical role in retinal physiology [81]. ICP-MS study of retinal pigment epithelium (RPE)/choroid and neuroretina demonstrated a mean Zn value of 292.1 ± 98.5 and 123.1 ± 62.2 μg/g, respectively [82]. Zn deficiency leads to ultrastructural changes in the retina and retinal pigment epithelium [83].

A high concentration of Zn^{2+} ions has been found in senile plaques of post-mortem AD brains [12,84]. Zinc ions interact with histidine residues at the N-terminal of Aβ [85,86]. However, the Aβ aggregation mechanism of Zn^{2+} is still elusive because assemblies of Zn-Aβ are variable and sensitive due to changes in pH, temperature, concentration, and buffer environment [87]. Accordingly, elevated levels of zinc have been reported in AD [88]. Interestingly, our results demonstrate lower Zn levels in the retina, hippocampus, and cortex of APP/PS1 compared with WT mice. A likely explanation for our finding is based on the fact that previous studies have shown that Zinc Transporter 3 (ZnT3) proteins are essential for loading Zn into synaptic vessels [89–91]. These proteins are abundantly available during the early stages of AD and its only with ageing that ZnT3 levels are depleted resulting in higher levels of free Zn. We did not evaluate ZnT3 expression in our study, as it fell outside the scope of the project, and therefore cannot determine whether an alteration in ZnT3 levels is driving the changes seen in Zn levels in our animals. A similar event could also be extended to the retina, as ZnT3 has been localised to the neural retina in regions which have been found reactive for Zn ions [92]. Zn suppression in senile plaques could also contribute to the smaller pool of synaptic zinc observed [93]. Further work is required to examine how Zn and its transporters are altered in the brain and the retina.

4.4. Limitations

Despite the current study being the first to visualise and objectively quantify transition metal levels in the retinas and brains of APP/PS1 and WT mice, our study has several limitations. First, as distribution of iron, zinc, and copper in the retina could be non-uniform, our approach of evaluating metal load in the retina as a whole may encompass biological bias. Second, we only studied animals at 9 months of age, which may have not captured the full continuum of age-associated changes in metal dyshomeostasis over the course of the disease. Third, transgenic animal models are artificially modified at a genetic level; hence, they may not represent the complex multifactorial origin of the most common form of AD, the sporadic variant [94]. AD progression in transgenic animal models happens in a very different time window than in AD patients [95]. As our study did not include a negative control group, it could be possible that the changes observed in our study may be a result of the genetic manipulation in our animal cohort. Whilst the distribution and metal load observed in our WT mice is broadly comparable to a study that used LA-ICP-MS to quantify metal load in 4–5 month-old C57BL/6 mice [96], further experiments using our findings are required. Future studies should also include analysis of the individual retinal layers as well as studying animals at various ages on the disease spectrum.

4.5. Conclusions

Whilst previous studies have used LA-ICP-MS to quantitatively image transition metals in the brain of AD mouse models as well as post-mortem donor human tissue samples [22,27,28], our study was the first to visualize and quantify the spatial distribution of Cu, Fe, and Zn in the retina in addition to the brain, and more specifically the hippocampus and the cortex of an AD mouse model. Our findings showed that the hippocampus, cortex, and retina of WT samples express significantly higher Cu, Fe, and Zn than APP/PS1 mice. Overall, our results suggest that an imbalance in retinal transition metal levels occurs in parallel in the hippocampus and cortex in AD, and this lays the groundwork for further work studying retinal metal load in the context of early onset Alzheimer's disease.

Author Contributions: Conceptualization, S.M.H.M., O.S., D.P.B. and S.M.G.; methodology, S.M.H.M., N.R.-R., D.P.B. and P.A.A.; formal analysis, S.M.H.M., D.P.B. and S.M.G.; resources, O.S., D.P.B., P.A.A. and S.M.G.; data curation, S.M.H.M., N.R.-R. and D.P.B.; writing—original draft preparation, S.M.H.M., D.P.B. and S.M.G.; writing—review and editing, S.M.H.M., N.R.-R., O.S., D.P.B., P.A.A. and S.M.G.; supervision, O.S., D.P.B., P.A.A. and S.M.G.; funding acquisition, O.S., D.P.B and S.M.G.; All authors have read and agreed to the published version of the manuscript.

Funding: S.M. Hosseinpour Mashkani was supported by a UTS international scholarship. S.M. Golzan and O. Shimoni were supported by an NHMRC-ARC Dementia Fellowship (GNT1105930). The study was supported in part by funding received from the Mason Foundation.

Institutional Review Board Statement: The protocol was approved by the the Florey Institute of Neuroscience Animal Ethics Committee prior to the commencement of experiments (19-060-FINMH, 2019).

Informed Consent Statement: Not applicable.

Data Availability Statement: The data presented in this study are available on request from the corresponding author.

Conflicts of Interest: The authors declare no conflict of interest.

References

1. Gupta, V.K.; Chitranshi, N.; Gupta, V.B.; Golzan, M.; Dheer, Y.; Wall, R.V.; Georgevsky, D.; King, A.E.; Vickers, J.C.; Chung, R.; et al. Amyloid β accumulation and inner retinal degenerative changes in Alzheimer's disease transgenic mouse. *Neurosci. Lett.* **2016**, *623*, 52–56. [CrossRef] [PubMed]
2. Atwood, C.S.; Huang, X.; Moir, R.D.; Tanzi, R.E.; Bush, A.I. Role of free radicals and metal ions in the pathogenesis of Alzheimer's disease. *Met. Ions Biol. Syst.* **1999**, *36*, 309–364. [PubMed]
3. Martic, S.; Rains, M.K.; Kraatz, H.-B. Probing copper/tau protein interactions electrochemically. *Anal. Biochem.* **2013**, *442*, 130–137. [CrossRef] [PubMed]
4. Gerber, H.; Wu, F.; Dimitrov, M.; Garcia Osuna, G.M.; Fraering, P.C. Zinc and Copper Differentially Modulate Amyloid Precursor Protein Processing by γ-Secretase and Amyloid-β Peptide Production. *J. Biol. Chem.* **2017**, *292*, 3751–3767. [CrossRef]
5. Craddock, T.J.; Tuszynski, J.A.; Chopra, D.; Casey, N.; Goldstein, L.E.; Hameroff, S.R.; Tanzi, R.E. The zinc dyshomeostasis hypothesis of Alzheimer's disease. *PLoS ONE* **2012**, *7*, e33552. [CrossRef]
6. Lovell, M.A.; Smith, J.L.; Xiong, S.; Markesbery, W.R. Alterations in zinc transporter protein-1 (ZnT-1) in the brain of subjects with mild cognitive impairment, early, and late-stage Alzheimer's disease. *Neurotox. Res.* **2005**, *7*, 265–271. [CrossRef]
7. Kitazawa, M.; Hsu, H.-W.; Medeiros, R. Copper Exposure Perturbs Brain Inflammatory Responses and Impairs Clearance of Amyloid-Beta. *Toxicol. Sci.* **2016**, *152*, 194–204. [CrossRef]
8. Zhang, X.; Surguladze, N.; Slagle-Webb, B.; Cozzi, A.; Connor, J.R. Cellular iron status influences the functional relationship between microglia and oligodendrocytes. *Glia* **2006**, *54*, 795–804. [CrossRef]
9. Liu, Y.; Nguyen, M.; Robert, A.; Meunier, B. Metal Ions in Alzheimer's Disease: A Key Role or Not? *Acc. Chem. Res.* **2019**, *52*, 2026–2035. [CrossRef]
10. Ozcelik, D.; Uzun, H. Copper intoxication; antioxidant defenses and oxidative damage in rat brain. *Biol. Trace Elem. Res.* **2009**, *127*, 45–52. [CrossRef]
11. Barnham, K.J.; Bush, A.I. Biological metals and metal-targeting compounds in major neurodegenerative diseases. *Chem. Soc. Rev.* **2014**, *43*, 6727–6749. [CrossRef] [PubMed]
12. Lovell, M.A.; Robertson, J.D.; Teesdale, W.J.; Campbell, J.L.; Markesbery, W.R. Copper, iron and zinc in Alzheimer's disease senile plaques. *J. Neurol. Sci.* **1998**, *158*, 47–52. [CrossRef] [PubMed]
13. Rottkamp, C.A.; Raina, A.K.; Zhu, X.; Gaier, E.; Bush, A.I.; Atwood, C.S.; Chevion, M.; Perry, G.; Smith, M.A. Redox-active iron mediates amyloid-beta toxicity. *Free Radic. Biol. Med.* **2001**, *30*, 447–450. [CrossRef] [PubMed]
14. Schubert, D.; Chevion, M. The role of iron in beta amyloid toxicity. *Biochem. Biophys. Res. Commun.* **1995**, *216*, 702–707. [CrossRef] [PubMed]
15. Mendez, M.F.; Mastri, A.R.; Zander, B.A.; Frey, W.H., 2nd. A clinicopathological study of CT scans in Alzheimer's disease. *J. Am. Geriatr. Soc.* **1992**, *40*, 476–478. [CrossRef] [PubMed]
16. Lipsman, N.; Meng, Y.; Bethune, A.J.; Huang, Y.; Lam, B.; Masellis, M.; Herrmann, N.; Heyn, C.; Aubert, I.; Boutet, A.; et al. Blood-brain barrier opening in Alzheimer's disease using MR-guided focused ultrasound. *Nat. Commun.* **2018**, *9*, 2336. [CrossRef]
17. Gao, F. Integrated Positron Emission Tomography/Magnetic Resonance Imaging in clinical diagnosis of Alzheimer's disease. *Eur. J. Radiol.* **2021**, *145*, 110017. [CrossRef]
18. Beach, T.G.; Monsell, S.E.; Phillips, L.E.; Kukull, W. Accuracy of the clinical diagnosis of Alzheimer disease at National Institute on Aging Alzheimer Disease Centers, 2005–2010. *J. Neuropathol. Exp. Neurol.* **2012**, *71*, 266–273. [CrossRef]
19. Yuan, A.; Lee, C.S. Retinal Biomarkers for Alzheimer Disease: The Facts and the Future. *Asia-Pac. J. Ophthalmol.* **2022**, *11*, 140–148. [CrossRef]
20. Savory, J.; Herman, M.M. Advances in instrumental methods for the measurement and speciation of trace metals. *Ann. Clin. Lab. Sci.* **1999**, *29*, 118–126.
21. Paul, B.; Hare, D.J.; Bishop, D.P.; Paton, C.; Nguyen, V.T.; Cole, N.; Niedwiecki, M.M.; Andreozzi, E.; Vais, A.; Billings, J.L.; et al. Visualising mouse neuroanatomy and function by metal distribution using laser ablation-inductively coupled plasma-mass spectrometry imaging. *Chem. Sci.* **2015**, *6*, 5383–5393. [CrossRef]
22. Matusch, A.; Depboylu, C.; Palm, C.; Wu, B.; Höglinger, G.U.; Schäfer, M.K.H.; Becker, J.S. Cerebral Bioimaging of Cu, Fe, Zn, and Mn in the MPTP Mouse Model of Parkinson's Disease Using Laser Ablation Inductively Coupled Plasma Mass Spectrometry (LA-ICP-MS). *J. Am. Soc. Mass Spectrom.* **2010**, *21*, 161–171. [CrossRef]
23. Ha, Y.; Tsay, O.G.; Churchill, D.G. ICP-MS for the neurodegenerative and brain sciences. In *Metal Ions in Neurological Systems*; Linert, W., Kozlowski, H., Eds.; Springer: Vienna, Austria, 2012; pp. 223–238.
24. Ackerman, C.M.; Weber, P.K.; Xiao, T.; Thai, B.; Kuo, T.J.; Zhang, E.; Pett-Ridge, J.; Chang, C.J. Multimodal LA-ICP-MS and nanoSIMS imaging enables copper mapping within photoreceptor megamitochondria in a zebrafish model of Menkes disease†. *Metallomics* **2018**, *10*, 474–485. [CrossRef]
25. Zoriy, M.V.; Dehnhardt, M.; Reifenberger, G.; Zilles, K.; Becker, J.S. Imaging of Cu, Zn, Pb and U in human brain tumor resections by laser ablation inductively coupled plasma mass spectrometry. *Int. J. Mass Spectrom.* **2006**, *257*, 27–33. [CrossRef]
26. Portbury, S.D.; Hare, D.J.; Finkelstein, D.I.; Adlard, P.A. Trehalose improves traumatic brain injury-induced cognitive impairment. *PLoS ONE* **2017**, *12*, e0183683. [CrossRef]

27. Bulk, M.; Abdelmoula, W.M.; Geut, H.; Wiarda, W.; Ronen, I.; Dijkstra, J.; van der Weerd, L. Quantitative MRI and laser ablation-inductively coupled plasma-mass spectrometry imaging of iron in the frontal cortex of healthy controls and Alzheimer's disease patients. *NeuroImage* **2020**, *215*, 116808. [CrossRef]
28. Cruz-Alonso, M.; Fernandez, B.; Navarro, A.; Junceda, S.; Astudillo, A.; Pereiro, R. Laser ablation ICP-MS for simultaneous quantitative imaging of iron and ferroportin in hippocampus of human brain tissues with Alzheimer's disease. *Talanta* **2019**, *197*, 413–421. [CrossRef]
29. Hare, D.J.; Raven, E.P.; Roberts, B.R.; Bogeski, M.; Portbury, S.D.; McLean, C.A.; Masters, C.L.; Connor, J.R.; Bush, A.I.; Crouch, P.J.; et al. Laser ablation-inductively coupled plasma-mass spectrometry imaging of white and gray matter iron distribution in Alzheimer's disease frontal cortex. *NeuroImage* **2016**, *137*, 124–131. [CrossRef]
30. London, A.; Benhar, I.; Schwartz, M. The retina as a window to the brain—From eye research to CNS disorders. *Nat. Rev. Neurol.* **2013**, *9*, 44–53. [CrossRef]
31. Georgevsky, D.; Retsas, S.; Raoufi, N.; Shimoni, O.; Golzan, S.M. A longitudinal assessment of retinal function and structure in the APP/PS1 transgenic mouse model of Alzheimer's disease. *Transl. Neurodegener.* **2019**, *8*, 30. [CrossRef]
32. Matei, N.; Leahy, S.; Blair, N.P.; Burford, J.; Rahimi, M.; Shahidi, M. Retinal Vascular Physiology Biomarkers in a 5XFAD Mouse Model of Alzheimer's Disease. *Cells* **2022**, *11*, 2413. [CrossRef] [PubMed]
33. Rao, S.S.; Lago, L.; Gonzalez de Vega, R.; Bray, L.; Hare, D.J.; Clases, D.; Doble, P.A.; Adlard, P.A. Characterising the spatial and temporal brain metal profile in a mouse model of tauopathy. *Metallomics* **2020**, *12*, 301–313. [CrossRef] [PubMed]
34. Guo, L.; Duggan, J.; Cordeiro, M.F. Alzheimer's disease and retinal neurodegeneration. *Curr. Alzheimer Res.* **2010**, *7*, 3–14. [CrossRef] [PubMed]
35. Gengler, S.; Hamilton, A.; Hölscher, C. Synaptic plasticity in the hippocampus of a APP/PS1 mouse model of Alzheimer's disease is impaired in old but not young mice. *PLoS ONE* **2010**, *5*, e9764. [CrossRef]
36. Lear, J.; Hare, D.J.; Fryer, F.; Adlard, P.A.; Finkelstein, D.I.; Doble, P.A. High-resolution elemental bioimaging of Ca, Mn, Fe, Co, Cu, and Zn employing LA-ICP-MS and hydrogen reaction gas. *Anal. Chem.* **2012**, *84*, 6707–6714. [CrossRef]
37. Lockwood, T.E.; Westerhausen, M.T.; Doble, P.A. Pew(2): Open-Source Imaging Software for Laser Ablation-Inductively Coupled Plasma-Mass Spectrometry. *Anal. Chem.* **2021**, *93*, 10418–10423. [CrossRef]
38. Atlas, A.M.B. Allen Mouse Brain Atlas. Available online: https://mouse.brain-map.org/ (accessed on 9 March 2023).
39. Moon, D.W.; Park, Y.H.; Lee, S.Y.; Lim, H.; Kwak, S.; Kim, M.S.; Kim, H.; Kim, E.; Jung, Y.; Hoe, H.S.; et al. Multiplex Protein Imaging with Secondary Ion Mass Spectrometry Using Metal Oxide Nanoparticle-Conjugated Antibodies. *ACS Appl. Mater. Interfaces* **2020**, *12*, 18056–18064. [CrossRef]
40. Danscher, G.; Jensen, K.B.; Frederickson, C.J.; Kemp, K.; Andreasen, A.; Juhl, S.; Stoltenberg, M.; Ravid, R. Increased amount of zinc in the hippocampus and amygdala of Alzheimer's diseased brains: A proton-induced X-ray emission spectroscopic analysis of cryostat sections from autopsy material. *J. Neurosci. Methods* **1997**, *76*, 53–59. [CrossRef]
41. Miller, L.M.; Wang, Q.; Telivala, T.P.; Smith, R.J.; Lanzirotti, A.; Miklossy, J. Synchrotron-based infrared and X-ray imaging shows focalized accumulation of Cu and Zn co-localized with beta-amyloid deposits in Alzheimer's disease. *J. Struct. Biol.* **2006**, *155*, 30–37. [CrossRef]
42. Magaki, S.; Raghavan, R.; Mueller, C.; Oberg, K.C.; Vinters, H.V.; Kirsch, W.M. Iron, copper, and iron regulatory protein 2 in Alzheimer's disease and related dementias. *Neurosci. Lett.* **2007**, *418*, 72–76. [CrossRef]
43. Ward, N.I.; Mason, J.A. Neutron activation analysis techniques for identifying elemental status in Alzheimer's disease. *J. Radioanal. Nucl. Chem.* **1987**, *113*, 515–526. [CrossRef]
44. Cornett, C.R.; Markesbery, W.R.; Ehmann, W.D. Imbalances of trace elements related to oxidative damage in Alzheimer's disease brain. *Neurotoxicology* **1998**, *19*, 339–345.
45. Leskovjan, A.C.; Kretlow, A.; Lanzirotti, A.; Barrea, R.; Vogt, S.; Miller, L.M. Increased brain iron coincides with early plaque formation in a mouse model of Alzheimer's disease. *Neuroimage* **2011**, *55*, 32–38. [CrossRef]
46. Scheiber, I.F.; Mercer, J.F.; Dringen, R. Metabolism and functions of copper in brain. *Prog. Neurobiol.* **2014**, *116*, 33–57. [CrossRef]
47. An, Y.; Li, S.; Huang, X.; Chen, X.; Shan, H.; Zhang, M. The Role of Copper Homeostasis in Brain Disease. *Int. J. Mol. Sci.* **2022**, *23*, 13850. [CrossRef]
48. Mathie, A.; Sutton, G.L.; Clarke, C.E.; Veale, E.L. Zinc and copper: Pharmacological probes and endogenous modulators of neuronal excitability. *Pharmacol. Ther.* **2006**, *111*, 567–583. [CrossRef]
49. Dingle, J.; Havener, W.H. Ophthalmoscopic changes in a patient with Wilson's disease during long-term penicillamine therapy. *Ann. Ophthalmol.* **1978**, *10*, 1227–1230.
50. Wills, N.K.; Ramanujam, V.M.; Kalariya, N.; Lewis, J.R.; van Kuijk, F.J. Copper and zinc distribution in the human retina: Relationship to cadmium accumulation, age, and gender. *Exp. Eye Res.* **2008**, *87*, 80–88. [CrossRef]
51. Maynard, C.J.; Cappai, R.; Volitakis, I.; Cherny, R.A.; White, A.R.; Beyreuther, K.; Masters, C.L.; Bush, A.I.; Li, Q.X. Overexpression of Alzheimer's disease amyloid-beta opposes the age-dependent elevations of brain copper and iron. *J. Biol. Chem.* **2002**, *277*, 44670–44676. [CrossRef]
52. James, S.A.; Volitakis, I.; Adlard, P.A.; Duce, J.A.; Masters, C.L.; Cherny, R.A.; Bush, A.I. Elevated labile Cu is associated with oxidative pathology in Alzheimer disease. *Free Radic. Biol. Med.* **2012**, *52*, 298–302. [CrossRef]
53. Xu, J.; Church, S.J.; Patassini, S.; Begley, P.; Waldvogel, H.J.; Curtis, M.A.; Faull, R.L.M.; Unwin, R.D.; Cooper, G.J.S. Evidence for widespread, severe brain copper deficiency in Alzheimer's dementia. *Metallomics* **2017**, *9*, 1106–1119. [CrossRef] [PubMed]

54. Bagheri, S.; Squitti, R.; Haertlé, T.; Siotto, M.; Saboury, A.A. Role of Copper in the Onset of Alzheimer's Disease Compared to Other Metals. *Front. Aging Neurosci.* **2017**, *9*, 446. [CrossRef] [PubMed]
55. Kong, G.K.; Miles, L.A.; Crespi, G.A.; Morton, C.J.; Ng, H.L.; Barnham, K.J.; McKinstry, W.J.; Cappai, R.; Parker, M.W. Copper binding to the Alzheimer's disease amyloid precursor protein. *Eur. Biophys. J.* **2008**, *37*, 269–279. [CrossRef] [PubMed]
56. Bayer, T.A.; Schäfer, S.; Simons, A.; Kemmling, A.; Kamer, T.; Tepest, R.; Eckert, A.; Schüssel, K.; Eikenberg, O.; Sturchler-Pierrat, C.; et al. Dietary Cu stabilizes brain superoxide dismutase 1 activity and reduces amyloid Abeta production in APP23 transgenic mice. *Proc. Natl. Acad. Sci. USA* **2003**, *100*, 14187–14192. [CrossRef]
57. Deloncle, R.; Guillard, O. Is brain copper deficiency in Alzheimer's, Lewy body, and Creutzfeldt Jakob diseases the common key for a free radical mechanism and oxidative stress-induced damage? *J. Alzheimer's Dis.* **2015**, *43*, 1149–1156. [CrossRef]
58. Reddy, P.H.; Beal, M.F. Amyloid beta, mitochondrial dysfunction and synaptic damage: Implications for cognitive decline in aging and Alzheimer's disease. *Trends. Mol. Med.* **2008**, *14*, 45–53. [CrossRef]
59. Sipe, J.C.; Lee, P.; Beutler, E. Brain iron metabolism and neurodegenerative disorders. *Dev. Neurosci.* **2002**, *24*, 188–196. [CrossRef]
60. Eisenstein, R.S. Iron regulatory proteins and the molecular control of mammalian iron metabolism. *Annu. Rev. Nutr.* **2000**, *20*, 627–662. [CrossRef]
61. Li, X.; Lei, P.; Tuo, Q.; Ayton, S.; Li, Q.X.; Moon, S.; Volitakis, I.; Liu, R.; Masters, C.L.; Finkelstein, D.I.; et al. Enduring Elevations of Hippocampal Amyloid Precursor Protein and Iron Are Features of β-Amyloid Toxicity and Are Mediated by Tau. *Neurotherapeutics* **2015**, *12*, 862–873. [CrossRef]
62. Cahill, C.M.; Lahiri, D.K.; Huang, X.; Rogers, J.T. Amyloid precursor protein and alpha synuclein translation, implications for iron and inflammation in neurodegenerative diseases. *Biochim. Biophys. Acta* **2009**, *1790*, 615–628. [CrossRef]
63. Silvestri, L.; Camaschella, C. A potential pathogenetic role of iron in Alzheimer's disease. *J. Cell. Mol. Med.* **2008**, *12*, 1548–1550. [CrossRef]
64. Chen, H.; Lukas, T.J.; Du, N.; Suyeoka, G.; Neufeld, A.H. Dysfunction of the retinal pigment epithelium with age: Increased iron decreases phagocytosis and lysosomal activity. *Investig. Ophthalmol. Vis. Sci.* **2009**, *50*, 1895–1902. [CrossRef]
65. He, X.; Hahn, P.; Iacovelli, J.; Wong, R.; King, C.; Bhisitkul, R.; Massaro-Giordano, M.; Dunaief, J.L. Iron homeostasis and toxicity in retinal degeneration. *Prog. Retin. Eye Res.* **2007**, *26*, 649–673. [CrossRef]
66. Zhao, L.; Li, Y.; Song, D.; Song, Y.; Theurl, M.; Wang, C.; Cwanger, A.; Su, G.; Dunaief, J.L. A high serum iron level causes mouse retinal iron accumulation despite an intact blood-retinal barrier. *Am. J. Pathol.* **2014**, *184*, 2862–2867. [CrossRef]
67. Campochiaro, P.A. Gene therapy for ocular neovascularization. *Curr. Gene Ther.* **2007**, *7*, 25–33. [CrossRef]
68. Song, D.; Dunaief, J.L. Retinal iron homeostasis in health and disease. *Front. Aging Neurosci.* **2013**, *5*, 24. [CrossRef]
69. Pamphlett, R.; Cherepanoff, S.; Too, L.K.; Kum Jew, S.; Doble, P.A.; Bishop, D.P. The distribution of toxic metals in the human retina and optic nerve head: Implications for age-related macular degeneration. *PLoS ONE* **2020**, *15*, e0241054. [CrossRef]
70. Hahn, P.; Ying, G.S.; Beard, J.; Dunaief, J.L. Iron levels in human retina: Sex difference and increase with age. *Neuroreport* **2006**, *17*, 1803–1806. [CrossRef]
71. Eckhert, C.D. Elemental concentrations in ocular tissues of various species. *Exp. Eye Res.* **1983**, *37*, 639–647. [CrossRef]
72. Lei, P.; Ayton, S.; Finkelstein, D.I.; Spoerri, L.; Ciccotosto, G.D.; Wright, D.K.; Wong, B.X.; Adlard, P.A.; Cherny, R.A.; Lam, L.Q.; et al. Tau deficiency induces parkinsonism with dementia by impairing APP-mediated iron export. *Nat. Med.* **2012**, *18*, 291–295. [CrossRef]
73. Acosta-Cabronero, J.; Betts, M.J.; Cardenas-Blanco, A.; Yang, S.; Nestor, P.J. In Vivo MRI Mapping of Brain Iron Deposition across the Adult Lifespan. *J. Neurosci.* **2016**, *36*, 364–374. [CrossRef] [PubMed]
74. Liu, J.L.; Fan, Y.G.; Yang, Z.S.; Wang, Z.Y.; Guo, C. Iron and Alzheimer's Disease: From Pathogenesis to Therapeutic Implications. *Front. Neurosci.* **2018**, *12*, 632. [CrossRef] [PubMed]
75. Connor, J.R.; Snyder, B.S.; Beard, J.L.; Fine, R.E.; Mufson, E.J. Regional distribution of iron and iron-regulatory proteins in the brain in aging and Alzheimer's disease. *J. Neurosci. Res.* **1992**, *31*, 327–335. [CrossRef] [PubMed]
76. Raha, A.A.; Vaishnav, R.A.; Friedland, R.P.; Bomford, A.; Raha-Chowdhury, R. The systemic iron-regulatory proteins hepcidin and ferroportin are reduced in the brain in Alzheimer's disease. *Acta Neuropathol. Commun.* **2013**, *1*, 55. [CrossRef] [PubMed]
77. Telling, N.D.; Everett, J.; Collingwood, J.F.; Dobson, J.; van der Laan, G.; Gallagher, J.J.; Wang, J.; Hitchcock, A.P. Iron Biochemistry is Correlated with Amyloid Plaque Morphology in an Established Mouse Model of Alzheimer's Disease. *Cell Chem. Biol.* **2017**, *24*, 1205–1215.e1203. [CrossRef]
78. Takeda, A.; Takada, S.; Nakamura, M.; Suzuki, M.; Tamano, H.; Ando, M.; Oku, N. Transient increase in Zn^{2+} in hippocampal CA1 pyramidal neurons causes reversible memory deficit. *PLoS ONE* **2011**, *6*, e28615. [CrossRef]
79. Roohani, N.; Hurrell, R.; Kelishadi, R.; Schulin, R. Zinc and its importance for human health: An integrative review. *J. Res. Med. Sci.* **2013**, *18*, 144–157.
80. Pal, A.; Prasad, R. Regional Distribution of Copper, Zinc and Iron in Brain of Wistar Rat Model for Non-Wilsonian Brain Copper Toxicosis. *Indian J. Clin. Biochem.* **2016**, *31*, 93–98. [CrossRef]
81. Ugarte, M.; Osborne, N.N. Zinc in the retina. *Prog. Neurobiol.* **2001**, *64*, 219–249. [CrossRef]
82. Erie, J.C.; Good, J.A.; Butz, J.A.; Pulido, J.S. Reduced zinc and copper in the retinal pigment epithelium and choroid in age-related macular degeneration. *Am. J. Ophthalmol.* **2009**, *147*, 276–282.e271. [CrossRef]
83. Grahn, B.H.; Paterson, P.G.; Gottschall-Pass, K.T.; Zhang, Z. Zinc and the eye. *J. Am. Coll. Nutr.* **2001**, *20*, 106–118. [CrossRef]

84. Frederickson, C.J.; Koh, J.Y.; Bush, A.I. The neurobiology of zinc in health and disease. *Nat. Rev. Neurosci.* **2005**, *6*, 449–462 [CrossRef]
85. Minicozzi, V.; Stellato, F.; Comai, M.; Dalla Serra, M.; Potrich, C.; Meyer-Klaucke, W.; Morante, S. Identifying the minimal copper- and zinc-binding site sequence in amyloid-beta peptides. *J. Biol. Chem.* **2008**, *283*, 10784–10792. [CrossRef]
86. Nair, N.G.; Perry, G.; Smith, M.A.; Reddy, V.P. NMR studies of zinc, copper, and iron binding to histidine, the principal metal ion complexing site of amyloid-beta peptide. *J. Alzheimer's Dis.* **2010**, *20*, 57–66. [CrossRef]
87. Lee, M.-C.; Yu, W.-C.; Shih, Y.-H.; Chen, C.-Y.; Guo, Z.-H.; Huang, S.-J.; Chan, J.C.C.; Chen, Y.-R. Zinc ion rapidly induces toxic off-pathway amyloid-β oligomers distinct from amyloid-β derived diffusible ligands in Alzheimer's disease. *Sci. Rep.* **2018**, *8*, 4772. [CrossRef]
88. Religa, D.; Strozyk, D.; Cherny, R.A.; Volitakis, I.; Haroutunian, V.; Winblad, B.; Naslund, J.; Bush, A.I. Elevated cortical zinc in Alzheimer disease. *Neurology* **2006**, *67*, 69–75. [CrossRef]
89. Adlard, P.A.; Parncutt, J.M.; Finkelstein, D.I.; Bush, A.I. Cognitive loss in zinc transporter-3 knock-out mice: A phenocopy for the synaptic and memory deficits of Alzheimer's disease? *J. Neurosci.* **2010**, *30*, 1631–1636. [CrossRef]
90. Hancock, S.M.; Portbury, S.D.; Gunn, A.P.; Roberts, B.R.; Bush, A.I.; Adlard, P.A. Zinc Transporter-3 Knockout Mice Demonstrate Age-Dependent Alterations in the Metalloproteome. *Int. J. Mol. Sci.* **2020**, *21*, 839. [CrossRef]
91. Xu, Y.; Xiao, G.; Liu, L.; Lang, M. Zinc transporters in Alzheimer's disease. *Mol. Brain* **2019**, *12*, 106. [CrossRef]
92. Redenti, S.; Chappell, R.L. Localization of zinc transporter-3 (ZnT-3) in mouse retina. *Vision Res.* **2004**, *44*, 3317–3321. [CrossRef]
93. Nuttall, J.R.; Oteiza, P.I. Zinc and the aging brain. *Genes Nutr.* **2014**, *9*, 379. [CrossRef] [PubMed]
94. Perez, C.M. Interplay between Synaptic GPCRs in Alzheimer's Disease. Ph.D. Thesis, University of Bristol, Bristol, UK, 2019.
95. Van Dam, D.; De Deyn, P.P. Animal models in the drug discovery pipeline for Alzheimer's disease. *Br. J. Pharmacol.* **2011**, *164*, 1285–1300. [CrossRef] [PubMed]
96. Hare, D.J.; Lee, J.K.; Beavis, A.D.; van Gramberg, A.; George, J.; Adlard, P.A.; Finkelstein, D.I.; Doble, P.A. Three-dimensional atlas of iron, copper, and zinc in the mouse cerebrum and brainstem. *Anal. Chem.* **2012**, *84*, 3990–3997. [CrossRef] [PubMed]

cells

MDPI

Review

Current and Future Landscape in Genetic Therapies for Leber Hereditary Optic Neuropathy

Hoda Shamsnajafabadi [1], Robert E. MacLaren [1,2] and Jasmina Cehajic-Kapetanovic [1,2,*]

1 Nuffield Laboratory of Ophthalmology, Department of Clinical Neurosciences, Oxford University, Oxford OX3 9DU, UK
2 Oxford Eye Hospital, Oxford University NHS Foundation Trust, John Radcliffe Hospital, Oxford OX3 9DU, UK
* Correspondence: enquiries@eye.ox.ac.uk

Abstract: Leber hereditary optic neuropathy (LHON) is the most common primary mitochondrial genetic disease that causes blindness in young adults. Over 50 inherited mitochondrial DNA (mtDNA) variations are associated with LHON; however, more than 95% of cases are caused by one of three missense variations (m.11778 G > A, m.3460 G > A, and m.14484 T > C) encoding for subunits ND4, ND1, and ND6 of the respiration complex I, respectively. These variants remain silent until further and currently poorly understood genetic and environmental factors precipitate the visual loss. The clinical course that ensues is variable, and a convincing treatment for LHON has yet to emerge. In 2015, an antioxidant idebenone (Raxone) received European marketing authorisation to treat visual impairment in patients with LHON, and since then it was introduced into clinical practice in several European countries. Alternative therapeutic strategies, including gene therapy and gene editing, antioxidant and neurotrophic agents, mitochondrial biogenesis, mitochondrial replacement, and stem cell therapies are being investigated in how effective they might be in altering the course of the disease. Allotopic gene therapies are in the most advanced stage of development (phase III clinical trials) whilst most other agents are in phase I or II trials or at pre-clinical stages. This manuscript discusses the phenotype and genotype of the LHON disease with complexities and peculiarities such as incomplete penetrance and gender bias, which have challenged the therapies in development emphasising the most recent use of gene therapy. Furthermore, we review the latest results of the three clinical trials based on adeno-associated viral (AAV) vector-mediated delivery of NADH dehydrogenase subunit 4 (ND4) with mitochondrial targeting sequence, highlighting the differences in the vector design and the rationale behind their use in the allotopic transfer.

Keywords: leber hereditary optic neuropathy; LHON; NADH dehydrogenase; retinal ganglion cells; gene therapy; mitochondrial inheritance; idebenone

Citation: Shamsnajafabadi, H.; MacLaren, R.E.; Cehajic-Kapetanovic, J. Current and Future Landscape in Genetic Therapies for Leber Hereditary Optic Neuropathy. *Cells* **2023**, *12*, 2013. https://doi.org/10.3390/cells12152013

Academic Editor: Hossein Ameri

Received: 25 June 2023
Revised: 27 July 2023
Accepted: 31 July 2023
Published: 7 August 2023

1. Introduction

Leber hereditary optic neuropathy (LHON) is the most common mitochondrially inherited disease that causes visual impairment. It affects ~1 in 30,000 to 1 in 50,000 individuals [1,2]. The symptoms of the disease were first described by the German ophthalmologist Theodor Leber in 1871 [3] and linked to mitochondrial DNA (mtDNA) variations by Wallace in 1988 [4] which led to mitochondrial dysfunction and ultimately cell death.

The hallmark of the disease is the selective vulnerability and premature degeneration of retinal ganglion cells (RGCs) leading to optic nerve atrophy and profound vision loss. The painless loss of vision occurs in children and young adults, and it typically affects the central visual field in one eye, followed by the loss of vision in the other eye within weeks to months. Sometimes the loss of vision is simultaneous in both eyes. There is a possibility that LHON patient can recover within a year, but this largely depends on the age of onset and mitochondrial genotype, and most patients with LHON become and remain legally blind, significantly impacting their quality of life [5–7].

Even though over 50 variations in mtDNA are associated with LHON, 95% of cases are caused by one of three mtDNA missense changes: m.11778 G > A (60%) [4,8], m.14484 T > C (14%) [9–11], and m.3460 G > A (13%) [12,13] encoding for nicotinamide adenine dinucleotide ND4, ND1, and ND6 subunits of the respiration complex I, respectively [9,14]. Due to the variable partitioning of the different mitochondrial haplotypes in each ovum, the penetrance is highly variable within and among families, making the condition difficult to predict [15]. Only a proportion of those that carry the mutation will thus develop the disease, with estimates for 50% of male and 10% of female carriers becoming affected [16].

Currently, there is no cure for LHON. Management of patients is limited to the best supportive care. Treatments based on antioxidant therapies such as idebenone, (licenced by EMA and available in some European countries) although promising, have shown limited overall efficacy dependent on disease stage and age of onset [17]. Advanced therapeutic modalities such as allotopic gene therapy, gene editing, mitochondrial biogenesis, mitochondrial replacement therapy, and stem cell therapy have been investigated, but an effective treatment for LHON is yet to emerge [18–21]. Researchers are faced with many difficulties when developing therapies for mitochondrial disease, the key challenges being our suboptimal approaches to deliver therapeutics including nucleic acids into the mitochondria and the lack of pre-clinical model systems. The nucleic acids are hydrophilic so even small naked DNA sequences do not cross the mitochondrial double membranes when unaided. Chemical approaches destabilise membranes with potential cytotoxic side effects [22]. Biological strategies to target mtDNA in mitochondrial intracellular space have, therefore, relied on mitochondrial targeting sequences (MTSs) added to therapeutic constructs via AAV capsids [23], mitochondrial proteins [24], or mRNA [25] to push them across the mitochondrial membranes. This sophisticated allotopic gene transfer where a mitochondrial gene is recoded into the nuclear codon, thereby allowing for protein expression in the cytoplasm and its subsequent transfer into the mitochondria [24,26–28] or a direct transfer of nuclear-transcribed mRNA to mitochondrial membrane for protein translation in the mitochondria [21,25,29,30], has formed the basis of three current LHON gene therapy trials [31–44].

These clinical trials are underway in Europe, the United States, and China and each involves the intravitreal injection of an AAV2 vector delivering normal copies of the ND4 complex I subunit in patients with LHON carrying m.11778 G > A variant. In this manuscript, we review the clinical phenotypes and complex molecular genetics of LHON disease, highlighting the peculiarities and challenges faced when developing new therapies with emphasis on the use of gene therapy in recent years.

2. Clinical Presentation of LHON

The typical LHON clinical presentation is the subacute onset of painless loss of central vision with sequential involvement in both eyes in 75% of cases or simultaneous presentation in 25% [16,45]. The time interval between the onset of sequential vision loss varies from a few weeks to a few months, with a median delay of 6 to 8 weeks. Since the majority of patients with LHON have bilateral involvement within one year (97% for m.11778 G > A variant) if the patient presents with unilateral optic neuropathy for more than one year, the patient is very unlikely to have the disease [6,46]. As the disease progresses, LHON goes through three phases: pre-symptomatic, acute, and atrophic. In the pre-symptomatic stage, patients typically do not present with obvious clinical symptoms but are diagnosed during family screenings. In the acute phase, the nerve fiber layer is swollen around the optic disc and circumpapillary telangiectatic microangiopathy affects visual acuity and colour vision. This stage ends with the disappearance of microangiopathy and the bundling of the first papillo-macular nerve fibers. As the disease progresses into the atrophic stage, optic nerve atrophy and thinning of the retinal nerve fiber layer become the pathological hallmark of the disease (Figure 1).

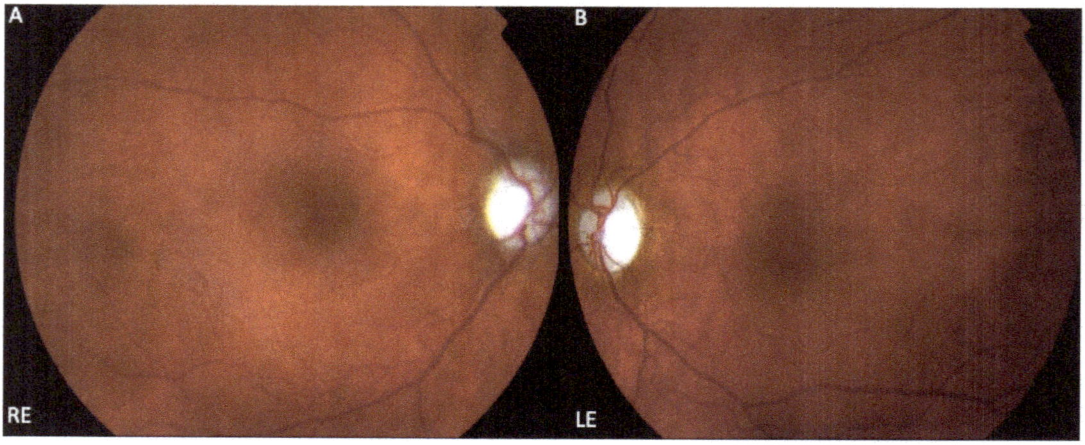

Figure 1. Clinical characteristics of Leber hereditary optic neuropathy (LHON). Colour fundus photographs of a 13-year-old girl with a loss of vision in both eyes (best corrected visual acuity 6/36 Snellen, right eye and left eye) showing bilaterally pale optic nerve discs (right eye (**A**), left eye (**B**)) and atrophy in the temporal part which is associated with the loss of papillo-macular bundle fibres and their associated astrocytes.

Although traditionally considered a disease that mostly affects young males, recent epidemiology studies show that LHON disease affects males and females of all ages with a 3:1 male-to-female ratio [7]. Females carrying an LHON pathogenic variant are, therefore, also at risk of losing vision, with the median age of onset of symptoms being slightly later at 30 years compared to 20 years for males [7]. Two major prognostic factors of visual outcome are the mitochondrial genotype and age at the onset of vision loss. There is, therefore, a possibility that visual field loss and colour vision can recover within a year, dependent on the specific variant of the mtDNA and the age of symptom onset. The reported rates vary depending on the criteria used to assess the rate of recovery. Overall, LHON patients harbouring m.11778 G > A variant represent the worst visual prognosis with only 4% visual recovery, although a recent prospective trial noted spontaneous recovery of visual acuity of 3 lines or more in 18% of patients occurring months to years after the initial loss of vision [47]. A total of 22% of patients carrying m.3460 G > A variant and 37% to 65% of patients with m.14484 T > C variant, and the best visual prognosis, recover some of their vision spontaneously [11,48] (Table 1). In addition, patients are more likely to recover when they are under 20 years of age [49], especially under 12 years of age, at the onset of the disease [50].

Table 1. Mitochondrial DNA variations associated with LHON disease and its characteristics.

mtDNA Variation/Complex I Subunit	NT Δ	AA Δ	Patients Carrying the Variant (%)	Variant Penetrance (% in Males)	Recovery of Vision (%)
m.11778 G > A/ND4	G > A	R340H	69	82	4–25
m.3460 G > A/ND1	G > A	A52T	13	40–80	22–25
m.14484 T > C/ND6	T > C	M64V	14	68	37–65

3. Molecular Genetics of LHON

In order to develop new therapeutic approaches, it is essential to fully understand both the pathogenic and natural compensatory mechanisms within the retinal ganglion cells (RGCs)—the primary cell type affected in LHON (Figure 2). A major challenge in understanding the mechanisms linking LHON variants to the selective neurodegeneration of RGCs is the lack of access to human tissue to study these cells and the limitations of existing animal models. In the retina, RGCs are a type of neuron located near the inner part of the retina which carry visual signals from photoreceptors to the brain in the form of action potentials. Approximately 1.2 million axons from RGCs run in the retinal nerve fibre layer and converge at the optic nerve head passing through the lamina cribrosa to form the optic nerve. These axons need high energy and mitochondria concentration to transmit visual information from the retina to the brain. RGCs produce adenosine triphosphate (ATP) due to oxidative phosphorylation of nutrients within the mitochondria. Several enzymes are involved in oxidative phosphorylation, including NADH dehydrogenase (complex I), Succinate dehydrogenase (complex II), Cytochrome b-c1 complex (complex III), Cytochrome c oxidase (complex IV), and ATP synthase. In the mitochondrial respiratory chain, electrons are shuttled from complex I to IV and provide energy by transferring protons across the membrane. As a protons channel, complex V conserves energy by catalysing ATP synthesis from adenosine diphosphate and inorganic phosphate. The mitochondrial respiratory chain comprises ~92 nuclear and mitochondrial DNA-encoded protein subunits. The presence of any related pathogenic variants mutation results in a reduction in ATP synthesis and increased production of reactive oxidative species [51]. A combination of these factors can lead to energy failure and cell death (Figure 2). In addition to high energy demand, the selective vulnerability to energy failure of RGCs could potentially be due to their narrow axons (especially for P-type RGCs at the papillo-macular bundle) and the lack of myelination necessary for retinal transparency and light penetration to photoreceptors.

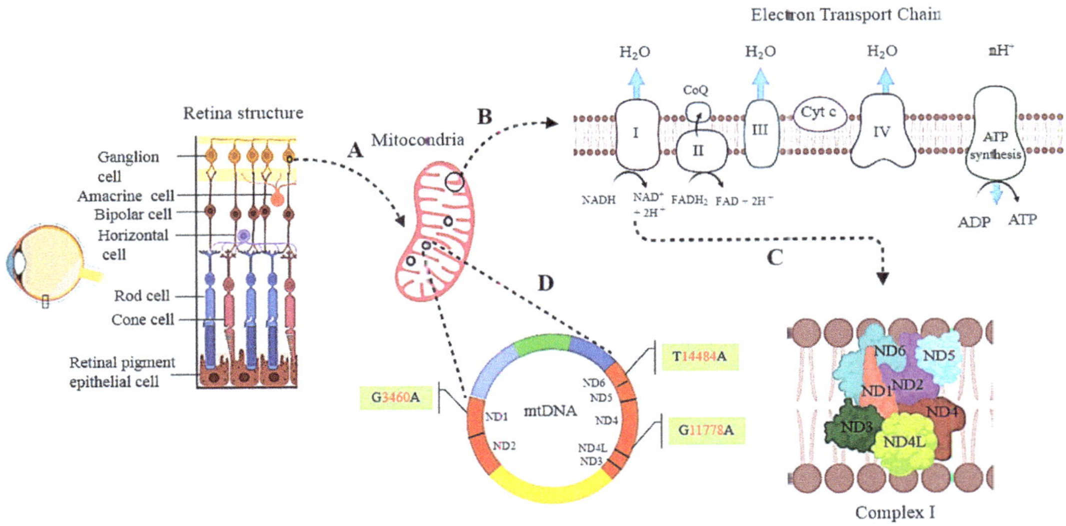

Figure 2. LHON is associated with mitochondrial DNA (mtDNA) variations leading to selective loss of retinal ganglion cells (RGCs). RGCs are high energy-demanding cells that prepare their energy using mitochondrial electron transport chains (**A**) where electrons are shuttled from complex I to IV in the mitochondrial respiratory chain to provide energy by transferring protons across complex V (**B**). Complex I is composed of ND1, ND2, ND3, ND4, ND4L, ND5, and ND6 subunits (**C**) encoded by mtDNA (**D**). The three most common primary mtDNA missense variants m.11778 G > A, m.14484 T > C, and m.3460 G > A in ND1, ND4, and ND6 subunits, respectively, lead to LHON disease. The figure was created using BioRender (www.biorender.com).

A mitochondrial genome consists of multiple copies of circular 16,569 base pair long mtDNA, encoding 37 genes (22 tRNAs, 2 rRNAs, and 13 subunits of the oxidative phosphorylation complex). mtDNA is characterised by a high copy number, maternal inheritance, as well as high levels of polymorphisms and variations. Variations in mtDNA are approximately 100 times higher than those in nuclear DNA, most likely due to a naked double-ring structure, insufficient mtDNA repair, and local environmental factors (possibly caused by oxidative radicals) [52]. Most variations associated with LHON are missense and found in complex I of the electron transport chain. Most of the patients with LHON (95%) are affected by either one of three primary mtDNA missense variants encoding for the respiratory chain subunits of the nicotinamide adenine dinucleotide ubiquinone- oxidoreductase (complex I) genes: 3460 G > A ND1, 11778 G > A ND4, and 14484 T > C ND6 [15]. In these patients, m.11778, m.3460, and m.14484 are present at a frequency of 69%, 13%, and 14% respectively (Table 1). It was estimated that 22%, 24%, and 31% of the points variations in m.11778, m.14484, and m.3460 represent a symptom onset peak in 21, 15, and 21 years old, respectively [7]. Adult patients with the m.11778 G > A variant in ND4 had the worst visual outcomes, as described above [8].

The link between the LHON variants and disease development remains poorly understood. The incomplete penetrance and male prevalence indicate that a primary variant is necessary, but not sufficient to develop the disease. Thus, the three primary mtDNA variations which alter the evolutionarily conserved amino acids remain clinically silent until an unknown trigger induces dysfunction of the respiratory chain of the RGCs, resulting in vision impairment. There is no evidence of these variations in control individuals. In addition to primary LHON variations, secondary mtDNA variations still exist and express with lower frequency in unaffected control individuals. The secondary mtDNA variations co-occur with primary variations or among each other but do not change evolutionarily conserved amino acids. The occurrence of these types of variants on specific mtDNA backgrounds increases the risk of LHON. Some secondary variations, such as ND1/4206, ND5/13708, CYTB/15257, and CYTB/15812, have co-occurred with the primary variants m.11778 and m.14484 in positive European families [53]. Carriers of LHON genotypes have variable and incomplete penetrance, and only a portion of the genotype carriers (males more than females) will manifest the disease with a low annual conversion rate. LHON reported penetrance is about 50% to 60% in males and 10% to 20% in females, although in larger cohorts, penetrance is also reported as low as 20% to 27% in males and 4% to 8% in females [1,54]. Although some carriers do not exhibit symptoms, it appears that a combination of genetic and environmental factors is strongly implicated in the onset of this condition. Disease penetrance is influenced by many genetic factors, including secondary mtDNA variations, the mtDNA copy number, and nuclear genes associated with correct maintenance and expression of mtDNA, heteroplasmy, and mtDNA haplogroups [55]. RGCs contain many mitochondria and, therefore, many copies of mtDNA. The levels of mutated mtDNA can vary from 0 to 100% and co-exist with wild-type mtDNA (heteroplasmy), or all mtDNA are either variant or wild-type (homoplasmy). In LHON, over 85% of the cases are homoplasmic states. Furthermore, LHON pathogenic variant carriers are at an increased risk of vision loss due to environmental factors such as tobacco smoking, excessive alcohol consumption, and toxic drug exposure [56]. The age factor is a major confounding factor contributing to the emergence of symptoms in older, asymptomatic individuals [57].

4. Treatment Modalities for LHON Disease

Several therapeutic strategies are in development to manage LHON, including genetic therapies, antioxidant and neurotrophic therapies, mitochondrial biogenesis, mitochondrial replacement therapy, and stem cell therapy (Figure 3). This review focuses on gene therapy for LHON, including the latest updates. In addition, we briefly discuss other potential therapies in development.

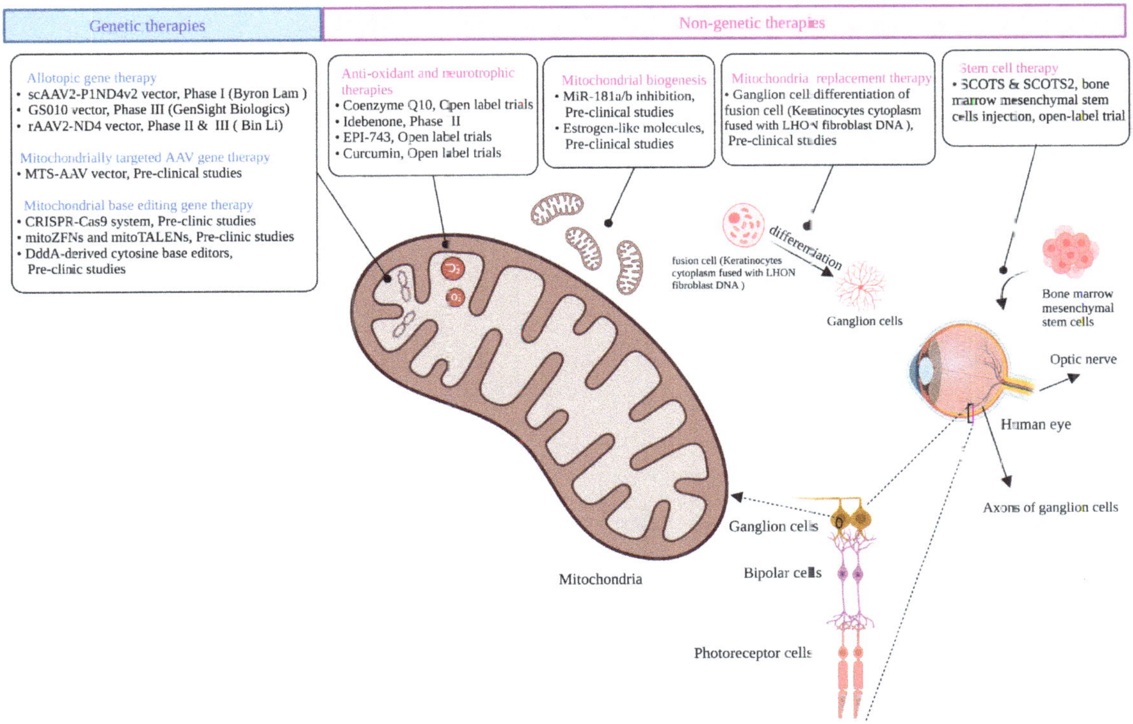

Figure 3. Schematic presentation of different therapeutic approaches in LHON. Current therapeutic strategies for LHON include genetic therapies, antioxidant and neurotrophic therapies, mitochondrial biogenesis, mitochondrial replacement therapy, and stem cell therapy. The figure was created using BioRender (www.biorender.com).

4.1. *Genetic Therapies in LHON*

Using gene therapy to replace the defective gene with the normal wild-type gene is an appealing treatment option for LHON, given that the RGC layer of the retina is relatively accessible for the delivery of corrected genes compared to other human tissue [58]. Among the potential gene therapies for mtDNA variations, allotopic gene therapy and mtDNA gene editing are the most promising approaches.

4.1.1. Allotopic Gene Therapy

Allotopic gene expression was developed to overcome the natural barrier to transporting molecules into the mitochondria—the mitochondrial double membrane. Applying this technique, mitochondrial genes are inserted into the nucleus, followed by the import of the resulting protein or mRNA with mitochondrial targeting sequence from the cytoplasm into the mitochondria. The efficiency of this method depends on the MTS arrangements and the protein assembly within the oxidative phosphorylation complexes (Figure 4).

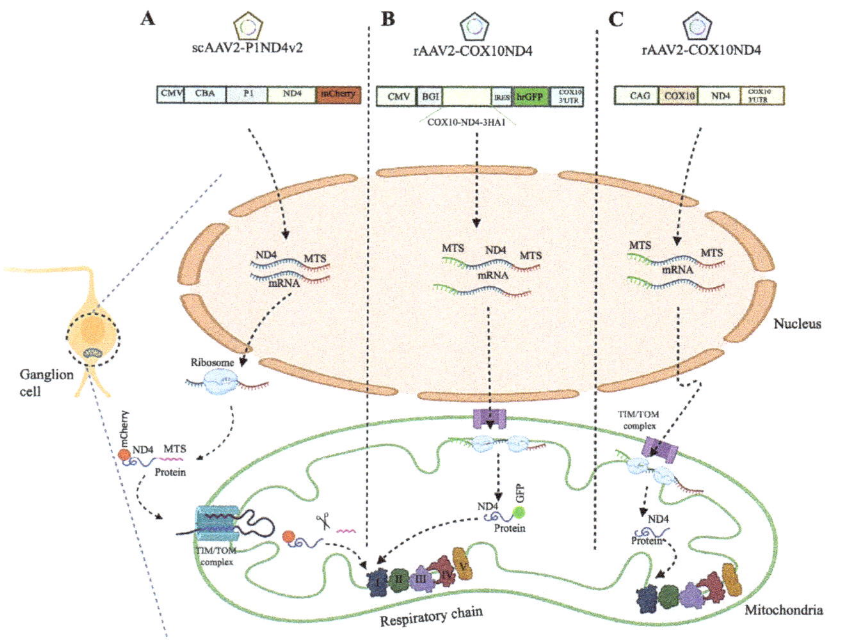

Figure 4. Allotopic gene therapy approaches for LHON disease. (**A**) The first allotopic gene therapy was performed using the scAAV2-P1ND4v2 viral vector carrying wild-type ND4 with the MTS nucleotide sequence from the ATP1 gene (P1). Following the delivery of the vector particles, genetic cargo is recorded in the nucleus and transcribed into mRNA which is transported into the cytoplasm for cytosolic ribosomal translation of the product protein containing the P1 MTS (P1-ND4). ND4-P1 is subsequently actively transported into mitochondria and integrated into complex I of mitochondrial respiratory chain (**B**) Second allotopic gene therapy was performed with rAAV2-COX10ND4 viral vector. This trial introduced a new MTS that allowed efficient allotopic expression of the human COX10ND4-cDNA in the nucleus followed by direct shuttling of mRNA into the outer membrane of the mitochondria. In turn, mitochondrial ribosomes translate mRNA into wild-type ND4 protein. (**C**) The third approach, also based on the AAV2-ND4 and the COX10 MTS (rAAV2-COX10ND4), allowed allotopic ND4 protein expression in the mitochondria in a similar manner to approach 2. The figure was created using BioRender (www.biorender.com).

Pre-Clinical Allotopic Gene Therapy Studies

In 2002, the first allotopic gene therapy approach was examined in human m.11778 LHON hybrids by a group of researchers led by the late John Guy and colleagues. As ND4 genes are expressed in the nucleus, a nuclear-encoded ND4 version should be prepared by converting the "non-standard" codons of the mitochondrial genetic system into the universal genetic code. Furthermore, the ND4 gene must contain enhancers, promoters, mitochondrial targeting sequences, polyadenylation signals, and a reporter gene. To direct the import of the recoded ND4 protein into the mitochondria from the cytoplasm, an MTS specifying the N-terminal region of either the P1 isoform of subunit c of human ATP synthase (ATPc) containing the entire 61 amino acid MTS plus the first five amino acids of the mature P1 polypeptide or the Aldh containing the first 19 amino acids MTS were used. Transduction of LHON cybrids with AAV-ND4 viral vectors resulted in effective expression of AAV-CMV-CBA-P1-ND4-FLAG and AAV-CMV-CBA-Aldh-ND4-GFP viral vector in the cytoplasm and import fusion protein into mitochondria. ATP synthesis was increased threefold in AAV-CMV-CBA-P1-ND4-FLAG complemented cybrids, while no effect was observed in AAV-CMV-CBA-Aldh-ND4-GFP treated cells. The first demonstration of allotopic respiration in vitro opened new prospects for future LHON gene therapy [26]. They produced the LHON mouse model by allotopic expression of the variant human R340H ND4 subunit of complex I in 2007, eliciting the hallmarks of human mitochondrial disease. According to their report in 2009, intravitreal injection of AAV2-CMV-CBA-P1-ND4-FLAG vector into LHON mouse model induced efficient expression in ganglion cells, while it did not induce the loss of RGCs, ATP synthesis, or PERG amplitude [24,27]. To follow up, they examined the safety and effects of scAAV2 (Y444,500,730F) -CMV-CBA-P1-ND4-FLAG viral vector intraocular injection into the eyes of mice, ex vivo human eye, rodents, and primates. They successfully reported that the human ND4 protein is expressed in almost all mouse RGCs and is integrated into complex I. In rodent models treated with this technique, vision is preserved, ATP synthesis is restored, and RGCs and optic nerve axons are prevented from being lost. In treated primates, a 3-month follow-up shows no adverse reactions. Ex vivo injection of ND4 into the human eye resulted in the expression of the protein in most retinal ganglion cells. Based on the results of this study, clinical testing followed for patients with the mitochondrial DNA variation 11778G > A [28].

In addition to targeting the ND4 protein to mitochondria, this group also worked on redirecting the AAV virions (carrying the ND4 gene) toward the mitochondria by the addition of the 23-aa cytochrome oxidase subunit 8 (COX8) sequence to the N terminus AAV2 capsid (COX8 MTS AAV capsid). The result demonstrated expression of the MTS AAV-delivered ND4 with the rescue of ATP synthesis in cultured LHON 11778G > A cybrids cells and visual loss and optic atrophy in rodents for almost their entire laboratory life span [59]. Using the same principle of MTS-tagged AAVs, the group recently reported that the MTS-AAV vector carrying wild-type human *ND1* restored mitochondrial respiratory complex I activity and ATP synthesis rate and protected RGCs and their axons from dysfunction and degeneration [23].

In 2007, another group of researchers led by Sahel and colleagues developed allotopic expression of *ND4* based on redirecting the mRNA to the mitochondria, in contrast to Guy's group approach where expressed ND4 protein is redirected. Through this MTS, derived from mitochondrial protein *COX10*, a modified complementary DNA (cDNA) encoding the human wild-type mitochondrial *ND4* gene was expressed allotopically in the nucleus by shuttling mRNA directly from the nucleus to the outer membrane of the mitochondria [25]. As an MTS (21 residues N-terminal and seven additional residues that ensure proper MTS cleavage (MAASPHTLSSRLLTGCVGGSVWYLERRT)), COX10 was introduced to a gene therapy vector in m.11778 LHON fibroblasts cells. As a result of transfection with the CMV-COX10-ND4-IRES-COX10 3′UTR-FLAG vector, the cells exhibited stably expressed ND4 protein and RNA, efficient mitochondrial translocation, as well as a significant restoration of oxidative phosphorylation activity [29,60]. The study was further extended by producing and injecting the pAAV-COX10-ND4-IRES-hrGFP viral vector into the rat model's vitreous body. Based on the findings of this study, the expression of human ND4 prevented the loss of RGCs and the impairment of visual function in the LHON rat model [21]. The safety and efficacy of this vector were confirmed in the LHON rat model by intravitreal injections of AAV2.2-CMV-beta globin intron-ND4-IRES-3HA1-hrGFP-COX10 3′UTR viral vector in 2015. In this study, human ND4 expression was demonstrated not to be harmful, with assembly in the respiratory chain complex I, and was found to significantly inhibit the degeneration of RGCs [25].

In 2013, a team of scientists led by Li and colleagues developed another mitochondrially targeted human *ND4* based on an amino aside protein sequence. The aside amine sequence of the produced ND4 protein was precisely the same as the native mitochondrial protein. An AAV2neo-CAG-COX10-ND4-COX103′UTR vector was designed based on the newly targeted gene and without a reporter sequence. The ND4 gene was confirmed to be translocated to the mitochondria in transduced HEK293 cells [30].

Clinical Allotopic Gene Therapy Trials

The results generated from the pre-clinical allotopic gene therapy studies formed the basis for launching the three major LHON clinical trials in the USA (NCT02161380, sponsored by NIH), Europe (NCT02064569, sponsored by GenSight Biologics), and China (NCT01267422, sponsored by NeurOphth) (Figure 5). Each trial was designed to supplement wild-type human *ND4* in patients carrying the most common LHON variant, the m.11778 G > A *ND4* variant.

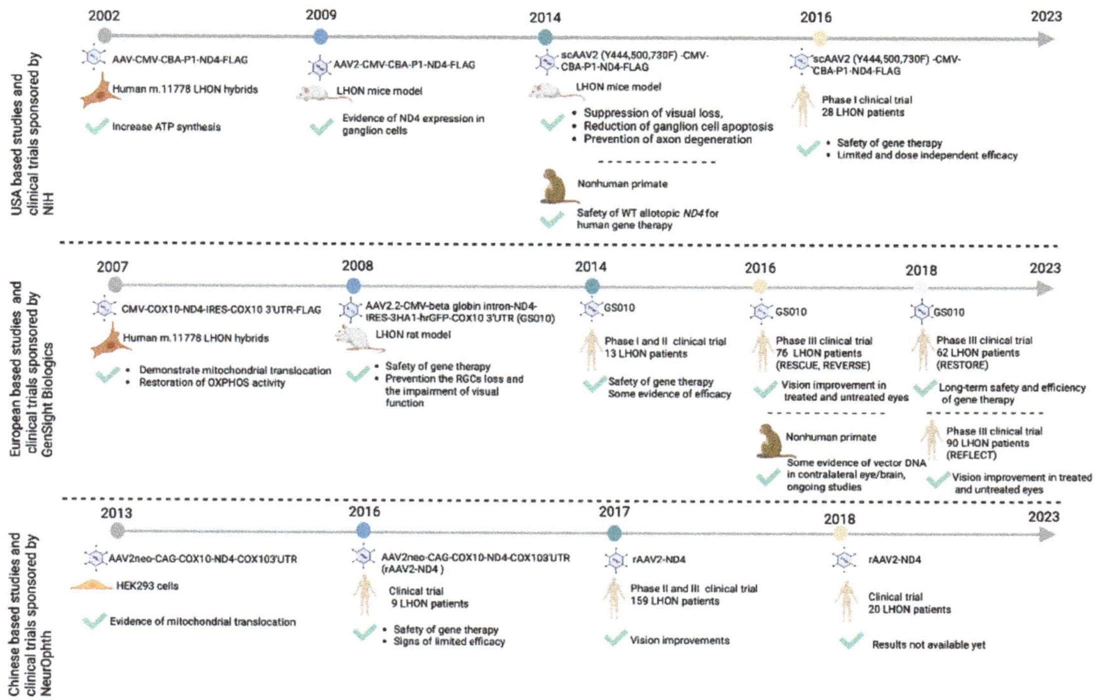

Figure 5. The development of LHON gene therapy from bench to bedside. Following proof-of-concept studies and having demonstrated safety and efficacy in pre-clinical models, three major clinical trials were launched, each aimed at testing safety and efficacy of gene supplementation of wild-type *ND4* in patients with mutant 11778 G > A *ND4*. The figure was created using BioRender (www.biorender.com).

An open-label dose escalation study (NCT02161380) was initiated in 2014 at the University of Miami, to demonstrate the safety of low and medium doses of scAAV2(Y444,500,730F)-P1ND4v2 viral vector for the treatment of LHON patients. The drug was injected intravitreally into one eye in five participants with vision loss secondary to G11778A LHON. A 90 to 180-day follow-up period was conducted on the treated participants, and ocular and systemic safety assessments were performed, as well as the visual structure and function examination. This initial phase I study revealed no serious safety concerns [34]. This trial was continued with intravitreal injection of vector into one eye of six participants with a chronic bilateral visual loss greater than 12 months (Group 1), six participants with bilateral visual loss less than 12 months (Group 2), and two participants with unilateral visual loss (Group 3) with a minimum of 12 months of follow-up. During the 12-month evaluation, it was confirmed that the average improvement in the injected eye was 0.24 logMAR, more significant than the 0.09 logMAR improvement observed in the fellow eyes. The average temporal RNFL thickness had not changed from baseline in the treated eyes, whereas it had reduced significantly in the fellow eyes 12 months after injection [31]. The longitudinal trial was completed in a total of 28 patients (5 females) after unilateral intravitreal injection of low, medium, high, and higher vector doses to patients with G11778A LHON and chronic bilateral visual loss > 12 months (Group 1, n = 11, worse eye), acute bilateral visual loss < 12 months (Group 2, n = 9, worse eye), or unilateral visual loss (Group 3, n = 8, better eye). The study result represents that incident uveitis (8 of 28, 29%) as a vector-related adverse event was related to vector dose and occurred in 71% of higher-dose eyes, and 14% of low, medium, or high-dose eyes. Improvements of ≥15-letter best-corrected visual acuity (BCVA) occurred in some treated and fellow eyes of Groups 1 and 2 and some surrogate study and fellow eyes of natural history subjects. Despite treatment, all study eyes (BCVA ≥ 20/40) in Group 3 lost ≥15 letters within the first year. The researchers concluded that the treatment efficacy, if any, is likely to be limited and dose independent. In addition, asymptomatic eyes with unilateral vision loss may be beyond the window of opportunity for neuroprotection given the reduced thickness of the ganglion cell layer and impaired pattern electroretinogram. Randomisation of patients to a control group that does not receive the vector in either eye would help to assess the treatment effect given confounders of subclinical disease and variable natural history [35].

In parallel, a Europe-based trial sponsored by GenSight Biologics (NCT02064569) was launched in 2014 to evaluate the safety of their vector AAV2.2-COX10ND4 (GS010). A single intravitreal injection of the vector was given to the worse-seeing eye of 19 LHON patients carrying the m.11778 variation. Initial follow-up results indicated no serious adverse events related to the treatment. Six of fourteen subjects with treated eyes had a clinically significant improvement in BCVA two years after injection [36]. A five-year follow-up revealed no serious adverse events related to the treatment or procedure. A clinically significant improvement in BCVA was observed in six of fourteen subjects at week 96 (similar proportions were observed at weeks 48 and 78) [37]. In 2016, two phase III clinical trials started to evaluate the clinical efficacy of GS010 in 76 LHON patients affected for up to 6 months (RESCUE, NCT02652767) and more than 6 months to 12 months (REVERSE, NCT02652780) by m.11778 variation in France, Germany, Italy, the United Kingdom, and the United States. Patients received a single dose of lenadogene nolparvovec (GS010) via a single intravitreal injection containing 9×10^{10} viral genomes in 90 µL in one of their randomly selected eyes [38]. The efficacy analysis of the RESCUE trial (multi-centre, randomised, double-masked, sham-controlled) included 38 subjects with a mean age of 36.8 years (82% male). Treatment started around 3.6 months and 3.9 months after the disease onset in the rAAV2/2-ND4 treated and sham-treated eyes, respectively. At week 48, the difference in BCVA from baseline between rAAV2/2-ND4 treated and sham-treated eyes were 0.01 logMAR, while it was 0.03 logMAR at week 96 [38]. The REVERSE trial (enrolled a total of 37 subjects) result showed that 25 subjects (68%) had a clinically relevant recovery in BCVA from baseline in at least one eye, and 29 subjects (78%) had an improvement in vision in both eyes (treated and sham). The difference in the change in BCVA between rAAV2/2-*ND4*-treated and sham-treated eyes at week 48 was −0.007 logMAR. Meanwhile, at week 96, there was an improvement in BCVA of −0.308 (0.068) logMAR, equivalent to a gain of 15 Early Treatment Diabetic Retinopathy Study (ET-DRS) letters ($p < 0.0001$ for change from baseline). The mean change from the baseline in BCVA increased continuously and bilaterally over the 96 weeks after treatment.

The possible mechanisms underlying the improvement in vision in the contralateral untreated eye were further investigated in nonhuman primates (NHPs). Three months after unilateral intravitreal injection of rAAV2/2-*ND4* in the NHP right eye, rAAV2/2-*ND4* DNA was detected in all the tissue and fluid samples tested for the three animals, including an anterior segment of the contralateral eye, retina, optic nerve, optic tract, lateral geniculate nucleus, and optic chiasm. In contrast, the contralateral visual cortex was not infected with the vector [39]. Further experiments are underway to explore this unexpected contralateral effect observed in clinical trials.

RESCUE subjects were treated on average 16 weeks after the onset of vision loss, whereas REVERSE subjects were treated 39 weeks after the onset of vision loss. On average, recovery of BCVA was observed 24 weeks after treatment in the RESCUE trial and 12 weeks after treatment in the REVERSE trial. Contrary to expectations, despite earlier treatment in RESCUE, visual outcomes at 96 weeks were less favourable than those among subjects treated at later stages of disease in REVERSE [40]. In 2018, the RESCUE or REVERSE long-term follow-up study assessed the safety and efficacy of GS010 in 62 (31 from REVERSE and 31 from RESCUE) patients with LHON five years after injection (RESTORE, NCT034061104). According to the 4-year data, objective visual acuity improvements are sustained and are associated with improvements in functional quality of life without any long-term safety concerns [33].

Another phase III, the global, multi-centre randomised, double-masked trial, was initiated in 2018 to evaluate the efficacy and safety of intravitreal injections of 9×10^{10} GS010 viral in 90 µL in 98 LHON patients up to 1 year from vision loss onset (REFLECT, NCT03293524). An intravitreal injection was administered in the first affected eye in all affected patients. Meanwhile, the second affected eye was randomised to either a second injection or a placebo injection. A bilateral treatment was administered to 48 patients, and a unilateral treatment to 50 patients. After two years, the mean BCVA in treated eyes was statistically significantly better than baseline, whereas improvement from baseline was not statistically significant in placebo-treated eyes. The results indicate a sustained treatment effect for all subjects, with the maximum improvement among bilaterally treated patients (https://www.gensight-biologics.com, accessed on 14 December 2021). GS010 currently remains an investigational compound and has not been registered in any country at this stage. A marketing authorization application is currently under review by the European Medicines Agency (EMA). GenSight also aims to resume its interactions with FDA to secure a regulatory pathway in the United States.

The third pre-clinical vector (pAAV2neo-COX10-ND4-COX10 3′UTR, or rAAV2-ND4) was taken into an investigator-led clinical trial (NCT01267422) in China. Nine patients were administered with the vector by intravitreal injection to one eye and then followed for 9 months During the 9-month follow-up period, no patients had local or systemic adverse events related to the vector, and the visual acuity (VA) of six patients improved by at least 0.3 logMAR. In these six patients, the VF was enlarged, but the RNFL remained relatively stable [41]. This trial's long-term results (3 years) confirmed the safety of intravitreal injection of rAAV2-ND4 for treating LHON. A significant improvement in BCVA, VF, and visual evoked potential (VEP) was also observed in treated eyes [61]. In 2017, a phase II and phase III clinical trial (NCT03153293) was initiated. The patients were treated intravitreally with a single injection of rAAV2-ND4 viral vector, with a 1×10^{10} virus dose in 0.05 mL. According to the results, higher baseline VFI, smaller baseline VF mean deviation, and higher baseline BCVA were associated with better VA outcomes. Long-term follow-up of these patients confirmed the strong association of age, the period between onset and treatment, and pre-treatment baseline BCVA with rapid and significant improvement in the VA of the injected eye [42,43]. It should also be noted that the severity of VF defects appeared independent of age, and it progressed within 6 months of disease onset remaining stable after 9 months. The differences in visual function between the eyes decreased as the disease progressed [44].

A new clinical trial (NCT03428178) was initiated in 2018 to evaluate the efficacy of gene therapy in patients with LHON symptom onset within 3 months. This clinical trial recruited 20 patients with the m.11778 variation onset within 3 months, 20 between 3 to 6 months, 20 between 6 to 12 months, 20 between 12 to 24 months, 20 between 24 to 60 months, and 20 over 60 months. All patients have been treated with a single injection of rAAV2-ND4 viral vector with dose 1×10^{10} viral 0.05 mL. A comparison will be made between the VA, VF, VEP, optical coherence tomography (OCT), electroretinograms (ERG), RNFL, and liver and kidney function in plasma before and after treatment at 1, 2, 3, 6, and 12 months. Results have not yet been published (Table 2).

Table 2. Summarises clinical trials targeting LHON disease by allotopic gene therapy.

Clinical Trial Identifier and Name	Phase	Vector	Age Range (Years)	Participants Population	Primary Endpoints	Secondary Endpoints	Results
NCT02161380	I	scAAV2-P1ND4v2	15+	28	IAE	Safety, efficacy, and VA change from baseline	No serious safety problems at low and medium doses; asymptomatic uveitis in 2 patients
NCT02064569	I and II	GS010	18+	19	IAE	-	Safety and tolerability of intravitreal
REVERSE, NCT02652780	III	GS010	15+	37	VA at Week 48	VA at Week 72 and Week 96, number of eye and subject responders, GCL macular volume, RNFL temporal quadrant and papillo-macular bundle thickness, ETDRS total macular volume, VF MD and foveal threshold, contrast sensitivity, and colour vision	Improvement of BCVA in treated and sham eye
RESCUE, NCT02652767	III	GS010	15+	39	VA at Week 48	VA at Week 72 and Week 96, number of eye and subject responders, GCL macular volume, RNFL temporal quadrant and papillo-macular bundle thickness, ETDRS total macular volume, VFI/MD and foveal threshold, contrast sensitivity, and colour vision	Improvement of BCVA
RESTORE, NCT03406104	III	GS010	15+	61	IAE up to 5 years	BCVA, HVF, and OCT, responder analysis, time course of the response, GCL, VFQ-25 and SF-36-v2 up to 5 years	No serious IAE
REFLECT, NCT03293524	III	GS010	15+	90	BCVA up to 1.5 year	BCVA, responder analysis, OCT, HVF, contrast sensitivity, VFQ-25 and SF-36-v2 up to 1.5 and 2 years	Improvement of BCVA
NCT01267422	Not Applicable	rAAV2-ND4	8 years to 60 years	9	BCVA up to 3 years	IOP, neutralizing antibodies, average RNFL thickness, VFI/MD, and VFI, VEP up to 3	Improvement of BCVA, VF, and VEPs, in both treated and untreated eyes, RNFL thickness relatively unchanged
NCT03153293	II and III	rAAV2-ND4	10 years to 65 years	159	BCVA up to 1 year	VEP, RNFL, and liver function in plasma up to 1 year	Results unavailable
NCT03428178	Not Applicable	rAAV2-ND4	8 years to 60 years	120	BCVA up to 1 year	VEP, RNFL, VF MD and VFI, liver, and kidney function in plasma	Results unavailable

Abbreviations: IAE, incidence of adverse events; BCVA, best corrected visual acuity; ETDRS, early treatment diabetic retinopathy study; IOP, intraocular pressure; OCT, optical coherence tomography; RNFL, retinal nerve fiber layer; VA, visual acuity; VEP, visual evoked potentials; VF, visual field; VFI, visual field index; VFI/MD, visual field index mean deviation; GCL, ganglion cell layer; VFQ-25, Quality of Life: Visual Functioning Questionnaire 25; and SF-36-v2, Quality of Life: 36-Item Short Form Health Survey, version 2.

4.1.2. Gene Editing

In mitochondrial disease caused by heteroplasmic variations in mtDNA, the clinical manifestations depend on the variation threshold resulting in a biochemical impairment. Therefore, eliminating some variant mtDNA molecules reduces the ratio of variant to wild-type mtDNA below the threshold for disease, decreasing the biochemical defect. Several endonucleases with localisation signals have been demonstrated to be capable of reaching mitochondria, identifying pathogenic variations as targets, and selectively causing double-strand breaks in heteroplasmic mtDNA, including mitochondria-targeted restriction endonucleases (mitoREs), zinc finger nucleases (mtZFNs), and mitochondrial-targeted transcription activator-like effector nucleases (mitoTALENs) [61,62]. In addition, both mitoZFNs and mitoTALENs approaches have been shown to restore oxidative phosphorylation function in a mouse model carrying a rare heteroplasmic variation of LHON [63,64].

In mtDNA editing, CRISPR-based systems have proven to be potent tools. The process uses guide RNAs (gRNAs) corresponding to the DNA sequences to be edited and nucleases that create double-stranded DNA breaks at specific sites. It has been demonstrated that the AAV2-mediated YFP/SpCas9 editing construct achieves the most efficient *YFP* reduction in the retinal cells of *YFP* transgenic mice [65,66]. Despite the potential of CRISPR/Cas it is restricted by the difficulty of delivering gRNA to mitochondria. Because of this reason, Hussain et al. propose a stem-loop element, called the RP-loop, as a potential enhanced mitochondrial transporter for CRISPR-Cas9 in 2021. The editing process was performed by adding a targeting guide RNA to an RP-loop and expressing the Cas9 enzyme containing mitochondrial localisation sequences. According to the results, RP-loop gRNA was colocalised with mitochondria, and transcription of the ND4 gene was significantly decreased in cells carrying 11205G. Therefore, CRISPR-Cas9 gene editing is a promising method for treating mitochondrial diseases [67]. Following the development of the CRISPR/Cas9 gene editing system, a promising candidate, mitoCas9 emerged, which targets mitochondria and is likely to be explored in future studies as a potential treatment for LHON [68]. Moreover, a CRISPR-free mitochondrial base editing with a bacterial cytidine deaminase toxin, the DddA-derived cytosine base editor (DdCBE) allowed precise manipulation of mtDNA in human cells, without the need to eliminate mtDNA copies that result from its cleavage by targeted nucleases [69].

4.2. *Antioxidant and Neurotrophic Therapies*

In the treatment of mitochondrial disorders, such as LHON, antioxidant therapies have played a crucial role. The first approach to antioxidant therapy involved nutritional supplements such as vitamins (B2, B3, B12, C, E, and folic acid), carnitine, creatine, l-arginine, dichloroacetate, or ketogenic diets. These treatments had limited and variable efficacy for patients with LHON [70]. Vitamin-like substance coenzyme Q10, or ubiquinone was also investigated for its antioxidant properties in LHON. By ubiquinone, electrons are shuttled from complexes I and II to complex III during the mitochondrial respiratory chain. A significant improvement in visual acuity was reported after 4 months of oral treatment with increasing coenzyme Q10 [71]. Treatment with combination therapies of CoQ10 and other nutritional supplements improved plasma lactate concentrations, body composition, ankle dorsiflexion strength, and oxidative stress in patients with LHON [72]. Subsequently, a synthetic analogue of coenzyme Q10 (CoQ10), idebenone became the most studied antioxidant drug for the treatment of LHON. There is evidence that ubiquinone-treated cell lines from patients with LHON can increase ATP production, decrease ROS levels, and prevent RGC deterioration in a mouse model of LHON [73]. In a 24-week multi-centre double-blind, randomised, placebo-controlled clinical trial (NCT00747487, RHODOS), 85 LHON patients with the three common mitochondrial DNA variations were treated with idebenone (300 mg three times per day) for 6 months (Canada, Germany, and the United Kingdom). Its treatment efficiency is controversial; effective in some patients but failed in others [74]. As a result of idebenone treatment, patients at imminent risk of further vision loss were able to maintain tritan colour vision and prevent any further loss of colour vision [75]. Following the conclusion of the previous trial, 39 LHON patients treated with idebenone were enrolled in a follow-up trial 30 months after discontinuation of the treatment (NCT01421381, RHODOS-OFU). Even after the drug was discontinued, RHODOS-OFU results indicate that VA continued to improve [76].

The results of a parallel study on 103 LHON patients revealed that earlier visual improvement was associated with younger age, the presence of the m.14484 variation, prompt initiation of the treatment, and a longer duration of the therapy [77].

In 2015, idebenone (Raxone®) was approved by the European Medicines Agency (EMA) in the European Union for treating LHON at 900 mg/day divided into three doses. For VA improvement, idebenone treatment should be initiated during the acute phase of the disease and continue for at least one year following the onset of the visual loss. As soon as improvements cease to be observed after one year, discontinuation should be considered [78]. Further clinical trials have confirmed that the duration of idebenone treatment needs to be increased to maximise its effectiveness. An open-label, multi-centre, retrospective, noncontrolled clinical trial (NCT04381091) examined the long-term VA and safety of idebenone (900 mg/day) treatment in 111 LHON patients (average treatment duration of 25.6 months). In 46% of LHON patients, clinically relevant recovery occurred between 2.5 and 26.5 months after treatment commenced, with a mean of 9.5 months. In addition to clinically relevant recovery, the magnitude of recovery increased with treatment duration.

Furthermore, idebenone treatment should be initiated early and maintained for more than 24 months to maximise its effectiveness, according to the results of this clinical trial [17]. From 2014 to 2020, a multi-centre, open-label clinical trial evaluated the efficiency of idebenone treatment on 72 Dutch LHON patients in the Netherlands. This study observed clinically relevant recovery (CRR) or clinically relevant stabilisation (CRS) in 53% and 11% of patients, respectively. Furthermore, RNFL and ganglion cell complex (GCC) thicknesses were irreversibly reduced [27]. Additional post-approval phase 4 clinical trials (NCT02774005, LEROS, and NCT02771379, PAROS) started in 2016 to examine the efficacy and safety of Raxone® in the long-term treatment of LHON patients. In LEROS, visual acuity outcomes following 24 months of idebenone treatment were compared to those of an external natural history cohort. A sub-analysis according to age at symptom onset and disease phase showed that treatment benefit was stronger in patients over 18 years. Subacute eyes of patients under 18 years of age showed limited treatment effect and had high recovery rate in the natural history cohort (ARVO 2023 Abstract no. 1953). In addition, a subanalysis of the potential treatment effects according to mitochondrial mutation show positive trends towards improved functional vision in eyes with m.11778 G > A and m.14484 T > C, whereas patients with m.3460 G > A had limited treatment effect and a mild disease course in the natural history cohort (ARVO 2023 Abstract no. 1954).

In addition to idebenone, several other antioxidants, such as EPI-743, elamipretide, and curcumin have been investigated. EPI-743 is a vitamin E base drug shown to have approximately 1000 to 10,000 fold greater antioxidant activity in vitro than coenzyme Q10 or idebenone [79]. The efficacy of EPI-743 was evaluated in a small, open-label clinical trial (NCT02300753) on five genetically confirmed LHON patients (four harbouring the m.11778 variant and one having the m.3460 variant) within 90 days of the onset of disease at a dose of 100–400 mg/day for at least one year. This trial confirmed that EPI-743 arrested disease progression and reversed vision loss in four patients. Based on the results of this trial, EPI-743 was found to arrest the progression of the disease and reverse vision loss in the participants [80]. Elamipretide is another promising candidate, and it was found to reduce reactive oxygen species (ROS) and increases adenosine triphosphate (ATP) levels [81]. As a result of a prospective, randomised, double-masked clinical trial (NCT02693119) conducted on 12 LHON patients with the m.11778 mtDNA variation, elamipretide (1 drop of elamipretide 1% topical ophthalmic solution) treatment for 52 weeks was found to be safe, tolerable, and efficacious. Continuing bilateral elamipretide therapy for at least

84 weeks is associated with significant improvements in best-corrected VA and colour discrimination in patients [82].

Moreover, an antioxidant drug known as curcumin (NCT00528151) has been clinically tested in a randomized, double-blind, placebo-controlled trial (NCT00528151). The trial (250 mg twice a day) was conducted in Thailand; however, the results still need to be published.

The newest antioxidant therapy is the near-infrared light-emitting diode therapy using low-energy lasers with a red to near-infrared range (630–1000 nm) to induce energy production and antioxidant protection [83]. The near-infrared light-emitting diode treatment is evaluated in a non-randomized, interventional clinical trial (NCT01389817) using a 630 nm wavelength and 4 J/cm^2 laser emissions to the closed eye (80 s) twice daily for 12 months. This clinical trial was conducted in 2011 involving four patients suffering from LHON, but the results are yet to become available.

Antioxidant therapies have been tested in clinical trials, but their efficacy and benefits remain unclear. Participants were instructed to maintain a regular diet, caffeine intake, fiber intake, and their normal levels of activity/exercise throughout the trial period. However, baseline differences in these participant attributes may impact the study's outcome. Furthermore, due to the relatively small number of participants, many of these trials have included subjects with defined mutations of various genes and phenotypic disorders. Also, it is difficult to determine whether genetic abnormalities or heteroplasmy levels caused differences in efficacy responses. Due to these reasons, a double-blind, crossover trial would be beneficial since it limits the number of patients required for statistical significance and reduces the effects of heterogeneity in the starting population.

4.3. *Mitochondrial Biogenesis and Replacement Therapies*

4.3.1. Mitochondrial Biogenesis

A compensatory mechanism for mitochondrial dysfunction in LHON occurs through mitobiogenesis or mitophagy. Many transcriptional activators peroxisome proliferator-activated receptor γ co-activator-1α, their regulators, and AMP-activated protein kinase include fibrates, rosiglitazone, metformin, and 5-aminoimidazole-4-carboxamide ribonucleoside are responsible for controlling mitobiogenesis [84].

Moreover, inhibiting the mammalian target of the rapamycin (mTOR) signalling pathway enhances mitophagy of dysfunctional mitochondria harbouring higher levels of variations. Rapamycin, as an mTOR inhibitor drug, induces selected mitophagy of dysfunctional mutated mitochondria and preserves the visual function in the LHON cybrid (G11778A) and Ndufs4 knockout mice model [85,86]. Several microRNAs, including miR181a/b, target Nrf1 to regulate mitochondrial biogenesis and mitophagy. In previous studies, the deactivation of MiR181a/b in LHON mouse models improved visual phenotypes. Therefore, this represents a gene-independent therapeutic target for the treatment of LHON [18].

In addition, since LHON has a gender bias, it is possible that different hormonal metabolisms between genders can affect mitochondrial function in LHON. Thus, a combination of natural estrogen-like molecules (i.e., phytoestrogens genistein, daidzein, and equol) or 17-norestradiol increased cell viability of LHON hybrid cells by reducing apoptosis, inducing mitochondrial biogenesis, and strongly reducing the levels of reactive oxygen species [87,88].

4.3.2. Mitochondrial Replacement Therapy

As part of LHON treatment, preventing mtDNA variation transmission to future generations has potential therapeutic benefits. This approach involves extracting nuclear DNA from the patient's egg or embryo and transplanting it into a donor's cytoplasm containing wild-type mitochondrial DNA. Pronuclear transfer, spindle transfer, and polar body transfer are possible methods of mitochondrial replacement therapy. In 2017, for the first time, a cybrid approach for correcting mtDNA variations in LHON patient-derived iPSCs with homoplasmic double variations m.4160 and m.14484 was proposed. In this study, the LHON fibroblasts' nuclear DNA was fused by wild-type keratinocytes cytoplasm. Successful mitochondrial replacement cells were isolated, expanded, and differentiated into RGC. Fused RGCs exhibited lower levels of mitochondrial superoxide than LHONs. They described the hybrid technique as a feasible approach for correcting mtDNA variations in LHON cell models, which could be applied to other mtDNA-related diseases [20]. However, this approach has been limited by active ethical considerations and unexplored long-term implications.

4.4. Stem Cell Therapy

The stem cell ophthalmology treatment study (SCOTS, NCT 01920867) and its follow-up study (SCOTS 2, NCT 03011541) are non-randomised, open-label efficacy trials for the treatment of retinal and optic nerve diseases by ganglion cell replacement. The mesenchymal stem cells, derived from the posterior iliac crest bone marrow, were administered through several delivery routes including retrobulbar, subtenon, intravitreal, intraocular, subretinal, and intravenous injections. The results indicated statistically significant improvements or vision stabilisations in most reported patients. Based on the results of the six LHON patients treated, 83.3% had improved their vision in both eyes within 24 months. This improvement is thought to have been achieved by transforming bone marrow mesenchymal stem cells into developed ganglion cells and transferring mitochondria and lysosomes to damaged cells [19,89]. However, claims of improvement in vision could not be verified because patients receiving the SCOTS treatment were discharged to local follow-up almost immediately after receiving their injections for which they paid up to USD 20,000 each. Hence, the SCOTS trial should not be considered seriously in any scientific debate and is only mentioned here to warn readers that its significance can be formally dismissed.

5. Summaries

LHON is the most common maternally inherited mitochondrial genetic disease and an important cause of blindness in children and young adults. The pathophysiology of LHON is particularly complex with peculiar specificity for RGCs, male bias, and incomplete penetrance. Research groups across the globe have made significant headways in improving our understanding of the disease mechanisms, natural history, and potential treatments. Several antioxidant therapies have been recommended, with varying outcomes, whilst idebenone became licenced by EMA in 2015 for the treatment of LHON. Post approval studies, however, have shown limited overall benefit to date with subgroup analyses showing dependence on the age at symptom onset, disease phase, and mitochondrial mutation. Currently, there are no recommendations to treat asymptomatic carriers.

Gene supplementation therapy is a logical treatment strategy for LHON, but the development of mitochondrial gene therapy has been hindered due to the difficulty in importing genes into the mitochondria and the lack of suitable pre-clinical models to carry out the studies. The development of allotopic gene transfer offered hope and its successful validation in cellular and animal models culminated in three major research groups leading multiple clinical trials. In 2015, GS010 (lenadogene nolparvovec, GenSight) received Investigational New Drug acceptance from the U.S. Food and Drug Administration (FDA) to enter phase III trials, and has indeed undergone phase III testing under three pivotal phase III trials (RESCUE, NCT02652767), (REVERSE, NCT02652780), and (REFLECT, NCT03293524). Unfortunately, despite visual improvements, beyond what would be expected based on the natural history of the disease, the overall efficacy was limited and complicated by the unexpected similar improvements in the untreated eye. This contralateral effect could be accounted for by the natural history of the disease, mechanical transfer of the viral vector DNA from one eye to the other, other potential local mediators, or simply a learning effect due to the clinical trial design and visual acuity measurements. NCT02161380 trial also suggests that if there is an efficacy effect in LHON, it is likely small and it was not uncommon in natural history patients. To ensure gene therapy efficac was picked up, some patients would need to be randomised into a group not receiving a injection in either eye. Overall, LHON clinical trial results have contributed greatly to ou understanding of LHON and provide valuable implications for future gene therapy clinica trial design and outcome measures.

6. Future Direction and Expert Opinion

Until now, promising progress has been seen in the gene therapy of LHON, but thi treatment is yet to be approved. The pre-clinical results are encouraging, but the expressio and function of ND4 in mitochondria require more evidence and more reliable assessment in pre-clinical models, which better recapitulate LHON disease. To this end, establishin induced pluripotent stem cells from patients with LHON and differentiating them int RGCs or full retinal organoids with preserved cellular microenvironments will open excit ing new research opportunities and therapeutic developments. Further expression studie of mitochondrial proteins post gene therapy in an NHP model will be critical in learnin the limitations of currently used vectors and the delivery method. The natural history o the LHON disease, the sequential nature of visual loss, and a varied degree of spontaneou improvement in affected individuals make the clinical trial design extremely challenging and improvements are needed in future study designs and clinical trial endpoints, wit the need to include a randomised untreated control group. Future trials should conside combination therapies, such as idebenone and gene therapy, which could result in a syner gistic, complementary therapeutic effect, especially in some patient subgroups. Ongoin pre-clinical research on genome editing and CRISPR-free mitochondrial base editing ha shown initial promising results and provides opportunities for further development specifi to LHON. The transition of LHON genetic treatments from bench to bedside will closel relate to the collaborative effort of molecular biologists and clinicians.

Author Contributions: Conceptualization, J.C.-K.; writing—original draft preparation, H.S. writing—review and editing, J.C.-K., H.S. and R.E.M. All authors have read and agreed to the published version of the manuscript.

Funding: This research was funded by UKRI MRC, NIHR Oxford BRC to J.C.K., UKRI MRC to H.S. and NIHR Oxford BRC to R.E.M.

Data Availability Statement: Data sharing not applicable.

Conflicts of Interest: The authors declare no conflict of interest. The funders had no role in the design of the study; in the writing of the manuscript; or in the decision to publish the results.

References

1. Puomila, A.; Hämäläinen, P.; Kivioja, S.; Savontaus, M.-L.; Koivumäki, S.; Huoponen, K.; Nikoskelainen, E.; Nikoskelainen E. Epidemiology and penetrance of Leber hereditary optic neuropathy in Finland. *Eur. J. Hum. Genet.* **2007**, *15*, 1079–1089 [CrossRef] [PubMed]
2. Yu-Wai-Man, P.; Griffiths, P.G.; Brown, D.T.; Howell, N.; Turnbull, D.M.; Chinnery, P.F. The epidemiology of Leber hereditary optic neuropathy in the North East of England. *Am. J. Hum. Genet.* **2003**, *72*, 333–339. [CrossRef] [PubMed]
3. Leber, T. Ueber hereditäre und congenital-angelegte Sehnervenleiden. *Albrecht Graefes Arch. Ophthalmol.* **1871**, *17*, 249–291 [CrossRef]
4. Wallace, D.C.; Singh, G.; Lott, M.T.; Hodge, J.A.; Schurr, T.G.; Lezza, A.M.; Elsas, L.J.; Nikoskelainen, E.K. Mitochondrial DNA mutation associated with Leber's hereditary optic neuropathy. *Science* **1988**, *242*, 1427–1430. [CrossRef]
5. Yu-Wai-Man, P.; Votruba, M.; Moore, A.T.; Chinnery, P.F. Treatment strategies for inherited optic neuropathies: Past, present and future. *Eye* **2014**, *28*, 521–537. [CrossRef]
6. Meyerson, C.; Van Stavern, G.; McClelland, C. Leber hereditary optic neuropathy: Current perspectives. *Clin. Ophthalmol.* **2015**, *9*, 1165–1176.
7. Poincenot, L.; Pearson, A.L.; Karanjia, R. Demographics of a Large International Population of Patients Affected by Leber's Hereditary Optic Neuropathy. *Ophthalmology* **2020**, *127*, 679–688. [CrossRef]
8. Yu-Wai-Man, P.; Newman, N.J.; Carelli, V.; La Morgia, C.; Biousse, V.; Bandello, F.M.; Clermont, C.V.; Campillo, L.C.; Leruez, S.; Moster, M.L.; et al. Natural history of patients with Leber hereditary optic neuropathy—Results from the REALITY study. *Eye* **2022**, *36*, 818–826. [CrossRef]

9. Mackey, D.A.; Oostra, R.-J.; Rosenberg, T.; Nikoskelainen, E.; Bronte-Stewart, J.; Poulton, J.; Harding, A.E.; Govan, G.; Bolhuis, P.A.; Norby, S. Primary pathogenic mtDNA mutations in multigeneration pedigrees with Leber hereditary optic neuropathy. *Am. J. Hum. Genet.* **1996**, *59*, 481.
10. Tonagel, F.; Wilhelm, H.; Richter, P.; Kelbsch, C. Leber's hereditary optic neuropathy: Course of disease in consideration of idebenone treatment and type of mutation. *Graefe's Arch. Clin. Exp. Ophthalmol.* **2021**, *259*, 1009–1013. [CrossRef]
11. Mackey, D.; Howell, N. A variant of Leber hereditary optic neuropathy characterized by recovery of vision and by an unusual mitochondrial genetic etiology. *Am. J. Hum. Genet.* **1992**, *51*, 1218–1228.
12. Johns, D.R.; Smith, K.H.; Miller, N.R. Leber's hereditary optic neuropathy: Clinical manifestations of the 3460 mutation. *Arch. Ophthalmol.* **1992**, *110*, 1577–1581. [CrossRef]
13. Huoponen, K.; Vilkki, J.; Aula, P.; Nikoskelainen, E.K.; Savontaus, M. A new mtDNA mutation associated with Leber hereditary optic neuroretinopathy. *Am. J. Hum. Genet.* **1991**, *48*, 1147.
14. Yu-Wai-Man, P.; Griffiths, P.G.; Chinnery, P.F. Mitochondrial optic neuropathies–disease mechanisms and therapeutic strategies. *Prog. Retin. Eye Res.* **2011**, *30*, 81–114. [CrossRef]
15. Jha, R.K.; Dawar, C.; Hasan, Q.; Pujar, A.; Gupta, G.; Vishnu, V.Y.; Thangaraj, K. Mitochondrial Genetic Heterogeneity in Leber's Hereditary Optic Neuropathy: Original Study with Meta-Analysis. *Genes* **2021**, *12*, 1300. [CrossRef]
16. Harding, A.; Sweeney, M.; Govan, G.; Riordan-Eva, P. Pedigree analysis in Leber hereditary optic neuropathy families with a pathogenic mtDNA mutation. *Am. J. Hum. Genet.* **1995**, *57*, 77.
17. Catarino, C.B.; von Livonius, B.; Priglinger, C.; Banik, R.; Matloob, S.; Tamhankar, M.A.; Castillo, L.; Friedburg, C.; Halfpenny, C.A.; Lincoln, J.A.; et al. Real-World Clinical Experience with Idebenone in the Treatment of Leber Hereditary Optic Neuropathy. *J. Neuro-Ophthalmol.* **2020**, *40*, 558. [CrossRef]
18. Indrieri, A.; Carrella, S.; Romano, A.; Spaziano, A.; Marrocco, E.; Fernandez-Vizarra, E.; Barbato, S.; Pizzo, M.; Ezhova, Y.; Golia, F.M.; et al. miR-181a/b downregulation exerts a protective action on mitochondrial disease models. *EMBO Mol. Med.* **2019**, *11*, e8734. [CrossRef]
19. Weiss, J.N.; Levy, S. Stem Cell Ophthalmology Treatment Study (SCOTS): Bone marrow derived stem cells in the treatment of Dominant Optic Atrophy. *Stem Cell Investig.* **2019**, *6*, 41. [CrossRef]
20. Wong, R.C.B.; Lim, S.Y.; Hung, S.S.C.; Jackson, S.; Khan, S.; Van Bergen, N.J.; De Smit, E.; Liang, H.H.; Kearns, L.S.; Clarke, L.; et al. Mitochondrial replacement in an iPSC model of Leber's hereditary optic neuropathy. *Aging* **2017**, *9*, 1341–1350. [CrossRef]
21. Ellouze, S.; Augustin, S.; Bouaita, A.; Bonnet, C.; Simonutti, M.; Forster, V.; Picaud, S.; Sahel, J.A.; Corral-Debrinski, M. Optimized allotopic expression of the human mitochondrial ND4 prevents blindness in a rat model of mitochondrial dysfunction. *Am. J. Hum. Genet.* **2008**, *83*, 373–387. [CrossRef]
22. Boddapati, S.V.; D'Souza, G.G.; Erdogan, S.; Torchilin, V.P.; Weissig, V. Organelle-targeted nanocarriers: Specific delivery of liposomal ceramide to mitochondria enhances its cytotoxicity in vitro and in vivo. *Nano Lett.* **2008**, *8*, 2559–2563. [CrossRef] [PubMed]
23. Liu, Y.; Eastwood, J.D.; Alba, D.E.; Velmurugan, S.; Sun, N.; Porciatti, V.; Guy, J.; Yu, H. Gene therapy restores mitochondrial function and protects retinal ganglion cells in optic neuropathy induced by a mito-targeted mutant ND1 gene. *Gene Ther.* **2022**, *29*, 368–378. [CrossRef]
24. Koilkonda, R.D.; Chou, T.H.; Porciatti, V.; Hauswirth, W.W.; Guy, J. Induction of rapid and highly efficient expression of the human ND4 complex I subunit in the mouse visual system by self-complementary adeno-associated virus. *Arch. Ophthalmol.* **2010**, *128*, 876–883. [CrossRef]
25. Cwerman-Thibault, H.; Augustin, S.; Lechauve, C.; Ayache, J.; Ellouze, S.; Sahel, J.A.; Corral-Debrinski, M. Nuclear expression of mitochondrial ND4 leads to the protein assembling in complex I and prevents optic atrophy and visual loss. *Mol. Ther. Methods Clin. Dev.* **2015**, *2*, 15003. [CrossRef]
26. Guy, J.; Qi, X.; Pallotti, F.; Schon, E.A.; Manfredi, G.; Carelli, V.; Martinuzzi, A.; Hauswirth, W.W.; Lewin, A.S. Rescue of a mitochondrial deficiency causing Leber Hereditary Optic Neuropathy. *Ann. Neurol.* **2002**, *52*, 534–542. [CrossRef]
27. van Everdingen, J.A.M.; Pott, J.W.R.; Bauer, N.J.C.; Krijnen, A.M.; Lushchyk, T.; Wubbels, R.J. Clinical outcomes of treatment with idebenone in Leber's hereditary optic neuropathy in the Netherlands: A national cohort study. *Acta Ophthalmol.* **2022**, *100*, 700–706. [CrossRef] [PubMed]
28. Koilkonda, R.D.; Yu, H.; Chou, T.-H.; Feuer, W.J.; Ruggeri, M.; Porciatti, V.; Tse, D.; Hauswirth, W.W.; Chiodo, V.; Boye, S.L.; et al. Safety and effects of the vector for the Leber hereditary optic neuropathy gene therapy clinical trial. *JAMA Ophthalmol.* **2014**, *132*, 409–420. [CrossRef]
29. Bonnet, C.; Kaltimbacher, V.; Ellouze, S.; Augustin, S.; Bénit, P.; Forster, V.; Rustin, P.; Sahel, J.-A.; Corral-Debrinski, M. Allotopic mRNA localization to the mitochondrial surface rescues respiratory chain defects in fibroblasts harboring mitochondrial DNA mutations affecting complex I or v subunits. *Rejuvenation Res.* **2007**, *10*, 127–144. [CrossRef]
30. Pei, H.; Wan, X.; Hu, W.-K.; Dong, X.-R.; Li, B. Construction and detection of a novel type of recombinant human rAAV2/2-ND4. *Eye Sci.* **2013**, *28*, 55–59.
31. Guy, J.; Feuer, W.J.; Davis, J.L.; Porciatti, V.; Gonzalez, P.J.; Koilkonda, R.D.; Yuan, H.; Hauswirth, W.W.; Lam, B.L. Gene Therapy for Leber Hereditary Optic Neuropathy: Low- and Medium-Dose Visual Results. *Ophthalmology* **2017**, *124*, 1621–1634. [CrossRef]
32. Yang, S.; Ma, S.-Q.; Wan, X.; He, H.; Pei, H.; Zhao, M.-J.; Chen, C.; Wang, D.-W.; Dong, X.-Y.; Yuan, J.-J.; et al. Long-term outcomes of gene therapy for the treatment of Leber's hereditary optic neuropathy. *EBioMedicine* **2016**, *10*, 258–268. [CrossRef]

33. Biousse, V.; Newman, N.J.; Yu-Wai-Man, P.; Carelli, V.; Moster, M.L.; Vignal-Clermont, C.; Klopstock, T.; Sadun, A.A.; Sergott, R.C.; Hage, R.; et al. Long-Term Follow-Up After Unilateral Intravitreal Gene Therapy for Leber Hereditary Optic Neuropathy: The RESTORE Study. *J. Neuro-Ophthalmol. Off. J. N. Am. Neuro-Ophthalmol. Soc.* **2021**, *41*, 309–315. [CrossRef]
34. Feuer, W.J.; Schiffman, J.C.; Davis, J.L.; Porciatti, V.; Gonzalez, P.; Koilkonda, R.D.; Yuan, H.; Lalwani, A.; Lam, B.L.; Guy, J.; et al. Gene Therapy for Leber Hereditary Optic Neuropathy: Initial Results. *Ophthalmology* **2016**, *123*, 558–570. [CrossRef] [PubMed]
35. Lam, B.L.; Feuer, W.J.; Davis, J.L.; Porciatti, V.; Yu, H.; Levy, R.B.; Vanner, E.; Guy, J. Leber Hereditary Optic Neuropathy Gene Therapy: Adverse Events and Visual Acuity Results of All Patient Groups. *Am. J. Ophthalmol.* **2022**, *241*, 262–271. [CrossRef]
36. Vignal, C.; Uretsky, S.; Fitoussi, S.; Galy, A.; Blouin, L.; Girmens, J.-F.; Bidot, S.; Thomasson, N.; Bouquet, C.; Valero, S. Safety of rAAV2/2-ND4 gene therapy for Leber hereditary optic neuropathy. *Ophthalmology* **2018**, *125*, 945–947. [CrossRef]
37. Vignal-Clermont, C.; Girmens, J.-F.; Audo, I.; Said, S.M.; Errera, M.-H.; Plaine, L.; O'Shaughnessy, D.; Taiel, M.; Sahel, J.-A. Safety of Intravitreal Gene Therapy for Treatment of Subjects with Leber Hereditary Optic Neuropathy due to Mutations in the Mitochondrial ND4 Gene: The REVEAL Study. *BioDrugs* **2021**, *35*, 201–214. [CrossRef]
38. Newman, N.J.; Yu-Wai-Man, P.; Carelli, V.; Moster, M.L.; Biousse, V.; Vignal-Clermont, C.; Sergott, R.C.; Klopstock, T.; Sadun, A.A.; Barboni, P.; et al. Efficacy and Safety of Intravitreal Gene Therapy for Leber Hereditary Optic Neuropathy Treated within 6 Months of Disease Onset. *Ophthalmology* **2021**, *128*, 649–660. [CrossRef]
39. Yu-Wai-Man, P.; Newman, N.J.; Carelli, V.; Moster, M.L.; Biousse, V.; Sadun, A.A.; Klopstock, T.; Vignal-Clermont, C.; Sergott, R.C.; Rudolph, G.; et al. Bilateral visual improvement with unilateral gene therapy injection for Leber hereditary optic neuropathy. *Sci. Transl. Med.* **2020**, *12*, eaaz7423. [CrossRef]
40. Newman, N.J.; Yu-Wai-Man, P.; Carelli, V.; Biousse, V.; Moster, M.L.; Vignal-Clermont, C.; Sergott, R.C.; Klopstock, T.; Sadun, A.A.; Girmens, J.F.; et al. Intravitreal Gene Therapy vs. Natural History in Patients with Leber Hereditary Optic Neuropathy Carrying the m.11778G>A ND4 Mutation: Systematic Review and Indirect Comparison. *Front. Neurol.* **2021**, *12*, 662838. [CrossRef]
41. Wan, X.; Pei, H.; Zhao, M.-J.; Yang, S.; Hu, W.-K.; He, H.; Ma, S.-Q.; Zhang, G.; Dong, X.-Y.; Chen, C.; et al. Efficacy and Safety of rAAV2-ND4 Treatment for Leber's Hereditary Optic Neuropathy. *Sci. Rep.* **2016**, *6*, 21587. [CrossRef]
42. Liu, H.-L.; Yuan, J.-J.; Zhang, Y.; Tian, Z.; Li, X.; Wang, D.; Du, Y.-Y.; Song, L.; Li, B. Factors associated with rapid improvement in visual acuity in patients with Leber's hereditary optic neuropathy after gene therapy. *Acta Ophthalmol.* **2020**, *98*, e730–e733. [CrossRef]
43. Zhang, Y.; Li, X.; Yuan, J.; Tian, Z.; Liu, H.; Wang, D.; Li, B. Prognostic factors for visual acuity in patients with Leber's hereditary optic neuropathy after rAAV2-ND4 gene therapy. *Clin. Exp. Ophthalmol.* **2019**, *47*, 774–778. [CrossRef]
44. Liu, H.L.; Yuan, J.J.; Tian, Z.; Li, X.; Song, L.; Li, B. What are the characteristics and progression of visual field defects in patients with Leber hereditary optic neuropathy: A prospective single-centre study in China. *BMJ Open* **2019**, *9*, e025307. [CrossRef] [PubMed]
45. Meunier, I.; Lenaers, G.; Hamel, C.; Defoort-Dhellemmes, S. Hereditary optic neuropathies: From clinical signs to diagnosis. *J. Fr. Ophtalmol.* **2013**, *36*, 886–900. [CrossRef]
46. Riordan-Eva, P.; Sanders, M.; Govan, G.; Sweeney, M.; Costa, J.D.; Harding, A. The clinical features of Leber's hereditary optic neuropathy defined by the presence of a pathogenic mitochondrial DNA mutation. *Brain* **1995**, *118*, 319–337. [CrossRef] [PubMed]
47. Lam, B.L.; Feuer, W.J.; Schiffman, J.C.; Porciatti, V.; Vandenbroucke, R.; Rosa, P.R.; Gregori, G.; Guy, J. Trial end points and natural history in patients with G11778A Leber hereditary optic neuropathy: Preparation for gene therapy clinical trial. *JAMA Ophthalmol.* **2014**, *132*, 428–436. [CrossRef] [PubMed]
48. Johns, D.R.; Heher, K.L.; Miller, N.R.; Smith, K.H. Leber's hereditary optic neuropathy. Clinical manifestations of the 14484 mutation. *Arch. Ophthalmol.* **1993**, *111*, 495–498. [CrossRef]
49. Nikoskelainen, E.K.; Huoponen, K.; Juvonen, V.; Lamminen, T.; Nummelin, K.; Savontaus, M.L. Ophthalmologic findings in Leber hereditary optic neuropathy, with special reference to mtDNA mutations. *Ophthalmology* **1996**, *103*, 504–514. [CrossRef]
50. Majander, A.; Bowman, R.; Poulton, J.; Antcliff, R.J.; Reddy, M.A.; Michaelides, M.; Webster, A.R.; Chinnery, P.F.; Votruba, M.; Moore, A.T.; et al. Childhood-onset Leber hereditary optic neuropathy. *Br. J. Ophthalmol.* **2017**, *101*, 1505–1509. [CrossRef]
51. Mukherjee, S.; Ghosh, A. Molecular mechanism of mitochondrial respiratory chain assembly and its relation to mitochondrial diseases. *Mitochondrion* **2020**, *53*, 1–20. [CrossRef] [PubMed]
52. Chial, H.; Craig, J. mtDNA and mitochondrial diseases. *Nat. Educ.* **2008**, *1*, 217.
53. Huoponen, K. Leber hereditary optic neuropathy: Clinical and molecular genetic findings. *Neurogenetics* **2001**, *3*, 119–125. [CrossRef] [PubMed]
54. Bianco, A.; Valletti, A.; Longo, G.; Bisceglia, L.; Montoya, J.; Emperador, S.; Guerriero, S.; Petruzzella, V. Mitochondrial DNA copy number in affected and unaffected LHON mutation carriers. *BMC Res. Notes* **2018**, *11*, 911. [CrossRef] [PubMed]
55. Bianco, A.; Bisceglia, L.; Russo, L.; Palese, L.L.; D'Agruma, L.; Emperador, S.; Montoya, J.; Guerriero, S.; Petruzzella, V. High Mitochondrial DNA Copy Number Is a Protective Factor from Vision Loss in Heteroplasmic Leber's Hereditary Optic Neuropathy (LHON). *Investig. Ophthalmol. Vis. Sci.* **2017**, *58*, 2193–2197. [CrossRef] [PubMed]
56. Kirkman, M.A.; Yu-Wai-Man, P.; Korsten, A.; Leonhardt, M.; Dimitriadis, K.; De Coo, I.F.; Klopstock, T.; Chinnery, P.F. Gene-environment interactions in Leber hereditary optic neuropathy. *Brain* **2009**, *132 Pt 9*, 2317–2326. [CrossRef]
57. Ajax, E.T.; Kardon, R. Late-onset Leber's hereditary optic neuropathy. *J. Neuro-Ophthalmol. Off. J. N. Am. Neuro-Ophthalmol. Soc.* **1998**, *18*, 30–31.
58. Yu-Wai-Man, P. Therapeutic Approaches to Inherited Optic Neuropathies. *Semin. Neurol.* **2015**, *35*, 578–586. [CrossRef]

59. Yu, H.; Koilkonda, R.D.; Chou, T.-H.; Porciatti, V.; Ozdemir, S.S.; Chiodo, V.; Boye, S.L.; Boye, S.E.; Hauswirth, W.W.; Lewin, A.S.; et al. Gene delivery to mitochondria by targeting modified adenoassociated virus suppresses Leber's hereditary optic neuropathy in a mouse model. *Proc. Natl. Acad. Sci. USA* **2012**, *109*, E1238–E1247. [CrossRef]
60. Bonnet, C.; Augustin, S.; Ellouze, S.; Bénit, P.; Bouaita, A.; Rustin, P.; Sahel, J.-A.; Corral-Debrinski, M. The optimized allotopic expression of ND1 or ND4 genes restores respiratory chain complex I activity in fibroblasts harboring mutations in these genes. *Biochim. Biophys. Acta (BBA)—Mol. Cell Res.* **2008**, *1783*, 1707–1717. [CrossRef]
61. Gammage, P.A.; Rorbach, J.; Vincent, A.I.; Rebar, E.J.; Minczuk, M. Mitochondrially targeted ZFNs for selective degradation of pathogenic mitochondrial genomes bearing large-scale deletions or point mutations. *EMBO Mol. Med.* **2014**, *6*, 458–466. [CrossRef]
62. Hashimoto, M.; Bacman, S.R.; Peralta, S.; Falk, M.J.; Chomyn, A.; Chan, D.C.; Sahel, J.-A.; Corral-Debrinski, M. MitoTALEN: A General Approach to Reduce Mutant mtDNA Loads and Restore Oxidative Phosphorylation Function in Mitochondrial Diseases. *Mol. Ther.* **2015**, *23*, 1592–1599. [CrossRef]
63. Gammage, P.A.; Viscomi, C.; Simard, M.-L.; Costa, A.S.H.; Gaude, E.; Powell, C.A.; Van Haute, L.; McCann, B.J.; Rebelo-Guiomar, P.; Cerutti, R.; et al. Genome editing in mitochondria corrects a pathogenic mtDNA mutation in vivo. *Nat. Med.* **2018**, *24*, 1691–1695. [CrossRef]
64. Bacman, S.R.; Kauppila, J.H.K.; Pereira, C.V.; Nissanka, N.; Miranda, M.; Pinto, M.; Williams, S.L.; Larsson, N.-G.; Stewart, J.B.; Moraes, C.T. MitoTALEN reduces mutant mtDNA load and restores tRNAAla levels in a mouse model of heteroplasmic mtDNA mutation. *Nat. Med.* **2018**, *24*, 1696–1700. [CrossRef] [PubMed]
65. Li, F.; Wing, K.; Wang, J.H.; Luu, C.D.; Bender, J.A.; Chen, J.; Wang, Q.; Lu, Q.; Nguyen Tran, M.T.; Young, K.M.; et al. Comparison of CRISPR/Cas Endonucleases for in vivo Retinal Gene Editing. *Front. Cell Neurosci.* **2020**, *14*, 570917. [CrossRef] [PubMed]
66. Li, F.; Hung, S.S.C.; Mohd Khalid, M.K.N.; Wang, J.H.; Chrysostomou, V.; Wong, V.H.Y.; Singh, V.; Wing, K.; Tu, L.; Bender, J.A.; et al. Utility of Self-Destructing CRISPR/Cas Constructs for Targeted Gene Editing in the Retina. *Hum. Gene Ther.* **2019**, *30*, 1349–1360. [CrossRef] [PubMed]
67. Hussain, S.-R.A.; Yalvac, M.E.; Khoo, B.; Eckardt, S.; McLaughlin, K.J. Adapting CRISPR/Cas9 System for Targeting Mitochondrial Genome. *Front. Genet.* **2021**, *12*, 627050. [CrossRef]
68. Jo, A.; Ham, S.; Lee, G.H.; Lee, Y.I.; Kim, S.; Lee, Y.S.; Shin, J.H.; Lee, Y. Efficient Mitochondrial Genome Editing by CRISPR/Cas9. *BioMed Res. Int.* **2015**, *2015*, 305716. [CrossRef]
69. Mok, B.Y.; de Moraes, M.H.; Zeng, J.; Bosch, D.E.; Kotrys, A.V.; Raguram, A.; Hsu, F.; Radey, M.C.; Peterson, S.B.; Mootha, V.K.; et al. A bacterial cytidine deaminase toxin enables CRISPR-free mitochondrial base editing. *Nature* **2020**, *583*, 631–637. [CrossRef]
70. Storoni, M.; Robert, M.P.; Plant, G.T. The therapeutic potential of a calorie-restricted ketogenic diet for the management of Leber hereditary optic neuropathy. *Nutr. Neurosci.* **2019**, *22*, 156–164. [CrossRef]
71. Chu, C.-C.; Huang, C.-C.; Kao, L.-Y.; Kao, H.-C.; Yu, T.-N.; Tso, D.-J.; Lee, H.-C.; Wei, Y.-H. Clinical phenotype and the G11778A mutation of mitochondrial DNA in patients with Leber's hereditary optic neuropathy in Taiwan. *Neuro-Ophthalmol.* **2001**, *26*, 207–216. [CrossRef]
72. Tarnopolsky, M. The mitochondrial cocktail: Rationale for combined nutraceutical therapy in mitochondrial cytopathies. *Adv. Drug Deliv. Rev.* **2008**, *60*, 1561–1567. [CrossRef]
73. Yu-Wai-Man, P.; Soiferman, D.; Moore, D.G.; Burté, F.; Saada, A. Evaluating the therapeutic potential of idebenone and related quinone analogues in Leber hereditary optic neuropathy. *Mitochondrion* **2017**, *36*, 36–42. [CrossRef]
74. Klopstock, T.; Yu-Wai-Man, P.; Dimitriadis, K.; Rouleau, J.; Heck, S.; Bailie, M.; Atawan, A.; Chattopadhyay, S.; Schubert, M.; Garip, A.; et al. A randomized placebo-controlled trial of idebenone in Leber's hereditary optic neuropathy. *Brain* **2011**, *134*, 2677–2686. [CrossRef]
75. Rudolph, G.; Dimitriadis, K.; Büchner, B.; Heck, S.; Al-Tamami, J.; Seidensticker, F.; Rummey, C.; Leinonen, M.; Meier, T.; Klopstock, T. Effects of idebenone on color vision in patients with leber hereditary optic neuropathy. *J. Neuro-Ophthalmol.* **2013**, *33*, 30. [CrossRef]
76. Klopstock, T.; Metz, G.; Yu-Wai-Man, P.; Büchner, B.; Gallenmüller, C.; Bailie, M.; Nwali, N.; Griffiths, P.G.; von Livonius, B.; Reznicek, L.; et al. Persistence of the treatment effect of idebenone in Leber's hereditary optic neuropathy. *Brain* **2013**, *136*, e230. [CrossRef]
77. Carelli, V.; La Morgia, C.; Valentino, M.L.; Rizzo, G.; Carbonelli, M.; De Negri, A.M.; Sadun, F.; Carta, A.; Guerriero, S.; Simonelli, F.; et al. Idebenone Treatment in Leber's Hereditary Optic Neuropathy. *Brain* **2011**, *134*, e188. [CrossRef]
78. Carelli, V.; Carbonelli, M.; Irenaeus, F.; Kawasaki, A.; Klopstock, T.; Lagrèze, W.A.; Sadun, F.; Carta, A.; Guerriero, S.; Simonelli, F.; et al. International consensus statement on the clinical and therapeutic management of Leber hereditary optic neuropathy. *J. Neuro-Ophthalmol.* **2017**, *37*, 371–381. [CrossRef]
79. Enns, G.M.; Kinsman, S.L.; Perlman, S.L.; Spicer, K.M.; Abdenur, J.E.; Cohen, B.H.; Amagata, A.; Barnes, A.; Kheifets, V.; Shrader, W.D.; et al. Initial experience in the treatment of inherited mitochondrial disease with EPI-743. *Mol. Genet. Metab.* **2012**, *105*, 91–102. [CrossRef]
80. Sadun, A.A.; Chicani, C.F.; Ross-Cisneros, F.N.; Barboni, P.; Thoolen, M.; Shrader, W.D.; Kubis, K.; Carelli, V.; Miller, G. Effect of EPI-743 on the clinical course of the mitochondrial disease Leber hereditary optic neuropathy. *Arch. Neurol.* **2012**, *69*, 331–338. [CrossRef]

81. Szeto, H.H. First-in-class cardiolipin-protective compound as a therapeutic agent to restore mitochondrial bioenergetics. *Br. Pharmacol.* **2014**, *171*, 2029–2050. [CrossRef] [PubMed]
82. Karanjia, R.; Coupland, S.G.; Garcia, M.; Sadun, A.A. Elamipretide (MTP-131) Topical Ophthalmic Solution for the Treatment c Leber's Hereditary Optic Neuropathy. *Investig. Ophthalmol. Vis. Sci.* **2019**, *60*, 2266.
83. Eells, J.T.; Wong-Riley, M.T.; VerHoeve, J.; Henry, M.; Buchman, E.V.; Kane, M.P.; Gould, L.J.; Das, R.; Jett, M.; Hodgson, B.D.; et a Mitochondrial signal transduction in accelerated wound and retinal healing by near-infrared light therapy. *Mitochondrion* **2004**, 559–567. [CrossRef] [PubMed]
84. La Morgia, C.; Carbonelli, M.; Barboni, P.; Sadun, A.A.; Carelli, V. Medical management of hereditary optic neuropathies. *Fron Neurol.* **2014**, *5*, 141. [CrossRef]
85. Dai, Y.; Zheng, K.; Clark, J.; Swerdlow, R.H.; Pulst, S.M.; Sutton, J.P.; Shinobu, L.A.; Simon, D.K. Rapamycin drives selectio against a pathogenic heteroplasmic mitochondrial DNA mutation. *Hum. Mol. Genet.* **2014**, *23*, 637–647. [CrossRef] [PubMed]
86. Yu, A.K.; Datta, S.; McMackin, M.Z.; Cortopassi, G.A. Rescue of cell death and inflammation of a mouse model of comple 1-mediated vision loss by repurposed drug molecules. *Hum. Mol. Genet.* **2017**, *26*, 4929–4936. [CrossRef]
87. Pisano, A.; Preziuso, C.; Iommarini, L.; Perli, E.; Grazioli, P.; Campese, A.F.; Maresca, A.; Montopoli, M.; Masuelli, L.; Sadun, A.A Targeting estrogen receptor β as preventive therapeutic strategy for Leber's hereditary optic neuropathy. *Hum. Mol. Genet.* **201** *24*, 6921–6931. [CrossRef]
88. Giordano, C.; Montopoli, M.; Perli, E.; Orlandi, M.; Fantin, M.; Ross-Cisneros, F.N.; Caparrotta, L.; Martinuzzi, A.; Ragazzi, E Ghelli, A. Oestrogens ameliorate mitochondrial dysfunction in Leber's hereditary optic neuropathy. *Brain* **2011**, *134*, 220–23 [CrossRef]
89. Weiss, J.N.; Levy, S. Stem Cell Ophthalmology Treatment Study (SCOTS): Bone Marrow-Derived Stem Cells in the Treatment o Stargardt Disease. *Medicines* **2021**, *8*, 10. [CrossRef]

cells

Review

Retinal Cell Damage in Diabetic Retinopathy

Jing Zhou and Bo Chen *

Department of Ophthalmology, Icahn School of Medicine at Mount Sinai, New York, NY 10029, USA
* Correspondence: bo chen@mssm.edu

Abstract: Diabetic retinopathy (DR), the most common microvascular complication that occurs in diabetes mellitus (DM), is the leading cause of vision loss in working-age adults. The prevalence of diabetic retinopathy is approximately 30% of the diabetic population and untreated DR can eventually cause blindness. For decades, diabetic retinopathy was considered a microvascular complication and clinically staged by its vascular manifestations. In recent years, emerging evidence has shown that diabetic retinopathy causes early neuronal dysfunction and neurodegeneration that may precede vascular pathology and affect retinal neurons as well as glial cells. This knowledge leads to new therapeutic strategies aiming to prevent dysfunction of retinal neurons at the early stage of DR. Early detection and timely treatment to protect retinal neurons are critical to preventing visual loss in DR. This review provides an overview of DR and the structural and functional changes associated with DR, and discusses neuronal degeneration during diabetic retinopathy, the mechanisms underlying retinal neurodegeneration and microvascular complications, and perspectives on current and future clinic therapies.

Keywords: diabetic retinopathy; diabetes mellitus; microvascular complication; neuronal dysfunction; neurodegeneration; clinic therapy

1. An Overview of Diabetic Retinopathy

DR is a leading cause of blindness in the world ranging from working-age adults to the elderly population (20–74 years old) [1]. It is estimated that the DR population worldwide will increase from 463 million in 2019 to 578 million in 2030 and to approximately 700 million by 2045 [1,2]. A study showed that more than 30 million people (~9.4% of the US population) have diabetes, and approximately one-third of them are diagnosed with diabetic retinopathy [3]. DR patients can suffer severe vision loss if left untreated. DR is also associated with the risks of systemic vascular complications of diabetes, including stroke, cardiovascular events and heart failure [4].

Diabetes affects all cells in the retina, though most studies have focused mainly on retinal microvascular pathology. Based on the presence of neovascularization, DR is classified into two stages, non-proliferative (NPDR) and proliferative diabetic retinopathy (PDR) [5–7]. NPDR is an early stage of DR. Early morphological signs of NPDR include basal membrane thickening, tight junction impairment, and blood–retina barrier (BRB) breakdown. Moreover, there is a cell loss of pericytes and endothelial dysfunction, resulting in fragile capillaries, formation of microaneurysms, small hemorrhages, cotton-wool spots, and capillary non-perfusion. These vascular lesions accumulate to induce ischemic conditions in some areas of the retina; as ischemia develops, proangiogenic factors such as vascular endothelial growth factor (VEGF) release, inducing the formation of neovascularization, a hallmark of proliferative DR [8]. The proliferative stage of DR is characterized by retinal neovascularization due to ischemia and hypoxia. Newly formed blood vessels are relatively fragile and susceptible to retinal and vitreous hemorrhage, leading to tractional retinal detachment and vision loss. Diabetic macular edema (DME) is characterized by macula thickening induced by abnormal accumulation of extravascular fluid in the macular, and hard exudates can occur at any stage [9]. The prevalence of DME in patients with DR

Citation: Zhou, J.; Chen, B. Retinal Cell Damage in Diabetic Retinopathy. *Cells* **2023**, *12*, 1342. https://doi.org/10.3390/cells12091342

Academic Editor: Hossein Ameri

Received: 27 March 2023
Revised: 5 May 2023
Accepted: 6 May 2023
Published: 8 May 2023

is 2.7–11% [10,11], and it depends on the type of diabetes and duration of the disease. DME and PDR are the main reasons for vision loss in patients with diabetic retinopathy and are increasing in prevalence around the world.

Several animal models of diabetes have been developed to study the pathogenesis of DR and to test therapies. These animal models were generated via genetic manipulation or induction. The most commonly used genetic models include db/db (Leprdb), Ins2^{Akita}, non-obese diabetic (NOD), Insulin2Q104del (Kuma), and Akimba mice; and the induced animal models include administration of drugs such as streptozotocin (STZ) and alloxan, applying laser or chemical damage to the eye, feeding high galactose diet, and surgical removal of the pancreas [12–16]. STZ administration is the most commonly used induced model as it results in the fastest rate of disease development. Neuronal apoptosis and reactive gliosis are the common histological features in the retina with DR. The complete picture for the DR disease mechanisms and the sequence of cellular events in the DR development is far from understood because each of the animal models can only mimic certain aspects of DR pathology in human patients. We will discuss in more detail various aspects of retinal cell damage in studies using some of these animal models.

2. Diabetic Retinopathy and Retinal Vasculature

The retina is a specialized tissue for light detection and vision formation, composed of several neuronal cell types (rods, cones, horizontal cells, bipolar cells, amacrine cells, and retinal ganglion cells) and glial cell types (Müller cells, astrocytes, and microglia). RGCs (retinal ganglion cells) are the sole output neurons located in the most inner layer of the retina, and they relay processed visual information from the retina and the brain for perception (Figure 1A,B).

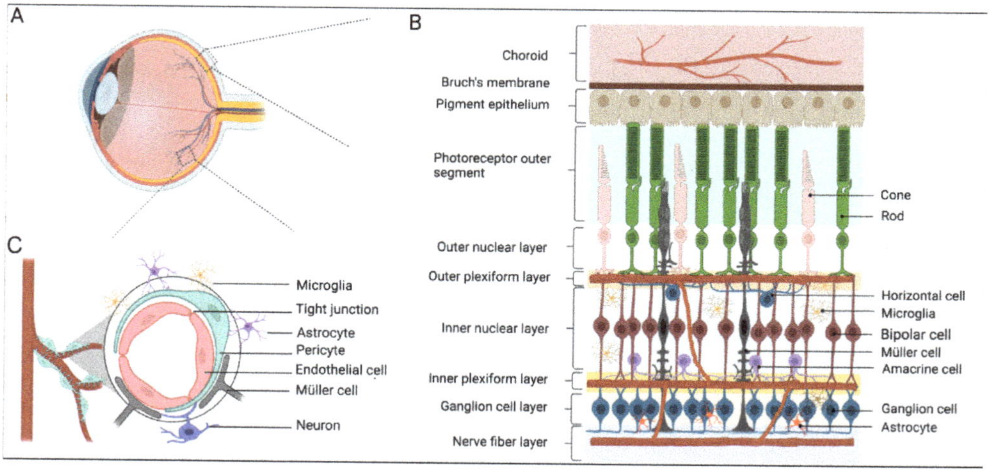

Figure 1. Retinal structure, iBRB, and neurovascular unit. (**A**,**B**) A drawing of a human eye with a schematic enlargement of the retinal structure. The retina lines the back of the eye, which consists of six types of neurons from the outer to the inner retina: rod/cone photoreceptors, interneurons (bipolar cells, horizontal cells, and amacrine cells), and ganglion cells as well as three types of glial cells (Müller cells, astrocytes, and microglia). The central retinal artery enters the eye through the optic nerve and branches into three vascular plexuses: superficial vascular plexus, intermediate vascular plexus, and deep vascular plexus localized in the ganglion cell layer, inner plexiform layer, and outer plexiform layer, respectively. (**C**) A schematic of the neurovascular unit and the inner blood–retina barrier (iBRB). iBRB is formed by tight junctions of adjacent endothelial cells. Pericytes, glial cells, and neurons surrounding the retinal vessel all together form a neurovascular unit to maintain the barrier function.

The retina relies on a well-functioning blood supply to maintain its functionality. The blood supply to the retina comes from two distinct vascular plexuses: the choroid that lines under the retina pigment epithelium to supply 90% of the total blood to the outer retina and the intraretinal blood vessel that primarily supports the inner retina [17,18]. The choroid blood vessels consist of fenestrated endothelial cells, whereas intraretinal blood vessels consist of non-fenestrated endothelia cells surrounded by pericytes, making the inner retinal vasculature form the blood–retina barrier (BRB) that regulates molecular traffic between the blood and the retina [19] (Figure 1B,C).

As one of the most energy-demanding tissues, the retina needs sufficient nutrients and high levels of oxygen to maintain its visual function [18,20]. The central retinal artery enters the eye through the optic nerve center and branches into superficial, intermediate, and deep layers. The highly specialized vasculature, together with neurons and glia, is integrated to form the retinal neurovascular unit to maintain homeostasis and modulate neuronal activities [20]. The components of the neurovascular unit of the retina include neurons (RGCs, bipolar cells, amacrine cells, and horizontal cells), glia (Müller cells, astrocytes and microglia), and vascular cells (endothelial cells and pericytes) [21–23] (Figure 1C). Neurovascular coupling is the process by which neural activity is linked to the blood flow and metabolism, allowing the retina to regulate the blood flow in response to neural activities or metabolic demands [24]. All components of the neurovascular unit work closely to maintain the integrity of the inner BRB and dynamically coordinate the local blood flow in response to metabolic demands. Retinal blood vessels dilate when flickering light stimulates them, whereas breathing 100% oxygen causes them to constrict. Retinal pathology is likely to involve all these components at varying degrees. Evidence shows that complications in diabetes impair the normal function of the retinal neurovascular unit. In the pathogenesis of DR, neurodegeneration and glial activation are widely observed, even occurring before the clinical signs of DR appear, in the experimental models of DR and from the retinas of diabetic donors [22].

BRB consists of two compartments: outer BRB (oBRB) is composed of retinal pigment epithelial cells, representing a highly selective barrier for molecules and solutes moving from the choroid into the retina, and inner BRB (iBRB), which is formed by adherents and tight junctions between adjacent retina capillary endothelia cells. iBRB, established by tight junctions between endothelial cells that are surrounded by pericytes and glial cells (Müller cells, astrocytes, and microglia), plays an important role in regulating the microenvironment and thus is crucial for proper vision. Distinctive characteristics of iBRB including tight junctions and lack of fenestrations in iBRB make it a highly selective barrier in nature. iBRB controls the transport of molecules, ions, and water between the vascular lumen and the retina. The transporters in iBRB play an essential role through the release of trophic factors and antioxidants into the retinal microenvironment [25]. Damage to the tight junctions of the iBRB causes disruption of the integrity of this inner barrier [26]. iBRB breakdown is a hallmark of diabetic retinopathy [27–29].

3. Neurodegeneration in Diabetic Retinopathy

For many years, diabetic retinopathy has traditionally been regarded as a microvascular disease and the current diagnosis of DR indeed relies on vascular alterations. However, emerging evidence suggests that retinal neurodegeneration is also involved in DR, even before clinical signs of DR appear [30]. During the early stage of diabetes, several morphological and electrophysiological alterations occur in neurons of the inner retina [26–28] before vascular changes, leading to the recommendation that precautions against DR should be taken immediately after diabetes is diagnosed. A deeper understanding of neurodegeneration during DR is essential for early detection and targeted therapies to prevent vision loss.

3.1. RGCs and DR

RGCs, located in the inner retina, are likely to be easily damaged because they are highly active in metabolism, making them particularly vulnerable to local and systemic metabolic stressors. Indeed, RGCs are the neurons in which the apoptotic process related to diabetes is first detected. The death of RGCs and the degeneration of the inner nuclear layer (INL) in the postmortem eye of people with diabetes were observed early back in 1961 [31]. Supporting the clinical data, experimental evidence also showed apparent RGC loss in diabetic rat models [32–34]. The first quantitative report of an increase in neural cell apoptosis in the diabetic rat retina shows that RGC degeneration is observed at the early stage after the onset of diabetes, even before the degeneration of retinal capillaries in diabetes [35], postulating that neurodegeneration may contribute to capillary degeneration. In the three-month streptozotocin (STZ)-induced diabetic rat, loss of RGCs is associated with morphological change; remaining RGCs in the diabetic retina, especially those in the RGA group (a subtype of RGCs, with a large soma and a large dendritic field), show significant enlargement of the dendritic field [36]. In STZ-induced diabetic rats, there are reduced axon numbers and diameters at the distal potion of the optic nerve at the early stage of diabetes mellitus without observing morphological changes of RGCs, which indicates the optic nerve is the early structural change in the diabetic visual pathway [37]. Studies in other species such as diabetic dogs show animals under moderate glycemic control have appreciable vascular alterations without apparent degeneration of RGCs [38].

Although neurodegeneration in RGCs in mice is less apparent compared to that in rats, the time-dependent loss of RGCs has been observed in many of the DR mouse models [39,40]. Yang et al. reported the apoptosis of RGCs in db/db mice, a model of spontaneous type 2 diabetes, whereas there was no obvious abnormality in the retinal vasculature [41]. RGC apoptosis and cell loss were also detected in Ins2^{Akita} mice within the first 3 months of retinal hyperglycemia, together with a reduced IPL thickness after 5.5 months with other marked alterations in the morphology of the surviving cells. In addition to the RGC loss, abnormal morphologic swellings were observed in the dendrites and axons, such as enlargement of the RGC soma and the extent and density of dendritic branches in ON-type RGCs [42–44]. In STZ-induced diabetic mice, RGC loss begins in 6~12 weeks after induction through apoptotic cell death. There is approximately a 30% RGC loss based on the studies of STZ-induced diabetic models [45,46]. In particular, the RGC complex, including the nerve fiber layer, the ganglion cell layer, and the inner plexiform layer, is affected, which is accompanied by a significant decrease in the density of RGCs and amacrine cells [47]. In STZ-induced diabetic mice, the morphology and passive membrane properties of the dendrites in ON-type RGCs are preferentially affected at the early stage, which is RGC subtype-dependent [48]. Interestingly, diabetes accelerates the longitudinal RGC dysfunction independently of elevated intraocular pressure, as evidenced in a combined model of STZ-induced diabetes in the DBA/2J model of glaucoma; diabetes exacerbates early progression of glaucomatous RGC dysfunction between 3 and 6 months of age, without influencing the intraocular pressure [49]. Nonetheless, other studies did not observe an obvious loss of RGCs in the STZ-induced models, likely due to different reactions of STZ in different animal strains.

Morphological changes in RGCs together with physiological alterations will influence how they process signals from the photoreceptors to the brain. A recent study showed an early response of RGCs in diabetic retinopathy, involving specific morpho-functional deficits in most RGC subtypes but there was no RGC loss [50]. Neuronal protective treatment using a chemical somatostatin analogue octreotide preserves the functionality of RGCs, highlighting the importance of neuronal protection in the early phase of diabetic retinopathy; RGC morphology can be preserved or adjusted to maintain RGC physiology [50]. This provides new therapeutic hopes for patients in the early phase of diabetic retinopathy.

3.2. Müller Cells and DR

Müller cells are the primary support glial cells in the retina. Müller cells span the entire thickness of the retina, come into contact with all types of retinal cells, and wrap around blood vessels, making Müller cells functionally unique and significant in maintaining retinal homeostasis. Müller cells express various voltage-gated channels and neurotransmitter receptors, which enable them to modulate neuronal activity by regulating the extracellular concentration of neuroactive substances, including K^-, glutamate, GABA, and H^+ [51]. In addition to supporting the structure and function of the retina, Müller cells also play an important role in the vascular function of the retina. The presence of Müller cells, astrocytes, and microglia, together with pericytes wrapping around the blood vessels, are all essential to maintain the homeostasis of the retina [29]. In the retina, astrocytes are restricted to the nerve fiber layer and RGCs in the vitreous side of the retina, and Müller cells play a predominant role in interacting with astrocytes and RGCs. Recent studies using diabetic animal models indicate that Müller cells are involved in dysfunctional aspects in diabetes by influencing BRB, especially at the early DR stage. Using a conditional Müller cell ablation model, a study shows that Müller glial deficiency may cause neuronal and vascular pathologies in retinal diseases [52]. Based on these diabetic animal models, depletion of Müller glia results in BRB breakdown and vascular alterations, hallmarks of DR pathology, indicating that Müller dysfunction may be a primary contributor to the vasculopathies of DR.

Reactive gliosis is one of the first responses to the inflammation in DR and is characterized by upregulation of various kinds of molecules, including the most well-established GFAP [53,54]. In the healthy retina, GFAP is mostly detected in astrocytes, with relatively low expression in Müller cells. GFAP activation is a key feature of any kind of impairment of the retina, such as mechanical damage, light damage, and injuries in various retinal diseases. In the STZ-induced diabetic rats, reactive changes in Müller cells, such as upregulation of GFAP, are observed within 3~4 months of the onset of diabetes while GFAP immunoreactivity in astrocytes is reduced, all of which precede the first signs of obvious vascular changes [55,56]. Another study pays particular attention to the first 4 weeks after STZ injection; induction of hyperglycemia occurs within 2 weeks and Müller cells undergo hyperplasia before GFAP expression, indicating that glial cells are the early targets of vascular hyperpermeability in the diabetic retina. In addition, microglia, the resident immune cells in the retina, also undergo highly dynamic morphological and functional alterations [54].

In addition, Müller cells produce high levels of pro-inflammatory molecules and neurotoxic factors, resulting in reactive gliosis under hyperglycemic conditions [31]. Single-cell transcriptomic analysis from DR shows that Müller cell subpopulation genes are involved in the lysosomal pathways, lysosomal membrane permeabilization, and leakage of lysosomal contents, all of which are known to lead to cell death [57]. It is noteworthy that DNA damage response 1 (REDD1) [58] is specifically expressed in Müller cells with the same expression pattern as Müller markers, contributing to diabetes-induced retinal pathology; REDD1 knockout mice exhibit reduced oxidative stress with no obvious retinal thinning or neurodegeneration in the RGC layer, providing a potential clinical therapeutic target for future DR therapy [58].

3.3. Photoreceptors and DR

Photoreceptors are the most abundant cells in the retina that convert light into neural signals. The photoreceptors and the retina pigment epithelium (RPE) in the outer retina normally function as a unit to maintain proper visual function. Phototransduction in photoreceptors involves a sequence of enzymatic reactions, including conversion of 11-cis retinal to all-trans retinal, causing the key conformational changes and activation of G proteins, followed by hydrolysis of cGMP to GMP, closure of cGMP-gated ion channels, hyperpolarization, and less glutamate release from these cells. RPE works with photoreceptor cells to regenerate 11-cis retinal through the classical visual cycle to support the function of both rods and cones [59,60].

To better understand the correlation between photoreceptors and DR, we need to understand O_2 distribution and usage in photoreceptors and the retina. The demand for energy in the retina is exceedingly high and much of the energy is derived from oxidative metabolism coupled with ATP synthesis, which is mainly dependent on continuous supply of oxygen [59,61,62]. The oxygen that is supplied to the retina cannot be stored in any region of the retina and must be in the vascular and choroidal vascular circulations of the retina. Photoreceptors use more energy at night when photoreceptor ion channels are open than in the daytime. In darkness, the surfaces of the outer segments of the rods are depolarized and leaky; water and sodium entering the retina must be extruded by pumps in the inner segment and thus more energy and oxygen are consumed within the inner segments to support ion pumping [63]. Therefore, photoreceptor activity in the dark makes the retina more hypoxic than the normal condition; this is especially the case for retinal vasculature diseases such as DR. Indeed, reduced or loss of dark adaptation is the first symptom in a variety of pathological conditions [64–66].

Growing evidence suggests that photoreceptors are affected in DR. Early studies on the association between poor color vison and signs of retinopathy date back to 1972 and 1973, indicating that blue-yellow and blue-green vision losses were much more severe in Scottish diabetic patients than normal controls [67,68]. Subsequently, other studies showed impairment of color vision, specifically blue-sensitive defects [69,70], as well as decreased contrast sensitivity in DR patients [71–73]. The rate of apoptosis increases in the outer nuclear layer with a reduction in photoreceptors between 4 and 24 weeks after the onset of diabetes [73]. Nevertheless, diabetes has not been reported to cause widespread degeneration of photoreceptors in patients and animal models. Up to now, conflicting results remain regarding whether DR induces photoreceptor death [74].

While emerging evidence shows that photoreceptors may be impacted by DR, many studies support that photoreceptors contribute to the development of retinal vascular lesions and early characteristics of DR by releasing inflammatory proteins [75–78]. Accumulation of inflammatory proteins in the diabetic retina accelerates the development of vascular lesions; when the production of proinflammatory proteins is inhibited, vascular lesions are also inhibited [79,80]. Using a model of retinitis pigmentosa (in particular rhodopsin knockout mice) in which the diabetic environment is induced via STZ, Gooyer et al. show that loss of the outer retina reduces the severity of diabetic retinopathy due to decreased hypoxia [78]. Photoreceptors are the main source of reactive oxygen species in the retina; deletion of photoreceptors inhibits the diabetes-induced increase in superoxide [75]. Photoreceptor cells themselves produce various proinflammatory and inflammatory proteins, such as interleukin-1β (IL-1β) [81], interleukin-1α(IL-1α), chemokine C-X-C motif ligand 1(CXCL1), monocyte chemoattractant protein 1(MCP-1), CXCL12a, chemokine ligand 25 (CCL25), TNF-a [76], inducible nitric oxide synthase (iNOS), intercellular adhesion molecule-1 (ICAM1), and vascular endothelial growth factor (VEGF) [75,78]. Some of these proteins have been reported to increase endothelial permeability or alter the tight junction and cell adhesion proteins [10,76,82].

4. Molecular Mechanisms Underlying Neurodegeneration in DR

Diabetic retinopathy was originally considered a microvascular disease; increasing evidence shows that it is a chronic inflammatory disease that leads to changes in the microcirculation of the retina. This raises a question whether the inflammation processs is involved in diabetic retinopathy; anti-inflammatory drugs should be able to alleviate various aspects of DR. In 1964, a study on a patient with rheumatoid arthritis showed a high regression rate of diabetic retinopathy as the patient had been taking a high dose of aspirin for 12 years [83]. Another study administered aminoguanidine and aspirin on a daily basis for 5 years in a dog model of DR, and both treatments were found to effectively inhibit the development of acellular capillaries in diabetes [84], indicating that inflammation is associated with DR. Using cDNA arrays to examine gene expression patterns in the diabetic retina in STZ-induced diabetic rats, upregulation of the inflammatory components was detected during the early stage of the disease onset [85]. DR is the manifestation of a mild chronic inflammation, during which inflammatory effectors including cytokines, pro-apoptotic molecules, and leukocytes are released and responsible for damages to the vascular endothelium of the retina. This type of damage is slow and cumulative over time. As diabetes progresses, acellular capillary forms and irreversible ischemia develops, leading to the discharge of certain vasoactive chemicals such as VEGF that promote the formation of new blood vessels and the transition to the proliferative stage of DR.

Hyperglycemia leads to considerable metabolic abnormalities. It causes non-enzymatic glycosylation as a result of the formation of complex cross-linked substances known as advanced glycation end products (AGEs). AGEs are a major consequence of sustained hyperglycemia during diabetes, which can trigger secondary complications. For example, increased levels of intracellular reactive oxygen species (ROS) [86,87] inflict oxidative damage to the retina. Multiple signaling pathways may have been altered in DR that include the polyol pathway [88], the protein kinase C pathway [89], and the protein kinase (MAP kinase) pathway [90], as well as abnormal activity of nuclear factors such as highly activated nuclear factor-κB (NF-κB). NF-κB can be activated via VEGF and translocated to the nucleus to promote the transcription and expression of VEGF itself; in the meantime, it also induces the expression of other pro-inflammatory mediators such as ICAM-1, vascular cell adhesion molecule-1 (VCAM-1), monocyte chemotactic proteins 1 (MCP-1), and cyclooxygenase-2 [91]. ICAM-1, VCAM-1, and VEGF in turn are implicated in BRB disruption that causes microaneurysms and leakage in the retina [92]. Nuclear factor erythroid 2-related factor 2 (Nrf2) is one of the primary regulators of cellular redox homeostasis, which controls the transcription of downstream antioxidant enzymes but its transcriptional activity is impaired in DR [93].

Oxidative stress is an imbalance between excessive generation of ROS and their removal [82]. ROS are oxidant molecules that contain an extra electron conferring on them great instability and reactivity. Common forms of ROS include hydrogen peroxide (H_2O_2), peroxyl radical (ROO·), superoxide anion ($O_2\cdot^-$), hydroxyl radical (·HO), and nitric oxide (NO·) [94]. Oxidative stress induces inflammation and mitochondrial dysfunction, leading to cell death through pyroptosis, apoptosis, or autophagy, and the resultant neurodegeneration causes neural vascular and retinal tissue damage. Oxidative stress is a critical contributor to the pathogenesis of diabetic retinopathy and also results from metabolic abnormalities induced by hyperglycemia. As we mentioned earlier, the retina is a highly energy-demanding organ and therefore is highly vulnerable to and easily damaged by high levels of ROS. Oxidative stress in DR has the ability to act as a trigger, modulator, and the link within the complex web of pathological events that occur in DR, including activation of protein kinase C (PKC), the hexosamine biosynthetic pathway, as well as the presence of increased amounts of AGEs and activation of receptors for AGE (RAGE).

5. Future Perspectives

Pathophysiological mechanisms underlying diabetic retinopathy are complex. Conventional treatment is based on vitreoretinal surgery and laser photocoagulation; however, traditional surgical treatment is only for proliferative diabetic retinopathy cases with hemorrhage or tractional retinal detachment. Drug treatment has been an emerging therapy to treat DR such as anti-VEGF or steroid drugs, which also target end-stages of the disease after damage has already occurred. Leading clinical anti-VEGF drugs including ranibizumab, bevacizumab, and aflibercept, have been widely utilized to treat DR patients [95–97]. Early anti-VEGF injection, before complications of DR have developed, can reduce further progression into severe stages [98]. However, anti-VEGF therapy shows low efficacy in certain patient populations [91]. A new technology of comparative ligandomics has identified Secretogranin III (Scg3) as a novel disease-associated ligand that selectively binds to the diabetic vessels but not healthy vessels to induce angiogenesis, and thus Scg3 may provide a new therapeutic target for antiangiogenic therapy of DR [99,100]. Although much effort has been made to investigate DR, currently there are no therapeutic strategies that can fully reverse the retinal damage caused by DR.

Neurodegeneration is an early event in diabetic retinopathy; therefore, one possible and targeted therapeutic strategy is to prevent or slow down neurodegeneration in the early stage of DR. The use of neuroprotective substances holds great potential for the treatment of DR. Endogenous neuroprotective agents, including insulin-like growth factor 1 [5], pigment epithelium-derived factor (PEDF) [101], somatostatin (SST), pituitary adenylate-cyclase-activating polypeptide (PACAP), glucagon-like peptide-1 (GLP-1), and neurotrophins such as brain-derived neurotrophic factor (BDNF) and nerve growth factor (NGF), are potential neuroprotective factors for DR [39,102]. Peroxisome proliferator-activator receptor alpha (PPARα) has been identified as a putative therapeutic target for retinopathy in type 1 and 2 diabetes [103,104]. Fenofibrate, a PPARα agonist, manifests unprecedented neuroprotective effects in DR with type 1 diabetes [103]. Further studies are needed to examine the protection and maintenance of RGC morphology and physiology. The contribution of neurotrophic factors and VEGF to diabetic RGC degeneration is controversial [105], as BDNF neuroprotection in the retina is concentration-dependent. Inadequate amounts of BDNF promote neuroretinal apoptosis and degeneration [106]. VEGF has neuroprotective effects at the early stage of DR, while at the late phase of DR, VEGF promotes diabetic neovascularization [105].

Current diagnosis for neurodegeneration in diabetic retinopathy can be measured by frequency domain optical coherence tomography (FD-OCT), allowing for detecting morphological changes such as thinning of the ganglion cell layer as well as the nerve fiber layer. Measurements of these two parameters using OCT provide valuable information about the extent of neurodegeneration in DR, representing the most practical way to monitor neurodegeneration. Functional abnormalities of retinal neurons at the early stage of DR can be measured via mfERG [107], standard automated perimetry, frequency doubling perimetry, or microperimetry, among which mfERG is the gold standard.

Gene therapy is a potential therapeutic strategy for DR and is designed to deliver genetic materials to patients or animal models for therapeutic purposes. Adeno-associated viruses (AAVs) have become the most popular gene therapy tool for the treatment of ocular disease for several of their advantages, including minimal toxicity, lower immune responses and the ability to sustain long-term treatment effects. Although many unanswered questions remain, a number of AAV serotypes have been tested, among which AAV serotype 2 is best characterized and has been commonly used for gene therapy in humans [108]. In 2017 an AAV-based gene therapy was approved for the human-inherited retinal disease, RPE65 mutation-associated Leber's congenital amaurosis (LCA), through delivering a functional copy of RPE65 cDNA to retinal pigment epithelial cells (RPEs). This is a historic landmark revolutionizing the therapies for retinal degenerative diseases [109,110]. Current possible gene therapy for DR mainly targets two pathological aspects: existing retinal neovascularization in the late phase of DR and neurodegeneration in the early phase of DR. VEGF is a prominent therapeutic target for DR. Various treatments have been attempted to interfere with VEGF pathways using AAVs, such as soluble VEGF receptor-1 (sFlt-1), Flt23k, an intraceptor inhibitor of VEGF containing VEGF-binding domains 2-3 of Flt-1, to suppress retinal neovascularization in animal models [101,111–113]. Erythropoietin (EPO) [114], a hematopoietic cytokine produced in the fetal liver and adult kidney, was reported to have a potent neuroprotective effect in the retina. AAV2-mediated delivery of EPO via subretinal injection was attempted in mouse diabetic retinas with reduced breakdown of the blood–retina barrier and less neuronal apoptosis in the outer nuclear layer [114]. Gene therapy holds promise to provide long-term treatment effects for DR, among many other desirable features compared to traditional drugs and surgical therapies. Clinical trial barriers are mainly due to the complexity of DR pathogenesis, as well as gene delivery safety and targeting.

Author Contributions: Conceptualization, J.Z. and B.C.; writing—original draft preparation, J.Z. and B.C.; writing—review and editing, J.Z. and B.C.; visualization, J.Z.; supervision, B.C.; funding acquisition, B.C. All authors have read and agreed to the published version of the manuscript.

Funding: This research was funded by National Institutes of Health (NIH) grants R01 EY024985 and R01 EY028921, an unrestricted challenge grant from Research to Prevent Blindness, the New York Eye and Ear Infirmary Foundation, and The Harold W McGraw, Jr. Family Foundation for Vision Research.

Institutional Review Board Statement: Not applicable.

Informed Consent Statement: Not applicable.

Data Availability Statement: Not applicable.

Acknowledgments: Figure 1 was created with BioRender.com (accessed on 18 April 2023).

Conflicts of Interest: The authors declare no conflict of interest.

References

1. Cheung, N.; Mitchell, P.; Wong, T.Y. Diabetic retinopathy. *Lancet* **2010**, *376*, 124–136. [CrossRef] [PubMed]
2. Cho, N.H.; Shaw, J.E.; Karuranga, S.; Huang, Y.; da Rocha Fernandes, J.D.; Ohlrogge, A.W.; Malanda, B. IDF Diabetes Atlas: Global estimates of diabetes prevalence for 2017 and projections for 2045. *Diabetes Res. Clin. Pract.* **2018**, *138*, 271–281. [CrossRef] [PubMed]
3. Thomas, R.; Halim, S.; Gurudas, S.; Sivaprasad, S.; Owens, D. IDF Diabetes Atlas: A review of studies utilising retinal photography on the global prevalence of diabetes related retinopathy between 2015 and 2018. *Diabetes Res. Clin. Pr.* **2019**, *157*, 107840. [CrossRef] [PubMed]
4. Kawasaki, R.; Tanaka, S.; Tanaka, S.; Abe, S.; Sone, H.; Yokote, K.; Ishibashi, S.; Katayama, S.; Ohashi, Y.; Akanuma, Y.; et al. Risk of cardiovascular diseases is increased even with mild diabetic retinopathy: The Japan Diabetes Complications Study. *Ophthalmology* **2013**, *120*, 574–582. [CrossRef] [PubMed]
5. Tarr, J.M.; Kaul, K.; Chopra, M.; Kohner, E.M.; Chibber, R. Pathophysiology of Diabetic Retinopathy. *ISRN Ophthalmol.* **2013**, *2013*, 1–13. [CrossRef] [PubMed]
6. Fong, D.S.; Aiello, L.; Gardner, T.W.; King, G.L.; Blankenship, G.; Cavallerano, J.D.; Ferris, F.L.; Klein, R. Diabetic retinopathy. *Diabetes Care* **2004**, *27*, 2540–2553. [CrossRef] [PubMed]
7. Hartnett, M.E.; Baehr, W.; Le, Y.Z. Diabetic retinopathy, an overview. *Vision Res.* **2017**, *139*, 1–6. [CrossRef] [PubMed]
8. Stitt, A.W.; Curtis, T.M.; Chen, M.; Medina, R.J.; McKay, G.J.; Jenkins, A.; Gardiner, T.A.; Lyons, T.J.; Hammes, H.-P.; Simó, R.; et al. The progress in understanding and treatment of diabetic retinopathy. *Prog. Retin. Eye Res.* **2016**, *51*, 156–186. [CrossRef] [PubMed]
9. Klein, R.; Moss, S.E.; Klein, B.E.; Davis, M.D.; DeMets, D.L. The Wisconsin Epidemiologic Study of Diabetic Retinopathy. *Ophthalmology* **1989**, *96*, 1501–1510. [CrossRef]
10. Yau, J.W.Y.; Rogers, S.L.; Kawasaki, R.; Lamoureux, E.L.; Kowalski, J.W.; Bek, T.; Chen, S.-J.; Dekker, J.M.; Fletcher, A.; Grauslund, J.; et al. Global Prevalence and Major Risk Factors of Diabetic Retinopathy. *Diabetes Care* **2012**, *35*, 556–564. [CrossRef] [PubMed]
11. Stewart, M.W.; Browning, D.J.; Lee, C. Diabetic macular edema: Evidence-based management. *Indian J. Ophthalmol.* **2018**, *66*, 1736–1750. [CrossRef]
12. Lai, A.K.W.; Lo, A.C.Y. Animal Models of Diabetic Retinopathy: Summary and Comparison. *J. Diabetes Res.* **2013**, *2013*, 1–29. [CrossRef]
13. Kumar, S.; Singh, R.; Vasudeva, N.; Sharma, S. Acute and chronic animal models for the evaluation of anti-diabetic agents. *Cardiovasc. Diabetol.* **2012**, *11*, 9. [CrossRef]
14. Rakieten, N.; Rakieten, M.L.; Nadkarni, M.V. Studies on the diabetogenic action of streptozotocin (NSC-37917). *Cancer Chemother. Rep.* **1963**, *29*, 91–98.
15. Grossniklaus, H.E.; Kang, S.J.; Berglin, L. Animal models of choroidal and retinal neovascularization. *Prog. Retin. Eye Res.* **2010**, *29*, 500–519. [CrossRef]
16. Sakano, D.; Inoue, A.; Enomoto, T.; Imasaka, M.; Okada, S.; Yokota, M.; Koike, M.; Araki, K.; Kume, S. Insulin2Q104del (Kuma) mutant mice develop diabetes with dominant inheritance. *Sci. Rep.* **2020**, *10*, 1–14. [CrossRef] [PubMed]
17. Nickla, D.L.; Wallman, J. The multifunctional choroid. *Prog. Retin. Eye Res.* **2010**, *29*, 144–168. [CrossRef] [PubMed]
18. Sun, Y.; Smith, L.E.H. Retinal Vasculature in Development and Diseases. *Annu. Rev. Vis. Sci.* **2018**, *4*, 101–122. [CrossRef] [PubMed]
19. Cunha-Vaz, J.G.; Shakib, M.; Ashton, N. Studies on the permeability of the blood-retinal barrier. I. On the existence, development, and site of a blood-retinal barrier. *Br. J. Ophthalmol.* **1966**, *50*, 441–453. [CrossRef] [PubMed]
20. Yu, D.-Y.; Cringle, S.J.; Yu, P.K.; Balaratnasingam, C.; Mehnert, A.; Sarunic, M.V.; An, D.; Su, E.-N. Retinal capillary perfusion: Spatial and temporal heterogeneity. *Prog. Retin. Eye Res.* **2019**, *70*, 23–54. [CrossRef]

21. Usui, Y.; Westenskow, P.D.; Kurihara, T.; Aguilar, E.; Sakimoto, S.; Paris, L.P.; Wittgrove, C.; Feitelberg, D.; Friedlander, M.S.; Moreno, S.K.; et al. Neurovascular crosstalk between interneurons and capillaries is required for vision. *J. Clin. Investig.* **2015**, *125*, 2335–2346. [CrossRef] [PubMed]
22. Metea, M.R.; Newman, E.A. Signalling within the neurovascular unit in the mammalian retina. *Exp. Physiol.* **2007**, *92*, 635–640. [CrossRef]
23. Hawkins, B.; Davis, T. The Blood-Brain Barrier/Neurovascular Unit in Health and Disease. *Pharmacol. Rev.* **2005**, *57*, 173–185. [CrossRef]
24. Kur, J.; Newman, E.A.; Chan-Ling, T. Cellular and physiological mechanisms underlying blood flow regulation in the retina and choroid in health and disease. *Prog. Retin. Eye Res.* **2012**, *31*, 377–406. [CrossRef] [PubMed]
25. Abbott, N.J.; Rönnbäck, L.; Hansson, E. Astrocyte-endothelial interactions at the blood-brain barrier. *Nat. Rev. Neurosci.* **2006**, *7*, 41–53. [CrossRef] [PubMed]
26. Leal, E.C.; Martins, J.; Voabil, P.; Liberal, J.; Chiavaroli, C.; Bauer, J.; Cunha-Vaz, J.; Ambrósio, A.F. Calcium Dobesilate Inhibits the Alterations in Tight Junction Proteins and Leukocyte Adhesion to Retinal Endothelial Cells Induced by Diabetes. *Diabetes* **2010**, *59*, 2637–2645. [CrossRef] [PubMed]
27. Cunha-Vaz, J.; de Abreu, J.R.F.; Campos, A.J. Early breakdown of the blood-retinal barrier in diabetes. *Br. J. Ophthalmol.* **1975**, *59*, 649–656. [CrossRef]
28. Navaratna, D.; McGuire, P.G.; Menicucci, G.; Das, A. Proteolytic Degradation of VE-Cadherin Alters the Blood-Retinal Barrier in Diabetes. *Diabetes* **2007**, *56*, 2380–2387. [CrossRef]
29. Fresta, C.G.; Fidilio, A.; Caruso, G.; Caraci, F.; Giblin, F.J.; Marco Leggio, G.; Salomone, S.; Drago, F.; Bucolo, C. A New Human Blood-Retinal Barrier Model Based on Endothelial Cells, Pericytes, and Astrocytes. *Int. J. Mol. Sci.* **2020**, *21*, 1636. [CrossRef]
30. Heng, L.Z.; Comyn, O.; Peto, T.; Tadros, C.; Ng, E.; Sivaprasad, S.; Hykin, P.G. Diabetic retinopathy: Pathogenesis, clinical grading, management and future developments. *Diabet. Med.* **2013**, *30*, 640–650. [CrossRef]
31. Wolter, J.R. Diabetic retinopathy. *Am. J. Ophthalmol.* **1961**, *51*, 1123–1141. [CrossRef] [PubMed]
32. Sima, A.A.F.; Zhang, W.-X.; Cherian, P.V.; Chakrabarti, S. Impaired visual evoked potential and primary axonopathy of the optic nerve in the diabetic BB/W-rat. *Diabetologia* **1992**, *35*, 602–607. [CrossRef] [PubMed]
33. Kamijo, M.; Cherian, P.V.; Sima, A.A.F. The preventive effect of aldose reductase inhibition on diabetic optic neuropathy in the BB/W-rat. *Diabetologia* **1993**, *36*, 893–898. [CrossRef] [PubMed]
34. Aizu, Y.; Oyanagi, K.; Hu, J.; Nakagawa, H. Degeneration of retinal neuronal processes and pigment epithelium in the early stage of the streptozotocin-diabetic rats. *Neuropathology* **2002**, *22*, 161–170. [CrossRef]
35. Barber, A.; Lieth, E.; Khin, S.A.; Antonetti, D.; Buchanan, A.G.; Gardner, T. Neural apoptosis in the retina during experimental and human diabetes. Early onset and effect of insulin. *J. Clin. Investig.* **1998**, *102*, 783–791. [CrossRef]
36. Qin, Y.; Xu, G.; Wang, W. Dendritic Abnormalities in Retinal Ganglion Cells of Three-Month Diabetic Rats. *Curr. Eye Res.* **2006**, *31*, 967–974. [CrossRef]
37. Fernandez, D.C.; Pasquini, L.A.; Dorfman, D.; Marcos, H.J.A.; Rosenstein, R.E. Early Distal Axonopathy of the Visual Pathway in Experimental Diabetes. *Am. J. Pathol.* **2012**, *180*, 303–313. [CrossRef]
38. Howell, S.J.; Mekhail, M.N.; Azem, R.; Ward, N.L.; Kern, T.S. Degeneration of retinal ganglion cells in diabetic dogs and mice: Relationship to glycemic control and retinal capillary degeneration. *Mol. Vis.* **2013**, *19*, 1413–1421.
39. Simo, R.; Stitt, A.W.; Gardner, T.W. Neurodegeneration in diabetic retinopathy: Does it really matter? *Diabetologia* **2018**, *61*, 1902–1912. [CrossRef]
40. Frydkjaer-Olsen, U.; Hansen, R.S.; Peto, T.; Grauslund, J. Structural neurodegeneration correlates with early diabetic retinopathy. *Int. Ophthalmol.* **2017**, *38*, 1621–1626. [CrossRef]
41. Yang, Q.; Xu, Y.; Xie, P.; Cheng, H.; Song, Q.; Su, T.; Yuan, S.; Liu, Q. Retinal Neurodegeneration in db/db Mice at the Early Period of Diabetes. *J. Ophthalmol.* **2015**, *2015*, 1–9. [CrossRef] [PubMed]
42. Barber, A.J.; Antonetti, D.A.; Kern, T.S.; Reiter, C.E.N.; Soans, R.S.; Krady, J.K.; Levison, S.W.; Gardner, T.W.; Bronson, S.K. The Ins2^{Akita}Mouse as a Model of Early Retinal Complications in Diabetes. *Investig. Opthalmology Vis. Sci.* **2005**, *46*, 2210–2218. [CrossRef] [PubMed]
43. Gastinger, M.J.; Kunselman, A.R.; Conboy, E.E.; Bronson, S.K.; Barber, A.J. Dendrite Remodeling and Other Abnormalities in the Retinal Ganglion Cells of Ins2^{Akita} Diabetic Mice. *Investig. Opthalmology Vis. Sci.* **2008**, *49*, 2635–2642. [CrossRef]
44. Gastinger, M.J.; Singh, R.S.J.; Barber, A.J. Loss of Cholinergic and Dopaminergic Amacrine Cells in Streptozotocin-Diabetic Rat and Ins2^{Akita}-Diabetic Mouse Retinas. *Investig. Opthalmology Vis. Sci.* **2006**, *47*, 3143–3150. [CrossRef] [PubMed]
45. Pitale, P.M.; Saltykova, I.V.; Adu-Agyeiwaah, Y.; Calzi, S.L.; Satoh, T.; Akira, S.; Gorbatyuk, O.; Boulton, M.E.; Pardue, M.T.; Garvey, W.T.; et al. Tribbles Homolog 3 Mediates the Development and Progression of Diabetic Retinopathy. *Diabetes* **2021**, *70*, 1738–1753. [CrossRef] [PubMed]
46. Sohn, E.H.; van Dijk, H.W.; Jiao, C.; Kok, P.H.B.; Jeong, W.; Demirkaya, N.; Garmager, A.; Wit, F.; Kucukevcilioglu, M.; van Velthoven, M.E.J.; et al. Retinal neurodegeneration may precede microvascular changes characteristic of diabetic retinopathy in diabetes mellitus. *Proc. Natl. Acad. Sci. USA* **2016**, *113*, E2655–E2664. [CrossRef]
47. Sergeys, J.; Etienne, I.; Van Hove, I.; Lefevere, E.; Stalmans, I.; Feyen, J.H.M.; Moons, L.; Van Bergen, T. Longitudinal In Vivo Characterization of the Streptozotocin-Induced Diabetic Mouse Model: Focus on Early Inner Retinal Responses. *Investig. Opthalmology Vis. Sci.* **2019**, *60*, 807–822. [CrossRef]

48. Cui, R.-Z.; Wang, L.; Qiao, S.-N.; Wang, Y.-C.; Wang, X.; Yuan, F.; Weng, S.-J.; Yang, X.-L.; Zhong, Y.-M. ON-Type Retinal Ganglion Cells are Preferentially Affected in STZ-Induced Diabetic Mice. *Investig. Opthalmology Vis. Sci.* **2019**, *60*, 1644–1656. [CrossRef]
49. Amato, R.; Lazzara, F.; Chou, T.-H.; Romano, G.L.; Cammalleri, M.; Monte, M.D.; Casini, G.; Porciatti, V. Diabetes Exacerbates the Intraocular Pressure-Independent Retinal Ganglion Cells Degeneration in the DBA/2J Model of Glaucoma. *Investig. Opthalmology Vis. Sci.* **2021**, *62*, 9. [CrossRef]
50. Amato, R.; Catalani, E.; Monte, M.D.; Cammalleri, M.; Cervia, D.; Casini, G. Morpho-functional analysis of the early changes induced in retinal ganglion cells by the onset of diabetic retinopathy: The effects of a neuroprotective strategy. *Pharmacol. Res.* **2022**, *185*, 106516. [CrossRef]
51. Newman, E.; Reichenbach, A. The Müller cell: A functional element of the retina. *Trends Neurosci.* **1996**, *19*, 307–312. [CrossRef]
52. Shen, W.; Fruttiger, M.; Zhu, L.; Chung, S.H.; Barnett, N.L.; Kirk, J.K.; Lee, S.; Coorey, N.J.; Killingsworth, M.; Sherman, L.S.; et al. Conditional Müllercell ablation causes independent neuronal and vascular pathologies in a novel transgenic model. *J. Neurosci.* **2012**, *32*, 15715–15727. [CrossRef]
53. Amaducci, L.; Forno, K.I.; Eng, L.F. Glial fibrillary acidic protein in cryogenic lesions of the rat brain. *Neurosci. Lett.* **1981**, *21*, 27–32. [CrossRef]
54. Rungger-Brändle, E.; Dosso, A.A.; Leuenberger, P.M. Glial reactivity, an early feature of diabetic retinopathy. *Investig. Opthalmology Vis. Sci.* **2000**, *41*, 1971–1980.
55. Barber, A.; Antonetti, D.; Gardner, T. Altered expression of retinal occludin and glial fibrillary acidic protein in experimental diabetes. The Penn State Retina Research Group. *Investig. Opthalmology Vis. Sci.* **2000**, *41*, 3561–3568.
56. Lieth, E.; Barber, A.J.; Xu, B.; Dice, C.; Ratz, M.J.; Tanase, D.; Strother, J.M. Glial reactivity and impaired glutamate metabolism in short-term experimental diabetic retinopathy. Penn State Retina Research Group. *Diabetes* **1998**, *47*, 815–820. [CrossRef] [PubMed]
57. Zhang, R.; Huang, C.; Chen, Y.; Li, T.; Pang, L. Single-cell transcriptomic analysis revealing changes in retinal cell subpopulation levels and the pathways involved in diabetic retinopathy. *Ann. Transl. Med.* **2022**, *10*, 562. [CrossRef] [PubMed]
58. Miller, W.P.; Toro, A.L.; Sunilkumar, S.; Stevens, S.A.; VanCleave, A.M.; Williamson, D.L.; Barber, A.J.; Dennis, M.D. Müller Glial Expression of REDD1 Is Required for Retinal Neurodegeneration and Visual Dysfunction in Diabetic Mice. *Diabetes* **2022**, *71*, 1051–1062. [CrossRef]
59. Okawa, H.; Sampath, A.P.; Laughlin, S.B.; Fain, G.L. ATP Consumption by Mammalian Rod Photoreceptors in Darkness and in Light. *Curr. Biol.* **2008**, *18*, 1917–1921. [CrossRef]
60. Meng, E.C.; Bourne, H.R. Receptor activation: What does the rhodopsin structure tell us? *Trends Pharmacol. Sci.* **2001**, *22*, 587–593. [CrossRef]
61. Vanderkooi, J.M.; Erecinska, M.; Silver, I.A. Oxygen in mammalian tissue: Methods of measurement and affinities of various reactions. *Am. J. Physiol. Physiol.* **1991**, *260*, C1131–C1150. [CrossRef] [PubMed]
62. Ames, A. Energy requirements of CNS cells as related to their function and to their vulnerability to ischemia: A commentary based on studies on retina. *Can. J. Physiol. Pharmacol.* **1992**, *70*, S158–S164. [CrossRef] [PubMed]
63. Arden, G.B.; Sidman, R.L.; Arap, W.; Schlingemann, R.O. Spare the rod and spoil the eye. *Br. J. Ophthalmol.* **2005**, *89*, 764–769. [CrossRef] [PubMed]
64. Havelius, U.; Berglund, S.; Falke, P.; Hindfelt, B.; Krakau, T. Impaired dark adaptation in polycythemia. Improvement after treatment. *Acta Ophthalmol. Scand.* **2000**, *78*, 53–57. [CrossRef] [PubMed]
65. Havelius, U.; Bergqvist, D.; Falke, P.; Hindfelt, B.; Krakau, T.I. Impaired dark adaptation in symptomatic carotid artery disease. *Neurology* **1997**, *49*, 1353–1359. [CrossRef] [PubMed]
66. Havelius, U.; Bergqvist, D.; Hindfelt, B.; Krakau, T. II. Improved dark adaptation after carotid endarterectomy. *Neurology* **1997**, *49*, 1360–1364. [CrossRef] [PubMed]
67. Kinnear, P.R.; Aspinall, P.; Lakowski, R. The diabetic eye and colour vision. *Trans. Ophthalmol. Soc. UK* **1972**, *92*, 69–78. [PubMed]
68. Lakowski, R.; Aspinall, P.A.; Kinnear, P.R. Association between Colour Vision Losses and Diabetes Mellitus. *Ophthalmic Res.* **1972**, *4*, 145–159. [CrossRef]
69. Daley, M.L.; Watzke, R.C.; Riddle, M.C. Early Loss of Blue-Sensitive Color Vision in Patients with Type I Diabetes. *Diabetes Care* **1987**, *10*, 777–781. [CrossRef]
70. Roy, M.S.; Gunkel, R.D.; Podgor, M.J. Color Vision Defects in Early Diabetic Retinopathy. *Arch. Ophthalmol.* **1986**, *104*, 225–228. [CrossRef]
71. Trick, G.L.; Burde, R.M.; Cordon, M.O.; Santiago, J.V.; Kilo, C. The Relationship between Hue Discrimination and Contrast Sensitivity Deficits in Patients with Diabetes Mellitus. *Ophthalmology* **1988**, *95*, 693–698. [CrossRef] [PubMed]
72. Sokol, S.; Moskowitz, A.; Skarf, B.; Evans, R.; Molitch, M.; Senior, B. Contrast Sensitivity in Diabetics with and without Background Retinopathy. *Arch. Ophthalmol.* **1985**, *103*, 51–54. [CrossRef] [PubMed]
73. Park, S.-J.; Park, J.-W.; Kim, K.-Y.; Chung, J.-W.; Chun, M.-H.; Oh, S.-J. Apoptotic death of photoreceptors in the streptozotocin-induced diabetic rat retina. *Diabetologia* **2003**, *46*, 1260–1268. [CrossRef] [PubMed]
74. Tonade, D.; Kern, T.S. Photoreceptor cells and RPE contribute to the development of diabetic retinopathy. *Prog. Retin. Eye Res.* **2020**, *83*, 100919. [CrossRef] [PubMed]
75. Du, Y.; Veenstra, A.; Palczewski, K.; Kern, T.S. Photoreceptor cells are major contributors to diabetes-induced oxidative stress and local inflammation in the retina. *Proc. Natl. Acad. Sci. USA* **2013**, *110*, 16586–16591. [CrossRef] [PubMed]

76. Tonade, D.; Liu, H.; Palczewski, K.; Kern, T.S. Photoreceptor cells produce inflammatory products that contribute to retin vascular permeability in a mouse model of diabetes. *Diabetologia* **2017**, *60*, 2111–2120. [CrossRef]
77. Tonade, D.; Liu, H.; Kern, T.S. Photoreceptor Cells Produce Inflammatory Mediators That Contribute to Endothelial Cell Death i Diabetes. *Investig. Opthalmology Vis. Sci.* **2016**, *57*, 4264–4271. [CrossRef]
78. de Gooyer, T.E.; Stevenson, K.A.; Humphries, P.; Simpson, D.A.; Gardiner, T.A.; Stitt, A.W. Retinopathy is reduced durin experimental diabetes in a mouse model of outer retinal degeneration. *Investig. Ophthalmol. Vis. Sci.* **2006**, *47*, 5561–556 [CrossRef]
79. Zheng, L.; Du, Y.; Miller, C.; Gubitosi-Klug, R.A.; Kern, T.S.; Ball, S.; Berkowitz, B.A. Critical role of inducible nitric oxide synthas in degeneration of retinal capillaries in mice with streptozotocin-induced diabetes. *Diabetologia* **2007**, *50*, 1987–1996. [CrossRef]
80. Joussen, A.M.; Poulaki, V.; Le, M.L.; Koizumi, K.; Esser, C.; Janicki, H.; Schraermeyer, U.; Kociok, N.; Fauser, S.; Kirchhof, B.; et a A central role for inflammation in the pathogenesis of diabetic retinopathy. *FASEB J.* **2004**, *18*, 1450–1452. [CrossRef]
81. Scuderi, S.; D'amico, A.G.; Federico, C.; Saccone, S.; Magro, G.; Bucolo, C.; Drago, F.; D'agata, V. Different Retinal Expressio Patterns of IL-1α, IL-1β, and Their Receptors in a Rat Model of Type 1 STZ-Induced Diabetes. *J. Mol. Neurosci.* **2015**, *56*, 431–43 [CrossRef]
82. Rochfort, K.D.; Collins, L.E.; McLoughlin, A.; Cummins, P.M. Tumour necrosis factor-α-mediated disruption of cerebrovascula endothelial barrier integrity in vitro involves the production of proinflammatory interleukin-6. *J. Neurochem.* **2015**, *136*, 564–57 [CrossRef] [PubMed]
83. Powell, E.D.; Field, R.A. Diabetic retinopathy and rheumatoid arthritis. *Lancet* **1964**, *2*, 17–18. [CrossRef] [PubMed]
84. Kern, T.S.; Engerman, R.L. Pharmacological inhibition of diabetic retinopathy: Aminoguanidine and aspirin. *Diabetes* **2001**, *5* 1636–1642. [CrossRef] [PubMed]
85. Joussen, A.M.; Huang, S.; Poulaki, V.; Camphausen, K.; Beecken, W.D.; Kirchhof, B.; Adamis, A.P. In vivo retinal gene expressio in early diabetes. *Investig. Opthalmology Vis. Sci.* **2001**, *42*, 3047–3057.
86. Brownlee, M.; Cerami, A.; Vlassara, H. Advanced Glycosylation End Products in Tissue and the Biochemical Basis of Diabeti Complications. *N. Engl. J. Med.* **1988**, *318*, 1315–1321. [CrossRef]
87. Hammes, H.-P. Diabetic retinopathy: Hyperglycaemia, oxidative stress and beyond. *Diabetologia* **2018**, *61*, 29–38. [CrossRef]
88. Lorenzi, M. The Polyol Pathway as a Mechanism for Diabetic Retinopathy: Attractive, Elusive, and Resilient. *Exp. Diabetes Res* **2007**, *2007*, 1–10. [CrossRef]
89. Ishii, H.; Koya, D.; King, G.L. Protein kinase C activation and its role in the development of vascular complications in diabete mellitus. *J. Mol. Med.* **1997**, *76*, 21–31. [CrossRef]
90. Mohammad, G.; Kowluru, R.A. The role of Raf-1 kinase in diabetic retinopathy. *Expert Opin. Ther. Targets* **2011**, *15*, 357–364 [CrossRef]
91. Cox, J.T.; Eliott, D.; Sobrin, L. Inflammatory Complications of Intravitreal Anti-VEGF Injections. *J. Clin. Med.* **2021**, *10*, 981 [CrossRef]
92. El-Remessy, A.B.; Al-Shabrawey, M.; Khalifa, Y.; Tsai, N.-T.; Caldwell, R.B.; Liou, G.I. Neuroprotective and Blood-Retina Barrier-Preserving Effects of Cannabidiol in Experimental Diabetes. *Am. J. Pathol.* **2006**, *168*, 235–244. [CrossRef] [PubMed]
93. Kowluru, R.A.; Mishra, M. Epigenetic regulation of redox signaling in diabetic retinopathy: Role of Nrf2. *Free Radic. Biol. Med* **2017**, *103*, 155–164. [CrossRef]
94. Poprac, P.; Jomova, K.; Simunkova, M.; Kollar, V.; Rhodes, C.J.; Valko, M. Targeting Free Radicals in Oxidative Stress-Related Human Diseases. *Trends Pharmacol. Sci.* **2017**, *38*, 592–607. [CrossRef] [PubMed]
95. Flaxel, C.J.; Adelman, R.A.; Bailey, S.T.; Fawzi, A.; Lim, J.I.; Vemulakonda, G.A.; Ying, G.-S. Diabetic Retinopathy Preferred Practice Pattern®. *Ophthalmology* **2019**, *127*, P66–P145. [CrossRef] [PubMed]
96. Yu, H.J.; Ehlers, J.P.; Sevgi, D.D.; Hach, J.; O'Connell, M.; Reese, J.L.; Srivastava, S.K.; Wykoff, C.C. Real-Time Photographic- and Fluorescein Angiographic-Guided Management of Diabetic Retinopathy: Randomized Prime Trial Outcomes. *Am. J. Ophthalmol* **2021**, *226*, 126–136. [CrossRef] [PubMed]
97. Lazzara, F.; Fidilio, A.; Platania, C.B.M.; Giurdanella, G.; Salomone, S.; Leggio, G.M.; Tarallo, V.; Cicatiello, V.; De Falco, S.; Eandi, C.M.; et al. Aflibercept regulates retinal inflammation elicited by high glucose via the PlGF/ERK pathway. *Biochem. Pharmacol* **2019**, *168*, 341–351. [CrossRef]
98. Maturi, R.K.; Glassman, A.R.; Josic, K.; Antoszyk, A.N.; Blodi, B.A.; Jampol, L.M.; Marcus, D.M.; Martin, D.F.; Melia, M.; Salehi-Had, H.; et al. Effect of Intravitreous Anti-Vascular Endothelial Growth Factor vs. Sham Treatment for Prevention of Vision-Threatening Complications of Diabetic Retinopathy: The Protocol W Randomized Clinical Trial. *JAMA Ophthalmol.* **2021**, *139*, 701–712. [CrossRef]
99. LeBlanc, M.E.; Wang, W.; Chen, X.; Caberoy, N.B.; Guo, F.; Shen, C.; Ji, Y.; Tian, H.; Wang, H.; Chen, R.; et al. Secretogranin III as a disease-associated ligand for antiangiogenic therapy of diabetic retinopathy. *J. Exp. Med.* **2017**, *214*, 1029–1047. [CrossRef]
100. Li, W.; Webster, K.A.; LeBlanc, M.E.; Tian, H. Secretogranin III: A diabetic retinopathy-selective angiogenic factor. *Cell. Mol. Life Sci.* **2017**, *75*, 635–647. [CrossRef]
101. Gehlbach, P.; Demetriades, A.M.; Yamamoto, S.; Deering, T.; Xiao, W.H.; Duh, E.J.; Yang, H.S.; Lai, H.; Kovesdi, I.; Carrion, M.; et al. Periocular gene transfer of sFlt-1 suppresses ocular neovascularization and vascular endothelial growth factor-induced breakdown of the blood-retinal barrier. *Hum. Gene Ther.* **2003**, *14*, 129–141. [CrossRef] [PubMed]

102. Hernández, C.; Monte, M.D.; Simó, R.; Casini, G. Neuroprotection as a Therapeutic Target for Diabetic Retinopathy. *J. Diabetes Res.* **2016**, *2016*, 1–18. [CrossRef] [PubMed]
103. Pearsall, E.A.; Cheng, R.; Matsuzaki, S.; Zhou, K.; Ding, L.; Ahn, B.; Kinter, M.; Humphries, K.M.; Quiambao, A.B.; Farjo, R.A.; et al. Neuroprotective effects of PPARα in retinopathy of type 1 diabetes. *PLoS ONE* **2019**, *14*, e0208399. [CrossRef] [PubMed]
104. Hu, Y.; Chen, Y.; Ding, L.; He, X.; Takahashi, Y.; Gao, Y.; Shen, W.; Cheng, R.; Chen, Q.; Qi, X.; et al. Pathogenic role of diabetes-induced PPAR-α down-regulation in microvascular dysfunction. *Proc. Natl. Acad. Sci. USA* **2013**, *110*, 15401–15406. [CrossRef]
105. Bikbova, G.; Oshitari, T.; Baba, T.; Yamamoto, S. Neurotrophic factors for retinal ganglion cell neuropathy—With a special reference to diabetic neuropathy in the retina. *Curr. Diabetes Rev.* **2014**, *10*, 166–176. [CrossRef]
106. Afarid, M.; Namvar, E.; Sanie-Jahromi, F. Diabetic Retinopathy and BDNF: A Review on Its Molecular Basis and Clinical Applications. *J. Ophthalmol.* **2020**, *2020*, 1–7. [CrossRef]
107. McAnany, J.J.; Persidina, O.S.; Park, J.C. Clinical electroretinography in diabetic retinopathy: A review. *Surv. Ophthalmol.* **2021**, *67*, 712–722. [CrossRef]
108. Wu, Z.; Asokan, A.; Samulski, R.J. Adeno-associated Virus Serotypes: Vector Toolkit for Human Gene Therapy. *Mol. Ther.* **2006**, *14*, 316–327. [CrossRef]
109. Smalley, E. First AAV gene therapy poised for landmark approval. *Nat. Biotechnol.* **2017**, *35*, 998–999. [CrossRef]
110. Morrison, C. Landmark gene therapy poised for US approval. *Nat. Rev. Drug Discov.* **2017**, *16*, 739–741. [CrossRef]
111. Zhang, X.; Das, S.K.; Passi, S.F.; Uehara, H.; Bohner, A.; Chen, M.; Tiem, M.; Archer, B.; Ambati, B.K. AAV2 delivery of Flt23k intraceptors inhibits murine choroidal neovascularization. *Mol. Ther. J. Am. Soc. Gene Ther.* **2015**, *23*, 226–234. [CrossRef] [PubMed]
112. Lai, Y.; Shen, W.; Brankov, M.; Lai, C.; Constable, I.; Rakoczy, P. Potential long-term inhibition of ocular neovascularisation by recombinant adeno-associated virus-mediated secretion gene therapy. *Gene Ther.* **2002**, *9*, 804–813. [CrossRef] [PubMed]
113. Bainbridge, J.W.; Mistry, A.; De Alwis, M.; Paleolog, E.; Baker, A.; Thrasher, A.J.; Ali, R.R. Inhibition of retinal neovascularisation by gene transfer of soluble VEGF receptor sFlt-1. *Gene Ther.* **2002**, *9*, 320–326. [CrossRef] [PubMed]
114. Xu, H.; Zhang, L.; Gu, L.; Lu, L.; Gao, G.; Li, W.; Xu, G.; Wang, J.; Gao, F.; Xu, J.-Y.; et al. Subretinal Delivery of AAV2-Mediated Human Erythropoietin Gene Is Protective and Safe in Experimental Diabetic Retinopathy. *Investig. Opthalmology Vis. Sci.* **2014**, *55*, 1519–1530. [CrossRef]

MDPI
St. Alban-Anlage 66
4052 Basel
Switzerland
www.mdpi.com

Cells Editorial Office
E-mail: cells@mdpi.com
www.mdpi.com/journal/cells

www.ingramcontent.com/pod-product-compliance
Lightning Source LLC
Chambersburg PA
CBHW061607220926
51850CB00005B/442